UTP-UMP Symposium on Energy Systems 2017 (SES 2017)

MATEC Web of Conferences Volume 131 (2017)

Perak, Malaysia
26 - 27 September 2017

Editors:

Hamzah Sakidin
Mohd Hazwan Yusof
Nordin Sa'ad
Dennis Ling Chuan Ching
Samsul Ariffin Abdul Karim

ISBN: 978-1-5108-5051-4

Printed from e-media with permission by:

Curran Associates, Inc.
57 Morehouse Lane
Red Hook, NY 12571

Some format issues inherent in the e-media version may also appear in this print version.

Printed by Curran Associates, Inc. (2017)

For additional information, please contact EDP Sciences – Web of Conferences at the address below.

EDP Sciences – Web of Conferences
17, Avenue du Hoggar
Parc d'Activité de Courtabœuf
BP 112
F-91944 Les Ulis Cedex A
France

Phone: +33 (0) 1 69 18 75 75
Fax: +33 (0) 1 69 28 84 91

contact-edps@webofconferences.org

Additional copies of this publication are available from:

Curran Associates, Inc.
57 Morehouse Lane
Red Hook, NY 12571 USA
Phone: 845-758-0400
Fax: 845-758-2633
Email: curran@proceedings.com
Web: www.proceedings.com

TABLE OF CONTENTS

edp sciences

MATEC Web of Conferences

Statement of Peer review

In submitting conference proceedings to MATEC Web of Conferences, I certify to the Publisher that I adhere to the **Policy on Publishing Integrity** of the journal in order to safeguard good scientific practice in publishing.

1. All articles have been subjected to peer review administered by the proceedings editors.
2. Reviews have been conducted by expert referees, who have been requested to provide unbiased and constructive comments aimed, whenever possible, at improving the work.
3. Proceedings editors have taken all reasonable steps to ensure the quality of the materials they publish and their decision to accept or reject a paper for publication has been based only on the merits of the work and the relevance to the journal.

Title, date and place of the conference

UTP - UMP Symposium on Energy Systems 2017 (SES 2017), September 26 - 27, 2017
UTP Perak, Malaysia and UMP Pahang, Malaysia, via tele-conference

Proceedings editor(s):

Hamzah Sakidin, Mohd Hazwan Yusof, Nordin Sa'ad, Dennis Ling Chuan Ching, Samsul Ariffin Abdul Karim

Date and editor's signature

11 October 2017

Numerical investigation of phase change materials thermal capacitor for pipe flow

Jundika Candra Kurnia[1,*], and *Agus Pulung* Sasmito[2]

[1]Department of Mechanical Engineering, Universiti Teknologi PETRONAS, 32610 Bandar Seri Iskandar, Perak Darul Ridzuan, Malaysia.
[2]Department of Mining and Materials Engineering, McGill University, 3450 University, Montreal, QC, H3A2A7 Canada

Abstract. This study addresses the performance of phase change material as thermal capacitor. A computational fluid dynamics (CFD) model is developed to take into account the conjugate heat transfer between water as the heat transfer fluid (HTF) and PCM as thermal capacitor. A pulsating inlet temperature with constant inlet velocity is prescribed to represent temperature variation. The performance of thermal capacitor is evaluated by closely monitoring outlet temperature and comparing it with inlet temperature to examine the reduction in temperature fluctuation. To intensify heat transfer between HTF and PCM, extended surfaces (fins) are installed on PCM side. The results indicate that PCM thermal capacitor can reduce temperature fluctuation by ~ 1 °C. This reduction can be improved further when extended surface is installed with ~ 1.5 °C reduction in temperature fluctuation is achieved. Moreover, it is found that the maximum temperature is delayed at the outlet due to slow conjugate heat transfer between HTF and PCM. Inlet velocity is found to have considerable influence of the temperature fluctuation reduction: Slower inlet velocity results in a better temperature fluctuation reduction. This study is expected to serve as a guideline in designing PCM-based thermal capacitor.

1 Introduction

Due to its capability of storing large amount of energy at minimum temperature rise, phase change materials have been widely adopted as energy storage medium [1]. Numerous phase change materials have been investigated as thermal energy storage ranging from organic materials over inorganic materials to eutectic materials [1-4]. In tandem, a number of PCM thermal energy storage configuration have been proposed, e.g. pipe storage, cylindrical storage, shell and tube storage, rectangular storage and triplex tube storage [2,5-7].

In addition to energy storage, there is a growing interest to utilize phase change materials as thermal capacitor or smoothing devices. Nardin et al [8] investigated the performance of aluminium as phase change materials to smooth the off-gas temperatures

* Corresponding author: jundika.kurnia@utp.edu.my

from the electric arc furnace process of steel industry for energy recovery purposes. They found out that the variation of the off gas temperature is significantly reduced. The concept was further developed for continuous charge electric arc furnace process in study by Dal Magro et al [9]. The findings highlight that reduction in off gas temperature fluctuation results in the increase of the whole recovery system efficiency. This in turn reduces size and investment cost of the recovery system and also eliminating the environmental issues due to dioxins generations. Novikov et al. [10] evaluate the performance of PCM powder to smooth temperature peak produced by the electronic component. It was found that PCM powder reduce the peak temperature. Among various PCM powder evaluated in their study, it was revealed that erythritol is the most promising. Verdier et al [11] experimentally investigated the PCM thermal energy storage to Hybrid Solar Gas Turbine (HSGT) systems for smoothing of temperature variation. In their study, the main goal is to keep the outlet air temperature from the collector above 650 °C.

These studies highlight the potential of PCM as thermal capacitor or temperature smoothing devices. Nevertheless, most of them focus on external flow, where the heat transfers fluid flow across bank of cylindrical tube container of PCM. It is therefore of interest to evaluate the effectiveness of phase change materials (PCM) as thermal capacitor or temperature smoothing device in a simple internal flow such as pipe flow. This study is conducted to evaluate the performance of PCM in reducing temperature variation of fluid flowing through straight pipe. To emulate the variation of temperature, a sinusoidal temperature profile with constant fluid velocity is prescribed at the inlet. The outlet temperature is monitored to evaluate the reduction in temperature variation. To enhance conjugate heat transfer between HTF and PCM and thus better smoothing, fin is installed in PCM side. The effect of inlet velocity will be evaluated as well.

2 Mathematical model

A thermal capacitor comprises HTF channel and pool of PCM is subjected to inlet temperature fluctuation, as illustrated in Fig.1. The PCM considered in this study is paraffin wax (n-octadecane) which has melting temperature around 28°C.

2.1 Governing equations

The mathematical model incorporates conjugate heat transfer between water as HTF and PCM as thermal capacitor. In HTF, fluid flow and convective heat transfer is taken into account. The conservation of mass, momentum and energy in HTF are given by [12].

$$\frac{\partial \rho_f}{\partial t} + \nabla \cdot (\rho_f \mathbf{u}) = 0 \tag{1}$$

$$\frac{\partial (\rho_f \mathbf{u})}{\partial t} + \nabla \cdot (\rho_f \mathbf{u}\mathbf{u}) = -\nabla p + \nabla \left[\mu_f \left(\nabla \mathbf{u} + (\nabla \mathbf{u})^T \right) \right] + \rho_f \mathbf{g} \tag{2}$$

$$\frac{\partial (\rho_f c_{p,f} T)}{\partial t} + \nabla \cdot (\rho_f c_{p,f} \mathbf{u} T) = \nabla \cdot (k_f \nabla T) \tag{3}$$

where ρ_f, μ_f, and $c_{p,f}$ are the density, dynamic viscosity and specific heat of the fluid, respectively. Meanwhile, p is the pressure, $\mathbf{u}$ is the fluid velocity and T is the temperature.

The conservation of mass, momentum and energy in PCM are given by [12]

$$\frac{\partial \rho_p}{\partial t} + \nabla \cdot (\rho_p \mathbf{u}) = 0 \tag{4}$$

$$\frac{\partial(\rho_p \mathbf{u})}{\partial t}+\nabla\cdot(\rho_p \mathbf{u}\mathbf{u})=-\nabla p+\nabla\left[\mu_p\left(\nabla\mathbf{u}+(\nabla\mathbf{u})^T\right)\right]+\rho_p\mathbf{g}+\mathbf{s}_{mom} \tag{5}$$

$$\frac{\partial(\rho_p H_p)}{\partial t}+\nabla\cdot(\rho_p \mathbf{u} H_p)=\nabla\cdot(k_p\nabla T) \tag{6}$$

where ρ_p and μ_f, are the density and dynamic viscosity of the PCM, respectively. H_p ans $\mathbf{s}_{mom}$ are additional terms added to momentum and energy equations to take into account reduced porosity in the mushy zone and latent heat of PCM.

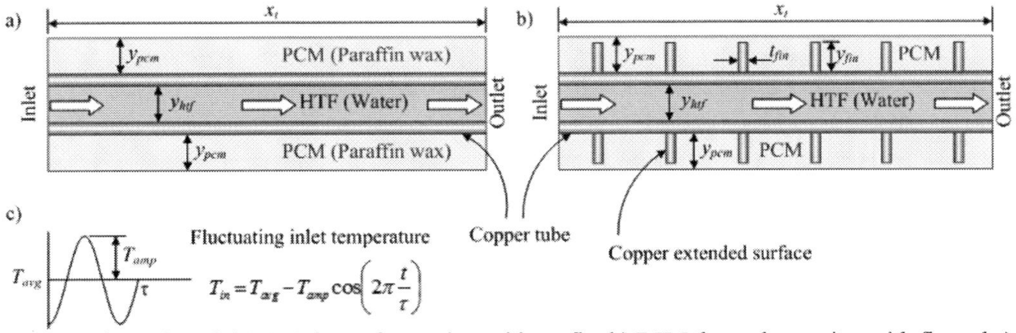

Fig. 1. Schematics of a) PCM thermal capacitor without fin, b) PCM thermal capacitor with fin and c) fluctuating inlet temperature.

2.2 Constitutive relations

Thermal properties of HTF (water) and PCM are strong function of temperature. Therefore a polynomial functions are adopted to represent this strong dependency (please refer to our previous publications [13-15] for details polynomial function adopted in this study).

2.3 Boundary conditions

The boundary conditions imposed on the mathematical model are summarized as follows:

- Inlet: at the inlet we specify constant inlet velocity and fluctuating temperature.
- Outlet: at the outlet, we set pressure and stream-wise gradient of temperature to zero.
- HTF channel wall: no slip condition and coupled temperature is prescribed.
- PCM outer wall: no slip condition and zero heat flux is implemented.

3 Numerical methodology

The mathematical model is incorporated to the computational domain, created, meshed and labelled in ANSYS workbench and the solved by using finite volume solver, ANSYS Fluent 15. User-defined-function is added to the computational model to account for thermo physical properties of water and PCM. The model is solved with semi-implicit pressure linked equation (SIMPLE) algorithm, Presto discretization for pressure and second order upwind discretization for other parameter together with algebraic multi-grid method. The convergence criterion of 10^{-3} is set for continuity and momentum while the convergence of energy is set at 10^{-6}. To ensure mesh independent result four different mesh amounting 2×10^4, 4×10^4, 6×10^4, and 8×10^4 are used and compared. The results indicate that no significant different is found between the result from 6×10^4 and 8×10^4 meshes. Hence mesh size of 6×10^4 is used for the remaining simulation. To assess the validity of the developed model, validation has been conducted for both PCM and HTF model, as has been presented in our previous work [13-16].

Table 1. Base case and operating parameters.

Parameter	Value	Units	Parameter	Value	Units
T_{avg}	300.65	K	T_m	300.65	K
T_{amp}	10	K	p_{out}	0	Pa (gage)
U_{in}	0.043, 0.085, 0.170	m/s	y_{pcm}	0.1	m
L	243,500	J/kg	y_{htf}	0.1	m
$c_{p,p}$	2160		y_{fin}, t_{fin}	0.08, 0.02	m
$k_p^{(s)}$, $k_p^{(l)}$	0.358, 0.148	W/m·K	t_{cu}	0.02	m
ζ	100,000	-	x_t	0.5	m

4 Results and discussion

The computational simulations are carried out for steady flow with fluctuating inlet temperature. The operating conditions and important parameters are listed in Table 1. The inlet temperature fluctuates in sinusoidal manner with maximum temperature of 290.65 and minimum temperature of 310.65 with a period of 10 minutes.

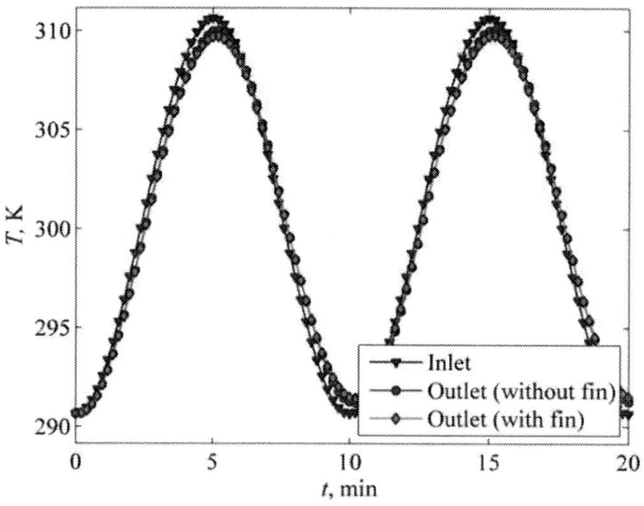

Fig. 2. Temperature fluctuation at inlet and outlet of PCM thermal capacitors at inlet Reynolds number of 1000.

The performance of investigated thermal capacitor is presented in Fig. 2. As can be seen, the investigated thermal capacitor is capable in reducing the temperature fluctuation as mirrored by the lower maximum outlet temperature as compared to the inlet temperature. The reduction is, however, relatively marginal. For PCM thermal without fin, the maximum temperature is reduced by ~ 1 °C. When the fin is added to the PCM to improve the heat transfer, only marginal improvement is observed where the maximum temperature is reduced by ~ 1.5 °C. Another interesting phenomenon that should be noted in Fig. 2 is the shifting of peak temperature: The maximum temperature at the outlet is delayed by 9 s and 7 s for PCM thermal capacitor without and with fin, respectively. This phenomenon is most likely attributed to the slow conjugate heat transfer between the heat transfer fluid and PCM. This is supported by the smaller delay for PCM thermal capacitor with fin as compared to that without fin.

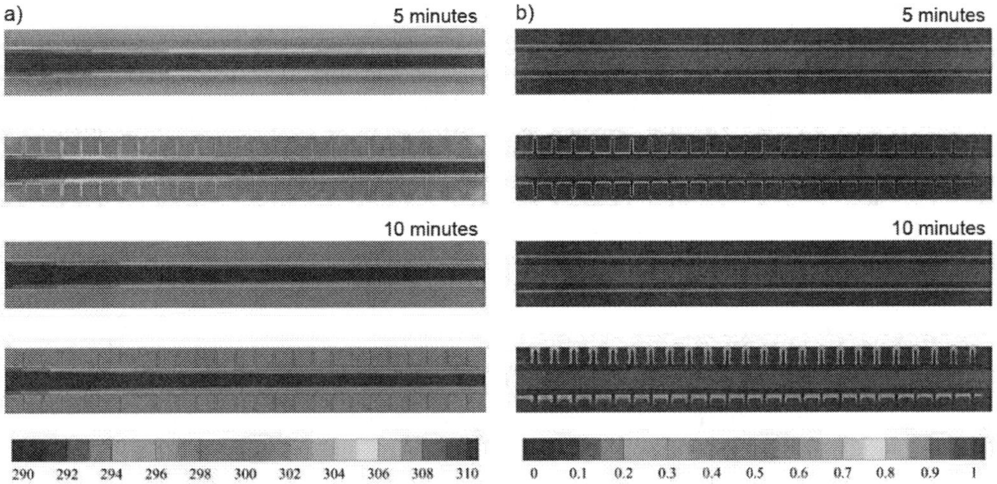

Fig. 3. a) Temperature distribution (K) and b) Liquid melting fraction within PCM thermal capacitor at time of 5 minutes and 10 minutes

Looking towards the temperature distributions within the PCM thermal capacitor during maximum and minimum inlet temperature, as presented in Fig. 3a, it is found that PCM temperature is relatively constant around the melting temperature despite large changing in heat transfer fluid temperature. This relatively constant temperature is due to phase change process on the PCM side as presented in Fig. 3b. As observed, phase change occurs only in the area near the metal surface. This is expected as the PCM considered in this study, paraffin wax, has low thermal conductivity. Therefore heat transfer from the HTF to PCM occurs slowly. This finding implies that higher heat transfer in PCM side is required to achieve high performance PCM thermal capacitor.

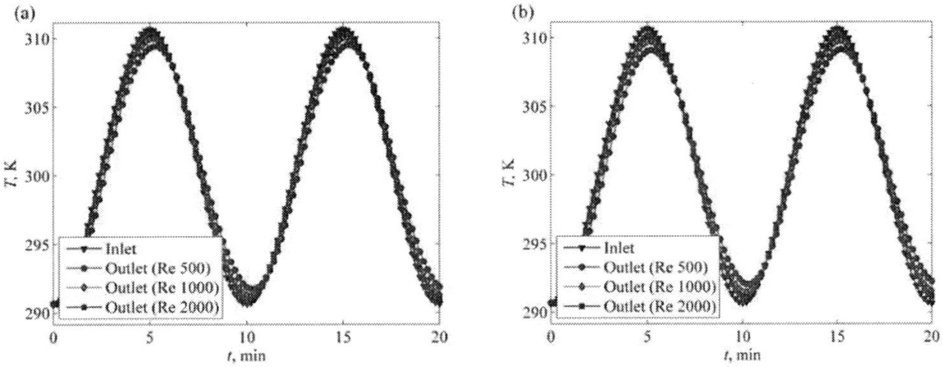

Fig. 4. Temperature fluctuation at inlet and outlet of PCM thermal capacitor a) with fin and b) without fin at various inlet Reynolds numbers

Next, the effect of inlet velocity on the performance of PCM thermal capacitor is evaluated. In this evaluation, three inlet velocities corresponding to inlet Reynolds number of 500, 1000 and 2000 is examined. Fig. 4 presents the temperature fluctuations at inlet and outlet for PCM thermal capacitor without and with fin at different inlet Reynolds numbers. As expected, higher Reynolds number yields in the lower temperature reduction. This is attributed to the fact that at lower velocity, the heat transfer fluid has longer heat transfer time.

5 Conclusions

A computational fluid dynamics (CFD) model which takes into account conjugate heat transfer between water as heat transfer fluid and phase change material as thermal capacitor is developed based on the validated model developed in previous study. Several thermal capacitor configurations were evaluated. Their performance is examined in term of reduction in temperature variation and induced pressure drop across the channel. The results indicate that PCM thermal capacitor without extended surface installed is capable to reduce temperature fluctuation by ~ 1°C. A larger reduction in temperature fluctuation is achieved when extended surface is installed in PCM. Subsequent study is currently underway to investigate various key factors affecting the performance of the proposed thermal capacitor. It should be noted that the mathematical model derived can easily be extended for other phase change material such as salt hydrate or metallic composite for high temperature application.

The authors gratefully acknowledge the facility and financial support from the Universiti Teknologi PETRONAS (UTP) through YUTP Fundamental Research Grant (YUTP-FRG) no 0153AA-E64.

References

1. A. Sharma, V. V. Tyagi, C. R. Chen and D. Buddhi. 2009. Renew Sustain Energy Rev **13**(2), 318345 (2009).
2. B. Xu, P. Li, and C. Chan. 2015. Appl Energy **160**, 286–307 (2015).
3. R. K. Sharma, P. Ganesan, V. V. Tyagi, H. S. C. Metselaar and S. C. Sandaran. Energy Convers Manage **95**:193–228 (2015).
4. P. Zhang, X. Xiao and Z. W. Ma. 2016. Appl Energy **165**, 472–510 (2016).
5. P. B. Salunke and P. S. Shembekar. 2012. Renew Sustain Energy Rev **16**(8), 5603–5616 (2012).
6. F. Agyenim, N. Hewitt, P. Eames and M. A. Smyth. Renew Sustain Energy Rev **14**(2), 615–628 (2010).
7. A. A. Al-Abidi, S. Mat, K. Sopian and M. Y. Sulaiman. 2013. Abdulrahman Th. Appl Therml Eng **53,** 147–156 (2013).
8. G. Nardin, A. Meneghetti, F. Dal Magro and N. Benedetti. Appl Energy **136**, 947–955 (2014).
9. F. Dal Magro, A. Meneghetti, G. Nardin and S. Savino. Energy Convers Manage **104**, 78–89 (2015).
10. A. Novikov, D. Lexow and M. Nowottnick. 2014. "Cooling of electronic assemblies through PCM containing coatings". In: *Proceedings of the 5th Electronics System-Integration Technology Conference* (Helsinki, Finland, 2014), pp. 1–5.
11. D. Verdiera, A. Ferrière, Q. Falcoz, F. Siros, and R. Couturier. Energy Procedia **49**, 1044–1053 (2014).
12. J.C. Kurnia, A.P. Sasmito, S.V. Jangam, A.S. Mujumdar, Appl Therm Eng **50**, 896–907 (2013).

13. J.C. Kurnia, A. P. Sasmito, T. Shamim, A. S. Mujumdar. Appl Therm Eng **102**,849–860 (2016).
14. J.C. Kurnia, A. P. Sasmito, A. S. Mujumdar. Heat Transfer Eng **37**,1341–1356 (2016).
15. J.C. Kurnia, A. P. Sasmito, S. I. Ping. Energy Procedia **105**, 4173–4178 (2017).
16. J.C. Kurnia, A. P. Sasmito. Appl Energy (doi: 10.1016/j.apenergy.2017.08.087)

Characterization of date palm frond as a fuel for thermal conversion processes

Hussain Sadig[1], *Shaharin Anwar* Sulaiman[2,*], *Mohamad Nazmi* Zaidi Moni[2] and *Lanisha Devi* Anbealagan[2]

[1]Department of Mechanical Engineering, Faculty of Engineeting and Technology, Nile Valley University, P.O Box 26, Atbara, Sudan
[2]Hybrid Energy Systems MOR, Universiti Teknologi PETRONAS, 32610 Seri Iskandar, Perak Darul Ridzuan, Malaysia

Abstract. Date palm fronds (DPF) have similar physical appearances to those of oil palm fronds and coconut palm fronds, which have been reported as having good potential as a source of energy through thermochemical conversion of biomass. However, nearly no report has been found pertaining to thermochemical properties of DPF. Hence, it has remained unclear whether DPF can become suitable feedstock for power generation. This study investigated the characteristics of DPF as a potential solid fuel for heat and power generation through various thermal conversion processes. DPF samples from selected sites in Sudan and Saudi Arabia were tested. The ultimate and proximate analyses and the calorific value of DPF were measured, and the results were compared with low to medium-rank coals and other common biomass materials. The calorific value range for DPF samples was found to be between 16.2 to 16.9 MJ/kg. The ultimate analysis of DPF samples revealed that more than 75% of their mass was composed of volatile materials, while the ash content in all samples was found to be less than 15%. The range of elementary carbon, hydrogen, nitrogen, sulfur and oxygen in DPF samples was found to be typical to that in biomass. The thermal decomposition trends the samples indicated the high reactivity of DPF with rising temperatures due to high holocellulose content. No distinctive differences in test results were observed between samples from Saudi Arabia and Northern Sudan. Overall, it was found that all DPF samples used in this study fulfilled the typical requirements for development and utilization as a solid fuel.

1 Introduction

Date is the fruit of date palm (Phoenix dactylifera L.). As one of the oldest known fruit crops, date has been cultivated in North Africa and the Middle East for at least 5000 years, with the earliest record of date culture discovered from the Mesopotamian region of Iraq dating back to 3000 BC[1]. Date cultivation has not only great spiritual and cultural significance but also became a sacred symbol especially to the early population of the arid regions in Middle East and North Africa. In today's world, date industry enjoys a worldwide courtesy due to the growing global demands. With reference to the online database maintained by the Food and Agriculture Organization of the United Nations (FAO), there are currently 40 countries in the world that cultivate date palms in a large scale for date trade. Most of these countries are located in the Middle East, while others are scattered at other parts of the globe. In 2014, around 7.76 million tons of dates were produced from over 1.14 million hectares of date palm plantation worldwide. The Middle Eastern countries contributed to around 40% of the total harvested area, with about 30% of the global production[2].Date palm frond (DPF) is the branching part of a date palm tree and is a waste residue of the date fruit industry. DPF is produced during pruning activity, along with other wastes such as foliar and thorns[3]. On average, a date palm tree sheds 14-20 kg of dry DPF annually, replacing with 12-15 new fronds[4-6]. Given that 1.14 million hectares of land were cultivated with date palm in 2014 and assuming an average density of 100-150 trees/ha, it is estimated that around 1.6 to 3.4 million tons of dry DPF are available each year[7]. From morphological perspective, DPF is quite similar to coconut frond (CF) and oil palm frond (OPF), and therefore is anticipated to have potentials as a source of power through thermochemical conversion process. However, little work has been done in evaluating the feasibility of utilizing DPF as a solid fuel, and thus a study is required before one can embark into power generation using DPF.

Corresponding author: shaharin@utp.edu.my

This paper examines the fuel properties of several samples of DPF collected from the North African and Middle Eastern regions for various thermal conversion process such as pyrolysis, torrefaction, gasification and as well as for general incineration purposes.

2 Methodology

Several dry DPF samples from Medina (Rothanah & Ajwah varieties) and Jeddah (Jeddah & Sukkari varieties) of Saudi Arabia and from Atbara, Northern Sudan (Mishriq variety) were received with removed leaflets via air courier. The samples were individually prepared to undergo three essential tests to evaluate their fuel properties: ultimate analysis, proximate analysis and calorific value analysis. Fig. 1 shows the as-received and the ground DPF samples.

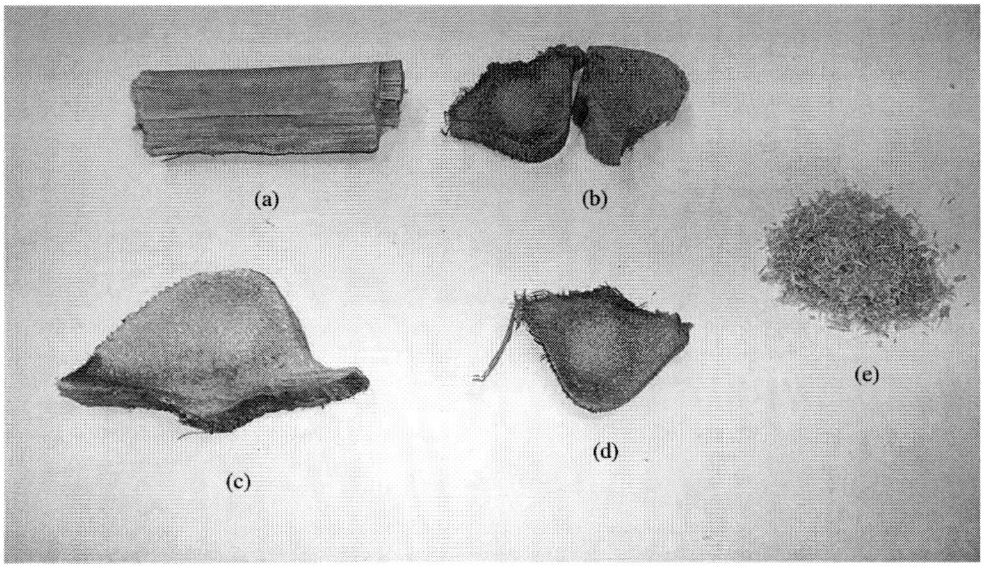

Fig. 1. As-received and ground DPF samples: (a) Mishriq variety, (b) Sukkari variety, (c) Jeddah variety, (d) Medina variety and (e) Rothanah variety (ground)

2.1 Sample preparation

Upon delivery, all DPF samples were dried in a convection oven at 105±0.5°C for 24 hours to remove remaining moisture and then stored in separated and labelled airtight containers. Around 50g of each sample were ground and sieved to obtain particles of 250 μm in apparent size for calorific value measurement and proximate and ultimate analyses. The sieved samples were stored in labelled airtight plastic containers and sent for analyses.

2.2 Calorific value

Calorific value (or heating value) is the amount of energy contained in a fuel and is determined by measuring the amount of heat produced by the complete combustion of a known quantity of it. The calorific value for each sample was determined using an IKA C-6000 isoperibol calorimeter with reference to ASTM D-5865-13 standard, and is expressed as the gross calorific value. Around 2 to 4 mg of sample was placed in the bomb vessel for calorimetric evaluation, which was conducted automatically by the device.

DOI: 10.1051/matecconf/201713101002

2.3 Proximate analysis

The proximate analysis reveals the amount of moisture, volatile matter, fixed carbon and ash content in a fuel. The proximate analysis of DPF samples was conducted using Pyris 1 TGA thermogravimetric analyzer by PerkinElmer with reference to ASTM E1131-08 standard test method. Each ground sample with a known mass was subjected to heating up to 850°C in an inert gas environment and then in oxidant gas as the temperature rose to 900°C. The TGA and derivative TGA (DTGA) trend generated by the test equipment for each sample was then examined for peaks to determine the amount of volatile matter, fixed carbon and ash based on mass loss, as well as the temperature peaks to determine the devolatilization stages during thermal degradation of DPF.

2.4 Ultimate analysis

The ultimate analysis is vital in assessing the elemental components in a biomass material in terms of carbon, hydrogen, nitrogen, sulfur and oxygen with regards to its suitability as a fuel. Otherwise known as the CHNS test, the ultimate analysis of the DPF samples was carried out using 2400 CHNS/O Series II System by PerkinElmer where the elementary carbon, hydrogen, nitrogen and sulfur components in the samples were measured by weight percentage. The oxygen content was determined by difference.

3 Results

3.1 Calorific value

Calorific value is the amount of energy contained in a specific quantity of fuel. Generally, the higher the calorific value of a fuel, the more energy will be released during its combustion. For this reason, fuel with high calorific value is more desirable for thermal conversion processes[8]. The calorific value for each of the DPF variety is shown in Table 1, along with other common biomasses and coals. As a baseline, the range of calorific value of biomass materials is typically between 15 to 20 MJ/kg, close to that of low rank coal such as peat and lignite. All DPF samples showed calorific value range of between 16.2 to 16.9 MJ/kg. The highest calorific values were obtained from the Jeddah and Sukkari varieties while the lowest was from Rothanah variety, although with no apparent difference.

Table 1. Calorific values of DPF samples and selected types of coal and biomass

Type	Sample	CV (MJ/kg)
DPF	Rothanah	16.23
	Jeddah	16.87
	Sukkari	16.86
	Ajwah	16.52
	Mishriq	16.62
	Sukkari, base[9]	15.47
	Sukkari, midrib[9]	17.57
Coal	Peat[11]	14.00
	Lignite coal[12, 13]	16.30
	Bituminous coal[12, 13]	23.25
Biomass	Corn stover[13]	17.65
	Walnut shell[13]	20.18
	Wheat straw[13]	16.71
	Cotton stalk[13]	18.26
	Coconut frond[14]	21.17
	Coconut shell[14]	21.17
	Palm kernel shell[14]	20.40
	Empty fruit bunch[14]	18.60
	Oil palm frond[14]	15.59
	Sugarcane bagasse[14]	15.25
	Rice husk[14]	14.79
	Sawdust[14]	18.19

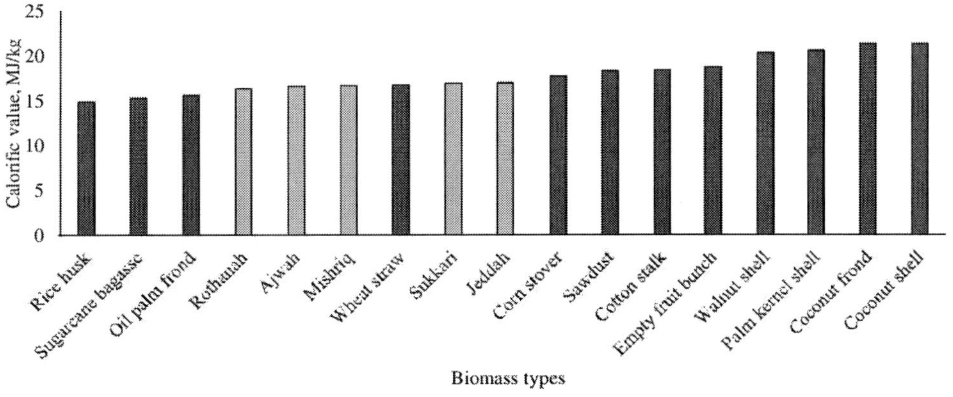

Fig. 2. Comparison of calorific value in DPF and selected biomass types

The calorific value of DPM samples were also verified with those from the work of Nasser et al, where the values were found to be in agreement with the reported figures[9]. The comparatively lower calorific values of DPF may be due to its low lignin content, as also revealed by Nasser et al in their work. Lignin is a natural polymeric fiber in plant matter that does not only contain the highest amount of energy but is also the hardest to thermally-decompose[10]. Compared to all parts of a date palm tree (coir, trunk, fruit stalk, empty fruit bunch and stone), DPF have the lowest content of lignin of around 25-26% by weight[9]. The calorific value of DPM samples was observed to be comparable with that of other biomass. All DPM samples were found to have lower calorific values compared to coconut frond, but nevertheless were higher than that of oil palm frond, as shown in Fig. 2. The calorific value of DPM samples was also found to be nearly similar to that of peat and lignite, thus confirming their potentials as solid fuels for home and industrial applications in terms of calorific value. Nevertheless, despite having comparatively lower calorific values than wood and several other biomasses, DPF is considered still suitable as a solid fuel for heat and power generation via thermal conversion processes.

3.2 Ultimate analysis

The ultimate analysis results for all DPF samples tested in this study are shown in Table 2. The range of carbon, hydrogen and oxygen in DPF was considered typical for biomass materials. The elemental carbon content in all DPF samples varied slightly with not so much differences. The same observation was made on hydrogen content, where its content was found to be between 5-6% in all samples, and was comparable with that in the selected types of coal and biomass.

The amount of oxygen in all DPF samples was found to be above 51% and was considerably higher than that in coal and most biomasses, indicating the comparatively smaller amount of stoichiometric air requirement for DPF combustion. The amount of nitrogen was highest in Rothanah and lowest in Sukkari and Ajwah varieties. The nitrogen content in all samples was found to be higher than that in peat and lignite coal, and was almost as much as in bituminous coal. The same observation was made in comparison to the DPF samples analyzed by Nasser et al and to all the featured common biomasses. Nitrogen content in fuel tends to produce oxides of nitrogen (NOx), a type of toxic and undesirable combustion byproduct[15]. The sulfur content in all DPF samples was also found to be higher than that in most biomasses, and was within the range of that in low to medium rank coals. In comparison to the DPF samples from other work, relatively low amount of nitrogen and sulfur was reported. Along with oxides of sulfur (SOx), which is the byproduct of combustion due to existence of sulfur in fuel, NOx emission has to be regulated below the allowable discharge limit by legal requirement in most countries due to their contributions to climate change and greenhouse gas effects[16]. For this reason, the utilization of fuels with high nitrogen and/or sulfur may require additional downstream filtration to reduce NOx and SOx emissions, and may pose a considerable impact to operating expenditures, and may be avoided by thermal power producers[17]. If not properly addressed, this would be the greatest drawback of utilizing DPF as a fuel material.

DOI: 10.1051/matecconf/201713101002

Table 2. Ultimate analysis of DPF samples and selected types of coal and biomass

Type	Sample	Ultimate analysis (wt. % db.)				
		C	H	N	S	O
DPF	Rothanah	32.44	5.11	3.50	0.75	58.20
	Jeddah	39.35	5.83	2.96	0.48	51.38
	Sukkari	37.78	5.79	2.88	0.34	53.20
	Ajwah	34.86	5.51	2.88	0.23	56.52
	Mishriq	36.61	5.82	3.17	0.26	54.13
	Sukkari, base[9]	40.48	5.63	0.28	0.00	53.61
	Sukkari, midrib[9]	45.65	5.95	0.27	0.00	48.13
Coal	Peat[11]	55.00	6.00	0.38	0.25	35.00
	Lignite coal[12, 13]	50.00	4.00	2.00	0.90	43.10
	Bituminous coal[12, 13]	68.00	4.00	3.00	0.90	24.10
Biomass	Corn stover[13]	43.65	5.56	0.61	0.01	50.17
	Walnut shell[13]	49.98	5.71	0.21	0.01	44.09
	Wheat straw[13]	43.20	5.00	0.61	0.11	51.08
	Cotton stalk[13]	43.64	5.81	0.00	0.00	50.55
	Coconut frond[14]	45.51	6.02	0.78	0.09	47.60
	Coconut shell[14]	43.00	6.30	0.75	0.05	49.90
	Palm kernel shell[14]	49.65	6.13	0.41	0.48	43.33
	Empty fruit bunch[14]	40.73	5.75	1.40	0.22	51.90
	Oil palm frond[14]	42.10	5.46	0.70	0.13	51.61
	Sugarcane bagasse[14]	42.93	5.82	0.68	0.06	50.51
	Rice husk[14]	38.74	5.83	0.55	0.06	54.82
	Sawdust[14]	43.68	6.65	0.23	0.04	49.40

Table 3. Proximate analysis of DPF samples and selected types of coal and biomass

Type	Sample	Proximate analysis (wt. % db.)		
		VM	FC	Ash
DPF	Rothanah	76.25	17.5	6.25
	Jeddah	88.61	9.28	2.11
	Sukkari	86.38	10.12	3.50
	Ajwah	86.10	6.26	7.63
	Mishriq	80.33	12.68	6.99
	Sukkari, base[9]	76.56	9.81	13.63
	Sukkari, midrib[9]	82.28	3.56	14.15
Coal	Peat[11]	65.00	26.50	8.50
	Lignite coal[12, 13]	34.41	48.39	17.20
	Bituminous coal[12, 13]	18.69	61.86	19.63
Biomass	Corn stover[13]	75.17	19.25	5.58
	Walnut shell[13]	78.28	21.16	0.56
	Wheat straw[13]	71.30	19.80	8.90
	Cotton stalk[13]	70.89	22.43	6.68
	Coconut frond[14]	80.24	13.06	6.70
	Coconut shell[14]	78.00	19.48	5.52
	Palm kernel shell[14]	81.03	14.87	4.10
	Empty fruit bunch[14]	84.61	9.89	5.50
	Oil palm frond[14]	83.19	12.91	1.26
	Sugarcane bagasse[14]	80.19	15.91	3.90
	Rice husk[14]	59.97	18.03	22.00
	Sawdust[14]	78.80	16.20	5.00

3.3 Proximate analysis

The proximate analysis results of DPF samples are shown in Table 3 along with the reported results for selected coal samples and biomasses. The range of volatile matter content in DPF was found to be between 76 to 89%; the highest volatile matter content was observed in Jeddah variety while the lowest was in Rothanah variety. The range was verified against those in Sukkari samples analyzed by Nasser et al and other in biomasses and was found to be in agreement with the published data. The significantly high content of volatile matter in DPF samples implies that they are highly suitable for pyrolysis and gasification processes, where the yields of solid to liquid and gas products through the thermal conversion processes may be more valuable than the heat it produces from direct incineration due to the combustion problems associated with biomass with high volatile content[15, 18]. The amount of fixed carbon in DPF samples was found to be comparatively lower than those in most biomasses. However, this observation is explained by the low amount of lignin component in DPF, which makes up for most of the fixed carbon in biomass fuel[19]. The ash content in DPF samples was found to be comparable with those in the featured biomasses, and was lower than those in peat, lignite and bituminous coals. This advantage makes DPF a highly appropriate fuel for continuous thermal processes, where ash removal and handling is a common technical barrier[20].

3.4 Thermal decomposition stages

The temperature-dependent devolatilization stages for each DPF samples were identified based on TGA trends and DTGA peaks. Four thermal breakdown regions – 1st stage of active pyrolysis, 2nd stage of active pyrolysis, passive pyrolysis and char combustion – were identified. The identified stages of thermal breakdown of DPF are shown in Table 4. It was found that the 1st active pyrolysis stage for all DPF samples began at around 141±4.5°C, where light hydrocarbon components and traces of moisture were released. The 1st active pyrolysis stage progressed until the temperature reached around 325°C for all samples, where the 2nd active pyrolysis stage began, resulting to the initial release of heavier hydrocarbon compounds and liquid products such as pyrolysis oil and light tar. Passive pyrolysis began at above 500°C for all samples except for Rothanah and Jeddah varieties, where the stage began at 457°C and 478°C respectively, signifying the high reactivity of the sample against rising temperature. Heavier carbon aromatic compounds were released and broken at this stage as the temperature rose. The solid residue at this point in all samples was majorly made up of char, where it underwent combustion at temperatures above 600°C except for Rothanah at 552°C. All samples were found to have lost half of their initial mass at around 320-340°C, where the 2nd stage active pyrolysis was ongoing, signifying that more than half of the tested DPF samples' mass was composed of hemicellulose and cellulose – the two most reactive natural polymeric fibers that break down at this particular thermal degradation stage. These findings were verified with the reported hemicellulose and cellulose content in Sukkari variety; hemicellulose content was reported to be between 43-45% by weight, and cellulose content was in the range of 28-31% by weight[9]. The high reactivity of DPF samples at low temperatures releasing high amount of volatiles revealed their potentials for pyrolysis and low temperature gasification purposes.

Table 4. Thermal decomposition temperatures by stages for DPF samples

DPF Sample	Temperature, °C				
	T_1	T_2	T_3	T_4	T_{hm}
Rothanah	135.57	325.59	457.29	551.96	329.23
Jeddah	138.24	329.41	478.11	637.21	322.41
Sukariah	141.27	324.59	507.77	654.52	329.88
Ajwah	147.11	327.86	515.41	681.71	333.14
Mishriq	146.25	330.62	519.79	679.77	336.63

T_1: initial 1st phase active pyrolysis;
T_2: initial 2nd phase active pyrolysis temperature;
T_3: initial passive pyrolysis temperature;
T_4: char combustion temperature;
T_{hm}: temperature where sample achieved half mass loss.

4 Conclusions

From the analysis results of DPF samples and in comparisons with the characteristics of selected low to medium rank coals and common biomass fuels, it was concluded that all DPF samples satisfied the typical requirements as solid fuel for thermal conversion processes in the heat and power generation sector. With its high and constant annual production amount, DPF has high potentials to be utilized as solid fuel for thermal applications in the Arab

Peninsula and North African countries. The arid origins of DPF make its moisture removal faster, easier and cheaper through sun drying technique, all the more reasons to utilize the otherwise undervalued and underused residues as fuel materials. However, the processing aspect in terms of sizing and storage needs to be addressed in order to fully review DPF full potentials. Such investigation is therefore recommended for the progression of this study.

References

1. C.T. Chao, R.R. Krueger, Hort. Sci. **42** (5), 1077-1082 (2007)
2. FAO Statistics website (www.fao.org/faostat)
3. M. Shoaib, H.A. Al-Swaidan, Asian J. of Chem. **26** (20), 7025-7028 (2014)
4. A. Chehma, H. F. Longo, Rev. Energ. Ren. : Production et Valorisation – Biomasse, 59-64 (2001)
5. J. J. Pascual, C. Fernandez, J. R. Diaz, C. Garces, J.J Rubert-Aleman, Arid Environ. **45**, 183-189 (2000)
6. S.A. Sulaiman, H.S. Bamufleh, S.N.A. Tamili, M. Inayat, M.Y. Naz, Bul. Chem. Soc. Ethiopia **30** (3), 465-472 (2016)
7. A. Zaid, P.F. de Wet, *Date Palm Cultivation* (FAO, Rome, 2002)
8. Kalyani A. Motghare, Ajit P. Rathod, Kailas L. Wasewar, Nitin K. Labhsetwar, Waste Mgmt. **47** Part A, J 40-45 (2016)
9. R.A. Nasser, M.Z.M. Salem, S. Hiziriglu, H.A. Al-Mefarrej, A.S. Mohareb, M. Alam, I.M. Aref, Energies **9** (374) (2016)
10. M.D. Bispo, J.A.S. Barros, D. Tomasini, C. Primaz, E.B. Caramao. C. Dariva, L.C. Krause, J. Earth Sci. Eng. **6**, 235-244 (2016)
11. FAO, *Energy use of peat* (http://www.fao.org/docrep/x5872e/x5872e0b.htm.)
12. A.G. Korotkokh, K.V. Slyusarskiy, K.B. Larionov, V.I. Osipov, J. Phy. Conference Series **754** (2016)
13. Y. Demirel, "Energy, energy types," in *Green Energy, Technology* (Springer-Verlag, London, 2012)
14. S.S. Abdullah, S. Yusup, J. App. Sci. **10** (24), 3301-3306 (2010)
15. B.M. Jenkins, L.L. Baxter, T.R. Miles Jr, T.R. Miles, Fuel Processing Tech. **54**, 17-46 (1998)
16. A.E.D.M.M. Ibrahim, "NOx, SOx emissions, climate changes," World App. Sci. J. **31** (8), 1422-1426 (2014)
17. C. Bibbiani, F. Fantozzi, C. Gargari, C.A. Campiotti, E. Schettini, G. Vox, Agriculture, Agricultural Sci. Proc. **8**, 637-645 (2016)
18. A.H. Tchapda, S.V. Pisupati, Energies **7**, 1098-1148 (2014)
19. A. Demirbas, Energy Conv. & Mgmt. **44** (9), 1481-1486 (2003)
20. Liang Wang, Johan E. Hustad, Øyvind Skreiberg, Geir Skjevrak, Morten Grønli, Energy Proc. **20**, 20-29 (2012)

Investigation of thermal behaviour, pressure drop, and pumping power in a Cu nanofluid-filled solar flat-plate collector

S. Reza Shamshirgaran[1, 2], *Morteza* K. Assadi[1,*], *Hussain H.* Al-Kayiem[1] and *K.V.* Sharma[3]

[1] Universiti Teknologi PETRONAS, Department of Mechanical Engineering, 31750, Tronoh, Perak, Malaysia.

[2] Shahid Beheshti University, Department of Mechanical and Energy Engineering, Tehran, 16765-1719, Iran.

[3] Jawaharlal Nehru Technological University, Department of Mechanical Engineering, Hyderabad, 500085, India.

Abstract. The evaluations of the performance of solar flat-plate collectors are reported in the literature. A computer program developed by MATLAB has been applied for modelling the performance of a solar collector under steady state laminar conditions. Results demonstrate that Cu-water nanofluid would be capable of boosting the thermal efficiency of the collector by 2.4% at 4% volume concentration in the case of using Cu-nanofluid instead of just water as the working fluid. It is noteworthy that, dispersing the nanoparticles into the water results in a higher pressure drop and, therefore, a higher power consumption for pumping the nanofluid within the collector. It has been estimated for the collector understudy, that the increase in the pressure drop and pumping power to be around 30%.

1. Introduction

Solar flat-plate collectors (FPCs) as an effective renewable-based device for the energy supply of domestic and industrial sectors, provide a potential for the performance improvement with the aid of nanofluids. Nanofluids would be able to change the thermophysical properties of a base liquid once they were dispersed as additives. Conductivity, amongst the other thermophysical properties, is most affected by nanoparticles; however, the other properties of base fluids, such as viscosity, heat capacity, and density are not excluded from the change. The interactions between these properties of nanofluids during the heat transfer phenomenon, and flowing in the pipes, will determine the overall performance of solar collectors. In fact, the conductivity enhancement will lead to an improvement in the performance of a FPC whilst the increase of viscosity and density will result in a higher pressure drop and, consequently, a higher pumping power in a forced circulation solar FPC. Nasrin et al. [1] conducted a numerical method to study the thermal behaviour of a FPC using four different water-based oxide nanofluids. Said et al. [2] analysed the pressure drop and heat transfer in a FPC with the SWCNT-water working

* Corresponding author: morteza.assadi@utp.edu.my

fluid. Lazarus et al. [3] examined the effect of the silver-water nanofluid on the thermal performance of a FPC by experimentation. Alim et al. [4] investigated the enhancement of the heat transfer performance and pumping energy required in the case of a water mixture with different oxide nanoparticles as the working fluid. Verma et al. [5] carried out an experimental research on the variety of oxide nanoparticles to realise the changes in efficiency and pumping power. Mahian et al. [6] undertook the study of heat transfer and pressure loss affected by nanofluids in a minichannel FPC.

This paper is concerned with evaluating the performance of a solar FPC working with a copper/water nanofluid. It is assumed that there is a constant velocity in the collector pipe to take into account the change of the mass flow rate by adding the nanoparticles. Furthermore, most of research studies have been carried out so far, have considered the same temperature for the inlet of the collector and the ambient temperature. Such a situation happens rarely in practice because it shows the conditions in which no thermal absorption takes place and only the optical performance of the collector plays the role of the heat transfer. This paper has considered a non-zero value for the difference of the inlet and ambient temperatures.

2. Problem description

The modelling and thermal behaviour evaluation of a solar FPC working with a Cu/water nanofluid has been considered in this study. The technical data of the collector understudy taken from the G-Series FPCs of Thermo Dynamics Ltd., are given in Table 1.

Table 1. The collector technical data

Collector Parameter	Value [unit]	Collector Parameter	Value [unit]
Collector tilt angle	25 [°]	Conductivity of bottom insulation	0.036 [W/m.K]
Optical efficiency	85 [%]	Conductivity of edge insulation	0.036 [W/m.K]
Tube spacing (pitch)	143 [mm]	Conductivity of bond	237 [W/m.K]
Number of risers	8	Length of bond	18 [mm]
Number of glass covers	1	Average thickness of bond	0.5 [mm]
Dimension [cm]	250×120×8.6	Emissivity of absorber plate	25 [%]
Riser tube diameter (in/out)	7 and 8 [mm]	Emissivity of glass cover	84 [%]
Header pipe diameter	14 [mm]	Wind velocity	3.2 [m/s]
Solar irradiance on collector	808 [W/m^2]	Conductivity of Al. absorber	237 [W/m.K]
Thickness of insulations	25 [mm]	Thickness of absorber plate	0.5 [mm]

3. Thermophysical properties of nanofluids

The main thermophysical properties of nanofluids consist of density, ρ, heat capacity, C_p, thermal conductivity, k, and viscosity, μ. The models and correlations for the estimation of the nanofluid properties have been presented in Table 2. In these equations, ϕ and d_p stand for the nanofluid volume concentration and diameter, and $\kappa_B = 1.381\times10^{-23}$ J/K is the Boltzmann constant.

4. Thermal efficiency of the solar collector

The thermal efficiency of a solar FPC is given by [7] as follows:

$$\eta_{eng} = \frac{Q_u}{A_c G_t} \quad (1)$$

where the useful heat gain, Q_u, from the available solar energy ($A_c . G_t$) is obtained from the following relations:

$$Q_u = \dot{m}C_p(T_{out} - T_{in}) = A_c[S - U_L(T_{pm} - T_a)] \tag{2}$$

which is related to the heat capacity and mass flow rate of the working fluid and its inlet and outlet temperatures. The first two are influenced by adding the nanoparticles and as a result, the outlet temperature will be affected. As observed from the last equation, the useful heat collection also depends on the effective radiation on the absorber of the FPC, $S = \eta_o G_t$, in which η_o represents the collector's optical efficiency; it is the product of the transmittance of the glazing and absorptance of the absorber, i.e., $\eta_o = (\tau\alpha)_{avg}$. Here T_{pm} and T_a are the absorber plate mean temperature and the ambient temperature in K.

Table 2. The correlations for prediction of the nanofluid properties

Property	Model	Correlation
Mass density	Principle of mixtures	$\rho_{nf} = \phi\rho_p + (1-\phi)\rho_{bf}$
Specific heat	Thermal equilibrium	$C_{p,nf} = \dfrac{\phi(\rho C)_p + (1-\phi)(\rho C)_{bf}}{\rho_{nf}}$
Thermal conductivity	Xuan	$\dfrac{k_{nf}}{k_{bf}} = \dfrac{k_p + 2k_{bf} - 2\phi(k_{bf} - k_p)}{k_p + 2k_{bf} + \phi(k_{bf} - k_p)} + \dfrac{\rho_p \phi C_p}{2k_{bf}}\sqrt{\dfrac{2\kappa_B T}{3\pi d_p \mu_{bf}}}$
Dynamic viscosity	Brinkman	$\dfrac{\mu_{nf}}{\mu_{bf}} = \dfrac{1}{(1-\phi)^{2.5}}$

5. Heat transfer criteria

The internal convection heat transfer, h_{fi}, appears in the collector's efficiency coefficient in the form of:

$$F' = \frac{1/U_L}{W[1/[U_L(D + (W - D)F] + 1/C_b + 1/\pi D_i h_{fi}]} \tag{3}$$

In fact, the factor F', as seen from Eq. (3), related the collector design parameters including tube pitch, W, diameter, D, fin efficiency, F, and bond conductance, C_b, to the operational parameters, such as the overall heat loss, U_L, from the collector enclosure. The overall heat loss consisted of three main components, i.e., from the back, edge and top sides, and the method of their calculation has been discussed in detail by [7] and [8]. On the other hand, the heat transfer coefficient is proportional to the Nusselt number and the conductivity of the nanofluid flow within the collector. The Nusselt number will be determined depending on the Reynolds number and Prandtl number. These dimensionless numbers are evaluated by means of the equations below:

$$h_{fi} = \frac{\mathrm{Nu}\, k_{nf}}{D_i} \quad (4) \qquad \mathrm{Re}_{nf} = \frac{\rho_{nf} V D_i}{\mu_{nf}} = \frac{4\dot{m}_{riser}}{\pi D_i \mu_{nf}} \quad (5) \qquad \mathrm{Pr}_{nf} = \frac{\mu_{nf} C_{p,nf}}{k_{nf}} \quad (6)$$

It can be seen that, the nanoparticles performed their duty by altering the properties of the working fluid and, consequently, via changing the aforementioned dimensionless numbers. In the laminar flow, i.e., Re < 2300, the heat transfer performance will be determined based on the thermal boundary condition. For the short circular tubes in which the hydrodynamic and thermal boundary layer are developing, the Heaton correlation is used to estimate the local Nusselt number with a constant heat flux:

DOI: 10.1051/matecconf/201713101003

$$\mathrm{Nu} = \mathrm{Nu}_\infty + \frac{a(\mathrm{Re}\,\mathrm{Pr}\,D_h/L)^m}{1 + b(\mathrm{Re}\,\mathrm{Pr}\,D_h/L)^n} \quad (7)$$

where D_h and L are the hydraulic diameter and tube length, respectively.

The Goldberg's correlation is essentially similar to Heaton's but with different coefficients of *a, b, m,* and *n*. It will be useful for the evaluation of the average Nusselt number under constant-wall temperature conditions. The values of the constant numbers are specified in Table 3.

Table 3. Constants of Heaton's and Goldberg's correlation for the Nu. Number

Pr.		*a*		*B*		*m*		*n*	
q = cte	**T= cte**	**q = cte**	**T= cte**	**q = cte**	**T= cte**	**q = cte**	**T= cte**	**q = cte**	**T= cte**
0.7	0.7	0.00398	0.0791	0.0114	0.0331	1.66	1.15	1.12	0.82
10	5	0.00236	0.0534	0.00857	0.0335	1.66	1.15	1.13	0.82
∞	∞	0.00172	0.0461	0.00281	0.0316	1.66	1.15	1.29	0.84
$\mathrm{Nu}_{\infty,q=cte} = 4.4$					$\mathrm{Nu}_{\infty,T=cte} = 3.7$				

The constants *a, b, m,* and *n* may be interpolated for the different Prandtl numbers from the values of Table 3.

6. Pressure drop and pumping power

The pressure drop of the nanofluid flow across the collector risers is calculated at a tilt angle β by applying the Bernoulli equation to the riser tube inlet and outlet. The overall pressure loss in the FPC, including the riser tubes, manifolds, and fittings is estimated as follows. Then, the pumping power can be predicted based on the overall pressure loss:

$$\Delta p = [f\frac{L}{D}\rho_{nf}\frac{V^2}{2}]_{in/out} + [\rho_{nf}g(L\sin\beta + h_L)]_{riser+fitting} \quad (8)$$

$$h_L = \frac{8\dot{m}_{riser}^2}{\rho_{nf}\,g\,\pi^2 D_i^4}(f\frac{L}{D_i} + \sum K_L) \quad (9) \qquad P = \frac{\dot{m}}{\rho_{nf}} \times \Delta p \quad (10)$$

where K_L is the coefficient loss taken as 2.0 for the laminar flow.

7. Results and discussion

In the present study, 25nm copper (Cu) nanoparticles were selected to analyse their effect on the performance of the collector under a steady state condition. The ambient temperature, its minimum difference with the inlet temperature, the sun's temperature, and the flow velocity in the header pipe were assumed to be 301 K, 2 K, 5770 K, and 0.3 m/s, respectively. An iterative method was required to obtain the FPC outlet temperature and its absorber temperature as well. MATLAB software was employed for solving the governing equations by a reasonable assumption for the plate's mean temperature as an initial guess. Assigning the given values, the desired parameters were evaluated and the results have been presented.

The addition of nanoparticles causes the specific heat of the fluid to decrease and its mass flow rate to increase, resulting from the density enhancement. The contribution of the specific heat to the change of the outlet temperature is more than density, which results in an increased outlet temperature according to Eq. (2). Therefore, the collector fluid's outlet temperature increases with the Cu nanofluid concentration. The mass flow innately influences the collector's performance via the mean temperature of the plate [8]; the higher the mass flow the lower the plate's temperature and, consequently, the lower the heat loss from the collector. As a consequence, a 3K drop in the plate's mean temperature is observed, leading to a lower heat loss and, hence, a higher heat gain as seen from Figure 1. As a result, the thermal efficiency of the FPC increased due to the greater amount of heat being absorbed by the collector. Figure 2 confirms this observation drawn for the different concentrations of the nanoparticles. The efficiency of the collector with a 25nm-copper-water working fluid improved from 70.3% for plain water to 72.7% with a 4% Cu/water nanofluid.

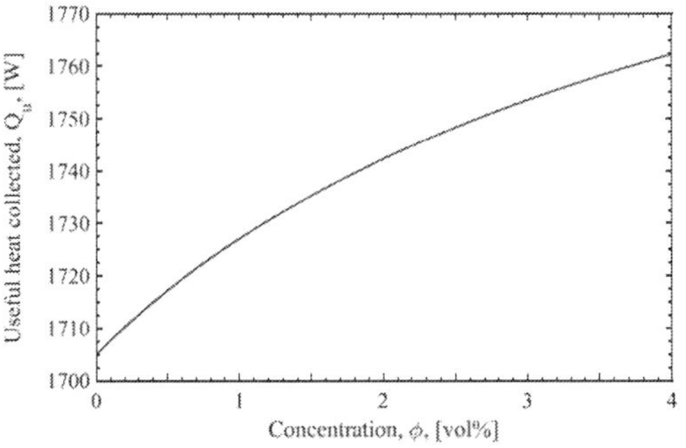

Fig. 1. Heat gain of the FPC with Cu nanofluid

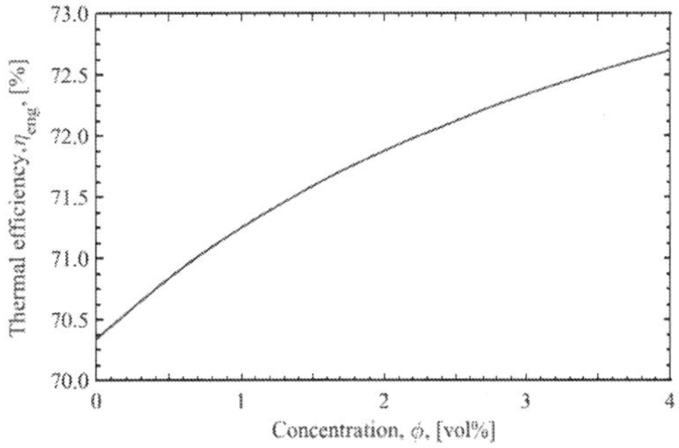

Fig. 2. Variations of the FPC thermal efficiency

The thermal behaviour of the collector in terms of the Nusselt number and heat transfer coefficient is depicted in Figures 3 and 4, respectively. Considering Eq. (6), the Prandtl number relates three important thermophysical properties of the water-copper nanofluid, i.e., specific heat, viscosity, and conductivity. With an increase of the volume fraction, the specific heat decreases, whilst the viscosity and conductivity increase. However, the values of the heat capacity and conductivity would be greater than the viscosity enhancement. The final interaction between the properties imposes a decrease of the Prandtl number with the volume concentration. The decrease of the Prandtl number overcomes the increase of the Reynolds number which causes the Nusselt number to be decreased, overall. The variation of the Nusselt number is not too much and sometimes it is assumed to be constant. Therefore, a small decrease in the Nusselt number is reported with the nanofluid concentration as shown in Figure 3.

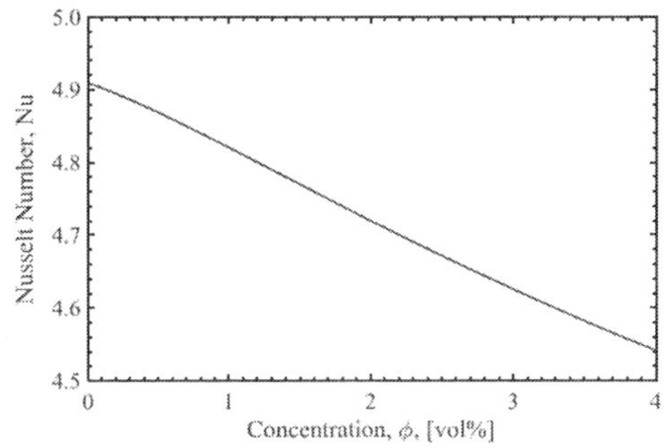

Fig. 3. Nusselt number of nanofluid in the FPC

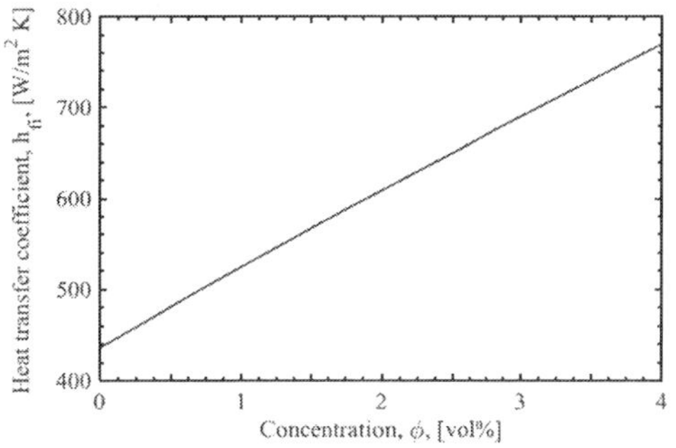

Fig. 4. Heat transfer coefficient of the FPC

Assuming either a constant or an insignificantly enhanced Nu number, it can be observed that the convective heat transfer coefficient shown in Figure 4 goes up based on Eq. (4). The increase would be more than 1.7 times with the 4vol% nanofluid concentration when compared to the values with water. The trends agree with [6].

Referring to Eq. (8) the pressure drop increases due to the addition of the copper nanoparticles and the resultant density enhancement when compared to the pure water. Figure 5 shows that the pressure drop could have been as high as 30% compared to the pure

 DOI: 10.1051/matecconf/201713101003

water. Owing to the pressure drop taking place in the collector pipes when using the nanofluid, the circulating pump consumed almost 30% more power at the maximum nanoparticle volume concentration according to Figure 6.

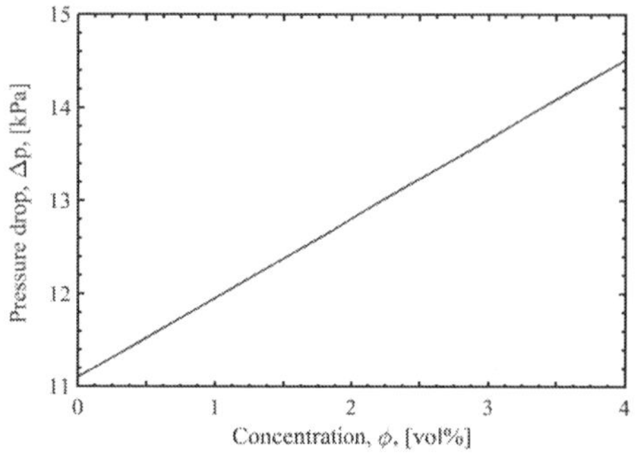

Fig. 5. Effect of nanoparticles on pressure drop

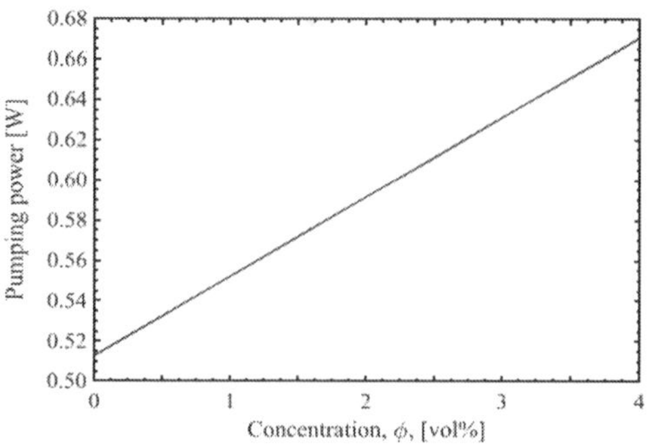

Fig. 6. Increased pumping power of the FPC

This affirms that although using nanoparticles, on one hand, improves the thermal behaviour of the collector by enhancing the conductivity, on the other hand, it worsens the pressure drop and pumping power via enhancing the viscosity and density of the flow. The quantities depend on the nanoparticle type and concentration.

Conclusions

In the current study, an analytical investigation has been undertaken using a computer program developed in MATLAB to determine the effect of copper nanoparticles on the thermal behaviour, pressure drop, and pumping power of a solar flat-plate collector under steady state laminar condition. The Results show that adding the copper nanoparticles into the water enhanced the energetic efficiency of the collector from 70.3% to 72.7%. Also, it can be deduced that using a copper-based nanofluid instead of water as the working fluid will lead to the increase of the collector's pressure drop and pumping power by approximately 30% at the maximum volume concentration.

The authors acknowledge Universiti Teknologi PETRONAS (UTP) for supporting the research through Yayasan University Technology PETRONAS (YUTP) grant number 0153AA-A96.

References

1. R. Nasrin, S. Parvin and M. Alim, Procedia Engineering **90**, 364-370 (2014).
2. Z. Said, R. Saidur, N. A. Rahim and M. A. Alim, Energy and Buildings **78**, 1-9 (2014).
3. G. Lazarus, S. Roy, D. Kunhappan, E. Cephas and S. Wongwises, Journal of Thermal Engineering **1** (2), 104-112 (2015).
4. M. Alim, Z. Abdin, R. Saidur, A. Hepbasli, M. Khairul and N. Rahim, Energy and Buildings **66**, 289-296 (2013).
5. S. K. Verma, A. K. Tiwari and D. S. Chauhan, Energy Conversion and Management **134**, 103-115 (2017).
6. O. Mahian, A. Kianifar, A. Z. Sahin and S. Wongwises, Journal of Heat Transfer **137** (6), 061011 (2015).
7. J. A. Duffie and W. A. Beckman, *Solar engineering of thermal processes*, 4th ed. (Wiley New York etc., 2013).
8. S. A. Kalogirou, *Solar energy engineering: processes and systems*. (Academic Press, 2013).

Numerical analysis of solar updraft power plant integrated with external heat source

Mohammed A. Aurybi[1, 2], *Hussain H.* Al-Kayiem[1,*], *Syed I. U.* Gilani[1] *and Ali A.* Ismaeel[1]

[1] Department of Mechanical Engineering, Universiti Teknologi PETRONAS, 32610 Bandar Seri Iskandar, Perak, Malaysia.
[2] Department of Mechanical Engineering, University of Baghdad, Baghdad, Iraq

Abstract. Many countries started to adopt clean energy sources, where solar energy is the first priority. Solar chimney power plant was selected in this study as a system which converts solar thermal energy to electrical power. In this research, hybrid solar chimney power plant model has been proposed as a day and night functional integrated system using external heat source to enhance the system performance and cover the setback of solar absence at night and cloudy days. With the help of ANSYS Fluent software, three-dimensional steady-state of Navier-Stokes and energy equations have been solved. Realized $k-\varepsilon$ two equation turbulent model equations and discrete ordinates (DO) radiation model equations were solved for the conventional model. The results were validated using experimental measurements. Afterward, the influence of thermal enhancing channels installation on the system performance was predicted and analysed in hybrid mode. The simulation results showed that this system could enhance the plant during the day. Percentage of enhancement was 10% and 14% for velocity and temperature respectively, and the power output percentage enhancement was 18% when the solar intensity was 1000 W/m^2, and external heat source mass flow rate and the temperature for each channels was 0.015 Kg/s and 100°C respectively, in comparison with the conventional model.

1 INTRODUCTION

Solar updraft power plant (SUPP) is considered as a promising technology for utilizing solar energy for power generation. The plant consists of three main components: collector, chimney, and turbine. The collector is circular in shape and composed of a transparent roof which allows the solar radiation to penetrate and then absorbed by the ground. The gap between the collector canopy and the ground allow the air movement underneath it. The chimney is installed in the center of the collector and the turbine is fixed at the chimney base. By the effect of solar radiation through the canopy, heat is absorbed by the ground which in turn heats up the air inside the collector by natural convection then the hot air flow and rise to the chimney by the bouncy effect.

* Corresponding author: hussain_kayiem@utp.edu.my

The updraft inside the chimney base rotate the turbine and then generate electricity by attached the turbine shaft with a generator. The main constraint of the SCPP is its low efficiency (less than 2% - total plant efficiency) and the interruption during the night, cloudy and rainy times [1].

Most of the researchers investigated the plant main parts and try to enhance the system. In the same line, [2] proposed a hybrid type of materials as thermal energy storage technique to enhance the SCPP power generation. [3], also proposed hybrid SCPP to enhance the power production at day and cloudy period operation or at night. The model was combined Geothermal/PV/SCPP. On the other side, there are many applied technologies which use and transform the waste energy to useful form. SCPP depending on the solar as the main energy source and some researchers suggested using another source with this system to enhance the performance and to survive the system from power decline at night. Ref. [4] introduced a mathematical model of hybrid SCPP by suggest installing thermal enhancing channels underneath the collector canopy. The proposed channels can be work as a heat exchanger.

[5] and [6] proposed hybrid SCPP model to enhance the plant performance by utilizing metal sheet as the absorber plate underneath the collector instead of the soil, the absorber plate stands as a heat exchanger which extract heat from the hot flue gas that flows through the bottom part of the collector while in the top/front of the absorber plate there was a collector which absorbed the solar energy to warm up the air inside it.

The main challenge of the SCPP is the poor electricity generation at night, so the researchers proposed a new power plant plan by using the SCPP with external heat source.

The main objective of this paper is to introduce a new enhancement technique to improve the performance of SCPP and ensure the continuity of its production for 24 hours a day throughout a week. Furthermore, the aim of this research is to study and analyze numerically the physical properties and heat transfer of air flow inside the collector and chimney to show the enhanced performance of the system with the help of thermal enhancing channels installations.

2 Proposed model description

This research presents a hybrid solar chimney power plant as a proposed model, as shown in figure.1. This model would not only increase the efficiency of the power plant during the day, but it would also provide power even when there is no at night time and cloudy days. The technique in this model utilizes the flue gasses or exhausts gasses from a nearby gas turbine power plant or any external waste heat source and converts the waste thermal energy in these exhausts to thermal energy in the SCPP, considering a new form of waste heat energy recapture. Thus, it is an ideal way to reduce global warming. In this way, the thermal energy of the air flow is increased in the solar collector of the plant by using hollow rectangular channels, known as Thermal Enhancing Channels (TECh), situated below the collector cover. The exhaust or flue gases are pushed through the channels by pipe holes and the gases passing through the channels cause heat transfer from walls of the channels to the air flowing below the collector (greenhouse). This phenomenon indicates that the air receives more kinetic energy in addition to that obtained from solar radiation, which in turn increases the temperature of air in an outlet collector, resulting in the enhanced performance of the plant. At the same time, this heat source helps the plant to operate at night. After that, the gases emerging from the channels enter the chimney and increase the updraft air velocity of the exit air flow from the chimney and that called "Co-generation".

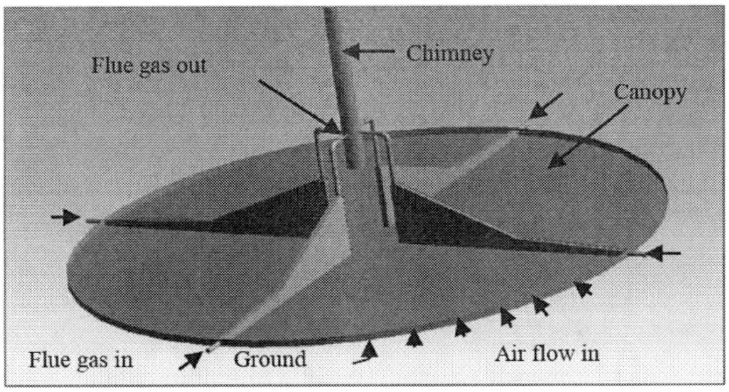

Fig. 1. Schematic diagram of proposed model.

3 Numerical solution

The fluid flow inside the model is solved through the conservation equation of mass, momentum, and energy. The governing equations, presented below, have been solved in three dimensions (3D).

Continuity equation:

No change in mass with time is assumed and when the density is constant, the conservation of mass equation can be written as:

$$\left(\frac{\partial u}{\partial x}+\frac{\partial v}{\partial y}+\frac{\partial w}{\partial z}\right)=0 \tag{1}$$

Conservation of momentum:

The x, y, z components of the momentum equation are determined by equations (2), (3), and (4), respectively:

$$\rho \nabla.(uV) = -\frac{\partial p}{\partial x}+\frac{\partial \tau_{xx}}{\partial x}+\frac{\partial \tau_{yx}}{\partial y}+\frac{\partial \tau_{zx}}{\partial z}+S_x \tag{2}$$

$$\rho \nabla.(vV) = -\frac{\partial p}{\partial y}+\frac{\partial \tau_{xy}}{\partial x}+\frac{\partial \tau_{yy}}{\partial y}+\frac{\partial \tau_{zy}}{\partial z}+S_y \tag{3}$$

$$\rho \nabla.(wV) = -\frac{\partial p}{\partial z}+\frac{\partial \tau_{xz}}{\partial x}+\frac{\partial \tau_{yz}}{\partial y}+\frac{\partial \tau_{zz}}{\partial z}+S_z \tag{4}$$

The source terms S_x, S_y, and S_z in (2), (3) and (4) represent the body force.

In equations (2, 3, and 4) the Boussinesq model is used by setting up the problem with fluid density as a function of temperature. This model treats density as a constant value in all solved equations, except for the buoyancy term in the momentum equation:

$$(\rho) = \rho_{\circ}(1-\beta\Delta T) \tag{5}$$

Equation (5) is valid when $(1\text{-}\beta\Delta T) \ll 1$

Conservation of energy:

The general flow equation after considered a constant thermal conductivity and specific heat of fluid elements as follows:

$$\rho C_p\left(\frac{\partial(uT)}{\partial x}+\frac{\partial(vT)}{\partial y}+\frac{\partial(wT)}{\partial z}\right) = k\left(\frac{\partial^2 T}{\partial x^2}+\frac{\partial^2 T}{\partial y^2}+\frac{\partial^2 T}{\partial z^2}\right)+S_E \tag{6}$$

3.4 Turbulence and radiation models

Realized $k-\varepsilon$ two equation turbulent model which considered one of the most widely used turbulent models is used in the numerical simulation. In this model the turbulent kinetic energy and dissipation rate can be computed as follows:

$$\rho\nabla.[\bar{v}k] = \nabla.[\mu_{eff,k}\nabla k] + S_k \tag{7}$$

$$\rho\nabla.[\bar{v}\varepsilon] = \nabla.[\mu_{eff,\varepsilon}\nabla\varepsilon] + S_\varepsilon \tag{8}$$

Where:
S_k and S_ε are the kinetic energy and dissipation rate source term.

The enhanced wall treatment ε-Equation is used in numerical solution as recommended by many literatures [10] to solve the near wall problem. Discrete Ordinates (DO) Radiation Model is used to solve the radiation problem. The DO model considers the radiation transfer equation for scattering, emitting and absorbing media at position $\vec{r}$ in the direction $\vec{s}$ after using the gray-band model and modeling the non-gray radiations as follows:

$$\nabla.(I_\lambda(\vec{r},\vec{s})\vec{s}) + (\alpha_\lambda + \sigma_s)I_\lambda(\vec{r},\vec{s}) = \alpha_\lambda n^2\, I_{b\lambda} + \frac{\sigma_s}{4\pi}\int_0^{4\pi} I_\lambda(\vec{r},\vec{s})\phi\,(\vec{s},\vec{s})d\acute{\Omega} \tag{9}$$

$\vec{r}$ = position vector, $\vec{s}$= direction vector, $\vec{s}$= scattering direction vector, α=absorption coefficient, n=refractive index, σ_s= scattering coefficient, σ= Stefan-Boltzmann constant ($5.669\times10^{-8}\ W/m^2.K^4$), I= radiation intensity, which depends on $\vec{r}$ and$\vec{s}$, T= temperature, ϕ = face function, $\acute{\Omega}$= solid angle.
λ = wave length, a_λ = spectral absorption coefficient, $I_{b\lambda}$= black body intensity.

4 Boundary conditions and dimensions

First, the Eqns. (1) – (9) were solved numerically with conventional model geometry which was implemented in the solar research site (SRS) at University Technology PETRONAS, as shown in figure 2. The dimensions of the model are given in Table 1 and the boundary conditions are presented in Table 2. The value of solar radiation adopted in the simulation was 1000 W/m^2 and ambient temperature was 307 K.

Table 1. Dimensions of the conventional model.

Parameter	Dimension(m)
Collector radius	6
Collector height at inlet	0.05
Collector height at outlet	0.65
Chimney height	6.5
Chimney Diameter	0.15

Table 2. Boundary conditions for conventional model

Zone	Type	Value
Ground	Interface wall	Interface
Canopy	Semi-transparent wall	h = 8 w/m^2 T_{amb}=307 k
Chimney	wall	Adiabatic
Collector inlet	Pressure inlet	ΔP=0, T_{amb}=307 k
Chimney outlet	Pressure outlet	ΔP=0, T_{amb}=307 k

DOI: 10.1051/matecconf/201713101004

Fig. 2. Implemented solar chimney conventional model.

Physical properties of the materials employed in the simulation are given in Table 3. A 3-D and steady state incompressible flow with second-order upwind discretization and pressure-based coupled algorithm were used as a solver in CFD simulation.

Table 3. Properties of materials used in the simulation

Physical property	**Perspex**	**Ground (black pebble)**	**Chimney (PVC)**
Absorption coefficient	0.06	0.9	0.04
Transmission coefficient	0.92	0	0
Density ρ (kg/m^3)	2700	2640	833
Specific heat (J/kg.K)	840	820	1170
Thermal conductivity (W/m.K)	0.78	1.73	0.19
Emissivity	0.9	0.9	0.91

4.1 Grid independence

It is logic to use fine mesh with a minimum mesh sizing to give a more accurate solution, but that means more computer cost and time. Grid independence is achieved by starting with coarse mesh which was defaults selection by mesh generation code, then medium and fine mesh. After that more reduction in grid size was conducted until the no change in output parameters is done. The structured mesh tetrahedral type was used to achieve the mesh generation for conventional SCPP model and mesh sizing (Maximum size 0.03 m and minimum size 0.001 m) were used to get the solution for the conventional model with percentage error 5% for the velocity at the chimney base comparing with the experimental value, and same mesh size with unstructured mesh hexahedral type were used for proposed model after mesh independence test as shown in figure. 3.

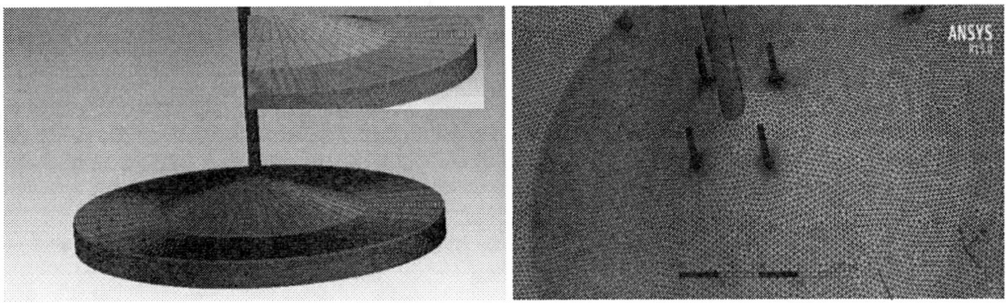

(a) Structured mesh tetrahedral type. (b) Unstructured mesh hexahedral type.

Fig. 3. Structure and unstructured mesh generation for the conventional and proposed model.

5 Result and discussion

Velocity and temperature contour maps were considered based on simulation results with no load condition and solar intensity of 1000 W/m^2. It could be observed that by simulation, the maximum velocity was observed at the chimney base and the temperature distributions increased when the solar insolation increased and this due to the solar intensity raised the temperature of storage media (pebble) inside the collector. Maximum temperature and velocity in simulation for the conventional model were in good agreement with the experimental readings depicting logical and reasonable error ratio. The maximum velocity for the numerical solution of the convenience model was 1.97 m/s and the temperature was 325K (ΔT= 18°C) as shown in figure. 4. Same boundary conditions are given in Table 2 with the addition of channels at hybrid mode with the external heat source is provided at this stage.

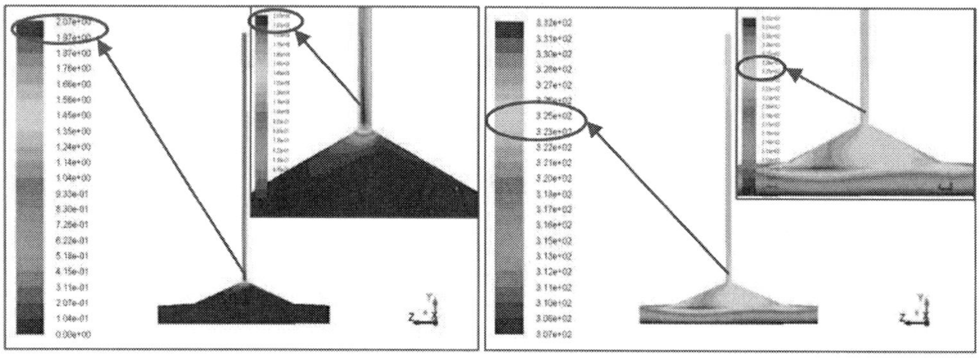

(a) Velocity contour distribution. (b) Temperature contour distribution.

Fig. 4. Velocity and temperature distribution for the conventional model.

Figure 5 illustrates the temperature and velocity distribution for the proposed model. The results showed that the percentage of enhancement of velocity and temperature for the working fluid of the model after adding channels and pushed flue gas with 0.015kg/s and 100°C for mass flow rate and temperature inside each channel were 10 % (v = 2.2 m/s) and 14% (T= 327K, ΔT = 20°C) respectively, with no load as compared with no channels case. The performance of the plant (collector efficiency and output power) was calculated according to Schlaich [9]. Table 4 shows the enhancement of the proposed model to the conventional one for velocity, temperature, and performance.

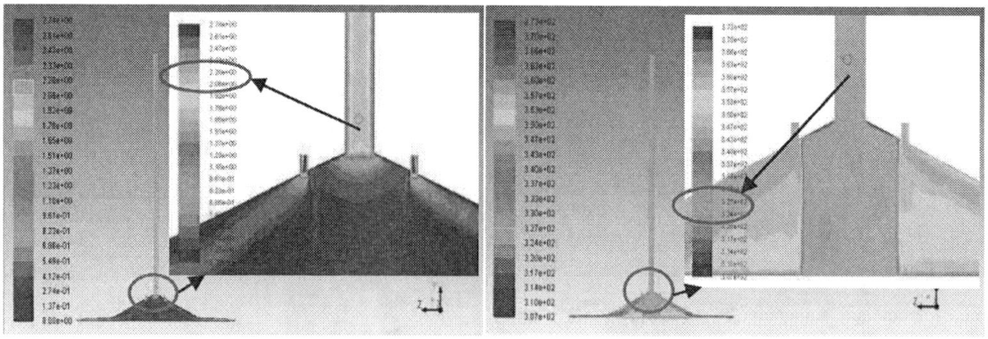

(a) Velocity contour distribution. (b) Temperature contour distribution.

Fig. 5. Velocity and temperature distribution for the proposed model.

Table 4. Experimental readings and simulation results

Model	Max. Velocity (m/s)	Temp. Diff ΔT (K)	Collector thermal efficiency (%)	Power output (W)
Conventional model (experimental)	1.89	17.6	3.7	0.07
Conventional model (simulation)	1.97	18	3.9	0.073
Proposed model with external heat source	2.2	20	4.8	0.09
Percentage of enhancement	10 %	14%	18.8%	18%

Conclusions

The simulation results of the conventional model showed a good agreement with the experimental results which confirm the accuracy of the simulation procedure. The numerical results disclose that the proposed technique has a momentous impact on the aero-thermal and performance characteristics. The velocity and temperature enhancements were 10% and 14%, respectively, and the power output percentage enhancement was 18%. Also, it can be concluded that there is a possibility to use the proposed model relying on solar energy (solar mode during the day) or an external heat source (during the night) or both (hybrid during the day) to produce electricity.

The logistic, technical and financial assistance provided by Universiti Teknologi PETRONAS is greatly acknowledged. In particular, the main author acknowledges Ministry of Industry and Minerals - Iraq (SIER-Company) for assistance to pursue his doctorate.

References

1. H. H. Al-Kayiem. WIT Transactions on Ecology and the Environment, Vol **179**, © WIT Press. (2013)
2. A. A. Ismaeel, H. H. Al-Kayiem, A. T. Baheta and M. A. Aurybi. International Energy Journal **16**, 11-24. (2016)
3. H. Alrobaei, Higher Institute of Engineering, P.O. Box: 61297; HOON; Libya. (2007)
4. M. A. Aurybi, H. H. Al-Kayiem, S. I. U. Gilani and A. A. Ismaeel. ARPN Journal of Engineering and Applied Sciences. **11**, 20, (2016).
5. H. H. Al-Kayiem, K. Y. Yin and C. Y. Sing. WIT Transactions on Engineering Sciences, **7**, © WIT Press. (2012)
6. A. O. Chikere, H. H. Al-Kayiem, and Z. A. Abdul Karim. Conference paper, 978-1-4577-1884-7/11/2011 IEEE. (2011)
7. J. Schlaich, and C. Maurer, Geislingen, Germany. (1995)
8. ANSYS Fluent Theory Guide. (2013)

Investigation on the performance of a prototype of thermo-electric generation with heat pipe-heat sink

Ali Elghool[1, 2], *Firdaus Basrawi*[1], *Hassan Ibrahim*[1] *and Thamir Ibrahim*[1]
[1]Energy Sustainability Focus Group, Faculty of Mechanical Engineering, Universiti Malaysia Pahang 26600 Pekan, Pahang, Malaysia
[2]Higher institute of sciences and technology/Gharian / Libya

Abstract. A significant problem in thermo-electric generators is the thermal design of the heat sink because it affects the performance of thermo-electric modules. As compared to conventional cooling systems, heat pipe heat sink have numerous advantages. Some of these advantages are: high heat-transfer rates; absence of moving parts and lack of auxiliary consumption (passive system). This paper presents the analysis of power generation using the combination of heat pipes and thermo-electric generators. The aim is to improve power output by an appropriate design of the heat sink. The average geometrical parameters of heat sink (fin height, fin space and fin thickness) were obtained from data collected from previous studies closely similar to this prototype. The prototype was tested and the temperature, voltage and current data were collected. All data were recorded by using a temperature data recorder, power meter and multimeter. It was found that the highest maximum power output was 1.925 watts at a temperature difference of 85°C. However, the prototype did not achieve the maximum output expected. This was a result of limitation of TEG model (where only one TEG was used) and the limitation of the performance of the prototype. The prototype successfully generated enough power to charge a cell phone and laptop when connected to two or three TEGs. Moreover the heat pipe heat sink needs optimization to meet the design output from the manufacturer of the TEG at hot side temperature and cold side temperature.

1 Introduction

Energy demand will keep increasing with the rapid development of technologies and industrialization. The current economy relies heavily on fossil fuels. Progress in the industrial sector and adoption of new technologies have further strengthened reliance on this fuel source. However, fossil fuels have inherent limitations since their reserves are increasingly depleting. In addition, there are a number of environmental issues related to fossil fuels production and utilization at both local and global scales. Furthermore, a significant portion of energy from fossil fuels is wasted during the transformation process and use. Owing to the issues related to fossil fuels, energy experts around the world have expressed concerns about the future of energy generation[1, 2]. Thermo-electric generator (TEG) has shown to offer a dependable and simpler way of thermo-electric energy conversion. Advantages of a TEG module include lack of moving parts, environmental safety, and silent operation. This generator can also be controlled in a seamless and accurate manner. Researchers have shown increased interest using thermo-electric technology in improving waste recovery efficiency over the last three decades. This has been made possible using a variety of heat-producing processes[3]. According to the Web of Science database, annual publications in thermo-electric technology had increased in the last 15 years from 500 to as high as 2000[4].

However, low efficiency in conversion has remained to be a primary challenge facing thermo-electric power production[5-7]. TEG heat exchanger's (heat sink) lack of design is one reason that causes shortcoming in the conversion efficiency of the TEG. These days, there is a profound research in heat sinks with heat pipes that enhances heat sinks significantly more, with a small temperature drop, because of the utilization of the latent heat of an internal fluid[8]. Heat pipe heat sink (HPHS) is the most suitable heat exchanger for the medium temperature range below 300°C with TEG (Energy Efficiency Guide for Industry in Asia)[9] which is the authors' work scope. Heat pipe has two ends, where one end works as evaporator placed on the TEG cold side to absorb the heat. This absorbed heat evaporates the fluid inside, which then flows to another end which works as condenser. At this end, it condensates, as it releases the heat to the ambient with the help of fins attached to it, and the fluid returns back to the evaporator. Thermo-electric generators have recently been investigated as a method of electricity generation with waste heat as heat source. Araiz.M [10] had studied thermo-syphons with phase change in thermo-electric generators in a waste-heat recovery application, stating that including thermo-syphons on the cold side of the generators improves the net thermo-electric production by 36% as compared to that obtained with finned dissipaters under forced convection.

Martinez, A[11], studied the combination of heat pipe exchangers and thermo-electric self-cooling and demonstrated its applicability to the cooling of power electronics. Experimental tests indicate that source-to-ambient thermal resistance reduces by around 30% when thermoelectric self-cooling system is installed, compared to that of the heat pipe exchanger under natural convection. Geometrical parameters of heat sink with heat pipe (fin height, fin space and fin thickness) have not been studied in previous work. In this paper, the main focus is to develop and test a prototype of thermo-electric-based power generator. The aim is to experimentally study the geometry of heat sink with heat pipe (fin height, fin space and fin thickness). The average geometrical size of heat sink was obtained depending on data collected from previous studies that were closely similar with this prototype.

2 Methodology

2.1 Description of the proposed system

Fig. 1 shows the design of the TEG prototype – one TEG was used in this experiment. The TEG hot side was placed on a heater to simulate the source of heat. Heat was transferred from the hot side to the cold side of the TEG. The cold side was attached to the aluminium block,

 DOI: 10.1051/matecconf/201713101005

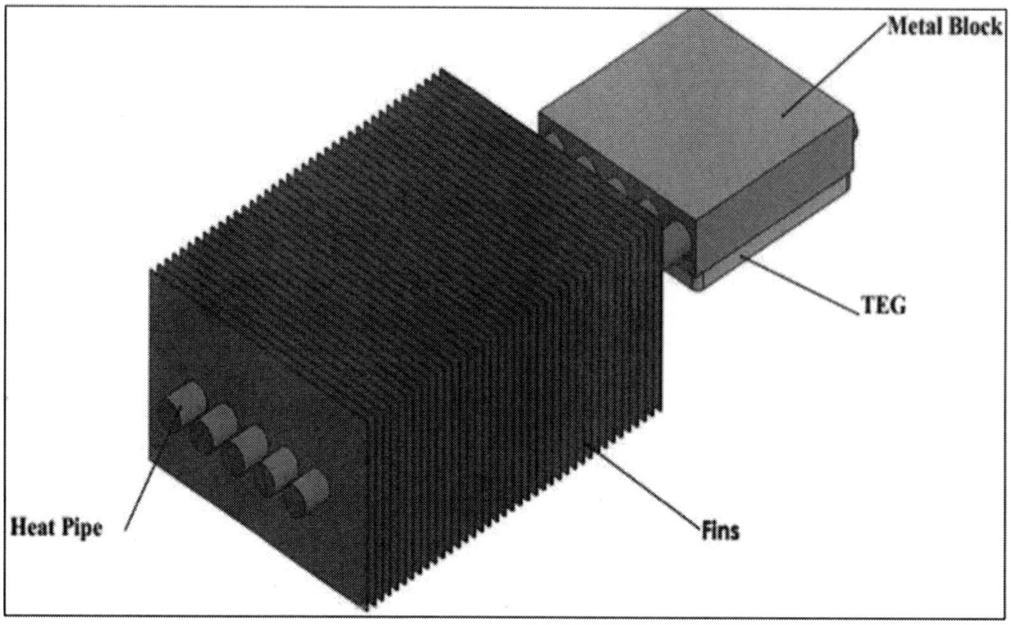

Figure 1. A schematic of the combined heat pipe- heat sink with thermoelectric generator.

where five heat pipes of 50W each were inserted in it. The TEG did not need any thermal paste as it was built with graphite foil as thermal interface material. On the other side of the heat pipes, there were fins to dissipate heat to the environment. Thirty-three aluminium (Al5052) were bounded mechanically on the heat pipes. The dimensions of each fin were 120×80×0.5mm. Interference fit method was used to assemble the fins to the heat pipes (also called press fit or friction fit), in which the fins had holes that were slightly smaller than the diameter of the heat pipe. By pressing the fins onto the heat pipes with pressure, the heat pipe and the fins were tightly assembled. When the temperature increased during working condition, thermal expansion of the fins and heat pipes will made the assembly even more secure. Table 1 describes the TEG specifications that were used in this study. Five pieces of copper heat pipes with 200 mm long and 8 mm outer diameter were used to dissipate heat from TEG cold side.When the hot side was supplied with 300°C of heat, whereas the cold side was maintained at 30°C, the TEG had the capability to generate 7.2 V, 3.0 amps, hence generates 21.6 W. This is the ideal case when the device is designed by the manufacturer.The heat pipes have fibre wick structure with heat flux of 50 W. Therefore, 250 W of heat can be transferred through the five heat pipes that were used.

2.2 Testing and data collection

Testing of the prototype was carried out as shown in Fig. 2. An electric heater was used as the heat source and a variable resistance was connected as electrical load. The TEG would deliver the maximum power output (Power = Volts x Amps) when the Load resistance equals the TEG internal resistance. (R of Load = R of TEG) by using the chart of "TEG resistance" form the manufacturer to determine the TEG's resistance at the temperatures that will provide it. If the TEG internal resistance could not match the load precisely, then the load resistance was made slightly higher than the TEG resistance rather than lower. The data collected were temperature, voltage and current. For recording of temperature, a 12-Channel Temperature

Recorder, BTM-4208SD was used. Thermocouples were placed at the hot side and cold side of the TEG. The date were recorded with intervals of 60 seconds for 1420 seconds (approximately 23 minutes). For voltage, it could be taken for both open and close circuits. But for the current, the circuit had to be loaded so the current can be measured. From the voltage and current measured, power (W) can be obtained by a conventional equation as shown below.

$$P = I \times V \tag{1}$$

Where I is current (amp), and V is voltage (Volt).The fins were cooled down by natural convection, with the average air speed about 1.6 m/s. It was simulated in the laboratory through a fan. Consequently, the cooling method was considered as passive. The average air velocity was obtained from data collected from previous studies at UMP.

Table 1. Thermoelectric Specifications

Maximum Hot side temperature (°C)	Maximum cold side temperature (°C)	Dimensions (mm)	Voltage (V)	Current (A)	Power (W)
320	180	56x56x4.45	7.2	3.0	21.6

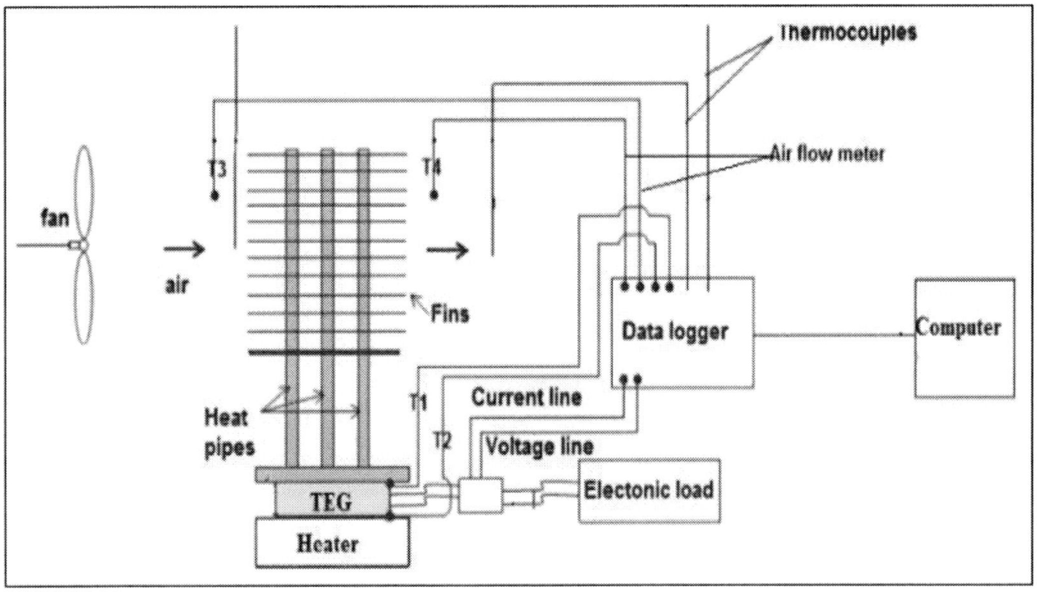

Figure 2. The schematic of the experimental setup

3 Results and discussion

Fig. 3, 4, 5 and 6 show results for voltage, current, power output and temperature difference, respectively. As shown in the figures, current and voltage show similar trend. The maximum voltage was 5.5 V, and current was 0.35 A. The maximum power obtained was 1.925 W as shown in Fig. 5. Subsequently, by connecting two or three TEGs to form a model, power output will double enough to run a mobile phone and laptop. The maximum power output of

1.925 W is much less than what was expected, which is 15W as per design from the manufacturer of TEG at hot side temperature 250°C cold side temperature 25°C. Fig. 6 shows the temperature difference between hot and cold sides of the TEG. It can be seen that the temperature difference increased constantly until 1200 s, then the temperature was almost constant which indicates that the operation has reached its steady-state.

It was found that the highest temperature difference was 83°C, For a low temperature difference, between hot side of TEG and cold side, the main cause was perhaps due to lack of heat sink design because it was observed that the temperature at the fins side also reached almost 100°C during the end of the testing. Thus, improvement in the heat sink is needed for better power output. As for the shortage of maximum power output, only one TEG was used due to the limitation of the performance of the heat pipe heat sink. For the power output limitation two or three TEGs should be connected to form one module. It can be concluded that the prototype successfully generated enough power to charge a cell phone and laptop with a multiple connection of two or three TEGs. Moreover heat pipe heat sink needs optimization to approach as close as possible from the ideal situation of the manufacturer's design of TEG, namely; with 250°C hot side and 25°C cold side, with the TEG generating 15W.

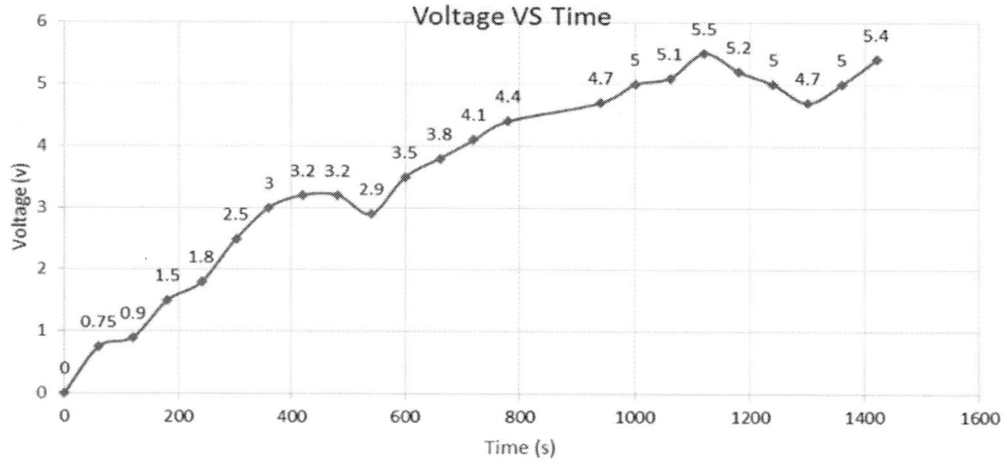

Figure 3. Chart for voltage vs time (S).

4 Conclusion

A prototype of thermo-electric generator (TEG) based power generation system was tested. The prototype consisted of a heat source, one TEG connected with heat pipe heat sink (passive system). It was found that the highest temperature difference was 83°C, maximum voltage was 5.5V, maximum current was 0.35A and maximum power obtained was 1.925 W. The maximum power output of 1.925 W was much less than what was expected, which was15 W as per design from the manufacturer of the TEG with hot side temperature 250°C and cold side temperature 25°C. This was a result of limitation of maximum power output (only one TEG was used) and the limitation of the performance of the heat pipe heat sink. For the power output limitation two or three TEG should be connected to form one module. For a low temperature difference between hot side of TEG and cold side, the main cause was possibly due to lack of proper heat sink design. Heat pipe heat sink needs optimization to meet the

 DOI: 10.1051/matecconf/201713101005

design output from the manufacturer of the TEG at hot side temperature and cold side temperature.

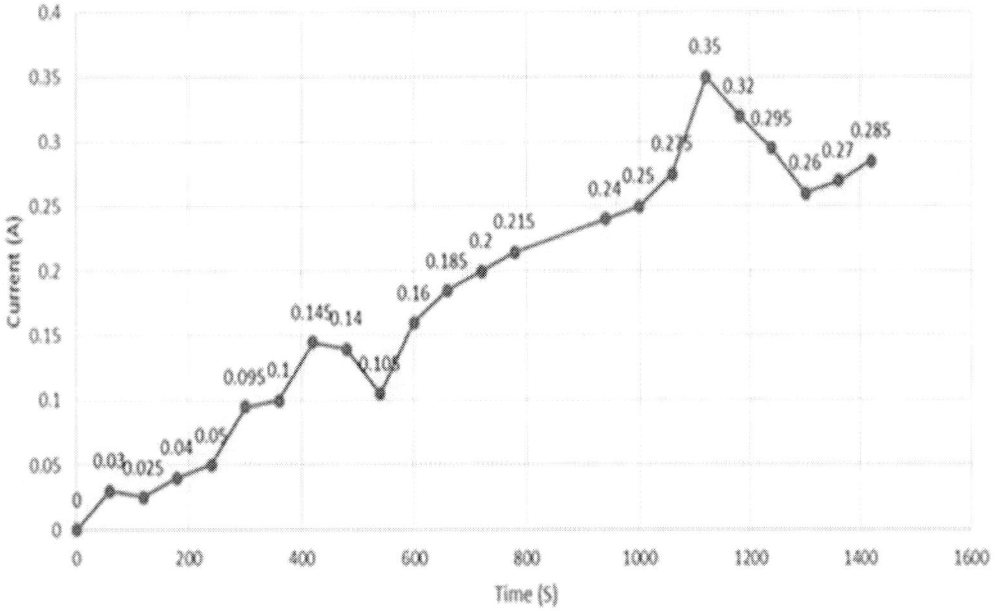

Figure 4. Chart for current vs time (S).

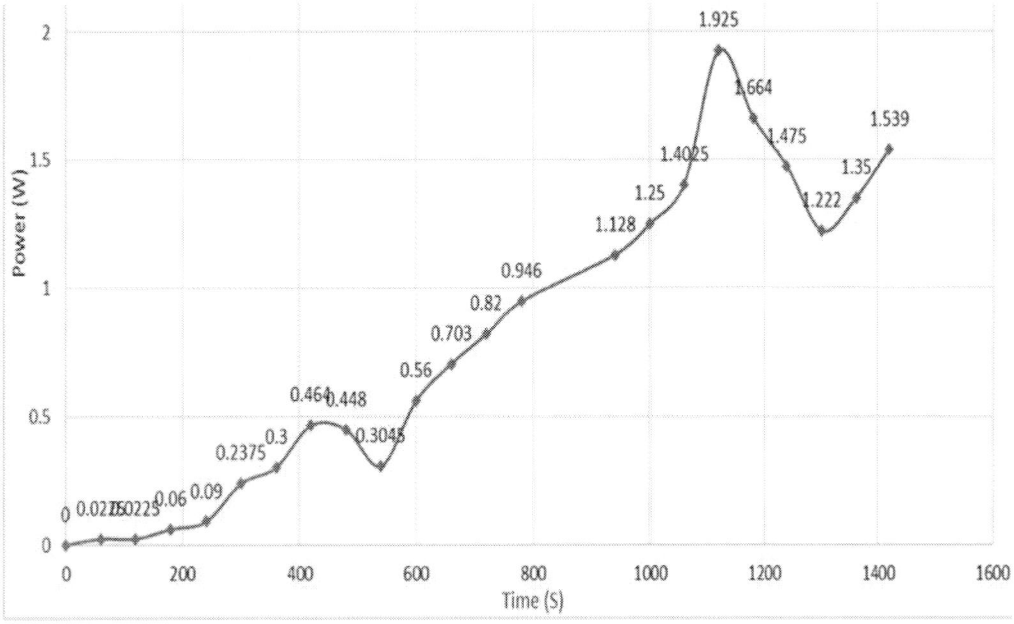

Figure 5. Chart for power vs time (S).

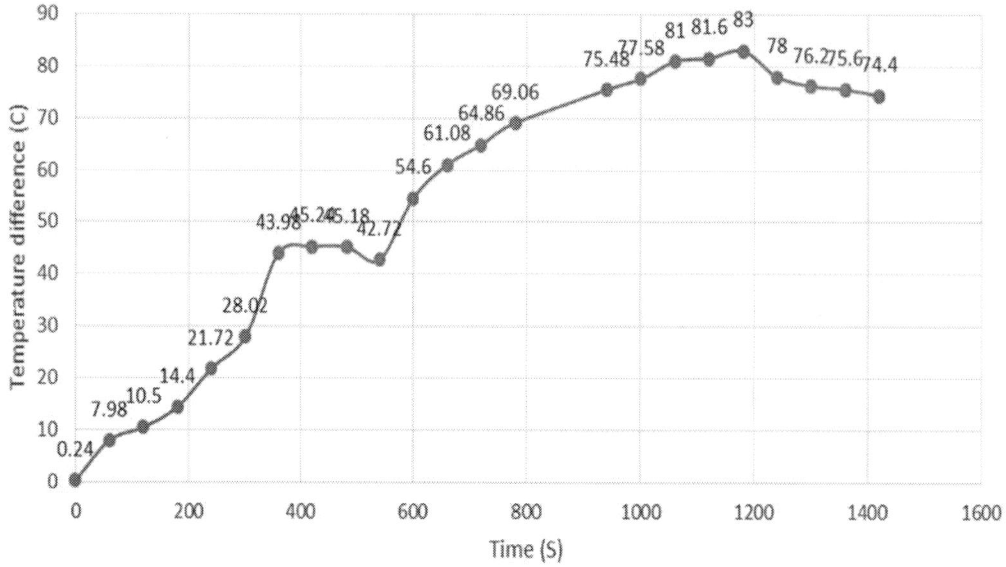

Figure 6.Chart for temperature difference vs time (S)

Acknowledgement

The authors gratefully acknowledge Universiti Malaysia Pahang for providing the internal grant under RDU160310.

References

1. A.Elghool, enrg.con.M., (2017). **134**: p. 260-277.
2. K.Yazawa, Y.R.K., and A. Shakouri, Appl. En, (2013). **109**: p. 1-9.
3. K.Yazawa, enrg.con.M., (2014). **84**: p. 244-252.
4. W.Liu, Acta .M, (2015). **87**: p. 357-376.
5. C.Gayner, K.K., (2016). **83**: p. 330-382.
6. H.Wu., Nano .E, (2015). **13**: p. 626-650.
7. M.Elsheikh, Ren,Sust. Ene. R, (2104). **30**: p. 337-355.
8. H.Shabgard., Inter. Jour.of Heat and Mass Trans., (2015). **98**: p. 138-158.
9. M.Remeli., enrg.conv.M. (2015). **91**: p. 110-119.
10. M.Araiz, enrg.con.M., (2017). **137**: p. .155-164.
11. A.Martinez, Energy., (2016). **112**: p. 1-7.

Development and implementation of Intelligent Soot Blowing Optimization System for TNB Janamanjung

Taneshwaren Sundaram[1], *Firas* Basim Ismail[1*], *Prem* Gunnasegaran[1], and *Pogganeswaren* Gurusingam[1]

[1]Power Generation Unit, Institute of Power Engineering (IPE), Universiti Tenaga Nasional, 43000 Kajang, Selangor, Malaysia.

Abstract. With an ever increasing demand for energy, Malaysia has become a nation that thrives on solid power generation sector to meet the energy demand and supply market. In a coal fired power plant, soot blowing operation is commonly used as a cleaning mechanism inside the boiler. There are many types of sequence available for this soot blowing operation. Hence, there is no efficient ways in utilizing the soot blowing operation to enhance the efficiency of boiler. Soot blowing optimization requires specific set of data preparation and simulation in order to achieve the best modal. Computational Fluid Dynamics (CFD) is used to model a 700MW super-critical boiler, whereby parameters with effect to soot blowing operation is studied. Two different boiler condition is studied to analyze parameters in a clean and faulty boiler. Artificial Neural Network (ANN) is used to train neural network modal with back propagation method to determine the best modal that will be used to predict soot blowing operation. Combination of neural network different number of neurons, hidden layers, training algorithm, and training functions is trained to find the modal with lowest error. By improving soot blowing sequence, efficiency of boiler can be improved by providing best parameter and model. This model is then used as a reference for advisory tool whereby a Neural Network Predictive Tool is suggested to the station to predict the soot blowing operation that occurs.

1 Introduction

The combustion of coal-fired power plant in Malaysia causes build-up of combustion deposits such as soot, ash and slag in the boiler. Combustion deposit generally decreases the efficiency of the boiler by reducing heat transfer. To maintain boiler efficiency, the surfaces in boiler are periodically cleaned by directing cleaning medium called soot blowers. Soot blowers are a cleaning mechanism that uses steam to blow of boiler flue gas circuit at different points. It carries out periodic cleaning to the parts that are exposed to ash deposits such as furnace, super heater, re heater, and economizer. There are many power plants in the world that uses an intelligent soot blowing system that is vastly becoming an important

* Corresponding author : Firas@uniten.edu.my

system during boiler operations. This is because an optimized soot blowing system will eventually increase the boiler efficiency [1, 2]. This optimized system only allows the soot blowers to function when it is needed and in necessary location. It does not follow a specific interval or period for it to work. An intelligent soot accumulation monitoring system will provide the updated data during boiler operations [3, 4]. There is a tool to predict and measure soot accumulation in boiler. Besides, by predicting soot accumulation, soot blowing can take place when slagging occurs. Hence, the soot blowers will be activated and soot blowing system will operate at the affected areas. The rate of heat flux and heat transfer, especially in the furnace indicates the level of soot accumulation in the boiler. The main idea is to reduce the frequency of soot blowing and hence, reducing the steam consumption during soot blowing that will eventually increase the efficiency of boiler and reduce maintenance cost [5].

2 Methodology

The methodology is built based on the research made on most related works in literature review. There are three main phases that is constructed in this methodology are as follows:

Data Preparation > Modelling > Analysis and Advisory Tool

2.1 Data preparation

Data collection is a key part to identify and characterize each data. In order to run simulation, a specific data set is required. All the necessary data is taken from the Plant Information (PI) station in TNB Janamanjung.

2.2 Modelling

The most important phase of this paper is modelling. There are two different types of software used for modelling phase. For CFD, ANSYS V15 Fluent is used to simulate two different boiler conditions, clean (with soot blowing) and faulty (without soot blowing). For ANN, MATLAB is used to run the neural network and to find the best neural network modal for soot blowing operation [6].

2.2.1 Computational fluid dynamics modelling

The main objective of simulating both boiler conditions is to study the behaviour of process. The pre-modelled boiler modal is used to simulate the boiler conditions. Two separate runs have been performed in orders to obtain two sets of different data. These sets of data are compared and the behaviour of the boiler can be analysed. The simulation is ran for 500-1000 iterations, hence the difference in data captured is tabulated.

2.2.2 Neural network modelling

The neural network basically contains 3 main layers, input, hidden layers and output. In feed forward network, the structure selection is done based on number of hidden layers, one hidden layer and two hidden layer with 1-10 number of neurons for each hidden layer. These different combination of structures will produce different output [7].

3 Results and Discussion

In this study, the results of CFD and ANN will be explained in more detail to get a clearer understanding about this project. The simulated data and contours from CFD is obtained with behaviour graphs of clean and faulty boiler. On the other hand, the results in form of RMSE values are tabulated and each input behaviour graph is portrayed for a clearer discussion.

3.1 Comparison of clean and faulty boiler

Through the analysis using Computational Fluid Dynamics (CFD), the factors that has been considered in soot blowing has been taken into account and simulated using this software. Throughout the analysis, two different sets of results have been obtained in clean and faulty boiler as shown in Table 1.

Table 1. Comparison of clean and faulty boiler

Boiler Parameter	Clean Boiler	Faulty Boiler
Pressure, Pa	359,023	419,055
Temperature, K	2708	2111
Heat Flux, w/m2	5,020,528	1,319,415
Heat Transfer, w/m2.K	7,052.88	1,961.11
Mass Imbalance, kg/s	0.008059487	0.02689896

When soot blowing happens, the boiler is assumed as a clean boiler. When the fireball is formed and flue gas is produced, soot accumulation and slagging will take place in boiler [8]. These two conditions are taken into account to validate the boiler condition by using CFD. Before soot blowing takes place, the pressure in the boiler is higher compared to faulty boiler. This is because, before soot blowing, the boiler is in dirty condition whereby soot accumulation and slagging occurs in all parts of boiler. By consuming space in boiler, the pressure will increase in presence of soot. After soot blowing, after all the slagging is removed, the pressure of boiler can be reduced as shown in Table 1. The temperature also plays a huge part in maintaining the boiler efficiency. In a clean boiler, the temperature of is relatively high which is important in relating with heat transfer and heat flux. However, in a faulty boiler, the temperature is reduced into almost half as the soot accumulated in the boiler absorbs the heat produced in boiler. This will directly reduce the performance of the boiler. Hence, soot blowing operation and ensuring the boiler is in clean condition is important. Heat flux and heat transfer relate to each other. Before soot blowing, slagging causes the rate of heat flux and heat transfer to decrease as the soot accumulation will disrupt the surface contact between the walls [9]. Hence with soot blowing, there will be no disruption at the walls as there is no soot accumulation. As shown in Table 1, heat flux and heat transfer increases after soot blowing happens in a clean boiler. Lastly, mass imbalance is a great way to measure the amount of soot accumulated in boiler. In a clean boiler, the mass imbalance is figuratively steady and low. However in a faulty boiler, the mass imbalance increases drastically as shown in Table 1. This is because of soot accumulation occurrence in various parts of boiler. By soot blowing operation, the mass imbalance can be kept in an optimum value.

3.2 ANN results and analysis

Construction of ANN consists of 2 different number of hidden layer, the first one would be with one hidden layer that has two different activation functions. The activation functions used in this project as stated previously are, Logsig (L), Tansig (T) and Purelin (P). Therefore the total combination that is present for one hidden layer with two activation functions is 9. L+L, L+T and L+P are some examples with 10 number of neurons per combination. Besides, four different types of training algorithms are used, 'trainscg', 'trainlm', 'trainbfg' and trainrp'. However, for 2 hidden layer, three activation functions are used hence a total of 27 combinations to possible for this modal. These different combination will produce different RMSE with each simulation ran in MATLAB. After training all the combination, the combination with the lowest RMSE is considered the best ANN modal to be used in latter stages.

Table 2. Result summary of ANN

Training Algorithm	Architecture	Activation Function	RMSE	Number of Iteration
trainscg	10HL1-6HL2	P+T+T	0.1561	86
trainlm	9HL1-8HL2	L+T+T	0.1277	87
trainbfg	7HL1-8HL2	P+T+T	0.1439	202
trainrp	7HL1-9HL2	P+T+T	0.1539	139

Table 2 summarizes the NN for 2 hidden layer for 4 different types of algorithm. The trainrp is the fastest training algorithm but produces an average RMSE. On the other hand, trainlm produces the best RMSE even though it took the longest to produce the RMSE. In addition, the 2 hidden layer NN produces better RMSE compared to 1 hidden layer. The graph in Figure 1 justifies the ANN combination modal which has the lowest RMSE value. Hence this modal will be used in future ANN predictive tool.

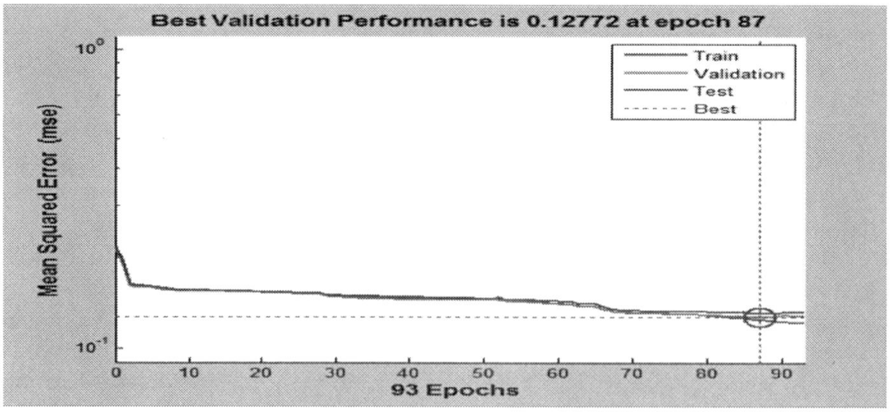

Figure 1. Best RMSE Performance Graph

4 Conclusion

This journal focuses on studying the pattern and coming up with a solution that will benefit the station. Firstly, the main parameters are identified and studied to ensure the best soot blowing sequence can be identified. CFD simulation is done using ANSYS V15-FLUENT to study two different boiler conditions, clean (with soot blowing) and faulty (without soot blowing). CFD allows the interpretation of real performance during boiler operation. By performing this, the soot blowing sequence will determine how it will affect the daily boiler operation. Furthermore, the best combination of ANN model is the main aim to determine the best soot blowing sequence through simulation, training and validation. The training algorithm, activation functions, number of hidden layer and number of neuron in each hidden layer are ANN topologies that will produce the best validation. Different combinations will produce different result. Root Mean Squared Error (RMSE) is used to measure how effective is the combination. The combination with the lowest RMSE is the best neural network modal.

4.1 Advisory tool

Modal Predictive Control (MPC) is a process model which is used to predict the future evolution of the process to optimize control signal [10]. By implementing MPC in soot blowing operation, the soot blowing sequence for each day can be predicted and set during boiler operation. MPC provides an excellent arrays of suggestion to the system in order not to interrupt the boiler operations frequently. By using the best neural network modal obtained from ANN, the modal is used in PMC to find the best sequence of soot blowing in power plant. The optimization happens after the neural network modal is being analysed and implemented in different conditions and hence providing a sequence to the plant to operate the soot blowing operation shown by the yellow arrows in Figure 2. If there is some error between the plant and neural network modal, the data is sent back to the optimizer in order for detailed optimization as shown by orange arrows in Figure 2. This process is continued until the error margin between the neural network modal and plant is zero. Hence, the soot blowing operation in the power station is fully optimized hence, allowing boiler efficiency to be improved.

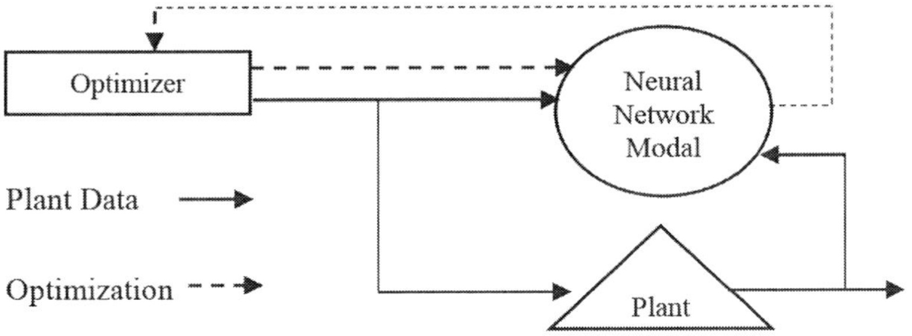

Figure 2. MPC modal

The authors would like to express our gratitude to the power generation unit, institute of power engineering Universiti Tenaga Nasional (UNITEN), TNB Janamanjung and Ministry of Higher Education (MOHE), Malaysia for the opportunity to sponsorship this corresponding research under the FRGS research fund with the project code 20160103FRGS.

References

1. Emerson Process Management (2011). Ensure Efficient Sootblowing [Online]. Available: http://www2.emersonprocess.com/siteadmincenter/PM%20Power%20and%20Water%20Documents/pws_003933.pdf
2. Powerclean Performance. Intelligent Sootblowing. [Online]. Available: http://www.babcock.com/library/documents/ps-388.pdf
3. Kansas City Power & Light (KCP&L) Hawthorn Generating Station.(2012) Boosting Efficiency with a Sootblowing Optimization System. [Online]. Available: http://www.powermag.com/boosting-efficiency-with-a-sootblowing-optimization-system/
4. B.Pena, L.I Diez and E. Teruel (May 2010). Soft-computing models for soot-blowing optimization in coal-fired utility boilers. [Online]. Available: http://www.sciencedirect.com.ezproxy.uniten.edu.my/science/article/pii/S156849461000102X
5. Shivaraj Kumar B. and Dr. N.S Venkatesh Gupta (2013). Effect of Smart Soot Blowing System in Boiler Furnace on Cycle Efficiency and Cost Saving. [Online]. Available: http://www.ijert.org/view-pdf/4396/effect-of-smart-soot-blowing-system-in-boiler-furnace-on-cycle-efficiency-and-cost-saving
6. Enrique Teruel, Cristobal Cortes, Luis Ignacio and Immaculada Arauzo (June 2005). Monitoring and prediction of fouling in coal-fired utility boilers using neural networks. [Online]. Available: http://www.sciencedirect.com.ezproxy.uniten.edu.my/science/article/pii/S0009250905003283
7. Hugo Calisto, Nelson Martins and Naim Afgan (2007) – CFD and Neural Network based expert system for boiler performance [Online] Available: http://dc.engconfintl.org/cgi/viewcontent.cgi?article=1059&context=heatexchanger2007
8. Bhatt Samik (2014) – Analysis of Clinker Formation Region & Soot Blower Optimization Using CFD Analyis [Online] Available: http://www.ijetae.com/files/Volume4Issue3/IJETAE_0314_111.pdf
9. Zhanhua Ma, Felicia Iman, Pisi Lu, Rod Sears, Lingbu Kong, A.S. Rokanuzzaman, Donald P. McCollar and Steven A. Benson (2007) – A comprehensive slagging and fouling prediction tool for coal-fired boilers and it's validation [Online] Available: http://www.academia.edu/5297900/A_comprehensive_slagging_and_fouling_prediction_tool_for_coal-fired_boilers_and_its_validation_application
10. Modal Predictive Control Toolbox [Online]. Available: https://www.mathworks.com/products/mpc.html [Accessed 29 January 2016]

DOI: 10.1051/matecconf/201713101007

Pre-treatment of oil palm frond biomass via extensive high temperature drying for gasification process

Nur Hazwani Mat Razali[1], *Shaharin Anwar* Sulaiman[1,*], *Mohamad Nazmi* Zaidi Moni[1] and *Mohamad Firdaus* Basrawi[2]

[1]*Department of Mechanical Engineering, Universiti Teknologi PETRONAS, 32610 Tronoh, Perak.*
[2]*Faculty of Mechanical Engineering, Universiti Malaysia Pahang, 26600 Pekan, Pahang*

Abstract. Oil palm frond has been utilized as a solid biomass fuel for gasification to produce synthesis gas or syngas to be used for heat and power generation. A fuel pre-treatment method by means of extensively-drying OPF blocks at 150°C and 200°C for 4 hours was implemented to investigate the effects of the fuel in terms of drying efficiency and gasification performances. Tar, pyrolysis oil and condensates were found to be squeezed out by heat during drying, signifying volatilization of fuel at temperatures between water boiling point at 100°C and fuel pyrolysis point at 280°C. Syngas produced from the updraft gasification of extensively-dried OPF blocks was analyzed and tested for sustainable gas flares. The syngas was found to be composed of 16.5% CO, 10% CO_2, 4% H_2 and 0.9% CH_4 and was produced at gasification temperatures lower than that exhibited by normal OPF blocks.

1 Introduction

Oil palm frond (OPF) is one of the highly-produced biomass in a typical oil palm plantation. Classified as a woody biomass, OPF has an average calorific value of 17.9 MJ/kg, which is fairly attractive as a source for solid biomass fuel1. Recent pattern in OPF fuel research showed that OPF has been utilized as a solid fuel for gasification to produce synthesis gas or syngas and has borne successful results1-9. To become a solid fuel, raw OPF has to undergo a series of processing stages including sizing, sieving, drying and storing. Drying is the most important criteria in solid biomass fuel processing. Normally, any biomass that is subjected for solid fuel production undergoes a drying process to remove excess moisture as a pre-treatment and preservation method to preserve its quality and ensure a longer storage life. For a small batch of feedstock, a drying temperature of 105°C for a period of 24 hours is used merely due to the purpose of removing only excess water by exposing the biomass to a temperature just slightly higher than that of boiling water1-9. If the drying temperature is set any higher, the biomass may undergo pre-pyrolysis state where its volatile matters will be released, thus slowly roasting the biomass in the heat, replicating the practice of charcoal-making (slow carbonization) although at a much lower temperature. By average, most plant materials volatilized at around 280°C11, whereby pre-pyrolysis state begins at slightly lower temperature, sometimes as low as 200°C in the case of wood12. With most of the volatile matters removed, the fixed carbon amount in the biomass will relatively be higher than that before extensive drying, hence the term carbonization to describe the process. The objectives of this study were to inspect the changes that occur to OPF blocks when undergoing an extensive drying process at elevated temperatures of 150°C and 200°C for 4 hours. The gasification of the blocks followed where the effects of using extensively-dried OPF blocks as gasifier fuel to syngas composition and overall quality were studied.

2 Materials and Methods

2.1 Raw material processing

Fresh OPF of less than two weeks old after pruned were collected from a nearby oil palm plantation. OPF were quickly processed to avoid decomposition due to prolonged storage as suggested by Moni and Sulaiman[1]. Each OPF was cut perpendicularly from its growing axis into 2.5 cm-thick sections using a power circular saw.

Corresponding author: shaharin@utp.edu.my

DOI: 10.1051/matecconf/201713101007

The sections were then further reduced into blocks of 2.5±0.5 cm in length using a machete, producing small blocks as shown in Fig. 1. The blocks were sun-dried at ambient conditions for 7 days to slowly and naturally remove the excess moisture content trapped within. The blocks were then artificially dried at 105±0.5°C in a convection oven for 24 hours to further remove all remaining moisture and to obtain the equilibrium moisture content value for OPF blocks by weight difference.

Fig 1. OPF blocks

2.2 Extensive drying

Oven-dried OPF blocks were directly heated to 150°C and 200°C in two separate tests in a convection oven after 24-hour drying at 105±2°C. Around 1 kg of OPF blocks was used for each test, with three repetitions for consistency. Both extensive drying sessions were conducted for 4 hours to inspect the changes during the extensive drying process. Continuous observations were made and recorded throughout the drying period.

2.3 Gasification

Extensively-dried OPF blocks were used as a solid fuel for gasification in an updraft gasifier. The arrangement of the updraft gasifier and its downstream equipment is shown in Fig. 2. Around 6 kg of OPF blocks were required for each gasifier operation. Ambient air was continuously supplied into the gasifier at a rate of 180–220 LPM using a centrifugal blower. Temperature profiles of the gasifier during gasification were taken and recorded using Type-N thermocouples and an 8-port thermocouple reader connected to a desktop computer. Syngas compositions were taken at 1-minute interval along the 60-minute gasifier operation using EMERSON X-STREAM gas analyser and syngas ignitability was constantly tested using a miniature jet burner.

3 Results and Discussions

3.1 Moisture content

Significant reduction in weight was observed in OPF samples after 7 days of sun-drying at ambient conditions. Averagely, each block lost around 50–60% of original moisture content. Artificial drying of OPF blocks in a convection oven at 105±2°C for 24 hours revealed that an average of 13±1% remaining moisture content was further removed, and this figure was taken as the equilibrium moisture content for OPF blocks when stored at ambient conditions.

After extensive drying, it was assumed that all remaining moisture in OPF blocks had been fully removed with

 DOI: 10.1051/matecconf/201713101007

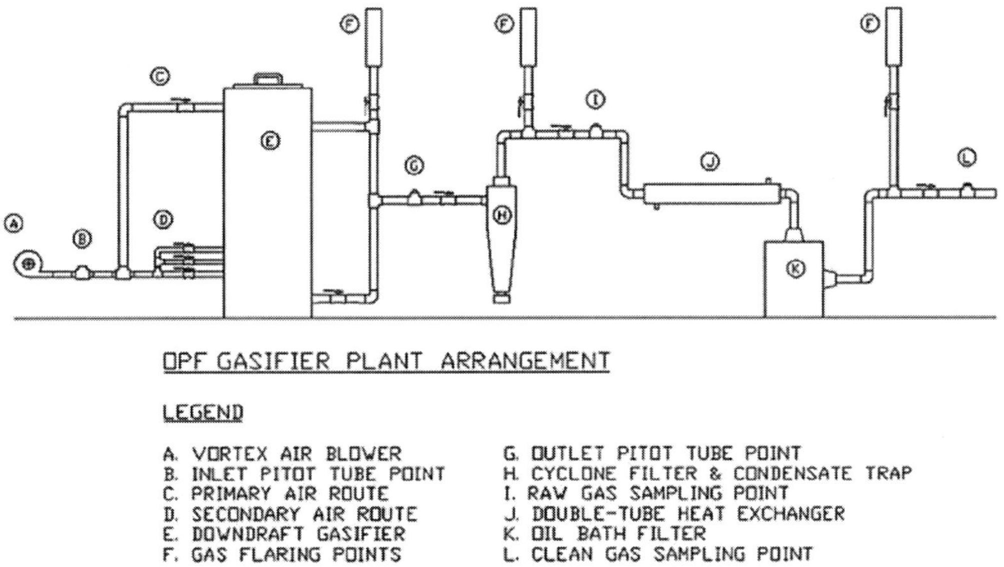

Fig. 2. OPF gasifier plant arrangement

averagely 4% more reduction in weight was measured, assumed to be due to removed moisture and the release of light volatiles as a result of the exposure to the elevated drying temperature.

3.2 Extensive drying

Oven-dried OPF blocks were subjected to an extended and extensive drying at 150°C and 200°C in two separate tests for 4 hours. Visual observations made suggested that the blocks underwent pre-pyrolysis process where traces of tarry condensate were detected to form on the internal wall and door surfaces of the convection oven during extensive drying process. Smoky and strong scent of tar was also noticed. The observation was consistent with the behaviour of biomass undergoing pyrolysis where its volatile matters and water contents are 'squeezed out' with applied heat above water boiling point temperature[1]. Some changes on the surface of OPF skin and flesh were observed where their colours changed from bright brown to dark brown darted with black spots indicating the formation of tar and condensates. Small traces of black, sticky substances composed of dehydrated tar and pyrolysis oil were found to form on OPF block surfaces and weakly tie contact surfaces of OPF blocks together. The traces of tar, pyrolysis oil and condensates produced from OPF blocks at 200°C drying temperature were observed to be higher than that at 150°C due to increased thermochemical activities at higher temperatures close to pyrolysis temperature.

3.3 Gasification of OPF Blocks

3.3.1 Gasification method

OPF blocks that have undergone the 4-hour extensive drying were fed into the updraft gasifier for gasification. Around 6 kg of OPF blocks were used in each gasifier operation that lasted for around 60 minutes. Hot gases from the gasifier were directed to a flare point via an exhaust pipeline for ignition tests and sampling.

3.3.2 Temperature profile

The hearth temperature during the updraft gasification of OPF blocks was found to be in the range of 400–650°C. The temperature range was relatively low when compared to the typical hearth temperature range during the gasification of OPF fuel that did not undergo any extensive drying process at 750–900°C and sometimes up to 1200°C[1,7]. The temperatures at drying and pyrolysis zones were found to be between 250-350°C respectively, which were also very low compared to typical OPF drying and pyrolysis zone temperatures of around 350-600°C.

The low temperature ranges at the given zones were suspected due to suffocation in which the supplied air was at an insufficient rate to keep a healthy combustion.

3.3.3 Gas ignition

Sustainable gas flares were obtained during gasification and lasted for a few minute before re-ignition was required. Syngas flares were observed to exhibit faint yellow and orangey flame colour, indicating low amount of unbroken hydrocarbon rings in the syngas stream. Traces of bluish flame that indicated the presence of H_2 was not observed. The length of the flare was estimated to reach around 1 m high from the flare tip and its temperature was measured to be between 400-500°C using a handheld thermocouple reader. The gas flare was able to sustain itself without the need for constant re-ignition, indicating the high concentration of flammable gases in the syngas.

3.3.4 Syngas composition

The syngas composition analysis showed that by average the syngas produced from extensively-dried OPF blocks contained 16.5% CO, 11.2% CO_2, 4% H_2 and 0.9% CH_4. The comparison of the syngas components with typical OPF syngas values is shown in Table 1 while the syngas components production trends are shown in Fig.3.

Table 1. Comparison of syngas components

OPF Fuel Type	Syngas composition (vo. %)			
	CO	CO_2	H_2	CH_4
Extensively-dried	16.5	10	4	0.9
Normally-dried	18.0	10	10	1.4
% difference	8.3	0	60	36

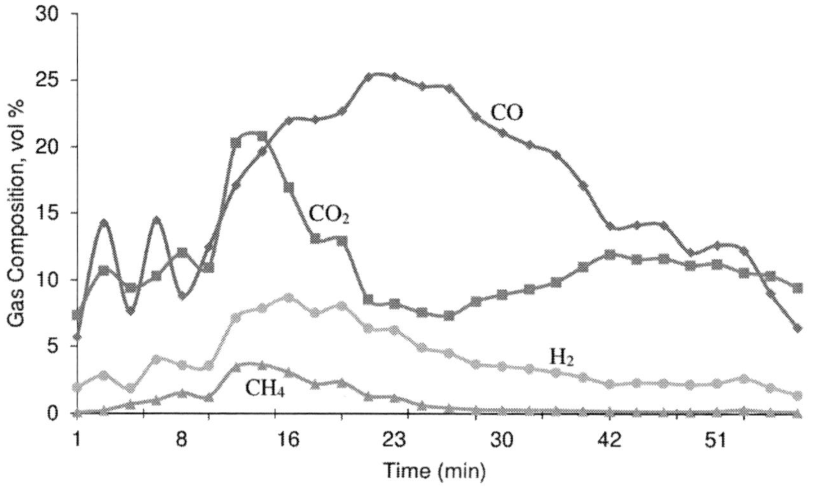

Fig. 3. Syngas composition as a function of gasifier operating time

The volume % of CO, H_2 and CH_4 were found to be lower than the typical values while the volume % of CO_2 was found to be equivalent. In Fig. 2, it can be observed that all syngas components experienced drops in production after reaching their peaks after around 15–20 minutes of gasifier operation. The slightly increasing amount of CO_2 after 30 minutes of operation showed that complete combustion was occurring, causing other syngas components to experience drops in concentrations. This may due to the decreasing amount of fuel after being consumed in combustion while the supplied air was maintained at the same flow rate, causing the air-fuel equivalence ratio to increase nearing to 1.0 where substoichiometric combustion ceased and stoichiometric combustion took place. The low productions of CO, H_2 and CH_4 can be explained by examining the thermochemical reactions that were responsible for their formations:

$$C + O_2 \rightarrow CO_2 \tag{3.1}$$

$$C + H_2O \rightarrow CO + H_2 \quad (3.2)$$

$$C + 2H_2O \rightarrow CO_2 + 2H_2 \quad (3.3)$$

$$CO + H_2O \rightarrow H_2 + CO_2 \quad (3.4)$$

$$C + H_2O \rightarrow \frac{1}{2}CH_4 + \frac{1}{2}CO_2 \quad (3.5)$$

Reactions (3.1) takes place at the oxidation stage where O_2 is consumed combustion to provide heat for fuel drying, to break up weak chemical bonds in pyrolysis stage and raise the overall reactor temperature for reactions (3.2) to (3.4) to occur. Reactions (3.2) and (3.3) are water-gas shift reactions, the principal gasification reactions, favouring high temperature and low pressure to occur. Reaction (3.4) is a part of water-gas shift and occurs at low temperature for H_2 production. Reaction (3.5) is methanation reaction where CH_4 is produced[1]. Almost all of the reactions require H_2O to occur – a condition that was scarcely achieved during the gasification of extensively-dried OPF blocks due to the removal of most moisture content from the OPF blocks during extensive drying. As a result, most of the reactions occur slowly depending to the available moisture content in the supplied ambient air into the gasifier, thus reducing the concentration of syngas components. In the case of CO_2, since air and hot char were available in high amount, the production of CO_2 was not affected. These observations suggested the important role of water in gasification especially for the production of H_2 from water-gas shift reaction, indicating that gasification feedstock that is too dry may not be beneficial for optimum syngas production. Therefore, feedstock with moisture content of less than 20% is recommended for quality syngas production, where the retained moisture content is low enough to maintain gasification temperature and to prevent condensates formation but high enough to initiate water-gas shift reactions from the exposure to hot char at the gasifier hearth to produce H_2 and CH_4.

4 Conclusions

Although pre-treating OPF blocks by extensive drying at high temperatures has reduced the volatile counts and increased the amount of carbon in fuel, the complete removal of water was found to cause gasification efficiency to drop and therefore affect the quality of syngas in terms of CO and H_2 fractions and gas ignitability. Furthermore, extensive drying of OPF blocks showed that tar, pyrolysis oil and condensates were 'squeezed' out by heat in gaseous form before reforming as liquid within the oven. This may contaminate the oven with hard stains and may affect the overall oven efficiency due to possible tar accumulation along the oven's ventilation points. In terms of gasification efficiency, mixing the extensively-dried OPF blocks with normally-dried OPF blocks may improve the overall fixed carbon amount to increase CO production while with adequate H_2O content in OPF block, the production of H_2 may also increase.

References

1. M.N.Z. Moni, S.A. Sulaiman, *Downdraft Gasification of Oil Palm Frond* (Lambert Academic Publishing, Germany, 2012)
2. S.A. Sulaiman, S. Balamohan, M.N.Z. Moni, S.M. Atnaw, A.O. Mohamed "Study on the Feasibility of Oil Palm-Fronds for Biomass Gasification" (5th IESE, Turkey 2010)
3. S.A. Sulaiman, M. Karim, M.N.Z. Moni, S.M. Atnaw, "Gasification Study on Different Plant-Based Biomass Materials" (3rd ICPER, Kuala Lumpur, 2012).
4. Atnaw S M, S.A. Sulaiman, S. Yusup, Jour. App. Sci. **11**, pp. 1913-1920 (2011)
5. S.M. Atnaw, S.A. Sulaiman, M.N.Z. Moni, "Experimental Study on Temperature Profile of Fixed – Bed Gasification of Oil- Palm Fronds" (4th IMAT, Melaka, 2011)
6. F.M. Guangul, S.A. Sulaiman, A. Ramli, Bioresour. Technol. **126**, pp 224-232 (2011)
7. M.N.Z. Moni, S.A. Sulaiman, "Downdraft Gasification of Oil Palm Frond: Effects of Temperature, Operation Time" (3rd ICPER, Kuala Lumpur, 2011)
8. F.M. Guangul, S.A. Sulaiman, M.N.Z. Moni, S.M. Atnaw, R.E. Konda, "Investigation of the Hygroscopic Property of Oil Palm Fronds for Gasification Process", (3rd ICPER, Kuala Lumpur, 2011)
9. M.N.Z. Moni, S.A. Sulaiman "Method of Processing Oil Palm Frond into Biomass Fuel for Downdraft Gasification," (3rd ICPER, Kuala Lumpur, 2011)
10. A.B. Nasrin, A.N. Ma, Y.M. Choo, S. Mohamad, M.H. Rohaya, A. Azali, Z. Zainal, American Jour. App. Sci. **5** (3) pp. 179-183 (2011)
11. Y. Shinogi, Y. Kanri, Bioresour. Technol. **90** pp. 241-247 (2003)
12. H. Pakdel, C. Roy, Energy & Fuels **5**, pp. 427-436, (1991)

Utilizing Proton Resonance Frequency of Isotopes Materials for Ultra-Precise Temperature Measurement: A Review

Monis Abdulmanan Abdullah [1,*], *Thar M. Badri Albarody*[1,†], and *Puteri Sri M. Bt Megat Yusoff* [1]

[1] Department of mechanical engineering, Universiti Teknologi PETRONAS, Malaysia

Abstract. High energy management in nuclear system and refractory metals productions are equipped with challenging procedures in terms of precise and remote controlling. In order to predict occurrence of contamination and avoidance of huge damages, there are often difficulties to access the equipment during their operation. In addition, estimating the precise and remote nucleation critical temperature of decay and growth of radioactive materials in the nuclear system has also proven to be a great challenge. Other than that, the eutectic crystallization temperature of the refractory metals during production also need to provide a precise estimation. However, it has been understood that the conventional temperature sensors are yet to be applicable to work precisely in such harsh environments. On the other hand, proton resonance frequency thermometry phenomenon have not been utilized or developed to serve as temperature sensors; despite the fact that they are capable to measure temperature in quantum level. Therefore, this article provides a review of the prior art on proton resonance frequency thermometry with its application and reliability, and elaborates on the trajectory of ultra-precise temperature measurement as the latest development.

1 Introduction

Temperature measurement for any system provides crucial information about how the system performs, how far the system has been under control, what is the temperature distribution in the whole system; despite the system's size or complexity. Nowadays, the available advanced technology of temperature measurements techniques at a level of sophistication accuracy have yet to be fulfilled by many applications. In order to measure the temperature in ultra-precise manner; as required by the quantum level of the system, there is a need to understand how the available techniques proven to be precise and remotely handled, can be implemented successfully. The solid-state nuclear magnetic resonance (NMR) technique is able to provide an ultra-precise temperature measurement through the proton of particular nuclei of specific

[*] Corresponding author: monis_g03669@utp.edu.my
[†] Corresponding author: dher.albarody@utp.edu.my

materials [1]. This paper aims to focus on temperature measurements using solid-state NMR technique, which has a real-time capability without any ionizing radiation occurrence.

2 Overview

NMR thermometry techniques are divided in to two main approaches: (1) the spin relaxations measurements (T_1, T_2), and (2) the chemical shifts through the magic angle spinning (MAS). Both techniques are based on the temperature dependence of the proton resonance frequency [1]. However, the fundamental relation of nuclear magnetic resonance is the proportionality between the angular resonance frequency of the proton ω and the magnetic-flux density β, $\omega = \beta\gamma$, where γ is the gyromagnetic ratio that nucleus specifies [2]. To make use of NMR thermometry isotopic chemical shifts of specific materials is where to begin with. Those materials are ^{1}H [3, 4] ^{13}C [5-7], ^{15}N [8], ^{31}P [9, 10], ^{119}Sn [11-13] and ^{207}Pb [14-16]. In solid-state, NMR technique is able to observe the temperature mapping of a particular sample; which also the temperature distribution of the sample. Comparing other temperature measurement techniques with solid-state NMR technique, a study has stated that it is able to provide a temperature distribution throughout the sample, not only the surface temperature of the sample as the probe is not in contact with the sample [17]. There has been a number of applications presented in the literature that represents the solid-state NMR temperature measurement techniques. In food industries, the precise reading of temperature in real time is required to ensure accurate output of the products is obtained; as a slight increment or decrement will affect the end-product. By a spatial measurements of the spin relaxations T_1 and T_2, the temperature profile can be mapped for cooking food, dairy products, and emulsified materials (a mixture of two or more liquids that are normally unmixable) [18]. Moreover, many researchers had done studies of chemical and physical characterizations of wide range of food which are categorized in terms of the amount of fat in canning meat, the percentage of water in daily dairy food process, storage of vegetables, and fruits and many others applications. With that, without adding any tracer materials, NMR is a powerful non-destructive technique which provides a specific type of information that relates to the product characterizations of chemical components and their percentage as well. The temperature mapping of the aforementioned applications can be examined by NMR technique, as shown below in Figure 1 and Figure 2, respectively.

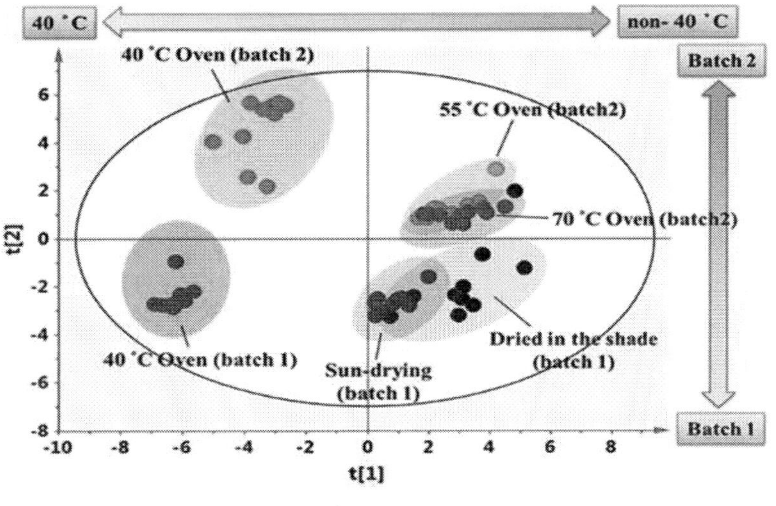

Fig. 1. Scatter plots of the PCA scores of the ^{1}H NMR dataset of extracts of Uraria Crinita that had been processed in different batches, by different drying methods, and at different oven temperatures [19].

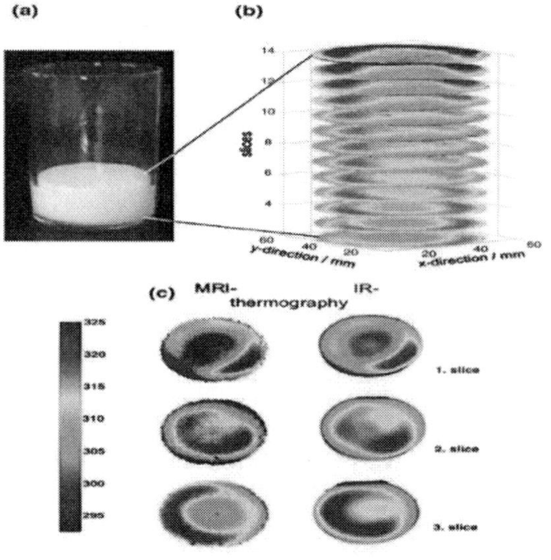

Fig. 2. (a) Heated model food with marked slices for (qualitative) comparison in (c). (b) MRI determined temperature distribution of heated model food (15 slices, slice thickness 1 mm, slice distance 1 mm). (c) Comparison of MRI and IR thermography of three selected slices [17].

However, NMR technique has been used widely in biotechnology and embryology applications. Studies on the components of biological members of humans and animals provide the exact information about their built. Doctors observed during the process of heating therapy temperatures measurement by NMR, it is able to provide directions towards where the heat should be applied and the exact amount that is required to stop additional heat, as compared with the normal tissues [20]. The temperature measurement ranges between -250°C to 47°C are identified in literature, respectively [1]. Other than that, in terms of material processing with NMR technique, there has been a wide range of application in many different industries. It is worth mentioning that the treatments process of medicine drugs is highly critical and sophisticated as it relates to individual's life. The precise temperature of syntheses of the materials is the most significant part of the process. This is due to the fact that any slight difference occurrence in the temperature can entirely change the chemical structure and provide a different output (medicine); hence leading to a side effect or inveracity of the medicine treatments [21] as illustrated in Figure 3.

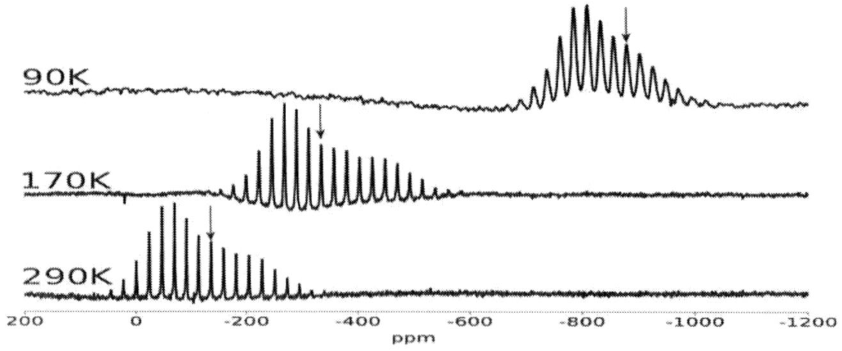

Fig. 3. ^{119}Sn MAS NMR spectra of $Sm_2Sn_2O_7$ at top 90K, middle 170K and bottom 290K [22].

Moreover, the NMR temperature measurement technique has also been used in processing the steel-making while it is being accompanied with complicated multi-component systems. NMR technique is a very powerful analysis tool to measure the

 DOI: 10.1051/matecconf/201713101008

temperature of the system and the structural analysis within all the stage of the process as well as covering a wide range of the temperatures. In situ high temperature NMR, temperature measurement could be recorded up to 1400°C, as shown in Figure 4. Varian infinity plus 400 (9.4 Tesla) spectrometer operating at 104.17MHz with a home-built static probe [23] has been used to perform in situ high temperature NMR experiments. Two different analysis techniques have been used at various temperature ranges. Relaxation time T_1, was used to measure by a saturation recovery pulse sequence and relaxation time T_2 value also calculated. It is essential to understand the fluctuations of local magnetic field with the spin lattice relaxation. However, these types of process include various types of components. Therefore, the measurement of MAS in solid state NMR plays a vital role with respect to different temperature NMR spectroscopy; once an observation of several MAS with specific line widths occurs, the data of temperature measurement is recorded. This is caused when the difference is detected due to the chemicals from one component to another component chemical shift. Each technique is complementary to the other technique, as it has to measure the temperature in ultra-precise manner by using the solid-state NMR instrument.

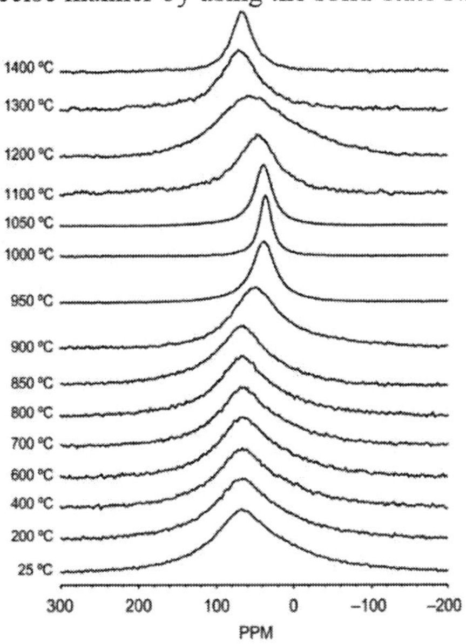

Fig. 4. ^{27}Al static spectra for the CAS glass and melt from 25°C to 1400°C [24].

Furthermore, the requirement to measure the temperature in NMR technique is not easy task in terms of acquisitions data and analysing it from the NMR spectroscopy. To this date, there is no standard formula for NMR spectroscopy data for temperature base. Many researchers have used different mathematical model techniques. Starting with a liner behaviour, the equation below shows the actual shift versus temperature relations [25].

$$\delta_{peak}(T) = -\{3670.6 \pm 1.0\, ppm\} + \{0.666 \pm 0.003\, ppm/K\}\, T \quad (1)$$

$$\delta_{iso}(T) = -\{3713.9 \pm 1.0\, ppm\} + \{0.758 \pm 0.002\, ppm/K\}\, T \quad (2)$$

The above two empirical equations vary due to the change of isotropic and the anisotropic components of the chemical shift. Equation (1) fits anisotropic components while, the equation (2) is comparatively based on the express of the temperature in an isotopic component. The uncertainties in these equations result from a slight linear fit of the data and reflects for the high signal-to-noise ratio of the spectra. Equation (1) provides the basis for measuring temperature in a static probe because the peak can be easily determined while

DOI: 10.1051/matecconf/201713101008

ensuring its accuracy in a limited number of scans without recourse to fitting or calculation [26]. Many other researchers have focused their studies on measuring the temperature variation by NMR solid-stat technique with the help of a known isotopic material. One researcher selected $Pb(NO_3)_2$, because it has a strong temperature dependence of the isotropic chemical shift [27].

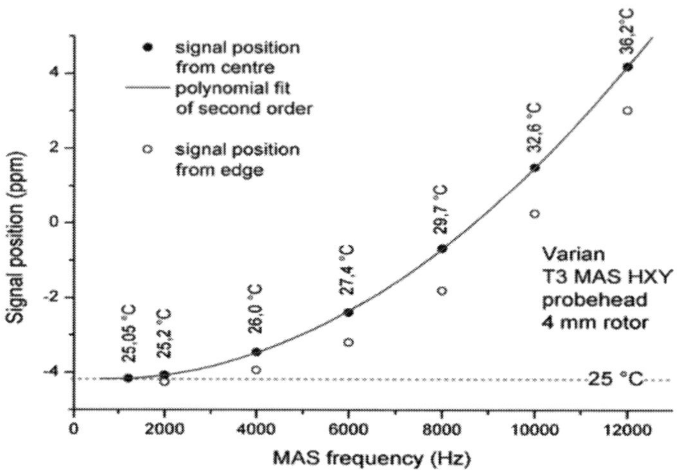

Fig. 5. ^{207}Pb NMR chemical shifts of $Pb(NO_3)_2$ and the sample temperature in the center (solid circles) and edge (open circles) of the rotor as a function of the MAS rate. The measurements were carried out at VT temperature set to 25 °C[28].

Special fitting curve has been used by the aforementioned researcher, where the second-order polynomial point method has been utilized. Temperature gradients can be calculated from the line width of the signals of NMR spectroscopy, which indicates the temperature of the sample in the base of chemical shift (figure 5). In addition, the variable temperature measurements (VT MAS NMR) is calculated when a specific temperature needs to be measured. In NMR solid-state, a rapid high precession temperature measurements can be obtained with fluctuations of 1s timescale as well as the temperature distribution that has wider ranges of temperatures of the product [29]. The authors highlighted that the demerits of this technique are very complicated in terms of result reading and analysis as it can be costly.

3 Conclusions

Solid-state NMR is a very powerful instrument that is utilized to analyze the chemical components of the material. Not only that, it is also able to study the physical state in spatial and temporal terms of the material at the quantum level. In addition, NMR technique does not affect the sample that is being used in the process, as it is a non-destructive testing technique. Temperature measurement of a system is remotely and instantaneously obtained in the solid-state NMR; which also covers a wide range of temperatures. Lastly, the NMR signals amplitudes and the line shapes of NMR spectroscopy has provided a meaningful insight.

Acknowledgment

The author acknowledges the support of Universiti Teknologi PETRONAS, Malaysia, for this work.

References

1. K.R. Thurber, R. Tycko, Journal of magnetic resonance, **196**, 84-87 (2009)
2. M.J. Duer, *Solid-State NMR Spectroscopy Principles and Applications*, Blackwell Science Ltd, USA, (2002)
3. Dvinskikh, Castro, Sandström, Magnetic Resonance in Chemistry, **42**, 875-881 (2004)
4. A.E. Aliev, K.D.M. Harris, Magnetic Resonance in Chemistry, **32**, 366-369 (1994)
5. A.N. Klymachyov, Dalal, Solid State Nuclear Magnetic Resonance, **7**, 127-134 (1996)
6. J.F. Haw, R.A. Crook, R.C. Crosby, Journal of Magnetic Resonance, **66**, 551-554 (1986)
7. G.C. Campbell, Crosby, J.F. Haw, Journal of Magnetic Resonance, **69**, 191-195 (1986)
8. Wehrle, Aguilar-Parrilla, Limbach, Journal of Magnetic Resonance, **87**, 584-591 (1990)
9. C.S. Zuo, K.R. Metz, Y. Sun, Sherry, Journal of magnetic resonance, **133**, 53-60 (1998)
10. H. Pan, B.C. Gerstein, Journal of Magnetic Resonance, **92**, 618-619 (1991)
11. T.F. Kemp, G. Balakrishnan, K.J. Pike, M.E. Smith, R. Dupree, Journal of magnetic resonance, **204**, 169-172 (2010)
12. B. Langer, Schnell, Spiess, A.-R., Journal of magnetic resonance, **138**, 182-186 (1999)
13. C.P., Cheetham, C.M., Journal of Magnetic Resonance, Series A, **101**, 299-306 (1993)
14. T. Takahashi, H. Kawashima, H. Sugisawa, B. Toshihide, Solid State Nuclear Magnetic Resonance, **15**, 119-123 (1999)
15. P.K. Isbester, A. Zalusky, D.H. Lewis, M.C. Douskey, M.J. Pomije, K.R. Mann, E.J. Munson, Catalysis Today, **49**, 363-375 (1999)
16. A. Bielecki, D.P. Burum, Journal of Magnetic Resonance, Series A, **116**, 215-220 (1995)
17. B. K. Knoerzer a, M. Regier b,c, E.H. Hardy d, H.P. Schuchmann b, H. Schubert, Innovative Food Science and Emerging Technologies, **10**, 537–544 (2009)
18. R.R.a. Xiaofei Ye a, Paul Chen a, Kehua Chang a,I.A.T.b. Ke Ning a, Chris Doona, Journal of Food Engineering, **59**, 143–150 (2003)
19. J. Chao, Y. Dai, H.Y. Cheng, W. Lam, Y.C. Cheng, K. Li, W.H. Peng, L.H. Pao, M.T. Hsieh, X.M. Qin, M.S. Lee, Scientific reports, **7**, 38763 (2017)
20. M. Shigeo Okuda, 1* Kagayaki Kuroda, PhD,1 Koichi Oshio, MD, PhD,1, P. Robert V. Mulkern, 2 Vincent Colucci, BA,1 Paul R. Morrison, MS,1, M. Osamu Kainuma, 1 and Ferenc A. Jolesz, MD1, Journal Of Magnetic Resonance Imaging, **12**, 330–338 (2000)
21. F.C.a.S. Grzesiek*, JMB, **715**, 739-752 (2002)
22. G.B. T.F. Kemp, K.J. Pike, M.E. Smith, R. Dupree, Magnetic Resonance, **204**, (2010)
23. J.F. STEBBINS, 1353-1373 (1991)
24. K. Kanehashi*, Nippon Steel Technical Report, Advanced Technology Research Laboratories , 20-1, Shintomi, Futtsu, Chiba, Japan, (2011)
25. C.D.b. Guenther Neue a, Solid State Nuclear Magnetic Resonance, **7**, 333-336 (1997)
26. A.C.D. Peter A. Beckmann*, Journal of magnetic resonance, **146**, 379–380 (2000)
27. J. Wu, Geological And Environmental Sciences, Stanford University, (2011)
28. V. Hronský, Acta Electrotechnica et Informatica, **13**, 95-98 (2013)
29. Jonathan f. Stebbins: nuclear magnetic resonance at high temperature. In *modern methods in solid-state NMR: practitoner's guide,* Stanford ca, (2017)

Thermal and strength performance of reinforced self-compacting concrete slabs mixed with basalt and PVA fibers in high intensity fire

Noraniza Mohd Jani[1,3], *Mohammad* Shakir Nasif [1, *], *Nasir* Shafiq[2], *and William* Pao[1]

[1]Mechanical Engineering Department, Universiti Teknologi PETRONAS, 32610 Bandar Seri Islander, Perak, Malaysia.
[2]Civil Engineering Department, Universiti Teknologi PETRONAS, 32610 Bandar Seri Islander, Perak, Malaysia.
[3]Fire Science Research Group, 41050 Jalan Meru, Klang, Selangor, Malaysia.

Abstract. Fibers addition to concrete and the innovation of self-compacting concrete technology lead to the development of high-performance concrete. However, high intensity fire may adversely affect the performance of this type of concrete. A series of fire resistance test experiments to evaluate the performance of fiber reinforced self-compacting concrete (FR-SCC) slabs consisting of various mix of basalt and PVA fibers were carried out by subjecting the concrete slabs as an element of construction to high intensity Hydrocarbon fire heating condition. The fire testing condition was in accordance with the standard time-temperature fire curve for 120 minutes up to 1100°C heating temperature. The temperatures on the surface and within the concrete slabs were recorded and the performance of each type of FRSCC slabs were evaluated. The performance of Basalt FR-SCC was found to be more resistant to fire in comparison to PVA FRSCC. There residual compressive strength of core samples were tested and SEM analysis were carried out to determine the effect of high intensity fire on the basalt and PVA FR-SCC slabs.

1 Introduction

The revolutionary and innovative development of self-compacting concrete (SCC) or self-consolidating concrete leads to saving in skilled labor cost and faster construction time [1]. The normal concrete basically need to be compacted by using vibrator which required skilled labor workers. In comparison with SCC, it is an innovative concrete which does not requires vibration for compacting process. Thus fresh concrete could flow on its own and can achieve full compaction by itself. Modern application of high performance self-compacting concrete (SCC) is focused on high performance for better and more reliable quality, dense and uniform surface texture, improved durability, higher strength, and faster construction time. SCC technology has taken a further development since its original development in Japan, it specifications and design for better practical application of the SCC technology on its physical and mechanical characteristics, including mixing of fibers to enhance the properties of the SCC the same way as normal concrete [2].

*Corresponding Author: mohammad.nasif@utp.edu.my

Under normal conditions, most concrete structures are subjected to a range of temperature no more severe than that imposed by ambient environmental conditions. Concrete is generally known for its capacity to endure high temperatures and fires, owing to its low thermal conductivity and high specific heat [3]. On the other hand, it does not mean that fire as well as higher temperatures does not affect the concrete properties. Characteristics such as color, compressive strength, elasticity, concrete density and surface appearance are affected by high temperature. Therefore, improving concrete's fire resistance is a field of interest for many researchers lately. According to their studies, it is possible to improve fire resistance of concrete in few ways including the addition of fibers to fresh concrete [3]. Basalt and polyvinyl alcohol (PVA) fiber-reinforced self-compacting concrete (FR-SCC) have been shown to enhance the mechanical properties at ambient conditions [4]; but its performance as a structural construction element in fire is yet to be assessed. There was no previous research done on the effects of basalt and PVA FR-SCCs' performance in accordance to standard fire i.e. for building fire (ISO Cellulosic Curve) and to rapid fire development (Hydrocarbon fire curve). The post-fire effects to the FR-SCC of basalt and PVA fibers also should be examined in order to assess the structural integrity of the FR-SCC which is the focus of this research.

4 Fire exposure characteristics

The type of fire, the fire size and heat output, have significant influence on the fire performance of a concrete structural system [5]. The researches pointed out that much of the current fire provisions contained in codes and standards are based on (standard ASTM E119 [6], BS 476-20 [7] ISO 834-1 [8] or similar) fire scenarios, which represents typical building fire. These provisions may not be directly applicable to fires resulting in infrastructure projects, offshore structures and tunnels due to a wide range of differences in fire characteristics. These fires, often referred to as hydrocarbon fires, typically represented by ASTM-E1529 Standard fire, much more severe than building fires and are characterised by fast heating rates or high fire intensities.

Considering, in a simplistic form, different fuel types and ventilation conditions, the following type of fire curves are considered for the testing of concrete slabs representing a typical fully developed compartmental building fire curve and a Hydrocarbon fire or liquid type fuel fire curve, referred to in ASTM E1529, BS 476-20: 1987 [6-7].

5 Experimental materials and test procedure

Considering the advantages offered by SCC compared to conventional concrete, experimental investigation have been carried out on the performance of fiber-reinforced self-compacting concrete (FR-SCC) slabs without fiber, with Basalt fiber and PVA fiber in high intensity fire. The mix design adopted in this experiment on fiber reinforced self-compacting concrete [4] is shown in table 1.

Fire resistance is a property of materials that prevents or retards the passage of excessive heat or flames under conditions of use. For structural elements, it is defined as the ability of the element to confine a fire or to continue its structural function, or both, for a stated period of time. Two important parameters for evaluating fire resistance of a construction element are load-bearing capacity and integrity [9]. The later means that no large crack is produced in the separating element so that the fire and large heat can pass through it. Therefore, the loss of strength and cracking of the specimens are the important factors which are to be considered for evaluating the fire behavior of the test specimens.

The self-compacting concrete (SCC) slabs were casted between May and August 2016. The slabs were casted using a portable concrete mixer. Each SCC slab dimension is 1600mm wide x 1600mm long x 300mm deep. The SCC slab was left to air- curing within the laboratory ambient condition for minimum 90 days prior to fire resistance testing.

Table 1. Design mix of fiber reinforced self-compacting concrete slabs text

*Code Mix**	*Cement*	*Fine Aggregate*		*Coarse Aggregate*		*Water*	*Super plasticizer*	*Fiber (PF - BF)*	
	kg/m³	*kg/m³*	*kg/m³*	*kg/m³*	*W/C*	*%*	*kg/m³*	*%*	*kg/m³*
M1- 0F	600	900	750	200	0.33	2.0	12.0	0.0	0.0
M2- 1.0PF	600	900	750	200	0.33	2.0	12.0	1.0	6.0
M3- 1.0BF	600	900	750	200	0.33	2.0	12.0	1.0	6.0
M4- 1.5PF	600	900	750	200	0.33	2.0	12.0	1.5	9.0
M5- 1.5BF	600	900	750	200	0.33	2.0	12.0	1.5	9.0
M6- 2.0PF	600	900	750	200	0.33	2.0	12.0	2.0	12.0
M7- 2.0BF	600	900	750	200	0.33	2.0	12.0	2.0	12.0

The fire resistance tests were carried out using an indicative fire resistance test furnace of nominal 1200mm x 1200mm x 1200mm chamber. The furnace temperature was measured by means of four (4) Type-K furnace thermocouples consisting of Type K thermocouple cable fixed to 12mm diameter copper discs; arranged symmetrically in the furnace in two rows of two thermocouples.

The SCC slab was placed to cover a horizontal opening of the furnace chamber of 1200mm x 1200mm. The temperature of the unexposed face i.e. non-fire side of the SCC slab was measured by means of five (5) fixed surface calibrated specimen thermocouples consisting of Type K thermocouple cable fixed to 12mm diameter copper discs and covered with 30mm square millboard pads. The internal temperatures of the slab was measured by means of eleven (11) specimen thermocouples consisting of Type K thermocouple cable fixed to 12mm diameter copper discs and placed within the slab during casting at every 25mm deep interval.

The fire resistance tests were carried out in accordance to the general requirements of BS 476-20 [7]; and adopting the specific requirements of non-loadbearing horizontal element of construction in accordance to BS 476-22 [9]. During the testing, observations were taken on the behavior of the SCC slab. The duration of the fire testing was 120 minutes.

A total of eight (8) slabs were tested. One of the plain SCC slab (Code Mix M1-0F) was tested under the heating conditions of ISO Cellulosic Curve. Whilst, all seven (7) other slabs (all Code Mix M1 to M7) were subjected to the test under Hydrocarbon heating conditions.

6 Test results and discussion

The fire resistance tests were carried out between October 2016 and March 2017. All slabs were tested tor a period of 120 minutes.

All the tested slabs satisfied the performance for insulation and integrity performance in accordance the standards [7, 10]. The standards describe the integrity as the ability of the test specimen to not collapse, not having any holes, cracks or fissures, and sustained flaming on the unexposed face which is the non-fire side. Whereas, insulation as the ability of the test specimen to restrict temperature rise of the unexposed face which is the non-fire side below the mean of 140 C above the initial value.

A plain SCC slab (Code Mix M1) was tested to the Hydrocarbon curve had a rapid rise from the commencement of heating, whilst another plain SCC slab was tested to the ISO Cellulosic curve had a gentle temperature rise throughout the duration of the test.

Table 2. Comparison on temperatures of SCC slabs (Code Mix M1-0F) subjected to fire exposure under ISO Cellulosic curve and Hydrocarbon curve.

Time (Mins)	ISO Cellulosic Fire Curve (Nominal) (°C)	Surface of Slab (Non-fire side) (Mean) (°C)	Mid-depth of Slab (Internal) (°C)	25mm from bottom slab (Internal) (°C)	Hydrocarbon Fire Exposure (Nominal) (°C)	Surface of Slab (Non-fire side) (Mean) (°C)	Mid-depth of Slab (Internal) (°C)	25mm from bottom slab (Internal) (°C)
0	20.0	31.9	30.1	30.5	44.0	30.94	29.5	30.6
15	738.6	32.1	30.3	1721.0	1070.7	31.2	32.3	256.1
30	841.8	32.3	37.5	312.0	1097.6	31.3	59	434.0
45	902.3	32.3	58.5	414.0	1099.8	32.6	82.0	544.0
60	945.3	33.3	73.5	476.0	1100.0	36.0	93.5	608.0
75	978.7	36.7	85.0	528	1100.0	40.6	104.5	644.0
90	1006.0	41.5	95.0	580	1100.0	45.6	115.5	672.0
105	1029.1	46.2	102.5	628	1100.0	50.5	127.0	696.0
120	1049.0	50.7	109.5	668	1100.0	54.9	139.0	716.0

Table 3. Comparison on temperatures of Basalt FR-SCC slabs subjected to Hydrocarbon fire exposure.

Code Mix		M3-1.0BF			M5			M7		
Time Mins	Hydro-carbon Fire Exposure (Nominal) (°C)	Surface of Slab (Mean) (°C)	Mid-depth of Slab (°C)	25mm from bottom slab (°C)	Surface of Slab (Mean) (°C)	Mid-depth of Slab (°C)	25mm from bottom slab (°C)	Surface of Slab (Mean) (°C)	Mid depth of Slab (°C)	25mm from bottom slab (°C)
0	44.0	28.6	27.8	28.5	28.54	28.7	29.4	30.46	29.6	30.9
15	1070.7	28.96	27.9	114	28.72	28.8	171	30.58	29.8	87
30	1097.6	28.86	30.9	191	28.88	33	316.1	30.26	32.8	146
45	1099.8	29.16	41.5	282	29.04	46.5	428	29.86	43	205
60	1100.0	30.86	55	352	30.42	60.5	504	29.98	54.5	260
75	1100.0	33.68	66	408	32.9	70.5	556	30.92	64.5	304
90	1100.0	37.86	74	452	36.3	78.5	600	32.9	72	340
105	1100.0	42.46	84	488	39.8	85	628	35.54	78	370
120	1100.0	46.92	92	516	43.46	90	656	38.56	83	396

Under ISO fire, after 8 minutes of fire testing the temperature at 25mm from the exposed face of slab only starting to reach 100°C. At the termination of test at 120 minutes of testing, the temperature at the depth of slab was 668°C. At termination of test, the temperature of the slab at 150mm deep from the unexposed face was still below 100°C. The surface temperature of the non-fire side was 53.3°C.

Under Hydrocarbon (HC) fire Table 2 shows the tabulation of temperatures for plain SCC under ISO Cellulosic and Hydrocarbon fire exposure. Table 2 shows the actual mean furnace temperatures, the slabs' temperatures on the surface of the non-fire side.

The plain SCC slab that was subjected to Hydrocarbon fire was found to be having higher temperature near to the bottom of the slab by slightly 48 C only. The non-fire side surface temperature also only showed slightly higher for the slab exposed to Hydrocarbon fire.

After these findings, all the Basalt and PVA fiber reinforced self-compacting concrete were subjected only to Hydrocarbon fire which have more severe effects to the slabs.

Tables 3 and 4 provide comparison of the Basalt and PVA fibers reinforced self-compacting concrete slabs when subjected to Hydrocarbon fire for 120 minutes.

All the Basalt and PVA fibers fiber reinforced self-compacting concrete slabs had surface temperature below 50° C at the termination of fire testing after 120 minutes of exposure to Hydrocarbon fire.

Table 4. Comparison on temperatures of PVA FR-SCC slabs subjected to Hydrocarbon fire exposure

Code Mix		M2-1.0PF			M4-1.5PF			M6-2.0PF		
Time Mins	Hydrocarbon Fire (Nominal) (°C)	Surface of Slab (Mean) (°C)	Mid depth of Slab (°C)	25mm from bottom slab (°C)	Surface of Slab (Mean) (°C)	Mid-depth of Slab (°C)	25mm from bottom slab (°C)	Surface of Slab (Mean) (°C)	Mid-depth of Slab (°C)	25mm from bottom slab (°C)
0	44	28.3	27.8	28.1	32.08	30.9	31.2	29.78	29.7	30.1
15	1070.6	28.36	28.2	229.2	32.38	31	162	29.84	30.1	428.2
30	1097.5	28.66	45.5	458.2	32.6	51.5	312.1	30.08	47.3	648
45	1099.8	29	72.5	560	32.68	85.5	430	30.14	70	736
60	1099.9	29.94	86	624	33.52	97	516	30.68	84.5	788
75	1099.9	31.88	94	688.1	35.16	99	584	31.92	94.5	844
90	1100	34.76	97	720	38.5	99.5	624	34.26	102.5	868
105	1100	38.02	99	672.7	42.42	100.5	656	38.04	109.6	880
120	1100	41.38	101	686.1	46.48	108.5	680	41.98	117.5	883.9

Table 5. Comparison of average compressive strength of various design mix.

Code Mix	Compressive Strength (MPa)		Reduction of Compressive Strength (*%)	Type of Fire
	Non-Fire	Fire Tested		
M1-0F	68.7	59.5	13.4	ISO Cellulosic
M1-0F	68.7	54.2	21.0	Hydrocarbon
M2-1.0PF	80.0	48.9	38.8	Hydrocarbon
M3-1.0BF	48.4	31.9	34.0	Hydrocarbon
M4-1.5PF	70.8	54.8	22.6	Hydrocarbon
M5-1.5BF	46.2	36.4	21.3	Hydrocarbon
M6-2.0PF	69.7	50.2	27.9	Hydrocarbon
M7-2.0BF	45.3	32.9	27.3	Hydrocarbon

After the fire tests, three (3) core samples of 100nn diameter were taken from each fire tested slabs and the control non-fire test slabs for the determination of compressive strength as shown in Table 5.

From the fire test results, Basalt FR-SCC Code Mix M7 with 2% Basalt fiber showed the best performance among all design mix when tested under high intensity Hydrocarbon fire.

It is found from the results presented that the best design mix when tested under Hydrocarbon fire was found to be basalt fibers reinforced self-compacting concrete with 2% basalt fiber percentage.

7 Conclusion

The fire resistance performance of fiber self-compacting concrete (FRSCC) with 2.0% basalt fiber (Code Mix M7 2.0BF) demonstrated the best among all tested design mixes when subjected to Hydrocarbon fire. The PVA FRSCC with 2% fibers (Code Mix M6 2.0PF) was found to have the least fire resistance performance compared to all FRSCC design mixes tested. Code Mix 6 was found to have higher temperature recorded by 167.9°C compared to the plain SCC when subjected to Hydrocarbon fire exposure. Whereas, basalt and PVA FRSCC of 1.5% were found to have slightly better fire resistance performance compared to plain SCC when subjected to exposure under Hydrocarbon fire. In terms of residual compressive strength when the basalt and PVA FR-SCCs' compressive residual strength compared to the plain SC exposed to Hydrocarbon fire, the 1.5% showed the most reduction in compressive strength i.e. by 22.6% and 21.3% respectively. Finding the balance of the fibers percentage and the fire resistance performance and its effects after fire in terms of residual compressive strength, both basalt and PVA mix of 1.5% were found to provide good fire resistance performance and higher residual compressive strength.

Acknowledgment

The authors are gratefully acknowledging Fire Science Research Group (FSRG) support to this research for the usage of the fire resistance test furnace. The authors also gratefully acknowledge the financial support by Promat (Malaysia) S/B.

References

1. H. Okamura, M. Ouchi, Jr. of Adv. Conc. Tech., **1(1)**, pp.5-15, (2003).
2. EFNARC, *Specification and Guidelines for Self-Compacting Concrete*, (EFNARC, 2002).
3. O. Arioz, Fire Safety Jr., **42**, pp. 516-522, (2007).
4. A. F. Mohamed Salih, *Performance of fibreorced self-compacting concrete containing different pozzolanic material*. (Phd. Dissertation., Universiti Teknologi Petronas, 2014).
5. V. Kodur, L. Phan, *Fire Safety Jr.* **42**, pp. 482-488., (2007).
6. ASTM, ASTM E119-98, *Standard Test Method for Fire Tests for Building Construction and Materials*, (American Society for Testing and Materials, 1998)
7. BSI, British Standard BS476, *Part 20: Method for determination of the fire resistance of elements of construction, (*British Standards Institution, London, 1987).
8. ISO, ISO 834-1, *Fire Resistance Tests Elements of Building Construction*, (International Organisation for Standardization, Geneva, 1999).
9. BSI, BS 476-22, *Part 22. Method for determination of non-loadbearing elements of construction*, (British Standards Institution, London, 1987).

Evaluations of thermocline and half cycle figure of merit of a thermal energy storage tank

Mohd Amin Abd Majid[1,*] and *Looi* Kar Kin[2]

[1]Mechanical Engineering Department, Universiti Teknologi PETRONAS.
[2]Mechanical Engineering Department, Universiti Teknologi PETRONAS.

Abstract. Two main criteria that are commonly used to evaluate thermal energy storage systems are thermocline thickness and half cycle figure of merit. For the thermocline thickness, the preference is to achieve as thin as possible the thermocline thickness. While the preference for half cycle figure of merit is to achieve the value of greater than 90 per cent. These two criteria were used to evaluate a thermal storage system at University Teknologi PETRONAS district cooling plant. The capacity of the thermal energy storage tank of the plant is 10,000 RTh. Operating data was used for the evaluation. The values of evaluated thermocline thickness ranges from 2.248 meters to 5.445 meters with an average of 3.251 meters. These values are very much higher in comparison to findings of other studies. One possible reason is due to higher flow rates. For the half cycle figure of merit the evaluated values ranges from 0.9469 to 0.9847, with the average of 0.9698, which are within the acceptable range. For future work a model should be developed which could automatically evaluate both the thermocline thickness and half cycle figure of merit. This would enable both of these parameters to be continuously evaluated.

1 Introduction

Thermal energy storage (TES) is an established technology. It is applied to correct the mismatch that occur between the supply and demand of electrical energy. This is to achieve more efficient utilization of electricity, whereby the off peak electricity is utilize to charge the TES. For effectiveness the TES tank should be stratified [1]. The most commonly used TES system is sensible TES system using water as storage medium.

TES system is widely used in USA, with estimated installed capacity of 353 MW

* Corresponding author: mamin_amajid@utp.edu.my

for chilled water [2]. In Asia TES systems are installed in Japan, Thailand, Singapore and Malaysia, with Japan having the most installations [3]. In Malaysia stratified TES system is installed at Universiti Teknologi PETRONAS (UTP) [4]. This plant is used as the case study.

The main principle of TES is storing of electrical energy for utilization at a later time. For the case of TES system using electrical energy, the electrical energy is used to cool the water at a thermal storage tank to a temperature of between 6°C to 7°C and the chilled water is then used for air conditioning during the day. Normally the initial temperature of the water is from 12°C to 13°C. The charging of cooling water at the thermal storage tank is done during off peak periods which are normally at night using electric chillers. The off peak is selected to take advantage of cheaper electrical rate, which is the case of district cooling plant for KLIA 2 [5]. For the case of district cooling with gas turbine the excess electricity during the night is used to operate the electric chillers to charge the thermal storage tank. This improved the utilization of electricity [4].

In terms of performance measures of stratified chilled water TES tank, there are three methods of evaluations which are commonly used. The methods are cycle thermal efficiency, half cycle figure of merit (FOM) and thermocline thickness [6, 7, 8, 9].

2 Performance measure

The performance measures used for this study are thermocline thickness and half cycle FOM. These measures were used for performance evaluations of a stratified TES tank at UTP district cooling plant.

2.1 Thermocline thickness

Thermocline thickness indicates the physical barrier that separates the warm and cool water in the TES tank. The thickness determine how well a TES is stratified. The temperature profile of the water during charging is as shown in **Fig. 1 (a)**. Due to the different temperature of the chilled water, the warm water with lesser density forms at the top layer of the tank while cool water with higher density settle at the bottom. Between these two regions, thermocline is formed in the middle [6]. **Fig. 1 (b)** shows the TES system with electric chillers.

The temperature profile of **Fig. 1** is represented by Sigmoid Dose Response (SDR) function as per equation (1) [10].

$$T(X) = T_c+\{(T_h-T_c)/[1+10^{(C-X)S}]\} \tag{1}$$

Equation (1) relates the temperature distribution in the TES to all the variables shown in **Fig. 1 (a)**. T_h and T_c are the warm and cool water temperatures in degrees C, respectively. X is the sensor elevation of the tank. Parameter C determines the thermocline position in the tank. While the parameter S is a constant defining the slope gradient of the thermocline profile.

 DOI: 10.1051/matecconf/201713101010

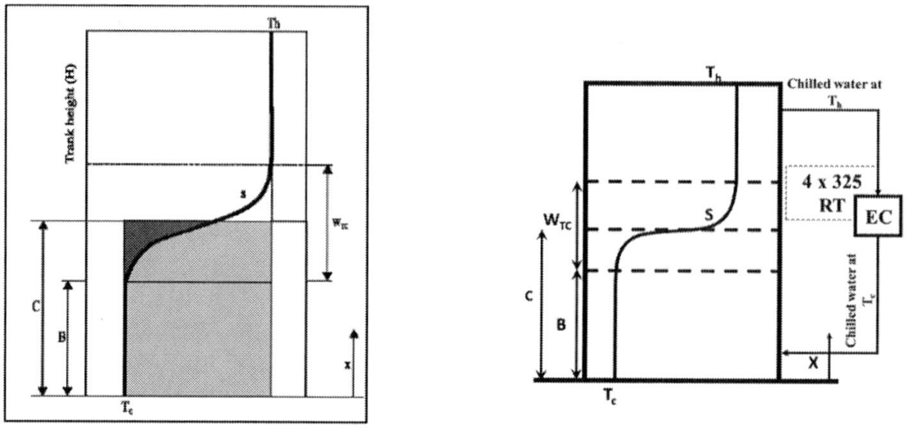

(a) **(b)**

Fig. 1. : (a) Temperature distribution profile of chilled water in TES tank. (b) TES tank with various parameters and electric chillers for the case study.

To quantify the thermocline thickness, limit points of the thermocline profile required to be analysed by using dimensionless cut-off temperature θ = (T-Tc)/(Th-Tc) [6]. Thermocline thickness, W_{TC}, is calculated using equation (2).

$$W_{TC} = \{[2log(1/\theta - 1)]/S\} \tag{2}$$

2.2 Half cycle figure of merit, ½ FOM

The half-cycle FOM, ½ FOM is the ratio of integrated capacity, C_{int} to theoretical capacity C_{max}, represented by an equal volume of water undergoing a temperature change equal to the difference between the average charged inlet temperatures. Derivation of the ½ FOM is as shown by equations (3) to (5) [9,10].

$$½\ FOM = C_{int}/C_{max} \tag{3}$$

Where,

$$C_{int} = \{[\rho AC_p(T_h\text{-}T_c)]/S\}[log(1+10^{SC})\text{-}log2] \tag{4}$$

$$C_{max} = C\rho AC_p(T_h\text{-}T_c) \tag{5}$$

The difference between the integrated capacity and theoretical capacity is the loss of cooling energy in TES, C_{loss}, and is represented by equation (6).

$$C_{loss} = C_{max} - C_{int} \tag{6}$$

Table 1. Hourly temperatures of 14 sensors (S1 to S14) in TES tank, 4th Jan 2016.

HR	S1	S2	S3	S4	S5	S6	S7	S8	S9	S10	S11	S12	S13	S14
	Temperature, °C													
0:00	8.0	7.9	7.9	7.9	7.8	7.8	7.7	7.8	7.8	7.7	7.7	7.4	7.5	7.3
1:00	8.0	7.9	7.9	7.9	7.8	7.8	7.7	7.8	7.8	7.7	7.7	7.3	7.5	7.2
2:00	8.0	7.9	7.9	7.9	7.8	7.8	7.7	7.8	7.8	7.7	7.6	7.3	7.4	7.1
3:00	7.9	7.9	7.9	7.9	7.8	7.8	7.7	7.8	7.8	7.6	7.5	7.3	7.3	7.0
4:00	7.9	7.9	7.9	7.9	7.8	7.8	7.7	7.8	7.8	7.6	7.5	7.2	7.2	7.0
5:00	7.9	7.9	7.9	7.8	7.8	7.8	7.6	7.8	7.7	7.5	7.4	7.2	7.1	7.0
6:00	7.9	7.9	7.9	7.8	7.8	7.8	7.6	7.7	7.7	7.4	7.4	7.1	7.1	6.9
7:00	8.1	7.9	7.9	7.8	7.8	7.8	7.6	7.7	7.7	7.4	7.4	7.1	7.1	7.0
8:00	12.2	7.9	7.9	7.8	7.8	7.8	7.6	7.7	7.7	7.5	7.5	7.2	7.2	6.9
9:00	13.8	10.2	7.9	7.8	7.8	7.8	7.6	7.8	7.8	7.7	7.6	7.3	7.4	7.0
10:00	14.0	13.8	10.5	7.9	7.9	7.8	7.7	7.8	7.8	7.7	7.8	7.5	7.5	7.1
11:00	14.0	14.0	13.9	10.2	7.9	7.9	7.7	7.8	7.9	7.8	7.9	7.7	7.7	7.4
12:00	14.1	14.1	14.0	13.9	11.3	8.0	7.8	7.9	7.9	7.8	7.9	7.8	7.9	7.5
13:00	14.3	14.2	14.2	14.1	14.0	12.5	8.3	7.9	8.0	7.9	7.9	7.8	8.0	7.7
14:00	14.5	14.5	14.4	14.2	14.1	14.0	13.4	9.9	8.0	7.9	8.0	7.8	8.0	7.9
15:00	14.6	14.5	14.5	14.4	14.2	14.1	13.9	13.9	11.6	8.2	8.0	7.9	8.1	7.9
16:00	14.6	14.6	14.5	14.5	14.4	14.3	14.0	14.1	14.1	13.2	9.5	8.0	8.1	8.0
17:00	14.6	14.6	14.6	14.5	14.4	14.3	14.1	14.2	14.1	13.9	12.3	8.2	8.1	8.0
18:00	14.6	14.6	14.5	14.4	14.3	14.2	14.0	14.1	14.0	12.4	8.6	8.2	8.2	8.1
19:00	14.6	14.5	14.5	14.3	14.2	14.2	13.9	14.0	12.2	8.5	8.3	8.1	8.3	8.1
20:00	14.5	14.4	14.3	14.2	14.1	14.0	13.8	11.8	8.5	8.2	8.2	8.0	8.2	8.1
21:00	14.4	14.3	14.2	14.1	14.0	13.9	11.4	8.3	8.2	8.0	8.1	8.0	8.1	8.0
22:00	14.3	14.2	14.2	14.1	13.9	11.2	8.1	8.1	8.1	7.9	8.0	7.9	8.1	8.0
23:00	14.2	14.2	14.1	14.0	12.2	8.4	8.0	8.0	8.0	7.9	8.0	7.8	8.0	7.9

3 Results and discussion

Thermocline thickness and ½ FOM were used for evaluations of TES at UTP District Cooling plant. The TES tank at the plant is with 10,000RTh capacity. There are 14 temperature sensors within the tank. January 2016 data was used for analysis. The hourly temperatures at the 14 sensors data for the 4th Jan 2016 is as shown in **Table 1**. The plot of hourly temperature profiles for the 14 sensors is shown in **Fig. 2 (a)** for 4th January 2016, while **Fig. 2 (b)** shows the plot for 6th January 2016. **Table 2** are the calculated SDR fitting temperature distribution on 4th Jan 2016 and 6th January 2016 respectively.

From **Fig. 2 (a)** and **Table 2** it is observed that the values of midpoint C of the thermocline on 4th January 206 were increasing as the charging hour increases. This indicates that the position of the thermocline was moving upward while the warm water drawn out and chilled water added into the TES tank. The increased in the portion of chilled water in the TES tank during the charging period resulted to increased cooling capacity of the TES tank. Since the values R2 in **Table 2** are greater than 0.99, this indicate the recorded temperature fitted well to SDR function.

For the case of 6th January, the C values increases from 18:00 to 12:00 as per **Table 2**. The plot for S-curve, fitted well during this period as shown in **Fig. 2 (b)**. However, it is not for the case of 01:00 and 02:00. The recorded temperatures did not fit the SDR function. The thermocline thickness for whole month of January 2016 were also evaluated. Plot of the values of the thermocline thickness is as shown in **Fig. 3 (a)**. The values range from 2.248 meter to 5.445 meter, with an average of 3.251 meter. These values are very much higher than that was obtained by [4] and [11], and the result close to that was obtained by [10]. One possible reason is due to higher flow rates in comparison to [4].

DOI: 10.1051/matecconf/201713101010

(a) (b)

Fig. 2. Plot of hourly temperatures from 6 pm to 2 am on **(a)** 4th January 2016 & **(b)** 6th January 2016 by 14 temperature sensors within the TES tank.

Table 2. Calculated SDR fitting of temperature distribution on 4th & 6th January 2016.

Initial Condition: ϴ= 0.01								
Date: 4th January 2016								
Hr	Tc	Th	C	S	R^2	Wtc(m)	1/2FOM	Remark
18:00	8.14	14.32	4.267	1.376	0.9959	2.90	0.9487	Accepted
19:00	8.18	14.29	5.32	1.437	0.9962	2.78	0.9606	Accepted
20:00	8.12	14.23	6.377	1.209	0.9978	3.30	0.9609	Accepted
21:00	8.05	14.19	7.456	1.335	0.9989	2.99	0.9697	Accepted
22:00	8.00	14.18	8.494	1.499	0.9992	2.66	0.9763	Accepted
23:00	7.95	14.15	9.274	1.422	0.9996	2.81	0.9771	Accepted
0:00	7.94	14.13	10.23	1.485	0.9993	2.69	0.9801	Accepted
1:00	7.92	14.06	11.02	1.738	0.9981	2.30	0.9842	Accepted
2:00	7.93	14.10	11.67	1.546	0.9965	2.58	0.9833	Accepted
Date: 6th January 2016								
Hr	Tc	Th	C	S	R^2	Wtc(m)	1/2FOM	Remark
18:00	8.10	14.36	5.921	1.285	0.997	3.11	0.9604	Accepted
19:00	8.28	14.38	7.132	1.219	0.9974	3.27	0.9653	Accepted
20:00	8.26	14.32	8.222	1.459	0.9977	2.74	0.9749	Accepted
21:00	8.20	14.27	9.256	1.322	0.998	3.02	0.9754	Accepted
22:00	8.18	14.25	10.23	1.333	0.9976	2.99	0.9779	Accepted
23:00	8.13	14.18	11.19	1.251	0.9966	3.19	0.9785	Accepted
0:00	8.14	13.98	12.14	1.37	0.9941	2.91	0.9819	Accepted
1:00	8.10	~	~	~	0.9851	Undefined	Undefined	N.P.S.C
2:00	7.67	~	~	~	0.896	Undefined	Undefined	N.P.S.C

***N.P.S.C = Not perfect S-Curve.*

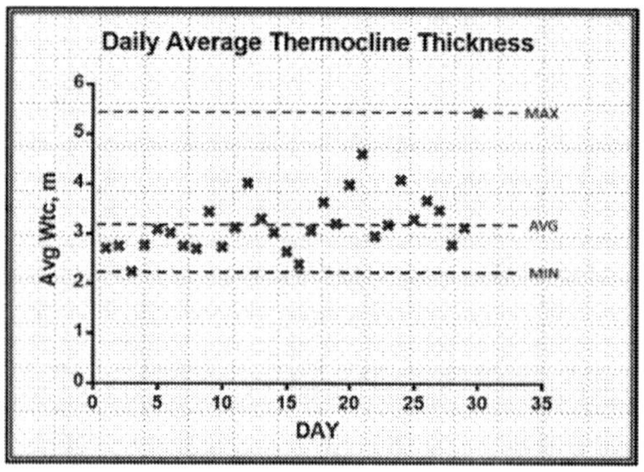

(a)

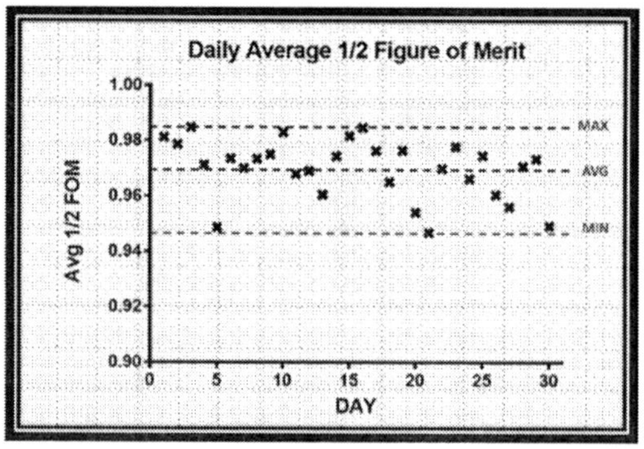

(b)

Fig. 3. Plot of **(a)** thermocline thickness & **(b)** ½ FOM for January 2016.

Since thermocline thickness might varies over a wide range, depending on tank dimension and operating temperatures, flow rates and diffuser designs, the magnitude of thermocline thickness could not be strictly compared.

For the half cycle FOM, the values obtained range from 0.9469 to 0.9847, with an average of 0.9698. The plot of the values is as shown in **Fig. 3 (b)**. Since the half cycle FOM are greater than 90%, the values are meeting the standards [6].

4 Conclusion

Thermocline thickness and ½ FOM were used for evaluations of TES at UTP District to assess the performance of the TES system. Results obtained indicate the TES tank performance is acceptable in terms of half cycle FOM but not for the thermocline thickness. The thermocline thickness are very much higher in comparison to other findings. One possible reason was due to higher flow rates. However, it is noted that due to a wide range of tank configurations and operating parameters, the magnitude of thermocline thickness could not be strictly compared. For continuously operating TES tank the performance monitoring should also be continuously evaluated. It is recommended that a model to be developed which could automatically evaluate both of these parameters. This would be enable continuous monitoring of these parameters.

References

1. Ibrahim Dincer, Marc R Rosen, *Thermal Energy Storage Systems and Applications*, (John Wiley & Sons Ltd., England, 2002)
2. M. Rutberg, M. Hastbacka, A. Cooperman, A Bauza, *Thermal Energy Storage, Emerging Technologies*, (ASHRAE Journal, June 2013)
3. SHINRYO Corporation, *District Heating and Cooling Systems*, https://www.shinryo.com/en/tech/bf_dhc.html.
4. A Majid, M Amin and W. Joko, (Journal of Energy and Power Engineering, **4** (2)): pp. 28-33 (2010)
5. Anmas Corporation, *Thermal Energy Storage Solutions*, http://www.anmascorp.com/pageKLIA2.html
6. A. Musser, W. Bahnfleth, (ASHRAE Transaction **104**(1)): pp. 55-67 (1998)
7. W. P Bahnfleth, J. Song, (Applied Thermal Engineering **25**): pp. 3067-3082 (2005)
8. J.E.B. Nelson 1, A.R. Balakrishnan, S. S. Murthy, (Applied Thermal Energy, **19**): pp. 89 -115 (1999)
9. Joko Waluyo, "*Determination of Performance Parameters of Hot Stratified Thermal Energy Storage Tank*", 6th International Annual Engineering Seminar (InAES), Yogyakarta, Indonesia, (2016).
10. Waluyo, J. and M.A.A. Majid, Journal of Applied Sciences. **11**(9): pp. 1642-1647 (2011)
11. Majid, M.A.A., M. Nasir, and J. Waluyo, Energy Procedia.14: pp. 1280-1285 (2012)

MATEC Web of Conferences 131, 01011 (2017)
UTP-UMP SES 2017 DOI: 10.1051/matecconf/201713101011

Experimental Investigation of Micro-Gas Turbine Performance at Reduced Air Inlet Temperature

Mior Azman Said[1, a)], Suhaimi Hassan[1] and M Faizairi M Nor[1]

[1]Mechanical Engineering Dept.,
Universiti Teknologi PETRONAS,
32610 Bandar Seri Iskandar, Perak, Malaysia

[a)]Corresponding author: miorazman@utp.edu.my

Abstract. Gas turbines have long been utilized as the main device for power generator. High temperature of ambient air flowing into inlet can reduce their performance. The objective of this study is to investigate the performance of a 1.5 kW micro-gas turbine at reduced inlet temperature using dry-ice as the cooling elements. Dry-ice is placed surrounding the outer wall of contained in a net-like holder at the inlet before the compressor and the inlet air temperature found to be decreased till 13°C from the ambient temperature. Comparison test was performed with a baseline result, which is the operation without the dry-ice cooling unit and the results shows that the power output decreases linearly with the increase of inlet air temperature. It was recorded that increase power output of 8.5% achieved from reduction of temperature of 7.5%. The power output was found also to be decreased with time, due to the inlet duct exposed to the heat form the combustion chamber. This indicates better insulation or adapted data recording approach is needed.

INTRODUCTION

Gas turbines is an important device in generating electricity, mobility and other engineering usage in industry. Higher thermal efficiency is one of the aspects that researchers look to achieve. In improving the gas turbine cycles, ambient temperature, pressure and humidity are the factors that can influence the gas turbine power output. However, according to El-Hadik [1], ambient temperature is the highest influence to the gas turbine power output, where as high as 1 K in temperature increase can produce 0.6% power increase and 0.18% thermal efficiency increase.

Due to the volatility of fossil fuel energy prices, researchers in inlet air cooling system for gas turbines to increase performance has escalated recently. Most research works on the inlet air cooling system can be categorized into these following classes based on their cooling systems: (1) Mechanical vapor compression or absorption chillers (2) Evaporation (3) Spray or fogging. For absorption chiller system, it is recorded that the gas turbine power output can be increased till 20% and efficiency by 2% but this has the highest cost compared to other two systems. Evaporation system consists of porous element like wood wool fibers where water inside it is evaporated when inlet air flows against it. For spray and fogging system, there are classified into two sub-category: (1) saturated air (2) overspray. The difference in both sub-category is the use of saturated air only before compressor for saturated air fogging system while for overspray system, is the use of high volume of water droplets in the spraying system [2].

LITERATURE REVIEW

Temperature plays a role in determining the efficiency of a gas turbine, therefore change in the inlet temperature will affect the efficiency and the power output of a gas turbine . Early et al. [3] mentioned that the amount of electrical power produced by the gas turbine or electric generator depends on the temperature of air at the inlet. In general the lower the temperature the higher the p is power output. Erdem and Sevilgen [4] have produced an equation of monthly electricity production:

$$E_m = N \cdot LF_m \cdot G \cdot 24 \tag{1}$$

where N is the power, LF is the monthly load factor and G is the number of days in the particular month. The power N, is calculated by the formula

$$N = \frac{P_1}{R_a} V \left[\frac{W_t}{T_1} - \frac{c_{pa}}{\eta_{cis}} \left(P_{rc}^{\frac{k_a - 1}{k_a}} - 1 \right) \right] \tag{2}$$

where P_1 is the inlet pressure and R_a is air ideal gas constant, V is the volumetric flow rate, W_t is the turbine work, T_1 compressor inlet temperature, c_{pa} is air specific heat at constant presssure, η_{cis} is the compressor isentropic efficiency. K_a is the air specific heat ratio and P_{rc} is the compressor pressure ratio.
Based from the study, it was shown that the electricity production, E_m decreases on the months that have a relatively higher T_1 temperature (the summer season between May to September), which shows that low temperature increases the power output.

The theory behind effects of temperature on the air is that temperature causes the change in air density. The density hot air is lighter, while cooled air is much denser. The dense air will give turbine a higher mass flow rate, which requires more power to compress the air but on the other hand it also results to an increase in turbine output and efficiency[5]. Ameri and Hejazi [5] stated in their study that the amount of increase in turbine power output due to higher mass flow rate outweigh the drop in power that may occur to compress the air. Previous studies have shown there are noticeable losses in terms of thermal efficiency and power output with the increase of ambient air temperature, where the latter's effect is much more substantial than the first. As documented by Erdem and Sevilgen [4], depending on the temperature variations, the production loss ranges from 1.32% up to 7.85%. In locations where the ambient temperature is above 15°C, especially in the summer, the production loss ranges between 1.67-7.22%. The increase in ambient temperature in hot regions will cause decrease in electricity production and also increases the fuel consumption rate per unit electricity produced. When the inlet air is cooled, the augmentation of power production is about 0.37 -7.59%.

In a study conducted by Ameri and Hejazi [5], it was found that there is relationship between air inlet temperature and the efficiency of a gas turbine. Due to the fact that cooler air is denser compared to hot air, cooler air will cause higher mass flow rate and results to an increase in turbine output and its efficiency. Thus, it was found that each 1°C increase of ambient air temperature, the power output will decrease by 0.74%. While the average power output, can be increased up to 11.3% when the inlet air inlet air cooled. The same pattern of deviation of performance also recorded by Sa and Zubaidy [6], in their study on the performance of gas turbine at varying ambient temperature, as shown in Figure 3 and Figure 4 respectively, stating for every K rise in ambient temperature above the ISO condition, the themal looses 0.1%, while the lost in power output is as high as 1.47 MW.

METHODOLOGY

The main equipment used in the study is a two shaft gas turbine unit, TGT .5 kW, placed at the energy laboratory in Universiti Teknologi PETRONAS. First, data collection is performed for the normal operating condition of gas turbine. This data will be used as a control before the modification of an air cooling unit is made. Next, the sizes of the inlet filters and air inlet will be measure for fabricating purposes.

 DOI: 10.1051/matecconf/201713101011

In order test the effects of cooling towards the system, net like-holder will be mounted to the air filter of the compressor during the fabrication stage. This net-like holder will be used to old dry ice while following air to pass through, hence allowing the process of heat transfer, making the air entering the gas turbine via air inlet of the compressor to cool down. Another cooling will be attached at the piping of the gas turbine, right after the first stage of compression. The test will be conducted by first using only one point of cooling – either the inlet or the piping cooling method – and it then both the points will be used to evaluate the cooling effects. In order to control the cooling load, the amount of dry ice used will vary in terms of mass for each run of experiment. This eases the process of finding the relation between the air temperature and power output of gas turbine. Hence, making it possible to observe the effect of decreased in air temperature on the performance of the gas turbine.

The data will be tabulated and compared with the control data, to observe the deviation in performance before the air is cooled and after the cooling system is installed.The amount of deviation and the data collected will also be compared to the ISO standards and the results from previous studies that have been conducted on the performance of gas turbine with varying air inlet temperature.

Equipment

The main equipment in the study is a two-shaft micro-gas turbine with maximum power output of 1.5 kW. The gas turbine is equipped with temperature probes, pressure gauge and speedometer. Data measured can be gathered through readings on the ,display panel. The gaseous fuel used is propane. The specifications of the gas turbine are displayed in TABLE 1.

TABLE 1. *Specifications of Two-Shaft Micro-Gas Turbine*

Parameter	Specification
Centrifugal ventilator flow rate	500m3/h
Compression ratio of compressor	1.2 – 3
Flow rate of compressor	100 – 700 kg/h
DC generator maximum speed	3750 rev/min
DC generator maximum power	1.5 kW

The cooling element that used in the study is solid carbon dioxide or known as dry-ice. Air flows against the dry-ice is cooled via sublimation process where the solid evaporates into gas. Dry-ice is placed around the outside diameter of the inlet pipe before the compressor. The properties of dry-ice is shown in **TABLE 2**.

TABLE 2. *Properties of Solid Carbon Dioxide*

Parameter	Specification
Molecular weight	44011 kg/kmol
Sublimation point	-78.9°C at 1 bar
Density	1977 kg/m^3
Triple point	-56.6°C at 5.18 bar bar
Critical temperature	31°C

Experimental Procedure

The procedure to perform the experimental investigation is as follows: (1) The main power supply, tap water and compressor are switched on. (2) LPG tank is connected to the gas turbine using a hose. (3) The gas turbine unit is turned on. Initial readings will be displayed on the display panel. (4) Air inlet from compressor and amount of fuel

 DOI: 10.1051/matecconf/201713101011

gas enters the gas turbine unit are adjusted. The pressure of gas inlet is set to 1 bar. (5) Manual restart button is turned on. (6) Air inlet and amount of fuel gas enters the gas turbine unit is adjusted to obtain stable combustion in the combustion by observing the flame. (7) Once, the flame is steady, all readings on the panel is taken.

RESULTS AND DISCUSSION

A test was conducted on the gas turbine unit without the dry-ice cooling system; to be used as baseline results. The angular speed of motor was set as constant and also the other parameters of gas inlet temperature and pressure. The result of the experiment is taken to be controlled parameter, taking the mean readings of ten runs. The result is plotted in **FIGURE 1**. The figure indicates that the inlet air temperature affects the performance of the gas turbine. As the gas turbine operates the inlet temperature become increased due to the heating from combustion. **FIGURE 1** shows that the increased temperature of the inlet air to the gas turbine decreases the gas turbine power output.

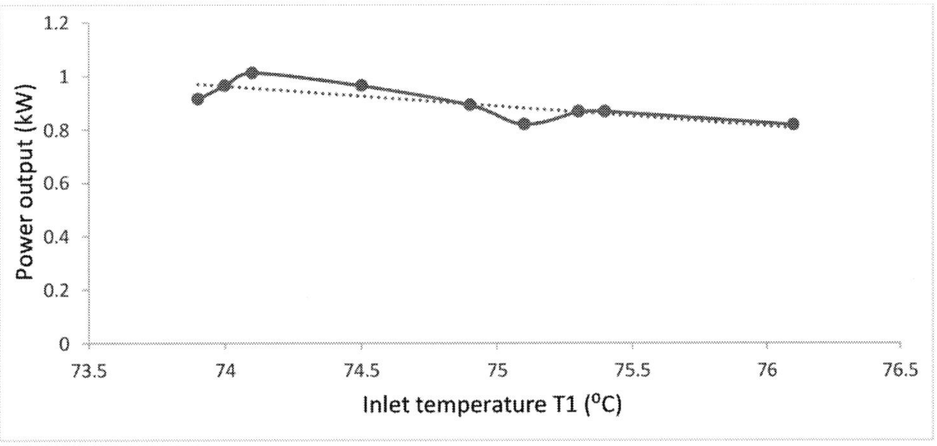

FIGURE 1. Power output vs inlet temperature for gas turbine test without dry-ice cooling.

With the installation of dry-ice cooling system in the inlet before the compressor of the gas turbine unit, the temperature decreases and the power output increases, as shown in **FIGURE 2**. The average temperature decreases from 74.9°C to 67.3°C and the power output increases from average power of 0.9 kW to 0.98 kW. These changes represent increase power output of 8.5% (0.8 kW) from reduction of temperature of 7.5% (15.6°C). This is coherent with the findings from [7] which states that Gas turbine output is a strong function of the ambient air temperature power output increase by 0.5-0.9% for every 1 °C decrement. The result also shows that the power output seems to be drecreasing with time, as the inlet air is being heated up due to the proximity to the combustion chamber.

DOI: 10.1051/matecconf/201713101011

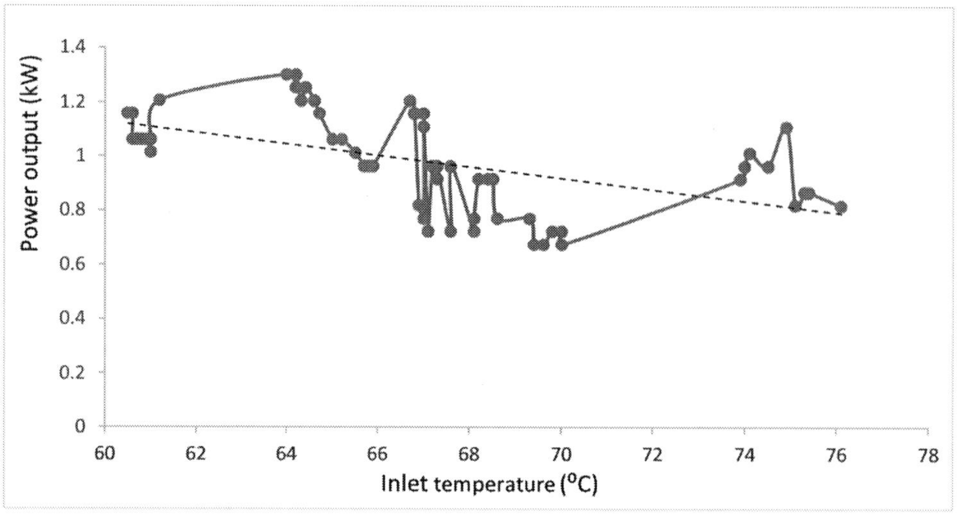

FIGURE 2. Power output vs inlet temperature for gas turbine test with dry-ice cooling.

The drop in power output as the temperature increases as mainly due to drop in air density. This can be shown by **FIGURE 3**, which describe a linear decrease of density of the inlet air with respect to the increase of inlet temperature.

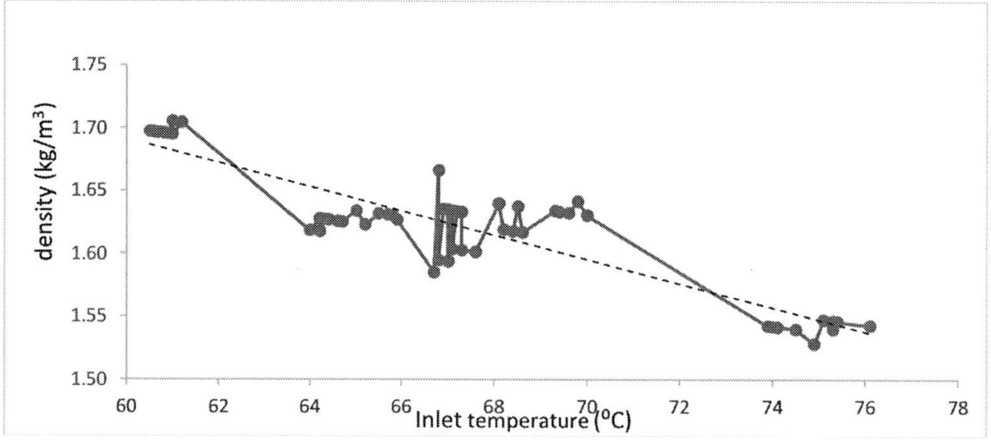

FIGURE 3. Density vs inlet temperature for gas turbine test with dry-ice cooling.

This is because as the hot air is lighter, lower mass flow rate, less amount of air goes to combustion. With same amount of volume, the air simply lighter compared to cold air, and thus tend to lower the combustion quality. This result concurs with other findings [4,8], which states that lower temperature resulted in higher air mass flow rate which correspond in higher density of the air.

CONCLUSION

This study was performed to investigate the performance of a 1.5kW micro-gas turbine with reduced inlet air temperature. The cooling mechanism employed is the use of dry-ice cooling through sublimation process. Some conclusions that can be drawn from the results:

(1) Using dry-ice as cooling medium proves a good potential as it can reduced the temperature up till 13°C from the ambient temperature.
(2) Power output decreases linearly with the increase of inlet air temperature. It was recorded that increase power output of 8.5% (0.8 kW) achieved from reduction of temperature of 7.5% (15.6°C).
(3) Power output decreases with time, due to the inlet duct exposed to the heat form the combustion chamber. This indicates better insulation or adapted data recording approach is needed.

ACKNOWLEDGMENTS

The study was supported and funded by the Mechanical Engineering Department of Universiti Teknologi PETRONAS. The authors wish to express thanks to the technology staff of Mechanical Engineering Laboratory for their commitment in setting up and maintaining the micro-gas turbine set-up.

REFERENCES

1. E. Hadik, Journal of Engineering for Gas Turbines and Power **112**, 590-596 (1993)
2. M. Jonsson and J. Yan, Energy 30, 1013-1078 (2005)
3. T.B.J. Early, R.D.J. Reens and M.R.S. Karnoff, U.S. Patent No. 5,463, 873 (7 November 1995)
4. H.H. Erdem and S.H. Sevilgen, Applied Thermal Engineering 26 320-326 (2006)
5. M. Ameri and S.H. Hejazi, Applied Thermal Engineering 24 59-68 (2004)
6. A.D. Sa and S.A. Zubaidy, Applied Thermal Engineering 31 2735-2739 (2011)
7. A. K. Shukla and O. Singh, International Journal of Scientific & Engineering Research **5**, 664-671 (2014)
8. A.M. Al-Ibrahim and V. Varnham, Applied Thermal Engineering **30**, 1879-1888 (2010)

MATEC Web of Conferences **131**, 02001 (2017) DOI: 10.1051/matecconf/201713102001
UTP-UMP SES 2017

Feasibility Study of Standalone PV-Wind-Diesel Energy Systems for Coastal Residential Application in Pekan, Pahang

Roziah Zailan[1,*], *Siti Nurzalikha* Zaini[2], *Muhammad Ikram* Mohd Rashid[2], *Amir* Abdul Razak[1]

[1]Faculty of Engineering Technology, Universiti Malaysia Pahang, Lebuhraya Tun Razak, 26300, Kuantan Pahang.
[2]Faculty of Electrical & Electronics Engineering, Universiti Malaysia Pahang, 26600 Pekan, Pahang Darul Makmur

Abstract. Techno economic study is feasible to optimize the usage of renewable energy components that targeting low cost of electricity generation system. The selected case study area is coastal area in Pekan, Pahang, Malaysia. The autonomous system designed in this study is hybrid standalone PV-wind-diesel energy system to fulfil a basic power demand of 20.1 kWh/day. Such power system was designed and optimized further to meet the power demand at a minimum cost of energy using energy optimization software, Hybrid Optimization Model for Electric Renewables (HOMER). The analysis focused on the operational characteristics and economics. The standalone PV-wind-diesel energy system has total net present cost about $61, 911 with cost of energy $0.66/kWh. Apparently, the generation of electricity from both wind turbine and PV can be inflated with the diesel generator. In comparison, return of investment (ROI) value turned out lower for Feed in Tariff (FIT) as compared to self-sustained house. Payback period also longer for FIT program that makes the selling back of electricity generated to Tenaga National Berhad (TNB) is considered not favourable.

1 Introduction

An abundant resources of solar and wind could be exploited to power up the coastal housing. However, the selection of location for suitable wind turbine and solar energy harvesting need to be studied in respect to the technology options, component costs, and resource availability. The study area located in Pekan, situated on the banks of Pahang River 50km South of Kuantan and near to popular beaches such as Legenda Beach and Air Leleh Beach. In this area, potential benefit to receive strongest wind and solar at coastal area will give advantages for this project. Therefore, this paper presents a feasibility study of hybrid standalone PV-

* Corresponding author: roziahz@ump.edu.my

wind-diesel energy system for Pekan, Kuantan Pahang. The analysis was done based on typical house power demand at about 20.1 kWh/day and 1.23kW peak demand [1]. Noted that all cost conversion in this study is 1$=RM4.28 as of current rate (August 2017).

Generally, total cost for establishing such generating unit in new location with unknown harvesting energy units is unclear without proper techno economic study. Therefore, this study is beneficial to optimize the usage of renewable energy components that targeting low cost of energy production. The arrangement of wind turbine, solar panel, diesel generator and storage systems were considered. Several configurations in the simulation include size of components, capacity, lifetime and so forth. The National Renewable Energy Laboratory's (NREL) optimization tool "HOMER" was utilized to identify feasible hybrid configurations and their applicability. HOMER simulates the operation of a system by making energy balance calculations and displaying a list of configurations, sorted by net cost that can be used to compare system design [1-4].

There are various previous studies using HOMER to initiate hybrid power system in the coastal areas. Sadrul Islam *et al.* [5] was attempted to model a hybrid electricity generation system for a small community of the St Martin's Island in Bangladesh that resulted lowest COE $0.345/kWh and total net present cost (NPC) of $137,927 with a renewable fraction of 31%. A study in the same study area by Hazra *et al.* [6] also stated that the depending on diesel only is not feasible but to integrate it with hybrid energy generation system. They concluded that PV-diesel-wind-battery hybrid generation system was found most feasible and optimized. Nevertheless, Ibrahim *et al.* [1] stressed that the hybrid renewable energy system is suitable to be applied in the coastal areas if current costs of the renewable energy components could be marked down.

In Malaysia, FIT is a policy mechanism which allows users to sell back the additional electricity generated from PV system to Tenaga National Berhad (TNB) based on TNB rate. This study also attempted to look into the possibility to participate in the FIT program. For this reason, the payback period ROI were calculated for both FIT program and self-sustained house to identify which one is more reliable to be applied. Matin *et al.* [7] also calculated payback period of the proposed system using different tariff based on their country, Bangladesh after they simulated hybrid PV-wind-diesel power system in HOMER to determine number of years required to recover the cost of investment. Therefore the economic analysis section will discuss more on the paybacks period and ROI analysis between self-sustained house and FIT program.

2 Material and Methods

2.1 Energy demand and resources

Electrical load: The electrical loads for the house are classified as domestic with basics electrical load components like fluorescent lamps, ceiling fan, television, refrigerator and also washing machine [1, 6]. The hourly load consumed by the house is presented in Figure 1. Electricity demand is the rate at which electric energy is required by the load, measured in kilowatts (kW) [8]. Average power demand was estimated at about 20.1 kWh/day and peak demand of 1.23 kW.

Solar radiation and Wind resources: Hourly solar radiation data and wind speed for year 2014 were collected from Malaysian Meteorological Department (MMD) as in Table 1.

 DOI: 10.1051/matecconf/201713102001

Table 1. Monthly Average Daily Radiation, Clearness Index & Wind Speed

Month	**Daily Radiation ($kWh/m^2/day$)**	**Clearness Index**	**Wind Speed (m/s)**
January	4.2	0.44	2.37
February	5.7	0.57	2.14
March	6.3	0.61	2.12
April	5.7	0.55	1.51
May	5.7	0.56	1.56
June	5.6	0.57	1.75
July	5.2	0.52	1.8
August	5.8	0.57	1.7
September	5.5	0.53	1.57
October	5.3	0.53	1.34
November	4.2	0.44	1.36
December	3.5	0.38	1.87

PV-wind-diesel power system: The schematic diagram of Photovoltaic (PV)-wind-diesel power system components are presented in Figure 1 consists of diesel generator, PV arrays, wind turbine, battery and power converters. Technical information to run the simulation using HOMER software were obtained from manufacturers of the equipment and previous studies [8-13].

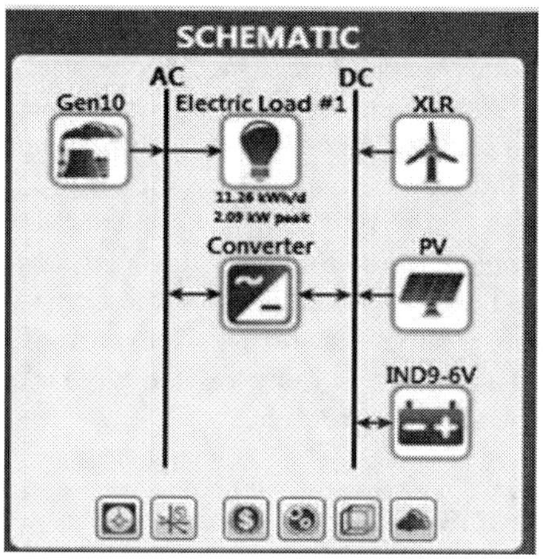

Fig 1. PV-Wind-Diesel Power System Configuration.

Table 2. Cost of Components.

Equipment	**Capacity/ Unit**	**Capital Cost ($)**	**Replacement Cost ($)**	**O&M Cost ($)**
Diesel Generator	16kW	6500	5800	0.15/h
Solar PV	5kW	11000	9500	-
Wind Turbine	1 Unit	19400	15000	75/year
Batteries (Trojan IND9-6V)	1 Unit	1325	1190	0.02/h
Converter	1kW	118	100	15/year

3 Result And Discussion

3.1 PV-wind-diesel system simulation

For hybrid PV-wind- diesel energy system, the equipment needed to build the system were diesel generator, PV array, wind turbine, batteries and power electronic converter with specified type and quantity. The HOMER simulation tool was used to optimize the sizes of different hardware components in the PV-wind-diesel system, taking into account the technical characteristics of system operation and minimizing total net present cost of the system. The optimization results of this power system are show in Figure 2.

Architecture							Cost				System
PV (kW)	XLR	DG (kW)	S6CS25P	Converter (kW)	Dispatch		COE ($)	NPC ($)	Operating cost ($)	Initial capital ($)	Ren Frac (%)
		5.00	6	2.00	CC		$0.584	$54,508	$3,585	$8,686	0.0
0.250		5.00	6	2.00	CC		$0.587	$54,800	$3,564	$9,236	0.0
	1	5.00	6	5.00	CC		$0.659	$61,496	$2,586	$28,440	22
0.250	1	5.00	6	5.00	CC		$0.663	$61,911	$2,575	$28,990	23
		5.00			CC		$0.787	$73,404	$5,703	$500.00	0.0
0.250		5.00		2.00	CC		$0.800	$74,660	$5,740	$1,286	0.0
	1	5.00		2.00	CC		$0.971	$90,571	$5,510	$20,136	0.0
0.250	1	5.00		2.00	CC		$0.975	$91,007	$5,501	$20,686	0.0
	24		36	5.00	CC		$6.54	$609,997	$7,518	$513,890	100

Fig 2. The simulation results for PV-wind-diesel energy system

The least Cost of Energy (COE), $0.58 kW/h resulted from the 5 kW diesel generator alone without contribution from renewable sources. But the integration with renewable energies fell into fourth least COE as $0.66 kW/h, resulted from the combination of 5 kW diesel generator, 0.25 kW of PV array, 1 unit of wind turbine, 6 unit of batteries and 5 kW converter. The strategy taken in this simulation is to ensure the power generator provide enough power to meet the demand. The renewable energy sources in collaboration with the diesel generator were evaluated to determine the feasibility of the system.

All values related to the electricity production and load served by the system are summarized in Table 3. The results of the simulation showed that the PV-wind-diesel hybrid system had a total annual electrical energy production of 10,046 kWh/yr. The electricity generation came mostly from diesel generator with 56.32% at about 5,658kWh/yr while wind turbine and PV-array generated energy at 42.82% (4,301 kWh/yr) and 0.87% (87 kWh/yr), respectively. Besides that, it is clear that approximately 8.7% of energy (873.5 kWh/yr) was neglected. Several amount of excess energy can be suggested to cater for PV-wind hydrogen energy system in the next study. The trend of monthly electricity production can be referred to Figure 3.

Table 3. Operational Characteristics of the PV-Wind-Diesel System

Annual Electricity Production	**kWh/year**	**%**
PV-array	87	0.87
Diesel generator	5,658	56.32
Wind Turbine	4,301	42.82
Total production	**10,046**	**100.00**
Annual electrical load served		
AC primary load served	7,300	100
Total	7,300	100
Other		
Excess electricity	873.5	8.7

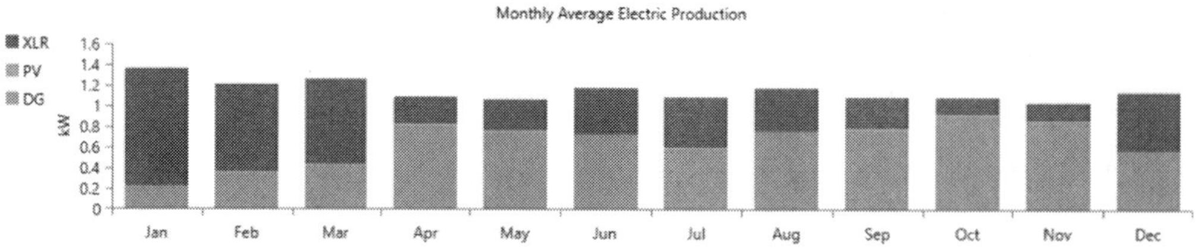

Fig 3. Monthly Electricity Production Trend of the PV-Wind-Diesel System

Through the economic analysis, the most expensive annualized cost is contributed from wind turbine at $1,518 followed by batteries ($622), converter ($46), PV ($43) and diesel generator ($39). Total annualized cost for overall system is $4,843. HOMER uses the total net present cost (NPC) to represent the life-cycle cost of a system (Figure 8). The highest NPC came out from the BWC Excel-R wind turbine $19,400, followed by batteries $7,950, converter $590, PV panel $550 and diesel generator $500. Total NPC is $61,911. F

From this study, diesel generator apparently contributed highest portion of electricity while wind speed is highly potential in three months, January, February and March. The contribution from solar is very minimal. Apparently, fully dependent on the consumption of diesel fuel was reduced with integration of those renewable resources. All in all, capital cost, total Net Present Value (NPC) and COE of the systems were recorded as $28,990, $61,911 and $0.66/kWh respectively. From the analysis, COE is considered high until price reduction be made to respective devices involved in such system.

3.2 Self-Sustained House and FIT Program Analysis

On the other hands, economic analysis was done to complete this study by introduced other alternative in order to promote FIT that currently stirred by government through SEDA Malaysia. Therefore, both payback period and ROI were calculated to determine which one is favourable option between self-sustained house and FIT program. Table 4 provide detail summary of both option. Noted that the cost of electricity for installation above 4kW and up to (and including) 24kW is $2.90/kWh as of January 2017 [14]. Payback period was calculated to identify specific duration to recover the cost of investment of the system. Meanwhile, the return on investment (ROI) expresses the benefit to the investor resulting from an investment of some resources. A high ROI means the investment is worthy for the investment cost. As a performance measure, ROI used to evaluate the efficiency of an investment [15]. In comparison, ROI value for FIT program is lower as compared to the self-sustained house. Thus, the selling back of electricity generated to Tenaga National Berhad (TNB) is not recommended for the proposed system.

Table 4. Comparison between Self-Sustained House and FIT Program

SELF-SUSTAINED HOUSE	FIT PROGRAM
Total capital cost= $28,990	Total capital cost= $28,990
Annualize income = Total production x HOMER COE = 10,046kWh/yr x $0.66/kWh = $6894.4/yr	Annualize income = Total production x TNB tariff rate ($) = 10,046 kWh/yr x $ 0.16/kWh = $1643.04/yr
Payback period = 4.2 years	Payback period =17.64 years
ROI= Annualize income/ Total capital cost = $6894.4/ $28,990 = 0.24 per annum (24%)	ROI= Annualize income/ Total capital cost = $1643.04/ $28,990 = 0.06 per annum (6%)

4 Conclusion

At the end of HOMER simulation, standalone PV-wind-diesel energy system presented total net present cost about \$61,911 that capable to generate electricity at cost of energy \$0.66/kWh. From the economic analysis, it has resulted that ROI for FIT program is lower than self-sustained house. Payback period also longer for FIT program that makes the selling back of electricity generated to Tenaga National Berhad (TNB) is not favourable. This study may be extended to involve the integration of other renewable resources system and environmental aspect for an extensive coastal area in the future.

References

1. M. Z. Ibrahim, R. Zailan, M. Ismail and A. M. Muzathik, *Pre-Feasibility Study of Hybrid Hydrogen Based Energy Systems for Coastal Residential Application*, Enrgy R. J **1**, 1:13-22 (2010)
2. Areef Kassam , *HOMER Software Training Guide for Renewable Energy Base Station Design.* (2010)
3. M. Shuhrawardy & K.T Ahmmed, *The Feasibility Study of a Grid Connected PV System to Meet the Power Demand in Bangladesh - A Case Study*, American J. Enrgy Eng **2**, 2 :59-64 (2014)
4. Z. Girma, Technical and Economic Assessment of solar PV/diesel Hybrid Power System for Rural School Electrification in Ethiopia, Int J. of Ren Enrgy R. **3**, 3 : 735-744 (2013)
5. A. K. M. Sadrul Islam, M. M. Rahman, M. A. H Mondal and F. Alam, *Hybrid energy system for St. Martin island, Bangladesh: An optimized model,* Procedia Eng **49**: 179–188 (2012)
6. S. R Hazra, K. S. Hossain, A. Al Jubaer and M. Rabby, *An Optimized Hybrid System Model : Solution for Coastal Area in Bangladesh*, Int. J. of R. in Eng & Tech **3**,12: 94–102 (2014)
7. M. A Matin, & A. Deb, *Optimum Planning of Hybrid Energy System using HOMER for Rural Electrification,* Int J. of Comp App **66**, 13: 45–52. 2013
8. Demiroren and U. Yilmaz, *Analysis of change in electric energy cost with using renewable energy sources in Gokceada, Turkey: An island example*. Ren & Sust Enrgy 14: 323-333. 2014
9. M.J. Khan and M.T. Iqbal, *Pre-feasibility Study of Stand-alone Hybrid Energy Systems for Applications in Newfoundland*. Ren Enrgy 30: 835-854 (2005)
10. E.I. Zoulias and N. Lymberopoulos, *Techno- Economic Analysis of the Integration of Hydrogen Energy Technologies in Renewable Energy-Based Stand Alone Power Systems*. Ren. Enrgy 32: 680-696 (2007)
11. W.C Goh and N.N. Barsoum, *Balancing Cost, Operation and Performance in Integrated Hydrogen Hybrid Energy System*, http://www.itee.uq.edu.au/~aupec/aupec06/htdocs/content/pdf/3.pdf (2006)
12. G.J. Dalton, D.A. Lockington and T.E. J. Baldock, *Feasibility Analysis of Renewable Energy Supply Options for a Grid-Connected Large Hotel*. Ren Enrgy 34: 955-964 (2009)
13. Bergey Wind Power. Small wind turbines: For homes, business and off-grid http://www.bergey.com (2017)
14. FIT Rates for Solar PV. http://seda.gov.my/ (2017)
15. K. Padmanathan, U. Govindarajan, V.K. Ramachandaramurthy, T.S.O. Selvi, *Multiple Criteria Decision Making (MCDM) Based Economic Analysis of Solar PV System with Respect to Performance Investigation for Indian Market.* Sust **9**, 6: 820. (2017)

Design and modelling of novel evacuated receiver tube of concentration solar power system by using heliostat stadium arrangement

Ayad Kadhim Khlief [1, 2], *Syed* Ihtsham- ul-Haq Gilani[1,*], *Hussain* H. Al-Kayiem[1] *and* *Sanan* T. Mohammad[1]

[1] Department of Mechanical Engineering, Universiti Teknologi PETRONAS, 32610 Bandar Seri Iskandar, Perak, Malaysia.
[2] Department of Electromechanical Engineering, University of Technology, Baghdad, Iraq

Abstract. A novel concentration solar evacuated receiver tube collector is proposed in this study. Details of design and simulation of the heliostat field are presented. The proposed system uses a heliostat stadium arrangement to focus sunlight towards a receiver (evacuated receiver tube). Heliostat field which looks like stadium arrangement is proposed an alternative system of the tower in the power tower plants to cancel the shading / blocking losses. In this study, simulation results showed that the energy input increases with increased solar irradiation; the increase ranges from 288 to 470W/m.The maximum amount of energy inputs that can be obtained in the receiver for the collector is around 470W/m for the maximum solar density of 688W/m^2. In the solar is 402 W/m^2, and the useful energy of collectors is about 288 W/m and the thermal efficiency of the receiver was around 23%.

1 Introduction

The receiver of the solar power tower plant absorbs the solar energy and transfers it to the Heat Transfer Fluid (HTF). Some of the incoming radiation is reflected due to the fact that absorbance of receiver is less than one. Surface temperature increases due to radiation absorption, causing loss by thermal emission and convection. On the edges of the receiver, some losses occur by thermal conductivity in the tower structure [1].

There are several ways to describe or model the thermal performance of an open receiver and cavity using molten salt as heat transfer fluid numerically. In a steady state model presented by Li et al. [2], there was internal flow in the tubes of the cavity receiver. The thermal performance was analysed considering the effects of several factors like surface of the receiver. Their model was experimentally validated. Lata et al. [3] reported that main heat loss was due to reflection and emissions, while quantifying the mechanism of open receiver. The heat transfer to molten salt was also simulated using a dynamic fluid code. They found that differences in the outer diameter of the tube have shown that efficiency improves when diameter is reduced. Yang et al. [4] performed the computational fluid dynamics (CFD) analysis to simulate heat transferred to the flow of molten salt in the tube. Their results were in good matching with experimental ones.

In this work, we report on the design and mathematical model of a novel concentrated evacuated receiver tube solar collector via heliostat stadium. The system is novel in the many regards. The purpose of this model is to understand the operation of the receiver

* Corresponding author: syedihtsham@utp.edu.my

 DOI: 10.1051/matecconf/201713102002

when exposed to solar radiation reflected from heliostat field. This paper also tries to find an approach to determine various collector parameters from measurements under non equilibrium conditions.

2 Design of evacuated tube receiver system

The main components of the proposed system are the (i) heliostat concentrator and (ii) evacuated tube receiver Figure 1. Evacuated tube absorbs which receives concentrated light from the heliostat and serves as the high-temperature fluid. An evacuated tube receiver comprising 4 tubes with aperture area 0.8 m^2 were installed on a steel structure. The receiver was installed on a North facing base. Initial investigations of the heliostat field led to a determination of 12 heliostats to be utilized in simulating the facility. The reflectivity established for the heliostat is 89%. Heliostats of collective reflective surface area 37.5 m^2 were selected for modeling of the facility, distributed as shown in Fig. 1, the distance between the tower and first row of heliostats is 10 meters, where as, the inter-row distance is 3.5m. The reflection area of each mirror is 3.75 m^2.

Fig. 1. Schematic diagram of proposed model.

3 The incident energy

The field efficiency is the factor that determines the energy of solar radiation that hits the receiver of the tower. This incident energy ($\acute{q}_{inci}$)on the receiver is defined as follows [5].

$$\acute{q}_{inci} = Q_{solar} \times \eta_{field} \tag{1}$$

$$\acute{q}_{inci} = DNI \times A_H \times N_H \times \eta_{cos} \cdot \eta_{sha} \cdot \eta_{blo} \cdot \eta_{att} \cdot \rho_{ref} \tag{2}$$

where DNI represents Direct Normal Irradiance, A_H the area of one heliostat, N_H the number of all heliostats in a field while $\eta_{cos}, \eta_{sha}, \eta_{blo}, \eta_{att}$ and ρ_{ref} are the cosine, shadowing, blocking, atmospheric attenuation and mirror reflectance losses, respectively.

4 Mathematical modeling for proposed model

Basic energy balance equations for the glass cover, the absorber pipe and the fluid are solved using an iterative procedure. To model the collector a number of simplifying assumptions are made. The assumptions are as follows: The collector is in steady state, thermal and radiation properties of the collector materials are independent of temperature and effects of incidence angle on the absorbed solar radiation are not considered.

DOI: 10.1051/matecconf/201713102002

Fig. 2 shows the one-dimensional steady-state energy balance for a cross-section of an HCE, with the glass envelope intact.

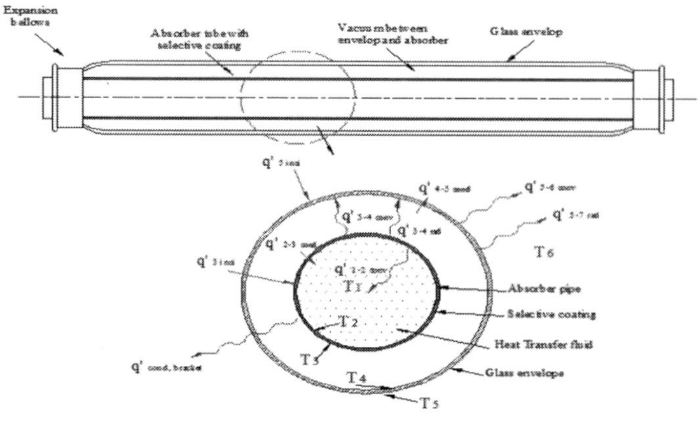

Fig. 2. One-dimensional steady-state energy balance

4-1 Energy Balance Equations between Glass Envelope and Atmosphere

The energy balance equations between the glass envelope and the atmosphere are represented by following equations presented in this sub-section.

$$\acute{q}_{4-5,cond} + \acute{q}_{inci} = \acute{q}_{5-6,conv} + \acute{q}_{5-7,rad} \tag{3}$$

$$2\pi k_{4-5}(\mathrm{T}_4 - \mathrm{T}_5)/ln(D_5/D_4) + \acute{q}_{s,inci}\,\alpha_{env} = \overline{Nu}_{D5}\,\pi k_{5-6}(\mathrm{T}_5 - \mathrm{T}_6) + \sigma\,\mathrm{D}_5\,\pi\,\varepsilon_5(\mathrm{T}_5^4 - \mathrm{T}_7^4) \tag{4}$$

where k_{4-5} (W/m.K) represents the glass envelope thermal conductance at the average glass envelope temperature given by $\{(T_2+T_3)/2\}$. The subscripts D_5 and D_4 (m) refer to the glass envelope's outlet and inlet respectively. $\acute{q}_{s,inci}$ is the solar irradiation per receiver length (W/m) and α_{env} is the absorptance of the glass envelope (Pyrex glass), T_5 and T_6 (°C), represent the glass envelope outer surface temperature and the ambient temperature, respectively, k_{5-6} (W/m.K) is the thermal conductance of air at temperature $\{(T_5 + T_6)/2\}$, $\overline{Nu}_{D5}$is the average Nusselt number based on the glass envelope outer diameter. ε_5 and σ respectively represents the glass envelope outer surface emissivity and Stefan Boltzmann's constant (5.670E-8) ($W/m^2.K^4$) while T_7 is effective sky temperature (K)

$$\overline{Nu}_{D5} = C\,Re_{D,max}^{m} Pr_6^{\;0.36}\left(\frac{Pr_6}{Pr_5}\right)^{1/4} \tag{5}$$

This correlation is valid for $0.7 \leq Pr_6 \leq 500$ and $1 \leq Re_{D,max} \leq 10^6$. All fluid properties are evaluated at the atmospheric temperature, T_6, except Pr_5, which is evaluated at the glass envelope outer surface temperature [6].

4-2 Energy Balance Equations between Absorber and Glass Envelope

This subsection contains the energy balance equations between the absorber and glass envelope.

$$\acute{q}_{3-4,conv} + \acute{q}_{3-4,rad} = \acute{q}_{4-5,cond} \quad (6)$$

$$\pi D_3 h_{3-4}(T_3 - T_4) + \frac{\sigma \pi D_3(T_3^4 - T_4^4)}{(1/\varepsilon_3 + (1-\varepsilon_4)D_3/(\varepsilon_4 D_4))} = 2\pi k_{4-5}(T_4 - T_5)/ln(D_5/D_4) \quad (7)$$

where D_3 is outer absorber surface diameter (m), T_3 is outer absorber surface temperature (°C), and h_{3-4} is convection heat transfer coefficient for the annulus gas at T_{3-4} (W/m^2-K), when the heat transfer is very small pressures (~< 0.0001 torr), the convection heat transfer coefficients (h_{3-4}) equals 0.000111 W/m^2-K [7].

4-3 Energy Balance Equations for Absorber

The energy balance equations for the absorber are:

$$\acute{q}_{inci} = \acute{q}_{3-4,conv} + \acute{q}_{3-4,rad} + \acute{q}_{2-3,cond} \quad (8)$$

$$\acute{q}_{s,inci} \cdot \tau_{env} \cdot \alpha_{abs} = \pi D_3 \cdot h_{3-4}(T_3 - T_4) + \frac{\sigma \cdot \pi \cdot D_3(T_3^4 - T_4^4)}{\frac{1}{\varepsilon_3} + \frac{(1-\varepsilon_4)D_3}{(\varepsilon_4 \cdot D_4)}} + \frac{2\pi k_{2-3}(T_2 - T_3)}{ln\left(\frac{D_3}{D_2}\right)} \quad (9)$$

where α_{abs} is the absorptance of the absorber and $\tau_{\_env}$ is the transmittance of the glass envelope.

4-4 Energy balance equations between the HTF and the Absorber

For the Heat transfer fluid and the absorber, the energy balance equations are given as:

$$\acute{q}_{2-1,conv} = \acute{q}_{2-3,cond} \quad (10)$$

$$Nu_{D2}\pi k_1(T_2 - T_1) = 2\pi k_{2-3}(T_2 - T_3)/ln(D_3/D_2) \quad (11)$$

where Nu_{D2} is the Nusselt number based on D_2 and k_1 is the thermal conductance of the HTF at temperature T_1 (W/m-K), such that:

$$T_1 = T_2 - \left[\frac{2\pi k_{2-3}(T_2 - T_3)/ln(D_3/D_2)}{Nu_{D2}\pi k_1}\right] \quad (12)$$

where D_2 is the inside diameter of the absorber pipe (m), T_1 is the mean (bulk) temperature of the HTF (°C), T_2 is the inside surface temperature of absorber pipe (°C), Nu_{D2} is the Nusselt number based on D_2 and k_1 is the thermal conductance of the HTF at T_1 (W/m-K)

$$T_o = 2T_1 - T_i \quad (13)$$

where T_o is the external heat transfer fluid temperature and T_i is the internal heat transfer fluid temperature.

$$\acute{q}_{loss} = \acute{q}_{5-6,conv} + \acute{q}_{5-7,rad} \quad (14)$$

$$\acute{q}_{useful} = \acute{q}_{inci} + \acute{q}_{loss} \quad (15)$$

$$\dot{m} = \frac{\acute{q}_{useful}}{cp\,(T_o - T_i)} \tag{16}$$

where $\acute{q}_{loss}$ *and* $\acute{q}_{useful}$ are heat loss per unit receiver length (W/m) and HTF heat gain per unit receiver length (W/m). $\dot{m}$ is the mass flow rate of HTF (Kg/s) and specific heat of HTF (J/kg.K).

The thermal efficiency can thus be calculated utilizing the following equation;

$$\eta_{thermal} = \frac{\acute{q}_{useful}}{\acute{q}_{inci}} \tag{17}$$

Receiver losses can be defined as the product of each efficiency in the loss mode.

$$\eta_{receiver} = \eta_{rad} \cdot \eta_{conv} \cdot \eta_{spil} \cdot \eta_{absorp} \tag{18}$$

where $\eta_{rad} \cdot \eta_{conv} \cdot \eta_{spil}$ and η_{absorp} are the efficiencies based on the receiver, radiation, convection, and spillage, absorption losses, respectively.

The above equations were used in MATLAB, and a boundary analysis was performed on the system (shown in Figure 1). For the evacuated tube receiver, the input variables needed for calculation are shown in Table 1.

Table 1. Specifications of solar collector

Model	**High temperature solar collector pipe**
Type	JJR-T-2140 collector tube
Length	2136 mm
Working temperature(Highest)	> 200°C(350°C)
Diameter of absorption tube	40 mm
Diameter of glass tube	100 mm
Absorptivity	≥ 0.95
Emittance	≤0.10at temperature 200°C
Transmittance of glass	≥ 0.91.5
Vacuum degree	3.0 × 10-4Pa

5 Result and discussion

The data uses direct solar radiation measurements to determine the radiation intensity of the facility located in UTP, Perak. More specifically, the amount of direct-use radiation available from the sun for any given hour of the year. Fig. 3 shows the average daily radiant values of the peak of each month, showing the highest solar radiation reached in February at 12:00. The calculations were carried out in February, where the sun was more visible than other months.

The optical losses in the heliostat field are shown in the bar chart in Fig. 4. The types of optical losses were cosine, shading and blocking, atmospheric attenuation, and mirror reflectance losses. All of those losses were depended on the heliostat field planning and the Sun's Position at any time of day.

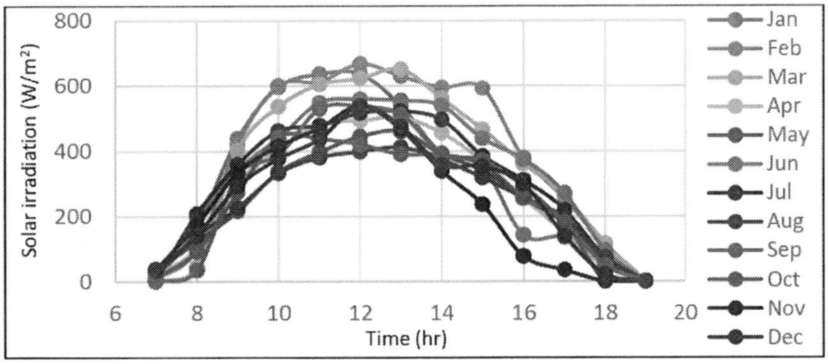

Fig. 3. Portrays an average of hourly values for peak daily radiation of each month.

The energy incident on the receiver depends on the heliostat efficiency directing the solar flux of the sun on the receiver. Fig. 4 shows the optical loss at 12 pm where it is observed that the biggest effect were the losses of the cosine.

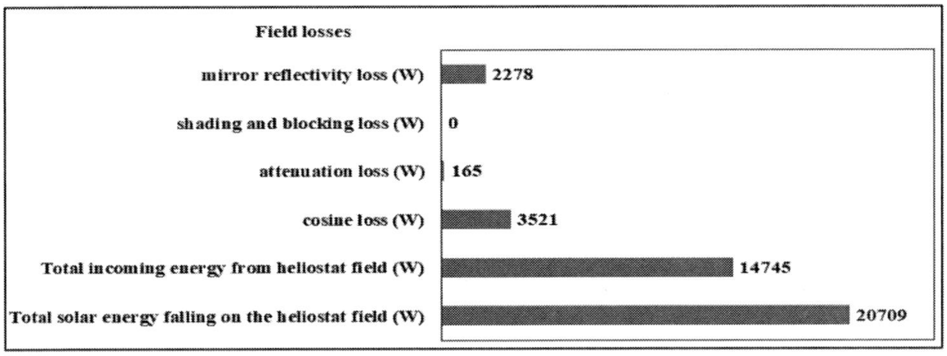

Fig. 4. Breakdown of the energy loss in heliostat field

A parametric analysis was performed to simulate the performance of the evacuated tube and change the above variables. Solar radiation in Perak, Malaysia varies throughout the day. This change is important because it allows us to study the impact of solar intensity on the energy input of the receiver. Fig. 5 shows this relationship. The energy input increases with increased solar irradiation with the increase ranging from 288 to 470W/m at 8 am and 12 pm. This implies that the maximum amount of energy inputs that can be obtained in the receiver for a collector is around 470W/m at 12 pm for the maximum solar density of 688W/m^2. The solar irradiation is 402 W/m^2 at 8 am, and the useful energy of collectors is about 288 W/m. These results show that the receiver will receive more energy at times of high solar intensity, thus increasing the useful energy. Increasing the flux of solar energy doesn't increase the efficiency much, as shown in Fig. 5, the efficiency increase is only from 22.7% to 23.8%, a total of 1.1%. This is reasonable since low solar flux also reduces the useful energy rate of the receiver and with the knowledge that the flux is also low, the efficiency will remain the same [8]. The efficiency of the receiver is low because spill losses are about 75 percent due to the small receiver space compared to the reflected image area of the heliostat field.

Increasing the flux of solar radiation leads to higher temperatures of heat transfer fluid in the receiver outlet from 92.8 to 151 °C with increasing mass flow rate from 0.0075 to 0.0122 Kg/s at 8 am and 12 pm, as illustrated in Fig. 6. These results show that the receiver will capture more energy in high solar intensity times. The mass flow rate of this HTF is decisive in the performance of the collector and will result in an increase in the

 DOI: 10.1051/matecconf/201713102002

useful energy of the receiver. This increase leads to an increase in the temperature of the HTF outlet in the receiver since more energy is captured and delivered to the HTF.

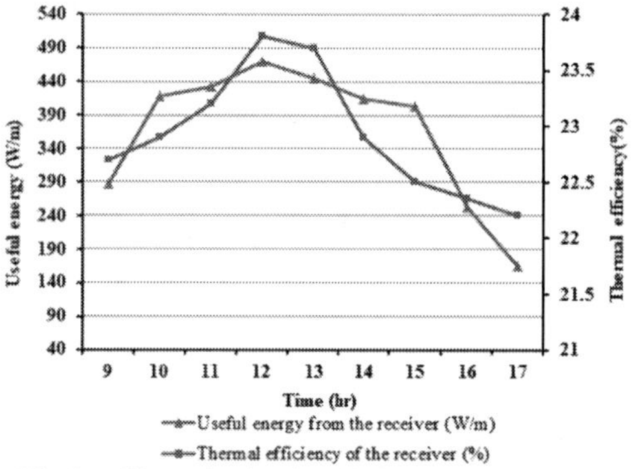

Fig.5. Effect of solar irradiation on the thermal efficiency and useful energy from the receiver.

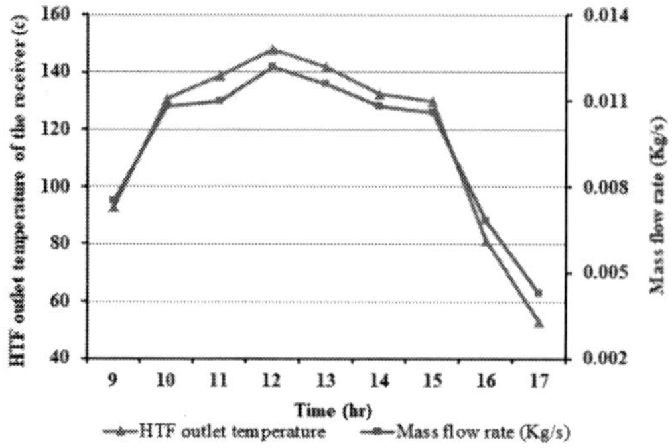

Fig.6. Effect of the solar irradiation on the mass flow rate of HTF and HTF outlet temperature from the receiver.

6 Conclusions

The novel concentration of evacuated solar tube receiver system has been developed to produce high temperature. In this paper, the proposed design details and field performance results are presented. The result showed the following:

- To reduce the loss of spill from the receiver, the number of tubes must be increased to intercept the maximum amount of solar energy.
- The biggest losses effect on the optical efficiency is the cosine and spillage.
- The thermal efficiency of the receiver was about 23%.

To increase the output temperature of the receiver as long as the receiver absorbs more heat, so the number of tubes must be increased.

The authors gratefully acknowledge the assistance provided by Universiti Teknologi PETRONAS for the financial and technical support to conduct this research work.

References

1. O. Garbrecht, F. Al-Sibai, R. Kneer, K. Wieghardt., Journal of Solar Energy **Vol** 90, 94-106.2013.
2. X. Li, W. Kong, Z. Wang, C. Chang, F. Bai., Renewable Energy **Vol** 35, 981–988.2010.
3. Lata, J.M., Rodrı´guez, M., A ´ lvares de Lara, M., Journal of Solar Energy Engineering 130, 021002 (5 pages).2008.
4. X. Yang, X. Yang, J. Ding, Y. Shao, H. Fan., Applied Energy **Vol** 90, 142–147.2012.
5. R. Pitz-Paal Lecture notes "Solartechnik", RWTH Aachen, 2008
6. R. Forristall., Report No. NREL/TP- 550-34169. CO, USA: NREL.2003.
7. Marshal, N., transl., New York, NY: Elsevier.1976.
8. M. S. Shahin , M. F. Orhan , F. Uygul., Solar Energy **Vol** 136, 183–196.2016

 DOI: 10.1051/matecconf/201713102003

Intelligent monitoring system of unburned carbon of fly ash for coal fired power plant boiler

Pogganeswaran Gurusingam[1], *Firas* Basim Ismail[1*], *Prem* Gunnasegaran[1], and *Taneshwaren* Sundaram[1]

[1]Power Generation Unit, Institute of Power Engineering (IPE), Universiti Tenaga Nasional, 43000 Kajang, Selangor, Malaysia.

Abstract. Coal fired power plant becoming preferable power plant type to support electricity demand mainly in Asia due to stable coal price and low maintenance. However, most coal fired plant operator struggle with condition where coal undergo incomplete combustion and produced unburned carbon where can be found in ashes especially in fly ash. Higher percentage of unburned carbon in fly ash reflects the lower efficiency of furnace and contributes to financial loses for plant operators. This problem also leads to technical issues such as slagging and clinkering and further reduces the efficiency of furnace. The plant operator determines the amount of unburned carbon by using conventional method and this proves be a challenge to identify and rectify the problem on day basis due time constraint to obtain results of unburned carbon. Thus in this paper, best Artificial Neural Network model was derived to develop intelligent monitoring system to predict unburned carbon level on more daily basis. By this model, the power producer can predict the unburned carbon level by using data in power plant to predict the unburned carbon level in short period of time.

1 Introduction

Malaysia's largest thermal power plant uses coal as fuel source to fuel their furnace. This type of power plant was considered due to the cheap coal prices and abundant of coal reserve in the world compare to other type of thermal power plant such as natural gas and oil based power plant. As by product of the coal combustion, two type of ashes were produced and known as bottom ash and fly ash. Bottom ash usually dropped to hopper of the boiler and fly ash will be collected by electrostatic precipitator or also known as ESP and will be collected in ash silo. The ash usually contains some percentage of unburned carbon that was produced due to incomplete combustion in the boiler. The higher concentration of unburned carbon usually found in fly ash compared to bottom ash. The higher percentage of unburned carbon proved not to be good in term of financial and technical factor for the power plant operators. Fly ash also poses some problem such as air pollution, groundwater contamination and also poses some health issue especially related to respiratory system. As energy demand increase

* Corresponding author: Firas@uniten.edu.my

and the world opted for greener environment, plant operator need to maintain higher environment friendly plant and a higher efficiency power plant. Lower carbon fly ash considered as high quality additives for cement manufacturing. High quality fly ash improve concrete strength and reduce heat of hydration of concrete [1]. Unburned carbon in the fly cannot be eliminating fully but can be reduced to significant amount. Best level of unburned carbon estimated to be around 2-5 % [2] from total fly ash. There is few method was tried such as usage on infrared light sensor to measure carbon intensity in fly ash [3], Computational Fluid Dynamic (CFD) analysis [4] and combination of ANN with other method to predict the presence of unburned carbon in fly ash [5,6,7,8]. Thus an intelligent monitoring system was designed to detect and predict the unburned carbon in short period of time. This system is based on Artificial Neural Network (ANN) theory. This approach was selected because it proves to produce prediction using real time data at faster and accurate compare to other existing prediction system.

2 Result and discussion

2.1 CFD simulation

Four set of simulation was carried out to find out the amount at different factor and the relation that lead to production higher unburned carbon. This simulation was done to show that the parameter that was selected for the input data influence the production of unburned carbon level. The simulation was carried out by using different rate that can lead to unburned carbon level. The simulation was done to four parameters as :

2.1.1 Load (MW) simulation

Two simulations were carried out to find the amount of unburned produced at two different load. The simulation was carried out for full load and half load. The simulation was done predict and study the behavior of the rate of unburned carbon produced. The simulation was done on full load and half load. The result of unburned carbon production was tabled in Table 1 below.

Table 1. Estimated unburned carbon in flue gas at different load.

Load (MW)	Unburned Carbon (ppm)
Full Load	0.064
Half Load	0.282

The unburned carbon amount was recorded in parts per million (ppm) during the simulation. Simulation for full load boiler operation shows that it produce 0.064 ppm and half load operation simulation shows the unburned carbon at 0.282.The data show that lower load produce more unburned carbon compare to full load. Thus, it was determine that load does affect the production of unburned carbon. The lower load does produce higher rate of unburned carbon compared to full load. This due to less heat produced in the boiler and less heat can't completely burn all the carbon because carbon needs high temperature for combustion. Thus the load does affect the unburned carbon in fly ash during boiler operation.

2.1.2 Tilt angle of Burner (°)

The second test was done to find the amount of unburned carbon produced at different tilt angle. Five different angle of tilt angle was determine with 0° was set as base angle with 20°, 10°, -10° and -20°. The result of simulation was presented in Table 2 below.

Table 2. Estimated unburned carbon in flue gas at different tilt angle of burner

Tilt Angle (°)	Unburned Carbon (ppm)
20	0.065
10	0.080
0	0.064
-10	0.041
-20	0.110

From Table 2 the result that was produced for 20°, 10°, -10°, 0° and -20° are 0.065, 0.080, 0.064, 0.041, 0.11 respective. This simulation shows result of different angle of tilt angle that influence the production of unburned carbon. The higher amount unburned carbon can be seen at -20°, the lower position of burner in the boiler. The lower tilt angle usually will influence the higher amount of unburned carbon because the coal from lower position need higher time to be getting burned completely [9]. Thus it can be seen that the position of burner does influence the production of unburned carbon although it does not show a clear pattern or relationship between tilt angle of burner and unburned carbon.

2.1.3 Coal Feeder Rate (t/h)

The third simulation was carried out by manipulating the coal feed rate per hour. This simulation was carried out because this parameter varies a lot during plant operation to meet demand power plant load. Thus three simulations were carried out by using 32.32 t/h, 30t/h and 26 t/h and the result was tabulated in Table 3below.

Table 3. Estimated unburned carbon in flue gas at different coal feed rate

Coal Feeder Rate (T/H)	Unburned Carbon (ppm)
32.32	0.064
30	0.080
26	0.124

The result that was obtained after the CFD simulation are 0.064 ppm, 0.080 ppm and 0.124 ppm respectively to 32.32 t/h, 30 t/h and 26t/h. The result shows the changes of unburned carbon produced during the simulation with coal feeder rate varies. Coal feeder rate was determined as references point due the feeder rate normal TNBJ operation was 30 t/h. From the table 3, it is known that 32.32 have lower unburned carbon compare to 30 t/h and 26t/h. The pattern shows that the unburned carbon in the fly ash decreases as the coal feeder rate increases. Feed rate is the rate of amount of coal that was feed into the boiler to support the combustion. Thus when less coal was feed in to the boiler, the efficiency of the fire reduces and the unburned carbon of the coal escapes through flue gas to outlet. The heat in flue gas also reduces due to the less heat was produced by fireball. Meanwhile for higher feed rate, more heat is produced and the carbon will be burnt and heat from flue gas assists to burn the remaining carbon that escape with fly ash [10]. Thus the factor can be concluded as input parameter that influences the production of unburned carbon. The coal feeder rate does influence the unburned carbon level in fly ash.

2.1.4 Excess Air (%)

The excess air in boiler also considered as influencing factor for unburned carbon level. Thus, the parameter was assessed by using CFD method by adjusting the percentage of excess air in the boiler. From the simulation, it was shown that the higher the percentage of excess air, the higher the unburned carbon at outlet. The result was tabulated in Table 4 below for easier understanding.

Table 4. Estimated unburned carbon in flue gas for different percentage of excess air in boiler

Excess Air (%)	Unburned Carbon (ppm)
5	0.094
5.26	0.064

The amount of unburned carbon produced at different percentage of excess air in the boiler. The 5 % produces 0.094 ppm of carbon oxide meanwhile 5.26% of excess air produces 0.064 which is less compare to 5% of excess air. From the result in table 4, it shows that more excess air reduces amount of unburned carbon produced in fly ash. This due to the more excess air will help the coal combustion to undergo more coal burning and more excess air will eventually increase burnout of the remaining unburned carbon during combustion. This eventually will produce less unburned carbon in fly ash. Although more excess proved to reduce unburned carbon in fly ash but if the excess air more than optimum level, it will reduce boilers efficiency and lead to other technical issues such as higher level of Nitrogen Oxide (NOx) and Carbon Dioxide (CO) [11]. The percentage excess air does influence the unburned carbon level.

2.2 Best ANN model

After all the simulation the lowest RSME from each hidden layer was compared to obtained best model to use for intelligent monitoring system of unburned in fly ash. The RSME from each hidden layer was listed in Table 3.5 below for comparison and best model selection.

Table 5. Summarized of the best model for each hidden layer

Training Algorithm	Architecture	Activation Function	RMSE
trainlm	9HL1	T+T	0.047
trainscg	5HL1-4HL2	P+T+T	0.148

The RSME obtained for each hidden layer can be observed in Table 5 above. The lowest RSME among both hidden layer are 0.047 from 1 hidden layer. The error was so minimal compared to 2 hidden layers which were at 0.148. Thus it is decided that the best model for this monitoring system will consists of 1 hidden layer, trained using trainlm as training function and consists of combination of tansig and tansig as its activation function[12]. Meanwhile it consists of 9 neuron in its hidden layer.

3 Conclusion

All the parameter was proved to influence the production of unburned carbon in fly ash. This was proved by using CFD simulation method and a basic study was done on the production of unburned carbon for each parameter and was discussed briefly. ANN model was developed using MATHLAB software and the model was decided to be use 1 hidden layer or 2 hidden layers. A total of the 9 input parameters were used to develop an intelligent monitoring system. From the 9 input parameters just 4 parameters was used for CFD simulation because the one parameter represent same type of input parameter. The model that was selected as the best model to be used for the intelligent monitoring system for unburned carbon in fly ash are made of 1 hidden layer with 9 neuron in it. The model uses combination of tansig and tansig as activation function and trainlm as training algorithm. The model produces RSME of 0.046637 compare to 2 hidden layer which produce RSME of 0.14797. Hence this model can be used for prediction of unburned carbon n fly ash. Meanwhile the training algorithm that was selected for this model are trainlm because it produced less error compare to other training algorithm using same activation function. By using this model, an intelligent monitoring system can be built and to predict the higher unburned carbon in fly ash and achieve more sustainable business and more environmentally friendly power plant.

The authors would like to express our gratitude to the power generation unit, institute of power engineering Universiti Tenaga Nasional (UNITEN) TNB Janamanjung and Ministry of Higher Education (MOHE), Malaysia for the opportunity to sponsorship this corresponding research under the FRGS research fund with the project code 20160103FRGS.

References

1. Why Use Fly Ash, [Online]. Available: http://www.proash.com / wordpress /?page_id=95. [Accessed 8 July 2015].
2. Katarzyna Styszko-Grochowick,Janusz Goles and Henryk Jankowski (2003), "Characterization of the coal fly ash for the purpose of improvement of industrial on-line measurement of unburned carbon", [Online]. Available: http://www.sciencedirect.com/science/article/pii/S0016236104000845 [Accessed 8 July 2015].
3. A.Schneider, R.Chabicosky and A.Aumuller (1998), "Optical sensor system for the on-line measurement of carbon in fly-ash", [Online]. Available: http://www.sciencedirect.com/science/article/pii/S0924424797017238. [Accessed 8 July 2015].
4. Javier Pallares, Inmaculada Arauzo, Enrique Teruel (2008), "Development of an engineering system for unburned carbon prediction", [Online]. Available: http://www.sciencedirect.com/science/article/pii/S0016236108002780. Accessed 10 November 2015].
5. Zhou Hao, Xinping Qian, Kefa Cen and Jian Fan (2003), "Optimizing pulverised coal combustion performance based on ANN and GA", [Online]. Available: http://www.sciencedirect.com/science/article/pii/S0378382003001553. [Accessed 14 September 2015].
6. S.Tronci,R.Baratti and A.Servida (2002), "Monitoring Pollutant emissions in a 4.8 MW Power Plant Through Neural Network", [Online]. Available: http://www.sciencedirect.com/science/article/pii/S0925231201006178. [Accessed 14 September 2015].
7. Jiejin Cai, Xiaoqian Ma and Qiong Li (2008), "On-line monitoring the performance of coal-fired power unit: A method based on support vector machine", [Online]. Available: http://www.sciencedirect.com/science/article/pii/S1359431108004651. Accessed 10 November 2015].

8. Weiping Yan and Jun Li (2009), "Modeling of the Unburned Carbon in Fly Ash", [Online]. Available: http://www.scirp.org/journal/PaperDownload.aspx?paperID=935. [Accessed 10 November 2015].
9. Jun-Li, Radoslaw Jankowski, Michal Kotecki, Weitong Yang, Dariusz Szewczyk, Artur Brdekiewicz, and Wlodzimier Blaisk (2012), "CFD Approach of Unburned Carbon Reducction on Pulverised Coal Boiler" [Online]. Available:https://pure.strath.ac.uk/portal/files/43323298/Li_etal_EAF_2012_CDF_approach_of_unburned_carbon_reduction_on_pulverized.pdf. [Accessed 31 January 2016].
10. Jong-Min Lee, Dong-Won Kim and Jae-Sung Kim (2011), "Unburned carbon fraction with operation variables in a commercial circulating fluidized bed during co-combustion of various antarcits". [Online]. Available: http://asiatec2013.spekorea.org/PDF/KJChE/KC29/KC29-4-0452.pdf. [Accessed 31 January 2016].
11. Combustion. [Online]. Available: http://www.e- inst.com/docs/Combustion-Booklet-2013.pdf. [Accessed 31 January 2015]
12. Neural Network Toolbox. [Online]. Available: http://www.mathworks.com/products/neural-network/features.html#network-architectures. [Accessed 31 January 2016]

DOI: 10.1051/matecconf/201713102004

Design and development of solar desalination plant

Thaneissha a/p Marimuthu [1], *Samson* Mekbib Atnaw [1,*], *Priyatharishini* a/p Mardarveran[1], *Sim* Shu Yi[1], *Mazlin Azura* binti Usop[1], *Muhammad Khairi* bin Md Gapar[1], *Siti Aishah* Binti Rusdan[1], *Roshahliza* Binti M Ramli[1] and *Shaharin* A. Sulaiman [2]

[1]Faculty of Engineering Technology, University Malaysia Pahang, 26300, Gambang, Pahang, Malaysia
[2]Department of Mechanical Engineering, Universiti Teknologi PETRONAS, 31750, Tronoh, Perak, Malaysia

Abstract. Direct sunlight has been utilized long back for desalination of water. The desalination process takes place in solar still. Solar still is a device that converts saline water to potable water. This process requires seawater and sunlight which are widely available on Earth. However, the current solar desalination generation capacity is generally low and has high installation cost. Hence, there is a need for the enhancement of the productivity which can be achieved through few modifications. This paper explores the challenges and opportunities of solar water desalination worldwide. It presents a comprehensive review of solar desalination technologies that have been developed in recent years which covers the economic and environmental aspects.

1 Background of solar desalination

Water and energy are two inseparable and vital commodities for human, plant and animal sustainability on earth. With total global water reserves of about 1.4 billion km^3, around 97.5% of it is in the oceans and the remaining 2.5% is fresh water present in the atmosphere, ice mountains, freshwater lakes and groundwater as shown in Figure 1 [1].

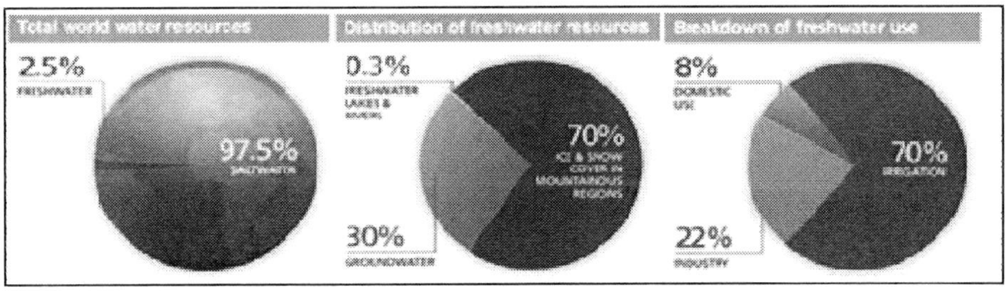

Fig 1. Distribution of world water resources

*Corresponding author: mekbibsams@ump.edu.my

 DOI: 10.1051/matecconf/201713102004

Since those days, humans survive with water from rivers, lakes, and recently aquifers where ground water can be extracted as potable water which is less than 1% of total amount of the water on Earth.

2 Classifications of solar desalination

Figure 2 shows the classification of solar still.

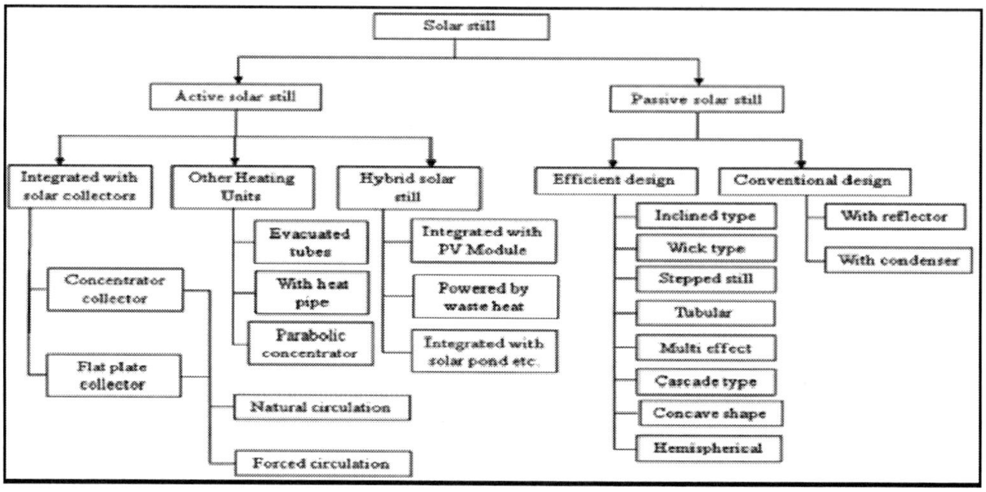

Fig 2. Classifications of solar still [2]

2.1 Passive solar still

Passive solar stills will be very simple in construction and operation. Passive solar stills only utilize available solar energy to remove the impurity in contaminated water, thus it is safe, clean, eco-friendly, and energy saving.

2.1.1 Basin type solar still

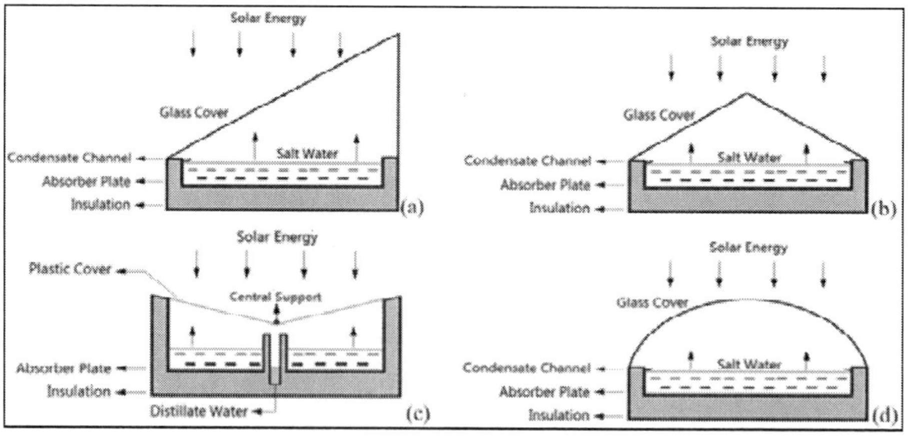

Fig 3. (a) single-slope basin still, (b) double-slope basin still, (c) V-type solar still, (d) Hemispherical type solar still [3]

As shown in Figure 3, basin type solar stills have been modified into several types according to their cover designs such as single slope, double slope, V-type and hemispherical. This basin type solar still is the most widely used nowadays due to its simplicity and low cost. Zaki et al. acclaims that the average distillate production rate of a standard single-basin still is between 2 L/m^2 depending on the insulation quality [4]. However, the productivity varies according to the design. The concept of this process is very simple where the still traps solar energy within the enclosure. This heats up the water causing evaporation and condensation on the inner face of the sloping transparent cover. This distilled water is generally potable because all the salts, inorganic and organic components and microbes are left behind in the bath. However, few modifications have been done in order to increase the productivity of basin type by integrating with solar panel or heater which refers as active solar still. After few improvements have been made, today's state-of-the-art single-effect solar stills have an efficiency of about 30–40% [5].

2.1.2 Weir type cascade solar still

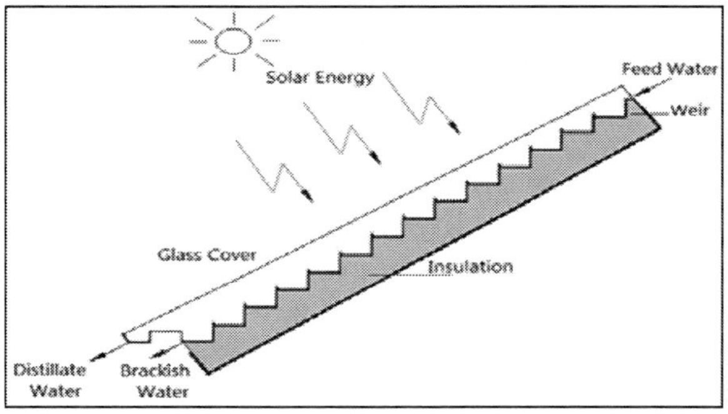

Fig 4. Weir type cascade solar still [6]

Figure 4 shows the weir type cascade solar still. The advantages of weir-type cascade solar stills is it do not suffer from dry spot or channelization problems since the brine is forced to flow each step one by one without leaving any dry surface on the absorber plate which increases the efficiency. Further study have been done to these weir-type cascade stills by including wick on each cascaded steps and phase change material beneath the absorber surface to store energy when it is abundant and give it back to the salt water when it is needed during cloudy days [7]. Table 1 shows the productivity of the weir-type still that is approximately 20% higher than conventional basin-type solar still. Besides, the average productivity of the still is approximately 5.5 L/m^2 in a day.

Table 1. Distillate productivity from the conventional basin and weir-type stills, September 2006 [8]

Day	Distillate productivity (l/m3/day)	
	Conventional basin-type still	Weir-type still (single-pane)
8 September	4.6	5.5
9 September	5.1	6.3
10 September	4.3	5.2
11 September	4.8	5.6
12 September	4.9	5.8
13 September	4.2	5.0
14 September	3.4	4.1

2.2 Active solar still

Active solar still is a still integrated with a solar panel or concentrator. Active solar stills receive additional thermal energy from an outer source to the water in the basin which improves the rate of evaporation.

2.2.1 Parabolic trough solar collector

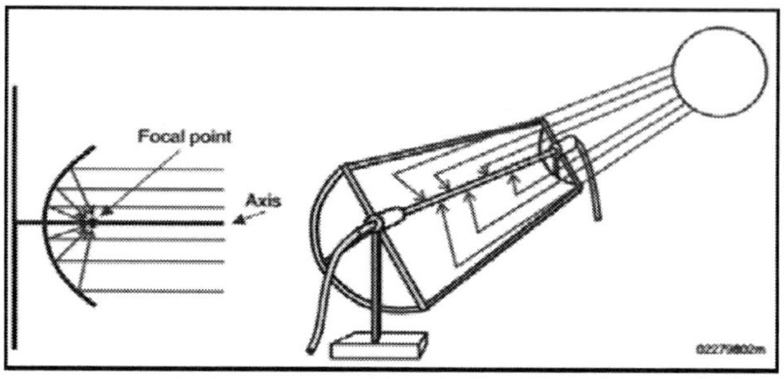

Fig 5. Parabolic trough solar collector on the right side and the concentrated focal point on the left side [9]

The parabola is an intriguing geometric shape with important practical uses including concentrating sunlight as shown in Figure 5. The curve of a parabola is set as such that light travelling parallel to the axis of a parabolic mirror will reflect to a single focal point from any place along the curve. Since the sun is far away, all light coming directly from it is essentially parallel, so if the parabola is facing the sun, the sunlight is concentrated at the focal point. A parabolic trough extends the parabolic shape to three dimensions along a single direction, creating a focal line along which the absorber tube is run [9]. Some design goals of a parabolic trough solar collector achieve high performance, reliability and durability. The drawback of this solar still is the cost is high and often used for large scale purpose. Improvements have been suggested in reflective and absorbent coatings which will incrementally improve economics, but still the cost is high.

2.2.2 Point-focus parabolic solar still

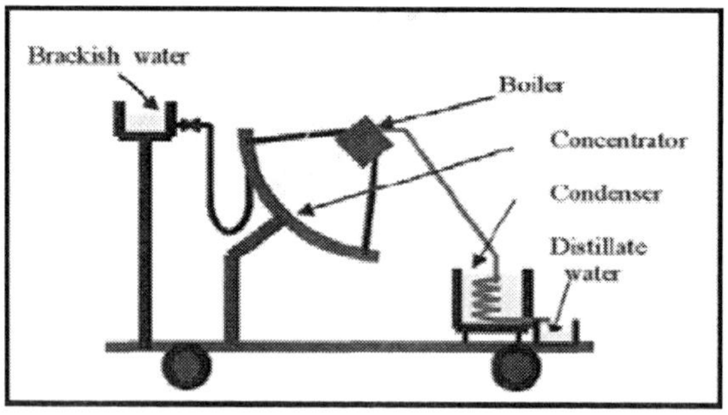

Fig 6. Point-focus parabolic solar still [10]

The point-focus parabolic solar still (PPSS) is the active solar still type as shown in Figure 6. This system consists of a parabolic dish concentrator, a single or two axis sun tracker based on programmable logic controllers (PLCs) and condenser. The concentrating solar energy on the focal line of a parabolic trough has line that usually a pipe that contains a thermal fluid which heats it up to high temperature [11]. Then, to improve the efficiency, heat exchanger can be added where the fluid is pumped to a heat exchanger which heats water. It eventually vaporizes and immediately condensed which is later withdrawn as desalinated water. The productivity of such units is high. Improvements have been made through high temperature and low heat loss area of the still basin or the absorber which is located on the focal point of parabolic concentrator greatly increases the efficiency of the still [12]. Despite this still's outstanding performance, it has many drawbacks; including the high cost of building and maintaining it, the need for strong direct sunlight, and its fragile nature. The total yield produced for this PPSS is 6.112 L per day for PPSS [11].

3 Summary table

Table 2 shows the comparison of productivity for different types of solar stills.

Table 2. Comparison of productivity for solar stills

Type of Solar Still	Methodology/ Modifications/ Analysis	Capacity (L/m2/day)	References
Basin Type Solar Stil	Simple basin that need for regular flushing of accumulated salts	2.00	Zaki et al.,1993 [4]
Wick Still	Effect of corrugated work by wick surface on the performance of the still	4.20	Kassem,2016 [13]
Weir type cascade solar still	Absorber plate, condensing cover, and insulation	3.85	Sadineni et al., 2006 [8]
Spherical Solar Still	Circular absorber basin coated with black paint	2.30	Arunkumar et al., 2012 [14]
Double basin still coupled to a collector	Double basin solar still integrated with flat plate solar collector	5.18	Tiris, 1998 [15]
Point-focus parabolic solar still	Concentrator (mirror + parabolic dish) and heat exchanger	3.56	Arunkumar et al., 2012 [14]

4 Conclusions

To sum up, water is essential to all living things in Earth as a nutritional requirement, agriculture and industrial processes. Since sunlight and seawater can be obtained abundantly and it is free, there would not be many challenges to get the resources for solar desalination system [16]. In order to come up with an effective solar desalination plant, the evaluation of the effects of environmental and operational parameters should be considered [17]. Therefore, there is a need for on-going research and demonstration projects to gain experience, knowledge and trust in new environmentally friendly technologies as well as a state incentives through the introduction of new polices and financial support.

References

1. T.A Mateo, Solar powered desalination system, 1 – 78 (2011)
2. H. Manchanda and Kumar, Renew. Energ. (2015)
3. T. AnfasMukram, P.U. Suneesh., Int. J. Innov. Res. Sci. Eng. Technol. 2(7) (2013)
4. G. Zaki, A. Radhwan and A. Balbeid, Sol. Energy. 51. 277–288. (1993)

5. G. Mink, M. Aboabbous and E. Karmazsin, Sol Energy 62. 309–317 (1998)
6. V. Sivakumar and E.G. Sundaram, Sust. Energ. Rev. 28. 246–64 (2013)
7. H.A Zoori, F.F Tabrizi, F. Sarhaddi and F. Heshmatnezhad, Desalination 325. 113–21 (2013)
8. S.B. Sadineni, R. Hurt, C.K. Halford and R.F. Boehm, J. Energy. 33. 71-80 (2008)
9. Federal Technology Alert. Parabolic-Trough Solar Water Heating. U.S. Department of Energy (DOE) (2000)
10. B. Chaouchi, A. Zrelli and S. Gabsi, Desalination 217. (2007)
11. S. Gorjian, B. Ghobadian, T.H. Teymour and B. Ahmad, Desalination. 352. 1-17(2014)
12. H.M. Qiblawey and F. Banat F, Desalination 220. 633–644 (2008)
13. T.K. Kassem, Int. J. Eng. Technol. Sci. 3 (2016)
14. T.Arunkumar, K. Vinothkumar, A. Ahsan, R. Jayaprakash and S. Kumar, Renew. Energ. (2012)
15. C. Tiris, M.Tiris, Y.Erdalli and M. Sohmen, Energ. Convers. Manage. 8, 853-856 (1998)
16. R. Zailan, A. Razak and M. Firdaus, Int. J. Eng. Technol. Sci. 3 (2015)
17. MNI Siddique and AW Zularisam, Int. J. Chem. Environ. Biol. Sci. (2014)

Microwave assisted pyrolysis of plastic waste for production of fuels: a review

Haroon Arshad [1,*], *Shaharin* A Sulaiman [1], *Zahid* Hussain [2], *Yasin* Naz [3], *Firdaus* Basrawi [4]

[1]Department of Mechanical Engineering, Universiti Teknologi PETRONAS, 32610 Seri Iskandar, Malaysia
[2] Department of Chemistry, Abdul Wali Khan University, 23200 Mardan, Pakistan
[3] Department of Physics, University of Agriculture, 38040 Faisalabad, Pakistan
[4] Department of Mechanical Engineering, Universiti Malaysia Pahang, 26600 Pekan, Malaysia

Abstract. This paper presents an overview of advantages of microwave assisted pyrolysis of waste plastics together with its limitations. It has been established that microwave induced pyrolysis can be used to get value added chemicals and fuels through its numerous noted advantages in contrast to conventional pyrolysis. The process has the potential for fast, volumetric and selective heating of plastics for the recovery of energy. The limitation in the use of dielectric material as absorbent in plastic pyrolysis has been highlighted. Special focus has been given to the constraints encountered in the accurate measurement of temperature and uniform heating in microwave assisted pyrolysis. A new alternative method based on microwave-metal interaction in the pyrolysis of plastic waste has been presented. Further it has been realized that proper investigation and understanding of microwave process shortcomings is fundamental to the successful implementation of the technology and at the same time provide a sustainable environment in the endeavor for waste to energy mission.

1 Introduction

Plastics are included among indispensable items of modern times. It has diverse applications due to its many excellent. Common types of plastics found in municipal solid waste (MSW) comprise high density polyethylene (HDPE), low density polyethylene (LDPE), polypropylene (PP), polystyrene (PS), polyethylene-terephthalate (PET) and polyvinyl chloride (PVC) [1], which are the principal contributors to the non-biodegradable class of waste. With rapid growth in the world population, there has been a drastic increase in the use of commodity plastics dominant in packaging, clothing, beverage and trash containers, and numerous other household items. It was observed that the global production of plastics rose to 229 million in 2013 with a 4% gain over that of 2012 [2]. At the same time, there was also huge accumulation of waste due to continuous rise in demand of plastics each year [3]. It was reported that the plastic waste reached to over 33 million tons in US according to 2013 statistic [4]. Moreover, in Europe a study revealed that a total of 25 million tons of plastics went to the wastestream in 2012. Further, it was found that 38% of

* Corresponding author: haroonarshad00@gmail.com

this waste was disposed to landfill, 26% were recycled and the remaining 36% was utilized for energy recovery [2]. It was also reported that plasticizers (additives used to enhance plastic properties) found in some specific plastics like PVC can also leach out from improperly designed landfills and may contaminate groundwater which is caused by the production of phthalate esters or PAEs [5]. Further it was explored in a study that open dumping of municipal solid waste poses environmental hazards through the transfer of heavy metals to the soil which adversely affect our vegetation [6]. Another problem was identified associated with disposal of plastic waste into oceans which results in formation of soap and garbage patch like the great Pacific garbage patch posing risk to health of aquatic animals [7]. Moreover, incineration of plastic waste may lead to environmental pollution through the toxic emissions [8].

Alternatively, the disposal of plastic waste through recycling has been considered as an unattractive option due to the extra cost involved in its recovery and intensive requirement of labor in the separation of its components [3]. The limitations on recycling method and ever-increasing load on plastic waste has created a need for the development of alternative technologies to solve the problems associated with disposal. Greater interest was observed on the conversion of waste plastic into value added chemicals and fuel products in a range of published papers on conventional as well as microwave assisted pyrolysis [3], [7], [9-12]. The conventional method also known as traditional method converts waste into fuel products through thermal or catalytic pyrolysis. Thermal pyrolysis which is uncatalyzed gives low quality products over a wide range [13]. However, application of suitable catalysts in catalytic pyrolysis has the potential to narrow the range of final products leading to outstanding selectivity for getting valuable products at lower temperatures [14]. Pyrolysis is a process that involves thermochemical breakdown of organic material at elevated temperatures 300-900°C in oxygen free environment, resulting in oil and gas yields with char as the solid residue. Based on heating strategies, pyrolysis can be classified as conventional pyrolysis and microwave assisted pyrolysis. The former involves surface heating of plastic waste using fuels like coal, whereas the latter works on the principle of microwave electromagnetic radiations conversion to thermal energy and has the potential for volumetric heating which is faster as compared to surface heating.

Though there are many advantages in microwave heating, there are also certain limitations. Accurate measurement of temperature within the reaction mixture is difficult [15]. Moreover, uneven distribution of scattered hotspots within the sample may result in non-uniform heating according to a simulation study on microwave design [16]. Another challenge has been the quantification of dielectric properties of the material to be treated since heating efficiency depends strongly on dielectric response of materials and there is insufficient data on the dielectric materials to date [3]. Problems related to microwave-metal discharges and scale-up of technology are elaborated. An alternative method of microwave-assisted pyrolysis of plastics and its future challenges have been highlighted. Therefore, this review focuses on the shortcomings of microwave assisted pyrolysis together with its advantages in contrast to conventional heating. The objective of this paper is to broaden the understanding of the limitations of microwave processing which is crucial to the success of any proposed work wherein relevant considerations could be made to get the desired results like efficient heating, accurate temperature measurement, minimum energy input and overall reduced cost.

2 Microwave heating technology

2.1 Microwave and conventional heating

Microwave is an electromagnetic form of energy found in the range of wavelengths from 0.001-1 m operating at a frequency band of 0.3-300 GHz. Most of the common microwaves used, particularly the domestic ovens operate at frequency of 2.45 GHz. Since microwaves have the ability to penetrate through solid material, it can accomplish volumetric heating in contrast to conventional method of heating in which only surface heating is achievable. The heat flow patterns showing the temperature distribution produced during the two heating schemes is illustrated in Fig. 1 [17]. The major difference between the microwave and conventional heating is based on the heat transfer mechanism. Conventional methods utilize furnaces to initially heat the surface, before any conductive heat transfer towards the inner core of the material takes place [18].

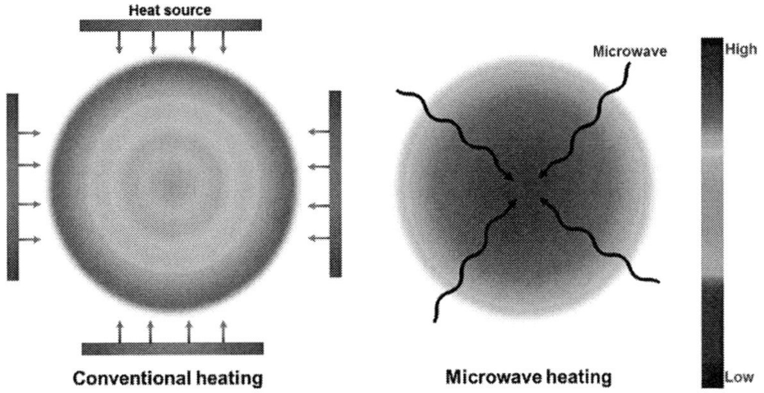

Fig. 1. Microwave and conventional heating patterns [17].

In case of microwave assisted pyrolysis of plastics, additional use of dielectric material called as absorbent is required. A study conducted on pyrolysis of plastic waste utilized carbon black as absorbent to convert the microwave energy to heat required for pyrolysis [12]. Contrarily, conventional heating does not need any absorbent due to direct mode of heat transfer to the material. Another criterion used to differentiate between microwave and conventional pyrolysis techniques is the distribution of products. An investigation led by (Khaghanikavkani, 2013) on variation in product composition was carried out for pyrolysis of plastics using microwaves and conventional method. The results revealed a very similar distribution of pyrolysis products. However, the heating uniformity was found to be considerably improved in contrast to conventional method as a result of microwave volumetric heating leading to faster pyrolysis. In case of plastics at a suitable scale (0.5 kg), the microwave heating has proved to be an efficient and cleaner process with faster and easier control [19].

There are several benefits of microwave heating in addition to volumetric heating [20], [21]. Selective material heating is a unique advantage of this technology that targets the specific material. In this process, the magnitude of heating strongly depends on the dielectric strength of the material. Since plastics have poor dielectric strength, they are mixed with dielectric absorbents like carbon which has been observed to convert microwave energy into thermal energy in a short span of time [22]. Thus, heating efficiency may vary for different materials which has also been a great challenge to industries [3]. Here, rapid heating can be easily promoted. It was pointed out that decomposition temperature up to 450°C in solid organic polymers can be achieved in a fraction of the time taken in conventional pyrolysis [23]. High temperatures can be reached in few minutes rather than hours in most cases, when the materials are heated at higher heating rates [23]. Another unique feature of this method is the non-contact heating. If microwaves can be controlled properly, it has the potential to treat the material without any physical contact with the heated sample [23]. Additionally, the control of the process is also simple. Plastic pyrolysis was achieved through efficient coupling and easy control of temperature by tuning and power control [24]. Microwaves have also been found to be effective for the upgrading of in-situ pyrolytic vapors into enhanced bio-oils, during biomass pyrolysis [25, 26]. Many studies have reported faster heating and in situ treatment of waste [22], [27-29]. The flexibility of operation and portability of equipment are the noted benefits of microwave technology [27, 28].

2.2 Limitations in microwave assisted pyrolysis

The potential of dielectric materials called as absorbents or susceptors to be used in the pyrolysis of polymers and plastics has been explored widely [22, 27, 28]. However, specific role of susceptor in the energy transfer or catalyzing effect on pyrolysis reaction is still not well known [15]. The effective use of dielectric material at industrial scale pyrolysis is a challenge since heating efficiency differs for different absorbents [3]. It is to be pointed out that not all materials have the ability to absorb microwaves [23]. For example, plastics have low dielectric constant and the mixture with carbon as microwave absorber during pyrolysis may enhance the energy absorbed to be converted to heat in shorter time [22]. Here, the challenge has been the quantification of properties of the dielectric materials to be treated since heating efficiency depends strongly on these properties and there is insufficient data on the dielectric materials to date [3].

To date, macroscopic temperature measurement methods, can only measure the overall or average value of temperature of the heating medium due to detection limitation of measurement devices at temporal and spatial scales, and it has been very difficult to measure temperature of isolated hotspots [30]. With the use of advanced detecting instruments like high-speed camera and optical temperature probe, existence of hotspot effect has been confirmed [31, 32]. Another difficulty is faced during accurate measurement of temperature within the reaction mixture. In a study led by Undri et al. [33], on microwave assisted pyrolysis of silicon oil, a temperature lag was observed in IR-based sensors in contrast to conventional use of thermocouples. The temperature monitoring of the process becomes difficult since the temperature goes beyond 1000°C and an instrument like thermocouple will melt down. In this scenario, there are two alternatives available for temperature measurement. One is the use of infrared thermometer and the other one being fiber optic thermometer. The infrared thermometer works without contact over a wide range of temperatures. But the IR thermometer sensors can only measure the surface temperature of the reactor which is lower than the temperature of the reaction mixture [15]. In this context only few studies have been able to measure the temperature within the reaction mixture [35-37]. On the other hand, the use of fiber optic device is not possible since it can measure temperatures only up to 400°C [38]. But there is a new variant of optical fiber temperature probe which is made of sapphire crystalline fiber and has been used effectively in microwave pyrolysis [39]. In another published work, a thermocouple was used for temperature measurement inside a domestic microwave oven operating under low power of magnetron [15]. This facilitated measurement of temperature inside the reaction mixture during the off-time which is greater than the on-time period in contrast to high power microwave. It was pointed that conventional thermocouple may lead to sparks and trips due to reflection of microwave. Therefore, in this experiment, a chromel-alumel thermocouple was used as a modified version of conventional thermocouple.

Formation of hotspot or micro-plasmas is another limitation that has been identified as a form of thermal instability associated with rapid heating in a microwave. According to a study, hotspot formation is a result of non-linear dependence of electromagnetic field on the thermal properties of the material [40]. The presence of such a hotspot may lead to non-uniform heating as materials not present in the hotspot region are not subjected to the same degree of microwave heating [16]. Further, it was reported that the design of microwave cavity plays a crucial role in the control or utilization of this hotspot for effective heating [40]. According to a simulation study on microwave design, uneven distribution of scattered hotspots within the sample may result in non-uniform heating [16]. A major problem is faced when dealing with microwave metal-discharge applications. Here, the main obstacle is that the discharge is highly unstable and difficult to sustain for a long time [30]. In microwave systems, discharge occurs when metals with sharp edges or tips are subjected to microwave irradiations. Spark discharge at the tip of metal electrode is an example where surrounding gas is broken down to conducting ions which may lead to formation of hotspots due to excessive generation of heat in the process. Another microwave-metal discharge effect can be observed in the use of thermocouples inside a microwave field, which can cause incorrect recording of temperature and may also damage the thermocouple in the extreme case of hotspots formation. Finally, application of microwave technology is still limited to only lab scale systems and is not effective at industrial scales. This drawback can be attributed to the limited understanding of microwave systems. Apart from material properties and characteristics, operating parameters such as radiation time and power, cavity design and material throughput have been found to determine the extent to which a successful treatment is achieved [23].

2.3 New trend in microwave assisted pyrolysis of plastics

Literature spans through numerous works that utilized the dielectric materials as absorbents to achieve microwave assisted pyrolysis [12, 15, 19, 41]. Carbonaceous materials like traditional coke, due to their excellent response to microwave absorption, heat tolerance and low cost has made them used widely in pyrolysis applications [20]. Later, in an experimental study led by Hussain et al. [42], pyrolysis was achieved effectively, based on microwave-metal interaction mechanism as a novel means of generating heat rather than the use of absorbents. Only few works have been explored principled on this mechanism [42-44]. The first experiment was conducted on a waste polystyrene (PS) plastic sample (20g), placed inside an iron mesh. The mesh acted as an antenna responsible for the generation of heat in a temperature range of 1100-1200°C. The products obtained were 80% liquid, 15% gas and 5% char residue [42]. In a similar work, PS waste sample was pyrolyzed using a copper coil instead of iron mesh [43]. The results gave 85% liquid, 10-12% gases and char residue. The process was achieved in a temperature range of 1000-1100°C. Both aforementioned studies were found to be effective in producing fuel like hydrocarbons showing major content of aromatics. The use of iron mesh [42] and copper coil [43], also demonstrated that the rate of reaction and reaction time were strongly influenced by the shape and nature of metal antenna. The mechanism of heat generation is based on the repeated reflections of microwaves trapped inside the coil shaped reactor. The metal coil or mesh initially acts as an antenna for reflections. Subsequently, when there is an interaction of free or conducting electrons, the metallic coil begins to act as a microwave absorbent. The rapid vibrations of the atomic core cause volumetric and rapid heating reaching high temperatures up to the melting point of metal [43].

Though the method is suitable, it still requires extended studies to develop proper control in getting the optimum yield and specific composition of products. Also, improper control of process parameters may result in non-uniform heating that can lead to formation of hotspots which in turn will reduce the overall treatment efficiency and lower the effective energy utilization of microwave systems [23]. Although, the initial capital cost of microwave system is high, this can be offset by economic benefits attained in operation by process time-savings, the yield of saleable by-products and environmental compatibility [23].

3 Conclusion

Microwave heating technology has evolved as a smart alternative for getting selective products from the pyrolysis of plastics. The process has the potential to deliver valuable chemicals and fuels. Despite so many advantages of this technology, limitations like inaccuracy in measurement of temperature, non-uniform heating caused by scattered hotspots, limited understanding of the specific role of dielectric materials in the pyrolysis reaction and gaps in efficient design of microwave systems has constrained the successful application of microwave induced pyrolysis of plastics at industrial scale. As such further investigations are required to push the technology to higher levels of performance where scale up of the technology together with cost reduction is not an issue.

References

1. P. Lettieri and S. M. Al-Salem, Thermochemical Treatment of Plastic Solid Waste, in *Waste*, Elsevier, 233–242, (2011)
2. Plastics Europe, *Association of Plastic Manufacturers Europe*. An analysis of European plastics production, demand and waste data. Belgium: *European Association of Plastics Recycling and Recovery Organisations*, Plastics Europe, 1–32, (2015)
3. S. D. Anuar Sharuddin, F. Abnisa, W. M. A. Wan Daud, M. K. Aroua. Energy Convers. Manag., **115**, 308–326, (2016)
4. U.S. *Environmental Protection Agency*, Common wastes and materials, U.S., (2014)
5. H. Liu, Y. Liang, D. Zhang, C. Wang, H. Liang, H. Cai. Waste Manag., **30**(8–9), 1569–1576, (2010)
6. S. M. Ali, A. Pervaiz, B. Afzal, N. Hamid, A. Yasmin. J. King Saud Univ. - Sci., **26**(1), 59–65, (2014)
7. B. Kunwar, H. N. Cheng, S. R. Chandrashekaran, B. K. Sharma. Renew. Sustain. Energy Rev., **54**, 421–428, (2016)
8. J. Aguado, D. P. Serrano, J. M. Escola. Ind. Eng. Chem. Res., **47**(21), 7982–7992, (2008)
9. R. Miandad, M. A. Barakat, A. S. Aburiazaiza, M. Rehan, A. S. Nizami. Process Saf. Environ. Prot., **102**, 822–838, (2016)
10. A. Undri, L. Rosi, M. Frediani, P. Frediani. *Microwave Heating*, 207–232, (2011)
11. A. Undri, L. Rosi, M. Frediani, P. Frediani. Fuel, **116**, 662–671, (2014)
12. K. N. Aishwarya and N. Sindhu. Procedia Technol., **25**(Raerest), 990–997, (2016)
13. W.-C. Huang, M.-S. Huang, C.-F. Huang, C.-C. Chen, K.-L. Ou. Fuel, **89**(9), 2305–2316, (2010)
14. A. López, I. de Marco, B. M. Caballero, A. Adrados, M. F. Laresgoiti. Waste Manag., **31**(8), 1852–1858, (2011)
15. D. V. Suriapparao and R. Vinu. J. Anal. Appl. Pyrolysis, **113**, 701–712, (2015)
16. J. P. Robinson, S. W. Kingman, C. E. Snape, S. M. Bradshaw, M. S. A. Bradley, H. Shang, R. Barranco. Chem. Eng. Res. Des., **88**(2), 146–154, (2010)

17. X. Zhang, K. Rajagopalan, H. Lei. Sustain. Energy Fuels, **0**, 1–36, (2017)
18. N. M. Mokhtar, R. Omar, A. Idris. Util. Environ. Eff., **34**(22), 2104–2122, (2012)
19. E. Khaghanikavkani. J. Chem. Eng. Process Technol., **4**(3), (2013)
20. B. Menéndez, J. A. Arenillas, A. Fidalgo, Y. Fernández, L. Zubizarreta, E. G. Calvo, J. M. Bermúdez. Fuel Process. Technol., **91**(1), 1–8, (2010)
21. K. E. Haque. Int. J. Miner. Process., **57**(1), 1–24, (1999)
22. S. S. Lam and H. A. Chase. Energies, **5**(10), 4209-4232, (2012)
23. T. J. Appleton, R. I. Colder, S. W. Kingman, I. S. Lowndes, A. G. Read. Appl. Energy, **81**(1), 85–113, (2005)
24. E. Khanikavkani. J. Chem. Eng. Process Technol., **4**(3), (2013)
25. Y. F. Huang, W. H. Kuan, S. L. Lo, C. F. Lin. Bioresour. Technol., **99**(17), 8252–8258, (2008)
26. A. Domínguez, J. Menéndez, Y. Fernández, J. J. Pis, J. M. Valente Nabais, P. J. M. Carrott, M. M. L. Ribeiro Carrott, J. Anal. Appl. Pyrolysis, **79**(1–2) SPEC. ISS., 128–135, (2007)
27. Y. Fernandez, A. Arenillas, J. Angel. Advances in Induction and Microwave Heating of Mineral and Organic Materials, InTech, (2011)
28. F. Motasemi and M. T. Afzal. Renew. Sustain. Energy Rev., **28**, 317–330, (2013)
29. S. Chandrasekaran, S. Ramanathan, T. Basak. AIChE J., **58**(2), 330–363, (2012)
30. J. Sun, W. Wang, Q. Yue, C. Ma, J. Zhang, X. Zhao, Z. Song. Applied Energy, **175**(August). Elsevier Ltd, 141–157, (2016)
31. S. Horikoshi, A. Osawa, S. Sakamoto, N. Serpone. Chem. Eng. Process. Process Intensif., **69**, 52–56, (2013)
32. W. Chen, B. Gutmann, C. O. Kappe. Chemistry Open, **1**(1), 39–48, (2012)
33. A. Undri, S. Meini, L. Rosi, M. Frediani, P. Frediani. J. Anal. Appl. Pyrolysis, **103**, 149–158, (2013)
34. X. Wang, W. Morrison, Z. Du, Y. Wan, X. Lin, P. Chan, R. Ruan. Appl. Energy, **99**, 386–392, (2012)
35. H. Lei, S. Ren, J. Julson. Energy & Fuels, **23**(6), 3254–3261, (2009)
36. N. Wang, J. Yu, A. Tahmasebi, Y. Han, J. Lucas, T. Wall, Y. Jiang. Energy & Fuels, **28**(1), 254–263, (2014)
37. D. Chen, L. Yin, H. Wang, P. He. Waste Manag., **37**(12), 116–136, (2015)
38. N. I. Djeu, Fiber Optic High Temperature Sensor, United States Patent, US6045259, US6045259, (2000)
39. C. Reimbert. IMA J. Appl. Math., **57**(2), 165–179, (1996)
40. D. A. Jones, T. P. Lelyveld, S. D. Mavrofidis, S. W. Kingman, N. J. Miles. Resour. Conserv. Recycl., **34**(2), 75–90, (2002)
41. A. Undri, L. Rosi, M. Frediani, P. Frediani. Fuel, **133**, 7–16, (2014)
42. Z. Hussain, K. M. Khan, K. Hussain. J. Anal. Appl. Pyrolysis, **89**(1), 39–43, (2010)
43. Z. Hussain, K. M. Khan, K. Hussain, S. Perveen. Util. Environ. Eff., **36**(18), 1982–1989, (2014)
44. Z. Hussain, K. M. Khan, N. Basheer, K. Hussain. J. Anal. Appl. Pyrolysis, **90**(1), 53–55, (2011)

MATEC Web of Conferences **131**, 02006 (2017)
UTP-UMP SES 2017
DOI: 10.1051/matecconf/201713102006

Flow control in s-shaped air intake diffuser of gas turbine using proposed energy promoters

Raed A. Jessam [1,2], *Hussain H.* Al-Kayiem [1,*], *Mohammed S.* Nasif [1]

[1] Mech. Eng. Dept., Universiti Teknologi PETRONAS, 32610 Bandar Seri Iskandar, Malaysia
[2] Electromechanical Engineering Department, University of Technology, Baghdad, Iraq

Abstract. This paper presents an experimental and numerical investigation of the flow control in an air intake S-shaped diffuser with and without energy promoters. The S-shaped diffuser had an area ratio 3.1and turning angle of 45°/45°. The proposed energy promoter was named as stream line sheet energy promoter. Computational Fluid Dynamics simulation was performed through commercial ANSYS-FLUENT 16.2 software. The measurements were made inside annular subsection, 45° from 360° of the complete annular shape of the diffuser, at Reynolds number 5.8×10^4 and turbulence intensity 4.1%. Results for the bare S-shaped diffuser (without energy promoters) showed the flow structures within the S-shaped diffuser were dominated by counter-rotating vortices and boundary layer separation especially in the outer surface. The combination of the adverse pressure gradient at the first bend of outer surface and upstream low momentum wakes caused the boundary layer to separate early. The combinations of proposed energy promoters were installed on the inner and outer surfaces at three installation planes. The use of energy promoters resulting in significantly decreased the outer surface boundary layer separation with consequential improving the static pressure coefficient and reduction of total pressure losses.

1 Introduction

Air intakes are main component of breathing aero-engines. The main task is to sustain high pressure recovery and reduce the inflow velocity to avoid presence of shock waves inside the diffuser part of the air intake. This part is strongly affecting the engine performance. Among the most commonly applied types of intakes is the S-shaped diffuser, which guides air from around a plane body to the engine inlet [1]. The S-shaped diffusers have centerline curvatures and cross-sectional areas that increase along the flow direction. Owing to this centerline curvature, cross-stream pressure gradients build up and increase secondary flows. This phenomenon creates cross-flow vortices within the boundary layer that generate a non-uniform pressure profile at the gas turbine engine face. This profile is called the aerodynamic inlet plane AIP. In addition, the adverse streamwise pressure gradient caused by the increasing cross-sectional area can also lead to flow separation, which is consider the major problem of S-shaped diffuser.

• Corresponding author: hussain_kayiem@utp.edu.my

To achieve acceptable performance, an S-shaped diffuser must incur minimal total pressure losses and facilitate almost uniform flow with small cross-flow velocity components at the AIP [2].

Many techniques have been used to improve the performance of S-shaped diffuser. One of the important technique is the passive flow control, vortex generators (VGs), which modifies the boundary layer fluid motion by bringing momentum from the outer flow region into the inner flow region of the wall bounded flow. Use of a set of VGs located on the outer and inner surfaces has been proposed for restructures the flow field [3]. Ref. [4] have been conducted experiments to study the effect of fences and VGs in terms of reducing exit flow distortion and improving total pressure recovery in two-dimensional, rectangular S-duct diffusers with different radius ratios. Ref. [5] investigated various low-profile VGs, counter-rotating vanes, single vane, forwards wedge, and backwards wedge referred to as sub boundary-layer vortex generators.

In the same direction, [6] investigated experimentally another type of VG named as Low-Profile Vortex Generators (LPVGs) to reduce the total pressure distortion in S-shaped diffuser. A low-profile wishbone type VG was used to improve the total pressure distortions and recovery performance of a diffusing S-duct by [7]. Ref. [8] used LPVGs within the first bend of this S-shaped intermediate turbine diffuser to study the influence of passive flow control devices on to separation. [9], have tested experimentally counter-rotating and co-rotating VGs configurations, installed on the casing of S-duct, on the flow development within an aggressive interturbine duct.

The objective of the present study is to investigate the contribution of different configurations of stream line EPs on the flow control in S-shaped diffuser by experiments through a test rig was set up for this purpose and numerically by three dimensional (3D) simulations by commercial ANSYS-FLUENT 16.2 software. Three proposed configurations of EPs have been investigated which have different numbers of EPs at three installation planes on the top and bottom surfaces of S-shaped diffuser. Total pressure loss coefficient, C_{TL}, static pressure recovery coefficient, C_{PR}, have been adopted as indicators to highlight the contribution of the EPs on controlling the flow and enhance the diffuser performance.

2 Experimental methodology

2.1 S-shaped Diffuser Model

The geometry of the S-shaped diffuser used in this study is shown in Figure 1. It was designed based on area ratio AR of 3.1. The radius of curvature, RC = 280 mm, the turning angle, β = 45°. The total length of the diffuser, L= 526 mm, and the hydraulic diameter, D_h = 65 mm. The modeled diffuser has a constant area conduit at the inlet with 65 mm long x 65 mm height, for smooth air inflow, and constant area conduit at the outlet with 65 mm long x125 mm height. The selected geometry of the model is representing 1/8 segment of the entire annulus. I.e., it is only 45° section of the 360° of the complete circular shape of the diffuser; is going to be named as annular subsection. Most if not all the previously reported studies, the investigator selected similar cross section, but with straight inner and outer surfaces. In the present work, the inner and outer surfaces have been simulated as curved inner and outer surfaces, which represent the actual case of the diffuser segment.

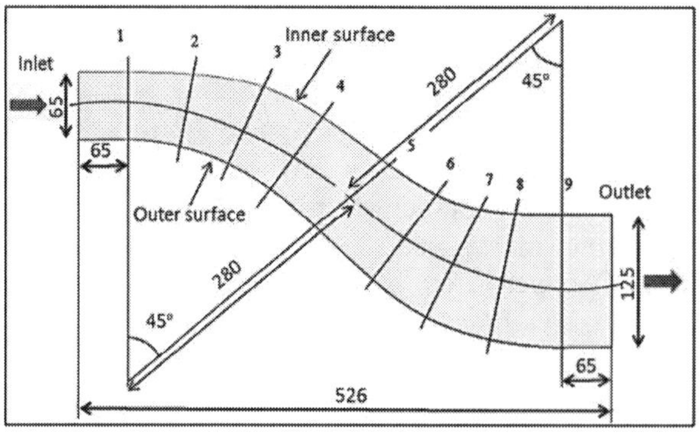

Fig. 1. Geometry of the modeled S-shaped diffuser and plane of interest (all dimensions in mm)

KIMO Multifunction AMI300 was connected with hot wire for measuring the velocity at these planes, also it has been used for measuring the turbulence intensity experimentally at the inlet plane. Also, it related to Pitot static tube for pressure measurements at 10 planes rated for ± 0.2 pa. At each plane, measurements are taken over a period of 10s which allows for the calculation of a mean value along with its standard deviation. The main parts of test rig are shown in Figure 2. The inlet velocity of 14.4 m/s and the Reynolds numbers of 5.8×10^4 obtained depending on the inlet hydraulic diameter of S-shaped diffuser. Also, the measured turbulence intensity by experiment at the inlet plane of S-shaped diffuser was 4.1%. These inlet velocity and inlet turbulence intensity have been used an inlet boundary conditions for CFD simulation.

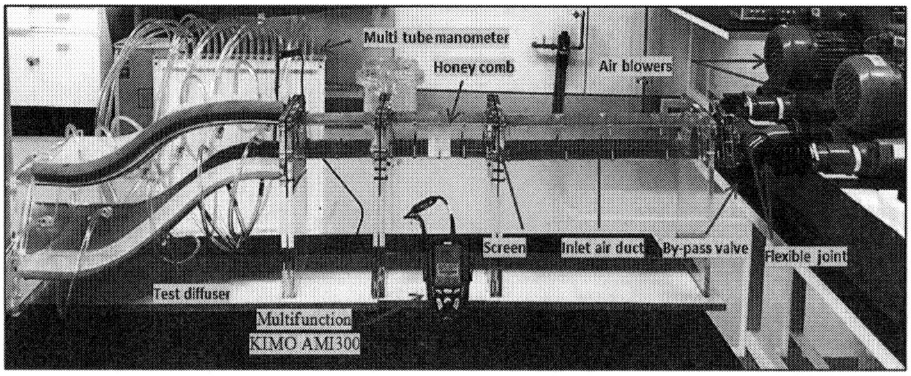

Fig. 2. S-shaped diffuser Test Apparatus

2.2 Proposed EPs Configurations

Three configurations of stream line sheet EPs were employed for elimination the flow separation region and delay separation starting. Also, to improve the flow pattern at the exit plane of the S-shaped diffuser by increase the static pressure recovery and reduce the total pressure loss. Figure 3 shows the investigated installation planes, while figure 4 shows the proposed EP geometry with the geometrical specifications. The EPs have been distributed on the inner and outer surfaces at plane 3,4 and 5 as bellow:

- At plane 3: 5 EPs on the inner surface and 7 EPs on the outer surface.
- At plane 4: 5 EPs on the inner surface and 7 EPs on the outer surface.
- At plane 5: 6 EPs on the inner surface and 8 EPs on the outer surface.
-

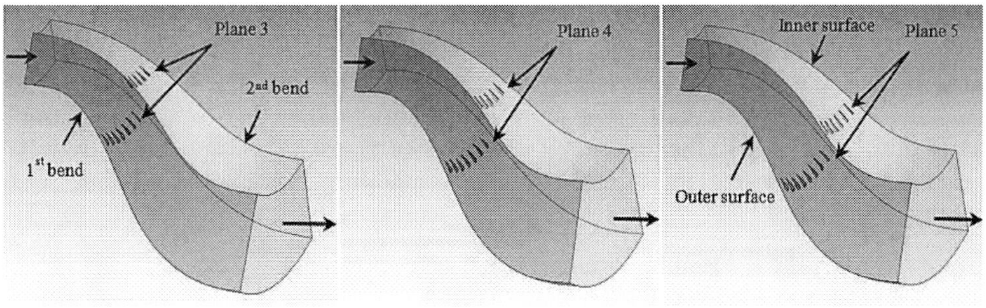

Fig. 3. Geometry of Co-rotating stream line sheet EP

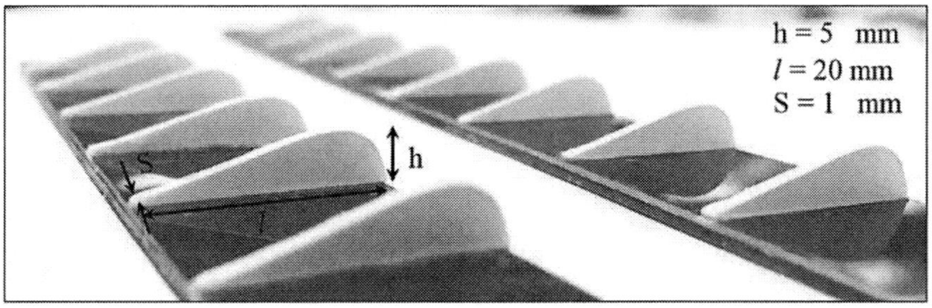

Fig. 4. Co-rotating stream line sheet EPs on the inner and outer surfaces of S-shaped diffuser at plane 3, 4 and 5.

3 Numerical methodology

This section provides information on the S-shaped diffuser grid generation, the governing equations, boundary conditions, mesh independency checking. Also, the validation of the turbulence models results with the experimental data has been presented.

3.1 Governing equations and boundary conditions

The governing equations for continuity and momentum, three–dimensional, steady state, Reynolds Average Navier Stokes (RANS) equations were numerically solved turbulent incompressible flows of air:
Continuity equation in coordinates (x, y, z)

$$\frac{\partial u}{\partial x}+\frac{\partial v}{\partial y}+\frac{\partial w}{\partial z}=0 \tag{1}$$

X - Momentum equation:

$$u\frac{\partial u}{\partial x}+v\frac{\partial u}{\partial y}+w\frac{\partial u}{\partial z}=-\frac{1}{\rho}\frac{\partial P}{\partial x}+v\left[\frac{\partial^2 u}{\partial x^2}+\frac{\partial^2 u}{\partial y^2}+\frac{\partial^2 u}{\partial z^2}\right]+\frac{1}{\rho}\left[\frac{\partial\left(-\rho\overline{u'^2}\right)}{\partial x}+\frac{\partial\left(-\rho\overline{u'v'}\right)}{\partial y}+\frac{\partial\left(-\rho\overline{u'w'}\right)}{\partial z}\right]$$

Y - Momentum equation:

$$u\frac{\partial v}{\partial x}+v\frac{\partial v}{\partial y}+w\frac{\partial v}{\partial z}=-\frac{1}{\rho}\frac{\partial P}{\partial y}+v\left[\frac{\partial^2 v}{\partial x^2}+\frac{\partial^2 v}{\partial y^2}+\frac{\partial^2 v}{\partial z^2}\right]+\frac{1}{\rho}\left[\frac{\partial\left(-\rho\overline{u'v'}\right)}{\partial x}+\frac{\partial\left(-\rho\overline{v'^2}\right)}{\partial y}+\frac{\partial\left(-\rho\overline{v'w'}\right)}{\partial z}\right]$$

Z - Momentum equation:

$$u\frac{\partial w}{\partial x}+v\frac{\partial w}{\partial y}+w\frac{\partial w}{\partial z}=-\frac{1}{\rho}\frac{\partial P}{\partial x}+v\left[\frac{\partial^2 w}{\partial x^2}+\frac{\partial^2 w}{\partial y^2}+\frac{\partial^2 w}{\partial z^2}\right]+\frac{1}{\rho}\left[\frac{\partial\left(-\rho\overline{u'w'}\right)}{\partial x}+\frac{\partial\left(-\rho\overline{v'w'}\right)}{\partial y}+\frac{\partial\left(-\rho\overline{w'^2}\right)}{\partial z}\right] \tag{2}$$

The flow field was predicted by solving the set of equations from (1) to (4). The 3D RNG k–ε model for simulating the turbulence quantities has been adopted. The enhanced wall treatment was selected to overcome with high complex three dimensional near wall flow phenomena of diffusers. In the present study, the first grid point of y+ ≈1.4 and the number cells around 25 cell within the inner layer of wall. The flow field was predicted by solving the set of equations from (1) to (4). The boundary conditions were used for simulations as summarized in Table1.

Table 1. Description of boundary conditions

Inlet boundary conditions		**Working fluid properties (air)**		**Outlet boundary conditions**		**Wall boundary conditions**	
Velocity inlet (m/s)	14.4	Density (kg/m^3)	1.1649	Pressure specified (Pa)	0	Share condition	No slip
Turbulence intensity	4.1%	Viscosity (kg/m.s)	1.85×10^{-5}				

3.2 Grid generation

The S-shaped diffuser used in the experiments has been modeled by modular design ANSYS 16.2, a grid generation by ANSYS - FLUENT 16.2. Figure 5 shows the structured mesh along the bare S-shaped diffuser (without EPs) and unstructured mesh of S-shaped diffuser with EPs. The S-shaped diffuser used in the experiments has been modeled by modular design ANSYS 16.2, a grid generation by ANSYS - FLUENT 16.2. Consistently with the experimental setup, a 65mm entrance length (upstream the S-shaped diffuser) and a 65mm long extension (downstream the S-shaped diffuser) are used in the numerical model.

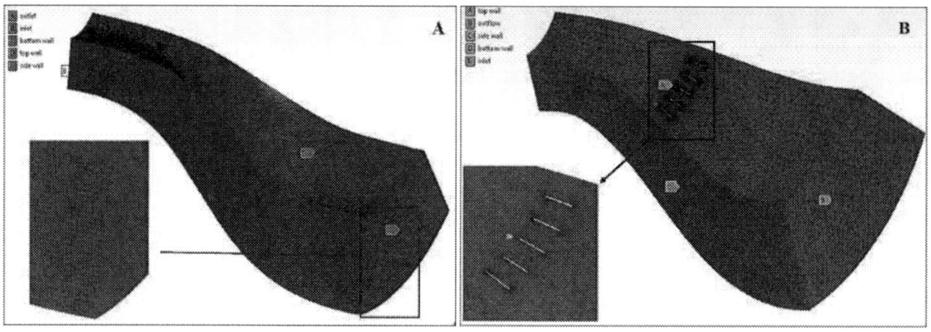

Fig. 5. Generated mesh of S-shaped diffuser: A. structured mesh without EPs, B. unstructured mesh with EPs.

3.3 Performance parameters

The main important parameters are using for evaluation the performance of S-shaped diffuser in the current study are:

Static Pressure Recovery Coefficient C_{PR}

It represents the magnitude by which kinetic energy has been converted into pressure energy due to diffusing action at any location along the S-shaped diffuser, it is represented by the equation:

 DOI: 10.1051/matecconf/201713102006

$$C_{PR} = \frac{P_s - P_{si}}{\frac{1}{2}\rho U_{avi}^2} \qquad (5)$$

Where: P_s static pressure (Pa)
P_{si} static pressure at the inlet plane (Pa)
U_{avi} average inlet velocity (m/s)

Total Pressure Loss Coefficient C_{TL}

It is a measure of how much total pressure is lost as a proportion of the mean inlet dynamic pressure due to viscous forces and turbulent mixing, this defined by the equation

$$C_{TL} = \frac{P_{ti} - P_t}{\frac{1}{2}\rho U_{avi}^2} \qquad (6)$$

where: P_{ti} inlet total pressure (Pa)
P_t total pressure (Pa)

4 Results and discussion

This section will be explaining the experimental and numerical results through the performance parameters C_{PR} and C_{TL}. The study cases were named as in Table2:

Table 2. Description of experimental and numerical study cases of S-shaped diffuser

Experimental Study		Numerical Study by CFD	
Name	Description	Name	Description
Exp.1	Bare S-shaped diffuser	CFD1	Bare S-shaped diffuser
Exp.2	EPs installed at plane 3	CFD2	EPs installed at plane 3
Exp.3	EPs installed at plane 4	CFD3	EPs installed at plane 4
Exp.4	EPs installed at plane 5	CFD4	EPs installed at plane 5

Figure 6-A shows the variation of static pressure recovery coefficient C_{PR} along the length of the S-shaped diffuser. The C_{PR} values that were computed experimentally by test rig measurements have been compared with those were computed numerically by CFD. The C_{PR} values, for CFD1, was increased in the initial phase up to plane 3 around $X/L \approx 0.35$, after that it decrease due to flow separation until flow reattachment occurs somewhere before plane 7 at $X/L \approx 0.68$ and then it increases steadily till the exit plane. For the experimental measurements, the trend of C_{PR} similar to the C_{PR} by CFD with some deviations could be observed in the location and length of the separated flow region. C_{PR} curve increase in the initial phase up to around $X/L \approx 0.3$ and then it decreases due to flow separation. The flow separation occurs early with experimental investigation, also flow reattachment occurs after plane 7. While, the values and the trend of C_{PR} were improved for both the experimental and numerical investigations with applying EPs on the inner and outer surfaces at the three installation planes of diffuser. Generally, the applying EPs at planes 3, 4 and 5 was improved the experimental and numerical values of C_{PR} of 28% and 30% respectively. The Exp.4 and CFD4 cases, give the best improvement of C_{PR} among the other cases, this improvement resulting from decrease the flow velocity and flow separation region especially on the outer surface as shown in Figure 6-B. The EPs were modified the boundary layer fluid flow by bringing momentum from the outer flow region into the inner flow region of the wall.

 DOI: 10.1051/matecconf/201713102006

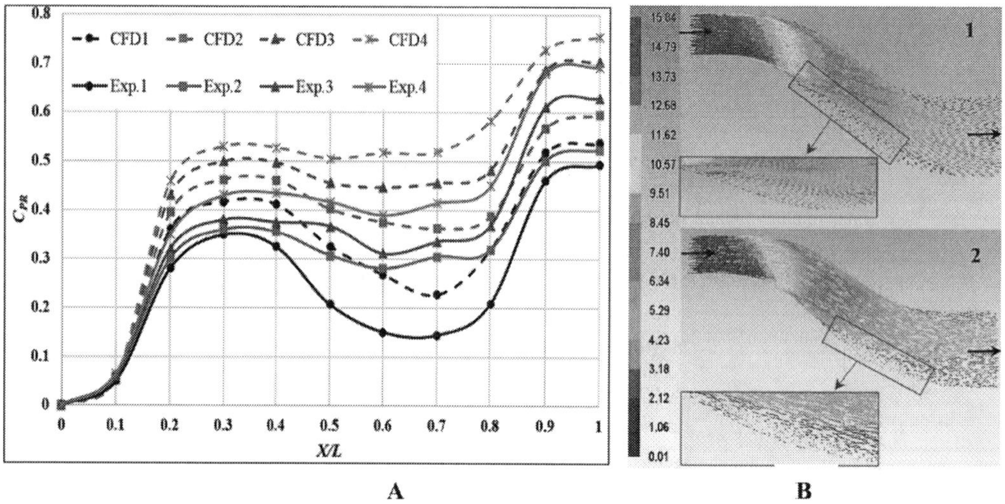

Fig. 6. (A)Variation of C_{PR} along the length of S-shaped diffuser, (B) Velocity vectors with separation regions of S-shaped diffuser: 1.Bare (without EPs), 2. With EPs

Figure 7-A shows that the C_{TL} decreases with the applying EPs on the inner and outer surfaces at planes 3, 4 and 5 as compared with bare S-shaped diffuser. Applying the EPs at plane 5 gives the best improvement in C_{TL} for CFD and experimental results of 18% and 11% as compared with bare S-shaped diffuser, respectively. The reduction in C_{TL} resulting from decrease of total pressure along the diffuser and separation region especially on the outer surface, as shown in Figure 7-B.

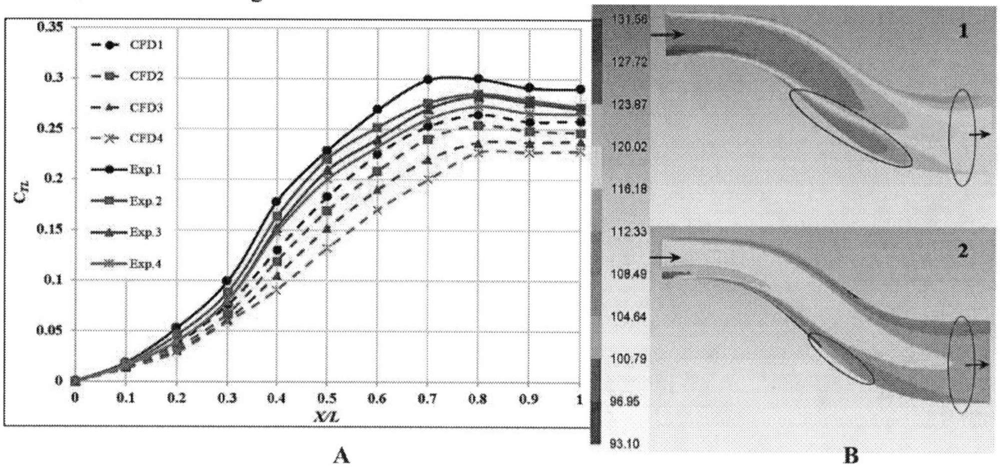

Fig. 7. (A) Values of C_{TL} along the length of bare S-shaped diffuser and S-shaped diffuser with EPs, B. Total pressure contours of S-shaped diffuser1.Bare (without EPs), 2. With EPs.

5 Conclusions

Experimental and numerical studies have been conducted on bare S-shaped diffuser (without EPs) and S-shaped diffuser with EPs. The following conclusions were drawn from current study:

- The results indicated that the adding energy promoters in planes 3,4 and 5 resulting in improve the S-shaped diffuser by increase the static pressure recovery and reduce the total pressure loss. Also, leading to delay the separation starting and eliminate the separation region.

- The highest performance enhancement when EPs installed at plane 5, the C_{PR} has been increased by 28%, the total loss coefficient C_{TL} reduced by 18% compared to the bare S-shaped diffuser.

The authors acknowledge Universiti Teknologi PETRONAS (UTP) - Malaysia for the logistic and technical support to produce this work. The first author appreciates UTP financial support of his PhD study under TFA scheme. Appreciations are due to Ministry of Higher Education – Malaysia (MOHE) for permitting the use of the measuring instruments purchased under FRGS grant FRGS/1/2017/STG02/UTP/01/1.

References

1. Madadi, A., M. J. Kermani, and M. Nili-Ahmadabadi. Journal of Engineering for Gas Turbines and Power **136**, no. 12, (2014)
2. Paul, A. R., Ranjan, P., Patel, V. K. and Jain, A. Aerospace Science and Technology **28**, no. 1, 332-343. (2013)
3. Anderson, B. H., Huang, P. S., Paschal, W. A., and Cavatorta, E. Journal of Propulsion and Power, Vol. **8**, No. 6, 1992, pp. 1266–1272. (1992).
4. Anderson, B. H., Huang, P.S., Paschal, W. A. and Cavatorta, E. Journal of Propulsion and Power **8**, no. 6, 1266-1272. (1992)
5. Ashill, P., J. Fulker, and K. Hackett. In 39th Aerospace Sciences Meeting and Exhibit, p. 887. (2001)
6. Anabtawi, A., Blackwelder, R., Lissaman, P. and Liebeck, R. In 37th Aerospace Sciences Meeting and Exhibit, p. 739. (1999)
7. Reichert, B. A, and Wendt B. J. In 31st Aerospace Sciences Meeting, p. 18. (1993)
8. Santner, C., GÃķttlich, E., Marn, A., Hubinka, J. and Paradiso, B. Journal of Turbomachinery **134**, no. 1, 011023. (2012)
9. Zhang, Y., Hu, S., Zhang, X. F., Benner, M., Mahallati, A. and Vlasic, E. Journal of Engineering for Gas Turbines and Power **136**, no. 11, 112604. (2014)

Free piston linear generator for low grid power generation

Izzeldin Abdalla[1], *Ezrann Z.* Zainal A.[2], *Nur Amalina* Ramlan[2], Firmansyah[2], *Abd. Rashid* A. Aziz[2,*], and *Morgan* Heikal[2]

[1]Department of Electrical and Electronic Engineering
[2]Centre for Automotive Research and Electric Mobility
[1,2]Universiti Teknologi PETRONAS, 32610 Bandar Seri Iskandar, Perak, Malaysia

Abstract. Generating power is of great importance nowadays across the world. However, recently, the world became aware of the climatic changes due to the greenhouse effect caused by CO_2 emissions and began seeking solutions to reduce the negative impact on the environment. Besides, the exhaustion of fossil fuels and their environmental impact, make it is crucial to develop clean energy sources, and efforts are focused on developing and improving the efficiency of all energy consuming systems. The tubular permanent magnet linear generators (TPMLGs) are the best candidate for energy converters. Despite being suffering problem of attraction force between permanent magnets and stator teeth, to eliminate such attraction force, ironless-stator could be considered. Thus, they could waive the presence of any magnetic attraction between the moving and stator part. This paper presents the design and analysis of ironless -cored TPMLG for low grid power generation. The main advantages of this generator are the low cogging force and high efficiency. Therefore, the magnetic field computation of the proposed generator has been performed by applying a magnetic vector potential and utilizing a 2-D finite element analysis (FEA). Moreover, the experimental results for the current profile, pressure profile and velocity profile have been presented.

1 Introduction

The climate change and fuel price both increase the stringent environmental policies have driven researchers to focus on finding technologies that are consistent with the environment, while the demand for electricity is forecasted to increase all over the world. One of the tools available to achieve the goal is to increase the share of renewable energy. Because of these commitments, the renewable energy industry is developing around the world and is one of the highest priorities of mankind. The new and renewable energy sources can guarantee a sustainable power production for the future energy needs [1, 2].

Most of the climatic phenomenon that occurred worldwide in the last few years that can be accounted is the global warming. The multiannual temperature reports indicated that the temperature is constantly rising. There is a close correlation between the temperature rise

* Corresponding author: rashid@utp.edu.my

and CO_2 as well as the other gasses emissions. Many gases besides CO_2 are accounted in different percentages for the greenhouse effect [3].

The free piston engine has a potential of minimizing of emission, as a result, more research has been undertaken in free piston engine to develop a more efficient system for vehicle propulsion and portable energy source application [2, 4]. The main problem concerning in electric vehicles is the performance of the batteries for the electric power storage system. Therefore, there are several solutions used to combine the power generated by the combustion engine and the electric motor inside a hybrid electric vehicle namely, parallel hybrid, series hybrid and parallel-series hybrid [3]. Notably, most of the systems use rotating generators and engines, therefore, linear electric generators connected directly to free piston engine is a viable solution. The vehicle traction is made by using two or four of electric motors mounted on the wheels. However, this system seems more complicated, but overall system efficiency is good [2, 3].

The free-piston linear generator engine (FPLG) offers many advantages over its conventional rotary counterpart. These features include high power-to-weight ratio and multiple fuel capability as well as the high efficiency with low emissions. This is achievable because of its free-piston characteristics whereby during combustion the piston stays a shorter time at top dead center (TDC) than a conventional crank engine. This reduces heat loss and lowers the combustion temperature. When the linear engine is made as a platform to convert mechanical to electrical energy through a particular arrangement of a linear generator for portable power generators for commercial and domestic use [4, 5]. Moreover, during the combustion stroke, the expansion of the free piston is faster compared with a conventional engine and the time spent around the TDC is shorter, meaning the speed is also higher around TDC [2, 3].

Figure 1 shows the cross-section of the major components of FPLG. This dual piston eliminates the need of a rebound device during functioning because the combustion stroke of one piston coincides with the compression stroke of the second piston. The external force that can be produced by the linear electrical machine needs to be utilized when the engine starts. The high vibration levels produced by such a system can be eliminated by using two or more systems that would work in anti-phase, requiring precise control [6-8]. This paper aims to design an appropriate linear generator for low grid power generation with low fuel. The idea is to use the TPMLG as the free-piston generator so that the mechanical energy can be directly converted to the electrical energy.

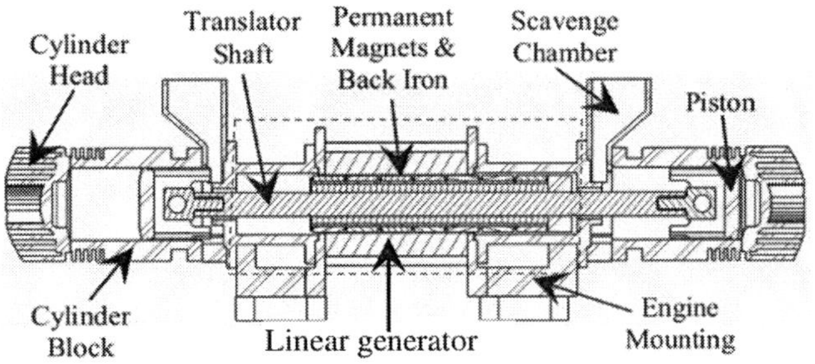

Fig. 1. Cross-section of the major components of free-piston linear generator engine [8]

2 Design of the linear electric generator

There are several linear electrical machines that could be used with a free piston engine, namely permanent magnet synchronous machines, permanent magnet variable reluctance machines, transverse flux machines, hybrid vernier machines, air-cored machines and moving coil machines. Each one of these technologies provide advantages and disadvantages that must be considered at the design stage, the moving coil structure has a light translator but uses sliding contacts or flexible connections that could reduce the system reliability, a moving magnet generator offers high air gap magnetic flux density and low moving mass but is sensitive to high temperatures near the combustion chambers and so on, and the air-cored structure eliminates most of the unwanted magnetic forces but requires a high magnetic flux densities to compensate the lack of the stator iron [6, 9].

In the selection of the linear generator for a free-piston engine power generation system, there are some aspects must be considered, namely the mover mass must be kept as low as possible because it has a negative influence on the entire mechanical system. Secondly, the proper insulation must be utilized to protect the permanent magnets because of very high temperature associated near the combustion chambers. Thirdly, rare-earth permanent magnets must be used to obtain high power densities that are resulting in a small size and easy way to place it inside the vehicle generating system. Lastly, for starting the internal combustion engine, the electrical machine must be able to work as a motor, thus the developed force must be high enough to compress the gas inside the cylinder for the first stroke.

However, the focus of this research is on a linear generator with permanent magnets that mounted on the moving part. The proposed design consists of a slotted core carrying six coils forming the stator part, and then another part is the moving part. The permanent magnets are quasi-Halbach magnetized are made of NdFeB. The shaft of the generator is made up of a nonmagnetic material, thus the leakage through that area is a magnetic field in the air gap.

When designing the generator some inputs were considered, such as rated power, rated voltage, rated efficiency, the linear velocity of a free piston engine, permanent magnet type, and others. Once the geometric dimensions of the linear generator were determined the structure was simulated using ANSYS, FEA based magnetic field computation software [8]. To reduce the simulation times only half of the generator was modeled and periodic boundary conditions were imposed. The optimal design of the linear generator used is a very important issue in the development of the entire power generation system. Based on Newton's second law of motion, the dynamic mechanical equation of linear generator is expressed as [5, 6]

$$\sum F_x = ma_x = m\frac{d^2z}{dt^2} \tag{1}$$

where m is the total mass of the moving part, including the shaft, pistons and magnets, and a is the acceleration. During the starting the motion along a single axis only (z), ignoring vibration along the other axes. Incorporating the mechanical forces above, therefore the following equation could be utilized to represent linear generator during starting for movement from TDC to BDC

$$m\frac{d^2z}{dt^2} = F_{mot} + F_{\exp} - F_{comp} - F_{cog} - f \tag{2}$$

$$= F_{mot}(x) + F_{\exp}(x) - F_{comp}(x) - F_{cog}(x) - f \tag{3}$$

Generally, the generators may be described by an equivalent electric circuit diagram, thus, can be simplified to *RLC* circuit. Therefore, the voltage equation of the linear generator is written as [6]

$$Ri + L\frac{di}{dt} + V = e \tag{4}$$

From the Faraday's electromagnetic law, e is expressed as

$$e = -\frac{d\phi}{dt} \tag{5}$$

By assuming that the flux-linkage is a function of z

$$e = -\frac{d\phi}{dt} = -\frac{d\phi}{dz} \cdot \frac{dz}{dt} \tag{6}$$

2 Analysis model of the linear generator

Based on the previous experiences and the given requirements a tubular permanent magnet linear generator (TPMLG) was designed for a low grid power generation. Then the field computation is performed by adopting a magnetic vector potential and a 2-D finite element analysis (FEA), thus, the electromagnetic performances of the proposed generator are simulated and analyzed.

Figure 2 shows the 3-D structure of the TPMLG and its main structure, that includes six coils and a mover consists of permanent magnet rings fixed on a linear shaft. The magnet topologies consist of quasi-Halbach magnetized PMs to create a magnetic flux path. Meanwhile, in the 2-D structure given in Figure 3 six coils are coupled together forming a single phase of the generator.

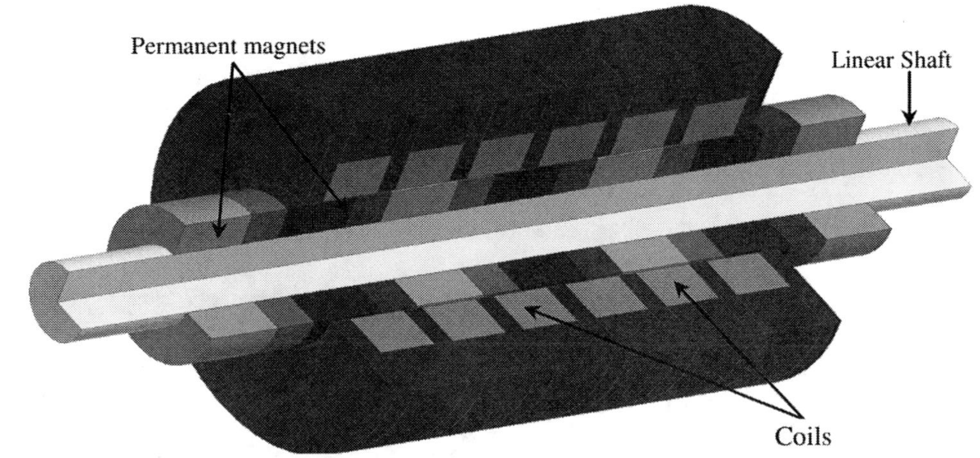

Fig. 2. The 3-D structure of the tubular permanent magnet linear generator

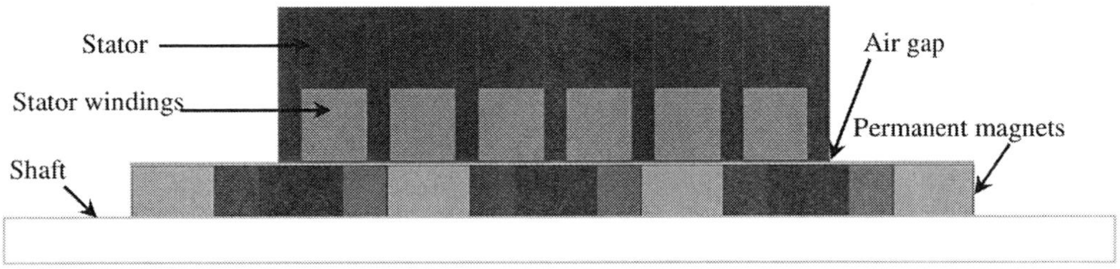

Fig. 3. The 2-D structure of the tubular permanent magnet linear generator

A 2-D axisymmetric finite element (FE) model of the generator was developed using commercial ANSYS software. The simulation results obtained are shown in Figure 4 is the magnetic flux plots in two positions of the mover, initial position and maximum stroke as shown in Figure 5. When the generator excited by the permanent magnets only, hence, to emphasize this statement, the shaded plot of the magnetic flux density distribution in the generator is presented in Figure 6, where high magnetic flux density values are noticed in the translator poles and stator teeth, while the average air gap magnetic flux density is 1.33 T. As it can be observed the flux density values are very small in the other parts of the machine.

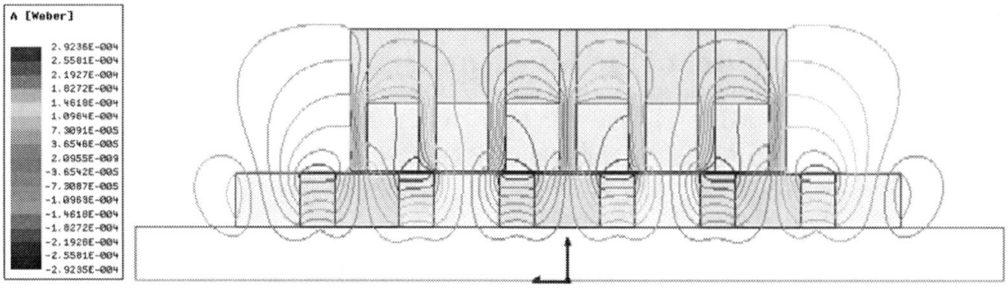

Fig. 4. The magnetic field lines obtained when the translator at the initial position

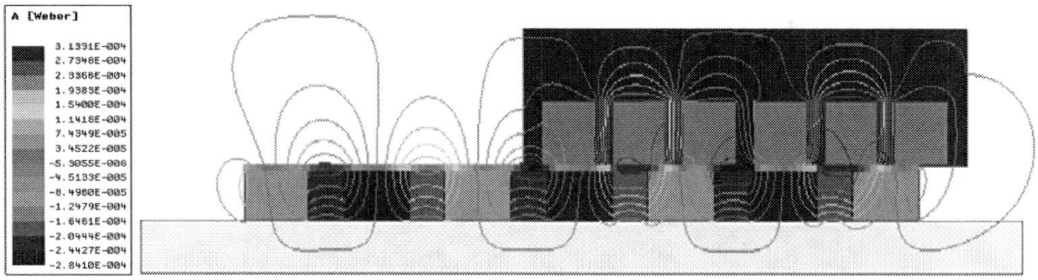

Fig. 5. The magnetic field lines obtained when the translator at the maximum stroke

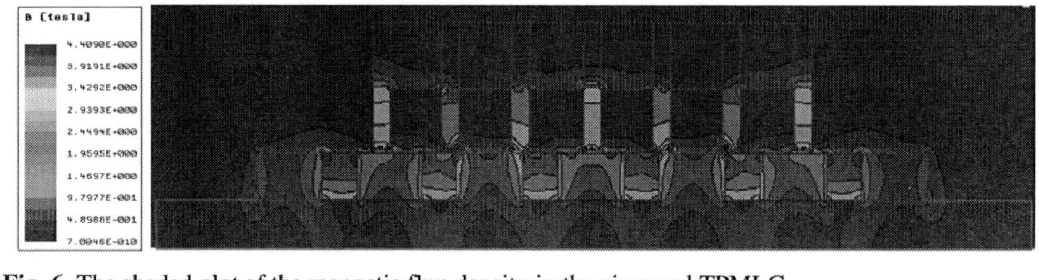

Fig. 6. The shaded plot of the magnetic flux density in the air-cored TPMLG

Figure 7 and Figure 8, respectively, show the distribution of magnetic flux and flux density produced by the PMs at the center of the air gap. The commercial software package ANSYS Maxwell is employed for non-linear 2-D FEA.

By moving the armature with the permanent magnets, a varying magnetic flux will pass through the windings, generating EMF. The main characteristics of the linear generator in this study were obtained by using FEA based numeric field computations, when the speed of the generator is constant. Therefore, for a dynamic simulation at the average constant linear speed of 6.0 m/s the obtained flux-linkage is shown in Figure 9 is plotted together with flux-linkage of the iron-cored generator with short-dash as can be observed, the iron-cored generate higher flux-linkage than the ironless-cored. The induced no-load EMF and the winding inductance are shown in Figure 10 and Figure 11, respectively. As can be

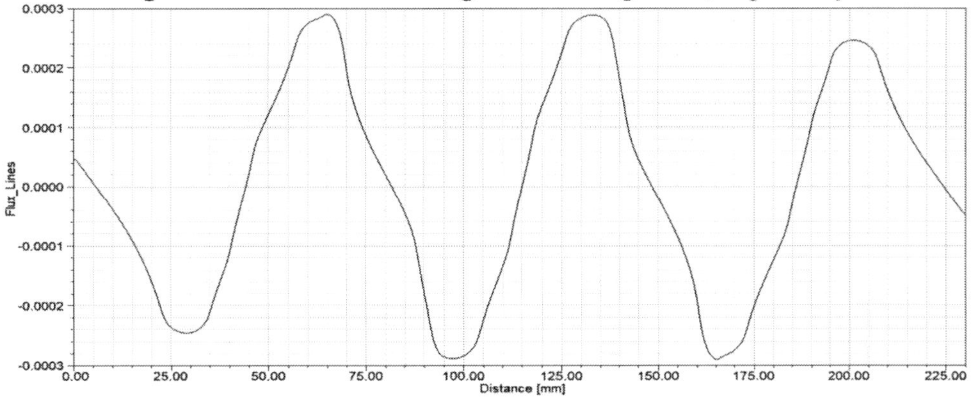

Fig. 7. Magnetic flux waveform of the air gap

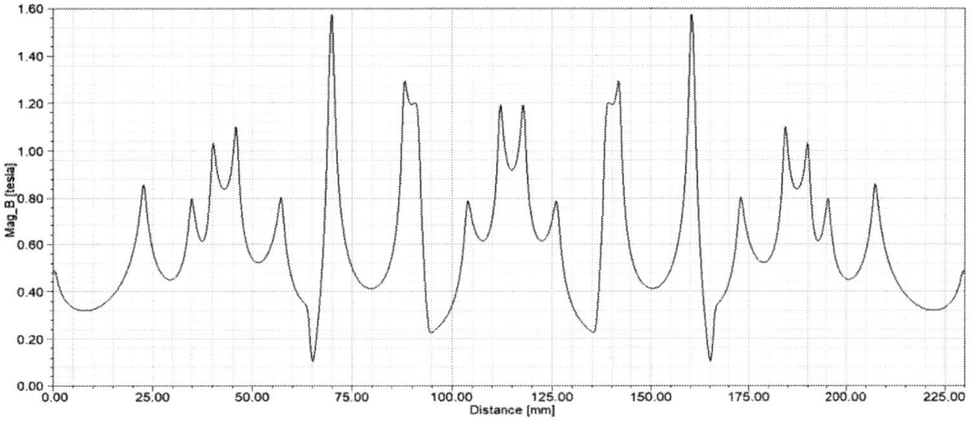

Fig. 8. Flux density waveform of the air gap

seen, also the iron-cord generator has a higher EMF and winding inductance and these indicate that it will posses higher power generation. For this type of linear generator, the effect that can be observed, only a small part of the magnetic flux generated by the magnets passes through the coil of the generator, hence EMF will be produced in its winding.

DOI: 10.1051/matecconf/201713102007

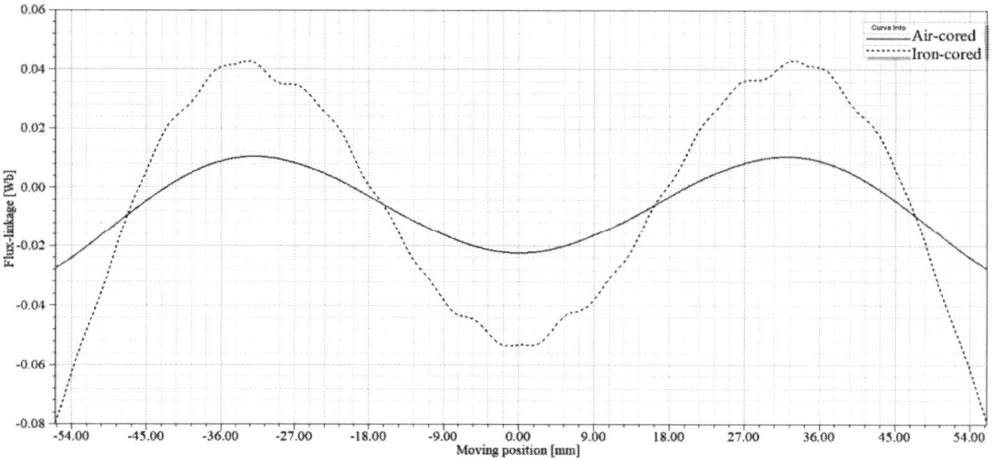

Fig. 9. Flux-linkage of air-cored and iron-cored TPMLG

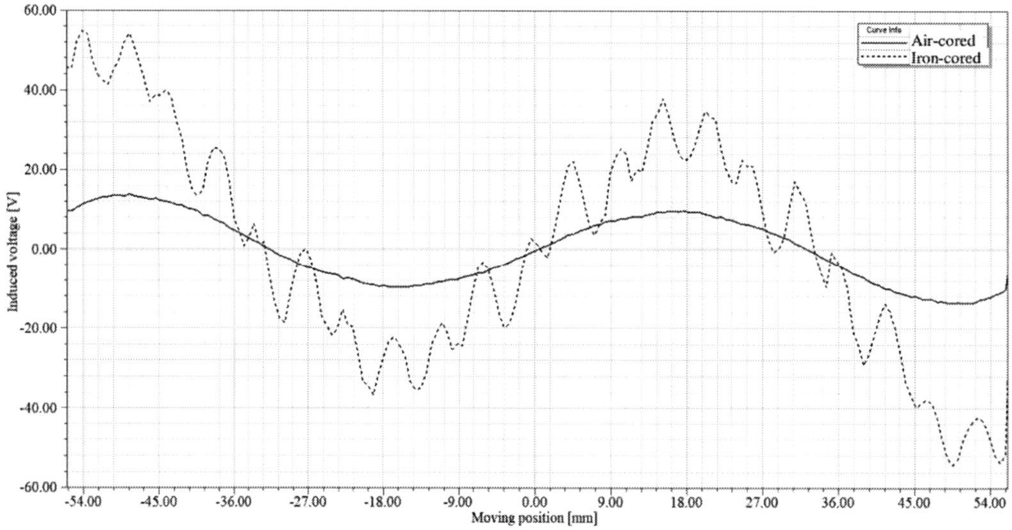

Fig. 10. Induced no-load EMF of air-cored and iron-cored TPMLG

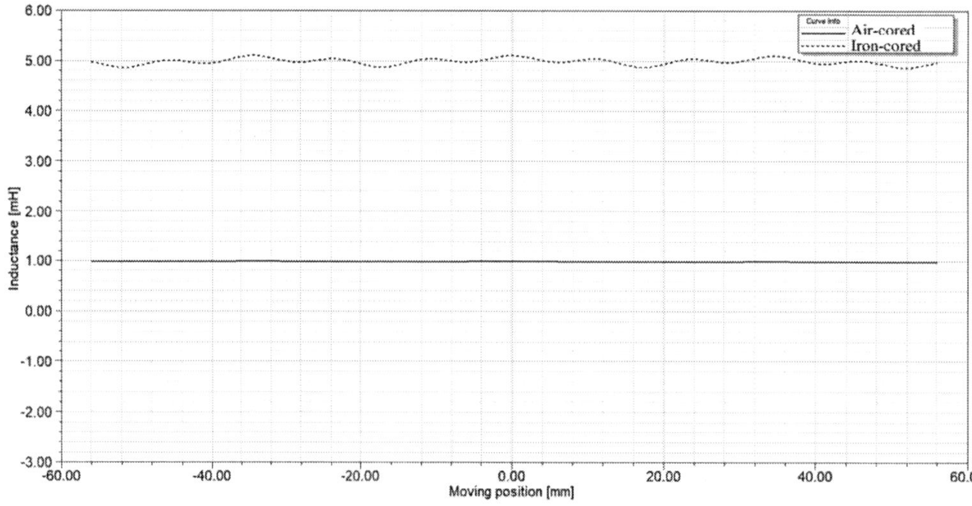

 DOI: 10.1051/matecconf/201713102007

Fig. 11. Winding inductance of air-cored and iron-cored TPMLG

The thrust force provides power generation it resulted related to the translator positions when the PMs field acts in the direction of motion, whereas the interaction of PMs with stator produces an attractive force it called cogging force. Compared this force to the thrust force, this force is an order of magnitude lower. Finite element results for the proposed generator indicate that this force expressed as a stress and clearly justifies the investigation of air-cored machines in terms of simplified design and reduced mechanical structure as well as the minimization of the cogging force as clearly observed from Figure 12 the solid line is for air-cored TPMLG.

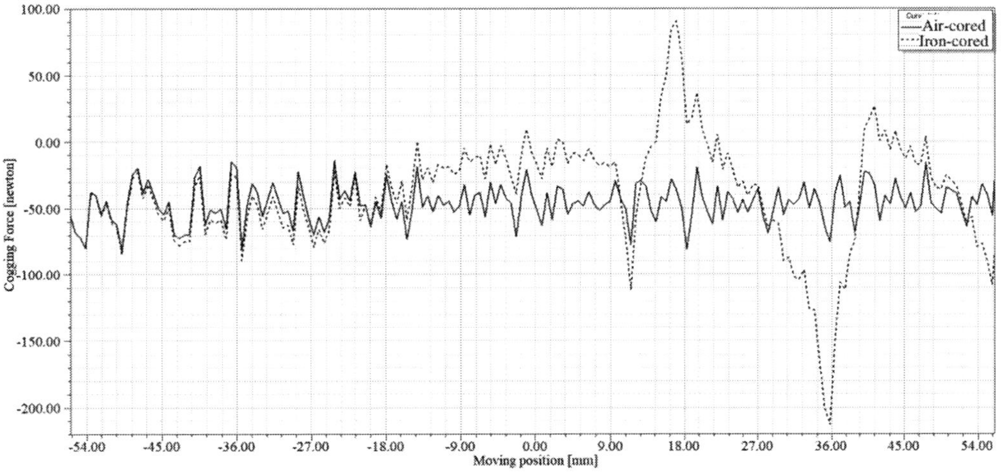

Fig. 12. The cogging force of air-cored and iron-cored TPMLG

The experiment was performed on a two-stroke, direct injection, and free-piston linear generator engine with two opposite combustion chambers as shown in Figure 13. The free-piston linear engine-generator (FPLG) involves a freely moving piston and a translator assembly between the two oppositely placed engine cylinders for combustion. The mover of FPLG is connected to the piston rods through flexible joints. This mover executes a reciprocating linear movement because of the gas pressure difference between the top of the two free pistons. The kinetic energy of the mover can be converted into electrical energy by the TPMLG. When the FPLG is an intake-expansion stroke, the piston moves from top dead center (TDC) to bottom dead center (BDC). The machine is operated as a brushless linear machine to produce the required motion. However, due to the very large peak compression force during the starting, thus, the limited current rating of stator coils and insufficient force have resulted in not possible to push the translator end to end in a single stroke. Hence, due to repeated compression-expansion of the engine cylinders and constant application of motoring force in the direction of natural bouncing motion, the amplitude and speed of the translator are expected to grow due to mechanical resonance.

DOI: 10.1051/matecconf/201713102007

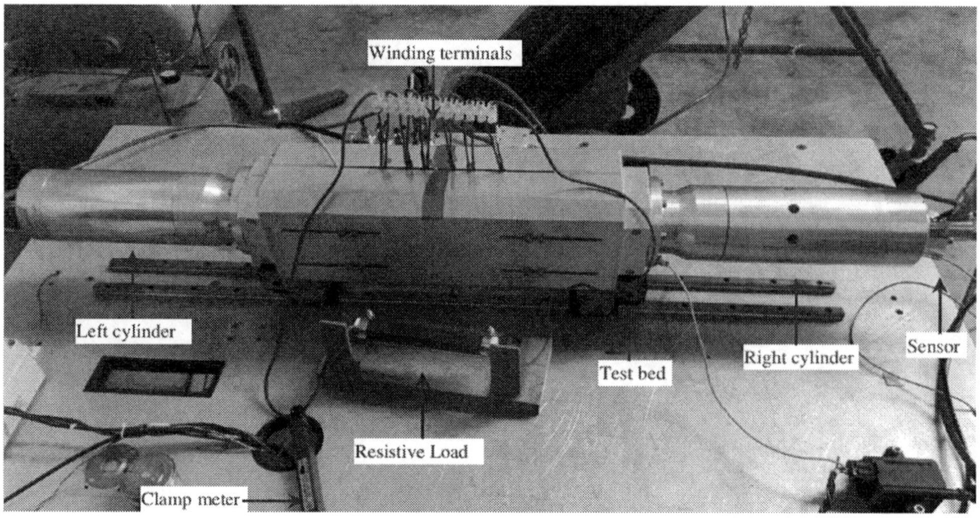

Fig. 13. Prototype of free-piston linear generator engine on the test bed

The TPMLG is the energy generation device in the FPLG, and thus, its appropriate selection at the conceptual design stage is crucial [3]. The Free piston internal combustion engines offer better dynamic profiles and can work at variable compressed ratio as well as has a less pollution. Figure 14 (a) and (b) shows pressure versus displacement of the left and right cylinder of the engine. It shows that the pressure in the left and right cylinder are not equal in both configurations. This can be attributed to the fact that there is a leakage in the left cylinder. The maximum pressure is observed in the left and right cylinders. Furthermore, the investigation showed that the compression profile is higher than the expansion profile. Hence, the region under the curve is a negative work due to the work was done by motoring force applied to the piston. This shows that the energy is supplied to the engine rather than extracting from it. Figure 15 shows the current profile versus time, whereas Figure 16 shows the current profile versus displacement, as can be observed, the maximum current obtained is around 9.0 A.

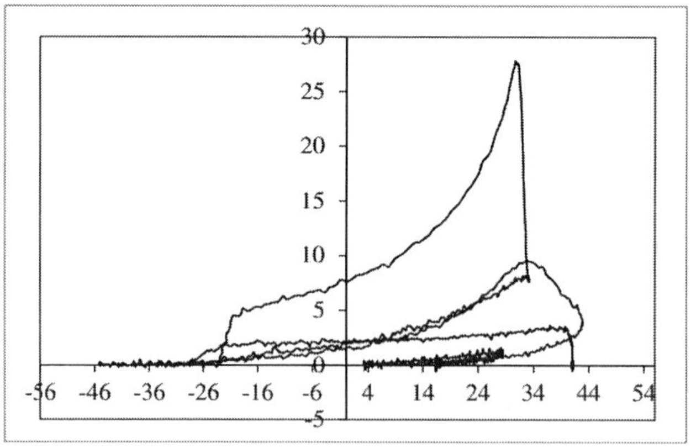

(a) In-cylinder pressure 1

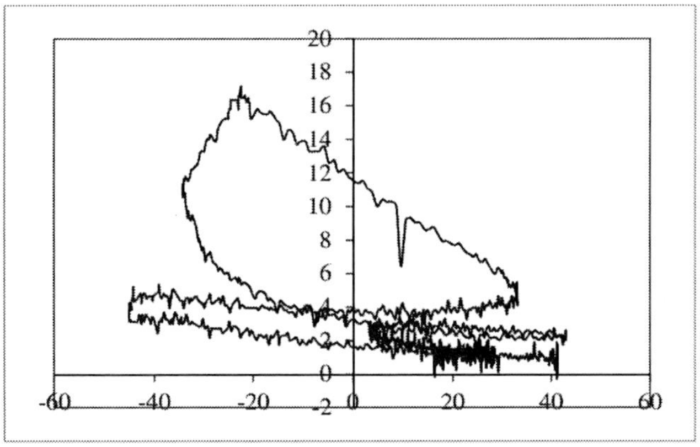

(b) In-cylinder pressure 2

Fig. 14. Engine compression pressure versus axial displacement

When the piston speed falls below the cut-off speed, power generation is stopped so reducing the electrical braking on the translator. At this speed, the voltage generated by the linear generator will be higher. Figure 17 shows the cut-off speed, the high speed will increase the compression ratio while lower speed will reduce it. The highest recorded piston velocity at 6.0 m/s, followed by 2.0 m/s. The velocity profile at both ends represents an expansion and compression profile whereby piston velocity decrees towards the end because of resistance during compression stroke and change of direction.

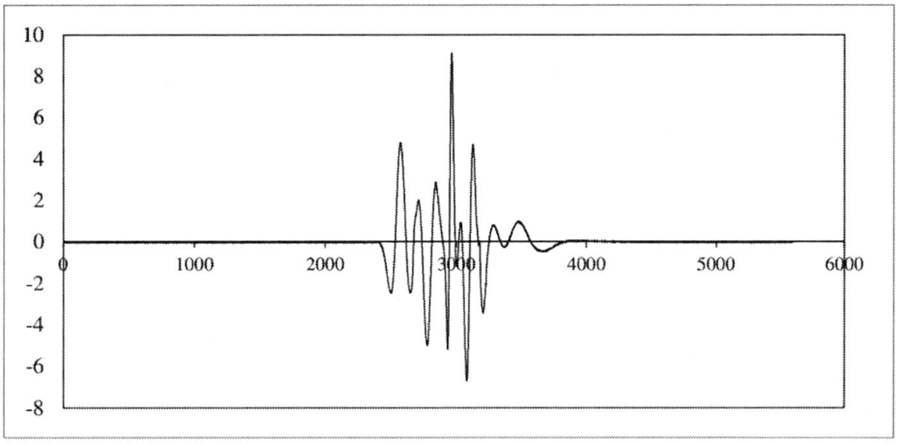

Fig. 15. Current profile versus time

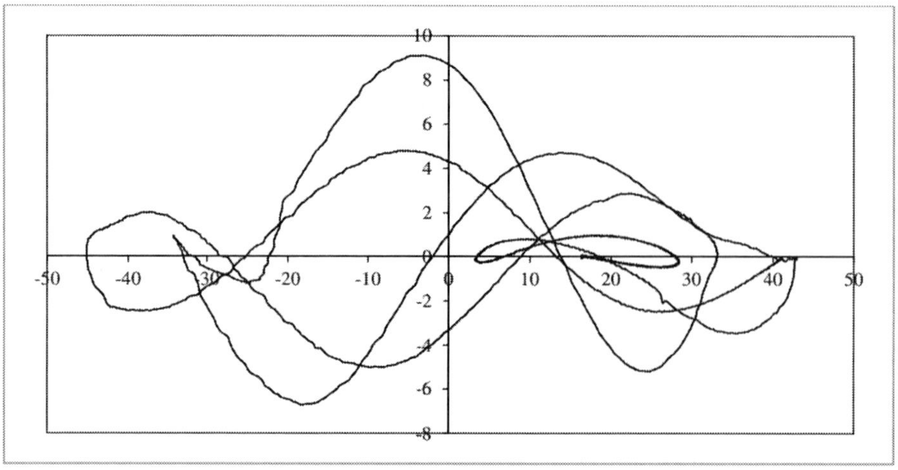

Fig. 16. Variation of current against displacement of the translator

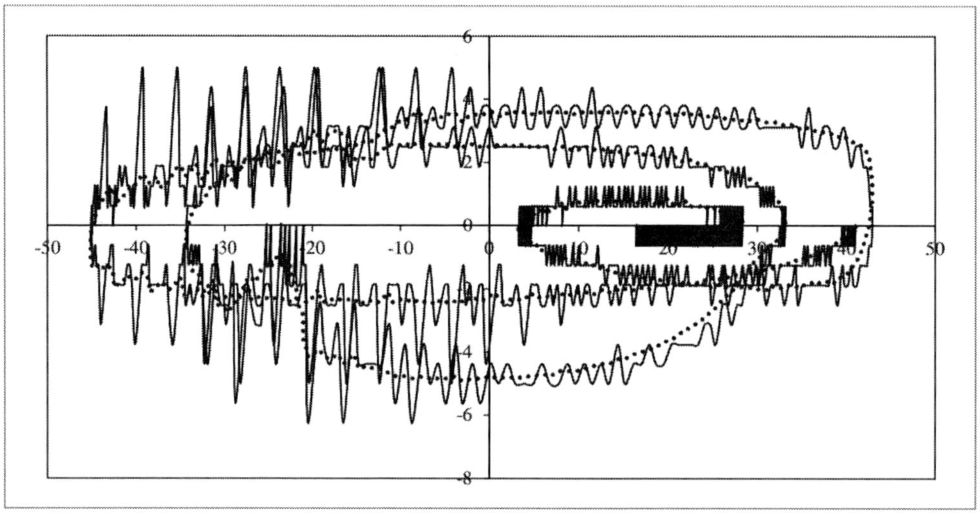

Fig. 17. Piston dynamics of a free-piston engine, piston velocity profiles versus displacement

3 Conclusion

The use of electric linear generators driven by internal combustion free piston engines offers better performance compared to conventional systems for clean energy converters. However, a tubular permanent magnet linear generator (TPMLG) for low grid power generation has been proposed and analyzed. This kind of linear generators has superior performance, nevertheless, it has a problem of high magnetic attraction force between the permanent magnets and the magnetic stator teeth. However, to eliminate this attraction force, this paper presents the design and analysis of ironless-stator single-phase TPMLG. It waives the use of any steel in a stator, therefore, the attraction force between the moving and stator part can be minimized as much as it is possible. The ironless-stator TPMLG has a low cogging force, despite, its power density is low. The magnetic field computation of the proposed generator was performed by applying a magnetic vector potential and a 2-D finite element analysis (FEA) and also some experimental results have been presented.

Authors gratefully would like to thank Yayasan Universiti Teknologi PETRONAS (UTP) (0153AA-A92) and the Petroleum Research Fund (0153AB-A34) for funding this research work.

References

1. L. Szabo and C. Oprea. Oradea University Annals pp. 120-125 (2006)
2. B. Jia, R. Mikalsen, A. Smallbone, Z. Zuo, H. Feng, and A. P. Roskilly. App. En., **179**, pp. 1166-1175 (2016)
3. J. Chen, Y. Liao, C. Zhang, and Z. Jiang. Design and ananysis of a permanent magnet linear generator for a free-piston energy converter. *IEEE 9th Conference on Industrial Electronics and Applications*, pp. 1719-1723 (2014)
4. Z. A. Ezrann Z., A. I. Abdulwehab, A. A. A. Rashid, Saiful A., and Zulkifli. J. of Mech. Eng. and Sci., **8**, pp. 1393-1400 (2015)
5. S. A. Zulkifli, M. N. Karsiti, and A. R. A. Aziz. Starting of a free-piston linear engine-generator by mechanical resonance and rectangular current commutation, *IEEE Vehicle Power and Propulsion Conference*, pp. 1-7 (2008)
6. H. Feng, Y. Song, Z. Zuo, J. Shang, Y. Wang, and A. P. Roskilly Energies **8**, pp. 765-785 (2015)
7. C. Oprea, L. Szabó, and C. Marţiş, Linear Permanent Magnet Electric Generator for Free Piston Engine Applications, *XXth International Conference on Electrical Machines*, pp. 691-696 (2012)
8. A. E. Z. B. Zainal. Int. J. of Aut. Eng. **4**, pp. 47-53 (2013)
9. I. I. Abdalla, T. Ibrahim, and N. B. M. Nor. Int. J. of Elec. Power & Energy Sys. **77**, pp. 263-270 (2016)

DOI: 10.1051/matecconf/201713102008

Design and modeling of solar parabolic trough power plant with MATLAB

Sanan T. Mohammad[1], *Hussain H.* Al-Kayiem[1*], *Morteza Khalaji* Assadi[1], *Syed I. U.* Gilani[1], *Ayad* K. Khlief[1]

[1]Mechanical Engineering Department, Universiti Teknologi PETRONAS, 32610 Bandar Seri Iskandar, Perak, Malaysia

Abstract. With the fact that Malaysia is one of the fast- growing countries, demand of energy increment is rapid. Malaysia is able to obtain ample amount of annual solar radiation due to its location at equator. If this is utilized proficiently and effectively, then, it can suffice the domestic needs as well as the industrial needs in terms of energy consumption. This article proposes a parabolic Trough Power Plant which is designed with 1.2 kW net electric output. Consequently, the results of theoretical calculations are detailed in the article, while, ensuring the analysing of design proposed through the MATLAB software. The results showed that by making use of aperture having an area of approximately 80 m2, maximum useful heat gain of 20701W at 13:00 pm was attained in March. The maximum net power is 11.84 kWh/day in February.

1 Introduction

The solar thermal power generation is closely similar to traditional ways of producing power based on fossil fuel combustion, which is also relying on the heat engines for conversion of energy to electricity. The implementation and experimentation begun in US, when the oil crisis occurred during 1970s and the first commercial plants were implanted at California in the late 1980s [1]. Due to compelling effect of global warming and volatile gas prices, both interest and investment are accelerated in solar thermal power technology, quickly. From the last two decades, investment is the biggest reason which lacked for the solar thermal power technology. If the investment were sufficient, this source of electrical power production will be the biggest source of electricity in economical way. It shows that there is a strong chance that this technology is one of the main stream of producing electricity, along with hydro, wind, and solar photovoltaic technologies; which is a key renewable energy for the future [2].

Solar energy is inexhaustible, locally available and renewable source of energy. It allows a local energy independence due to its clear and clean energy source. As the sun's power of reach earth at the annual rate of approximately 1000 W/ m^2 [3], it still varies its availability with the location and time of year. In order to capture the solar energy, it requires equipment with demands a high initial capital cost.

*corresponding Author: hussain_kayiem@utp.edu.my

DOI: 10.1051/matecconf/201713102008

Nevertheless, certain cases were reported where a lifetime of the solar equipment is mostly to state that these systems are able to prove a cost competitive market, when it is compared with the conventional energy technologies [4]. The solar thermal power technologies are distinguished to concentrate solar radiation by systems, such as, (1) parabolic trough, (2) solar tower, and (3) solar dish. The direct radiation from the sun is concentrated with the use of reflectors and the concentrated energy is then transformed into steam, which can be used to drive conventional electricity generators [5]. Therefore, this study explores the conceptual design and performance analysis of solar thermal power plant using Direct Steam Generation (DSG) criteria in a parabolic trough solar field by using MATLAB software. The location of the study is in Universiti Teknologi PETRONAS, Perak Malaysia. This study also proposes case study which assumes the total output of solar thermal power plant at 1.2 kW.

2 Mathematical modeling

A basic Rankine cycle with a variety of reference values and inputs are incorporated in the design shown in Figure 1. Firstly, there is a different state points which are determined through a theoretical concept by analysing the heat transfer that occurs in each component of the cycle. Consequently, these state points are then used to obtain steam flow rate, condenser water flow rate and other related values.

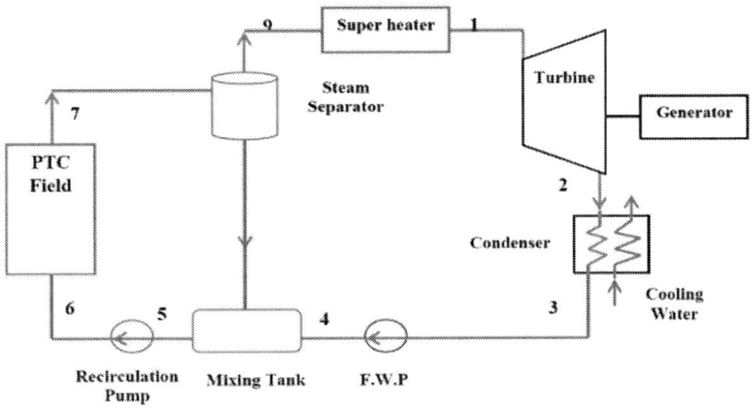

Fig. 1. Simplified process flow diagram a PTC based solar thermal power plant.

This system present mathematical modeling, where the energy analysis of the PTC is referred to the equations presented in Refs [6, 7, 8]. In presentation, the modeling of solar system is displayed first. Next, the performance assessment equations to represent the entire system are presented, respectively. For the calculations, assumptions are made. One of the assumptions is that the system is at a steady state. Other than that, the pressure change is negligible except in the turbines and pumps. The equation to define useful energy rate from the collector are:

$$Q_{gain} = \dot{m}_{CL}\ (h_7 - h_6) = \eta . I . A_p\ (1); \quad \eta = \eta_o - U_L . \left(\frac{\Delta T}{I}\right)\ (2); \quad \Delta T = T_m - T_a\ (3)$$

Where: m_{CL} is the mass flow rate of steam passing through Collector (kg/s), I is direct normal incidence radiation (W/m^2, A_P is aperture area of the collector (m^2), $T_m = (T_6 + T_7)/2$, denotes mean temperature (°C), Ta is ambient temperature (°C), and U_L is loss co-efficient based on aperture area (W/m^2.K). If the work input Feed Water Pump (*F.W.P*) and Recirculation Water Pump (*RC.W.P*) as well as heat losses through piping are neglected then:

$$Q_{gain} = \dot{m}_{CL}\ (h_7 - h_6) = .(h_1 - h_3) \quad (4)$$

From Eqs. (1) and (4):

$$\dot{m}_{CL}\,(h_1 - h_3\,) = \eta\,.I.A_p \qquad (5)$$

It may be noted that the PTC inlet mass flow rate (m_{cl}) will generally remain fixed during the operation and as a result the PTC outlet dryness fraction as well as mean temperature difference (ΔT) will vary at part load. The value of ΔT can be calculated by neglecting the pump work and assuming design point outlet dryness fraction of the PTC field (x_7). The value of PTC field inlet temperature (T_6) can be calculated from inlet pressure (P_6) and inlet enthalpy (h_6) which could be obtained from:

$$h_6 = (1 - x_7).\,h_8 + x_7.\,h_3 \qquad (6)$$

It may be noted that h_i and T_i denote the enthalpy and temperature at *i*-th state point, respectively. η_o is optical efficiency which is defined as:

$$\eta_o = \rho_c\,\gamma\,\tau\,\alpha\,K_\gamma \qquad (7)$$

where ρ_c, γ , τ , α and $K_\gamma$are the reflectance of the mirror, intercept factor, transmittance of the glass cover, absorbance of the receiver, and incidence angle modifier, respectively and their values are given in Table 1. The aperture area is defined as:

$$A_p = (w - D_{co}\,)L \qquad (8)$$

Where L, w, and D_{co} are collector (module) length, collector width, and receiver cover outer diameter, respectively. The solar collector heat loss coefficient between ambient and receiver is defined as:

$$U_L = \left[\frac{A_r}{(h_{c,ca}+h_{r,ca})A_C} + \frac{1}{h_{r,cr}}\right]^{-1} \qquad (9)$$

The radiation heat coefficient between ambient and the cover is defined as:

$$h_{r,ca} = \varepsilon_{Cv}\,\sigma\,(T_c + T_a)(T_c^{\,2} + T_a^{\,2}) \qquad (10)$$

Where ε_{Cv} is the emittance of the cover and σ is Stefane-Boltzmann constant. The radiation heat coefficient between the receiver and the cover is defined as:

$$h_{r,cr} = \frac{\sigma\,(T_c + T_{r,av})(T_c^{\,2}+T_a^{\,2})}{\frac{1}{\varepsilon_r}+\frac{A_r}{A_c}\left(\frac{1}{\varepsilon_{cv}}-1\right)} \qquad (11)$$

Where ε_r is the emittance of the receiver and the subscript *av* is refers to average. The convection heat loss coefficient between the cover and ambient is defined as:

$$h_{c,ca} = \left(\frac{Nu\;k_{air}}{D_{c.o}}\right) \qquad (12)$$

Where Nu is the Nusselt number and k_{air} is the thermal conductivity of the air. The subscript r refers to the receiver. The temperature of the cover can be calculated using this equation:

$$T_c = \frac{h_{r,cr}\,T_{r,a}+(h_{c,ca}+h_{r,ca})\,To\,\frac{A_C}{A_r}}{h_{r,cr}+(h_{c,ca}+h_{r,ca})\,\frac{A_C}{A_r}} \qquad (13)$$

The amount of the solar radiation that shines upon the collector, considered as heat into the system, is defined as:

$$Q_{solar} = m_{cl}.\,(h_7 - h_6) + m.\,(h_1 - h_9) \qquad (14)$$

The performance equations of the overall system are presented next. The power produced by the steam turbine is defined as [8, 9]:

$$W_{st} = \dot{m}\,(h_1 - h_2) \qquad (15)$$

Where h is the enthalpy and the subscript st indicates steam. The net power of the steam Rankine cycle is defined as:

$$W_{cyc} = \eta_g\,W_{st} - (W_{F.W.P} + W_{Rc.P.W}) \qquad (16)$$

$W_{F.W.P}$ is power for feed water pump and $W_{Rc.P.W}$ is power for Recirculation pump water are defined as:

$$W_{F.W.P} = \dot{m}\,(h_4 - h_3) = \dot{m}\,\vartheta\,(P_4 - P_3) \quad (17)$$

$$W_{Rc.P.W} = \dot{m}\,(h_6 - h_5) = \dot{m}\,\vartheta\,(P_6 - P_5) \quad (18)$$

The net electrical efficiency for the steam Rankine cycle system is defined as:

$$\eta_{el} = \frac{W_{cyc}}{Q_{solar}} \quad (19)$$

The considered PTC specifications and characteristics are shown in Table 1.

Table 1. Characteristics of PTC used for simulation

Parameters	Value
Length of the receiver (L)	6.4 m
Width of the collector (W_a)	2.6 m
Focal length (F)	0.87 m
Receiver external diameter (Dao)	40 mm
Receiver internal diameter (Dai)	38 mm
Concentration ratio (C)	20
Rim angle (°)	72.5

3 Result and discussion

The collection of data was measured at every 5 minutes direct normal irradiation (DNI) by a device located in the solar side zone of Universiti Teknologi PETRONAS. A typical day of the month is selected to represent the month as per [5] criterion of recommended average day of the month. It accumulates total solar radiation starting at sunrise hour and ending at sunset hour. The input parameters of PTC field, turbine and condenser used for the simulation are given in Table 1. Simulation of the plant, based on selected typical day of each month average hourly DNI, has been done to see the daily variations of collector efficiency and heat gain. For these calculations, the typical day monthly average hourly DNI and ambient temperature data are taken from EKO Pyranometer located in the solar side zone of UTP. Hourly solar insolation of a direct normal irradiance on a specific day each month throughout the year is depicted in Figure 2.

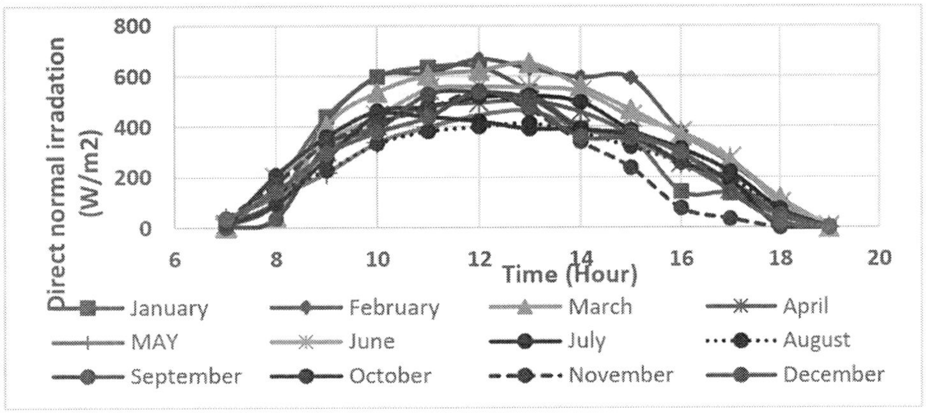

Fig. 2. Monthly average hourly direct normal irradiance in Universiti Teknologi PETRONAS during the year.

It was observed that the hourly maximum direct normal beam insolation on the collector was recorded around 668 W/m^2 in the month of February at noon. However, there are more fluctuations occurring in the direct normal beam for all months because of cloud covers in Malaysia. Figure. 3 shows the variation of useful heat gain by the PTCs against the hourly time on a specific day of every month. It is evidently shown in the figure that the highest useful hourly heat transferred by PTCs considerably fluctuates during the year.

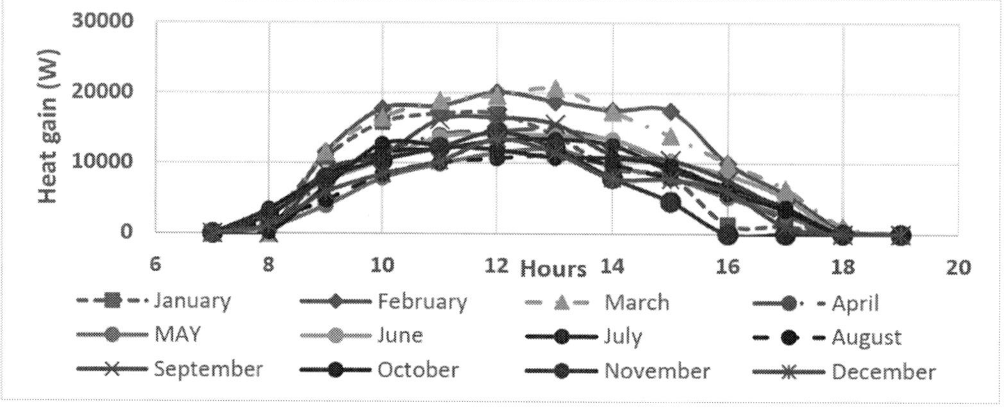

Fig. 3. Hourly collectors useful heat gain during the year.

It can touch maximum 20701.7 W at 13:00 pm in the month of March due to high direct normal irradiance (DNI) approximately 650 W/m^2 and approximately 11137.2 W (minimum) in the month of August on plane. This is a considerable fluctuation in useful heat gain occurs due to cloudy weather and rainfall (mainly between August and November) as a contrast to the driest season (generally between December to April), where insolation is attained by PTCs in Ipoh. As a result, higher values of useful heat gain were obtained throughout the low rainfall season as compared to various high rainfall seasons and it has high values on the plane according to the earlier expectation of system.

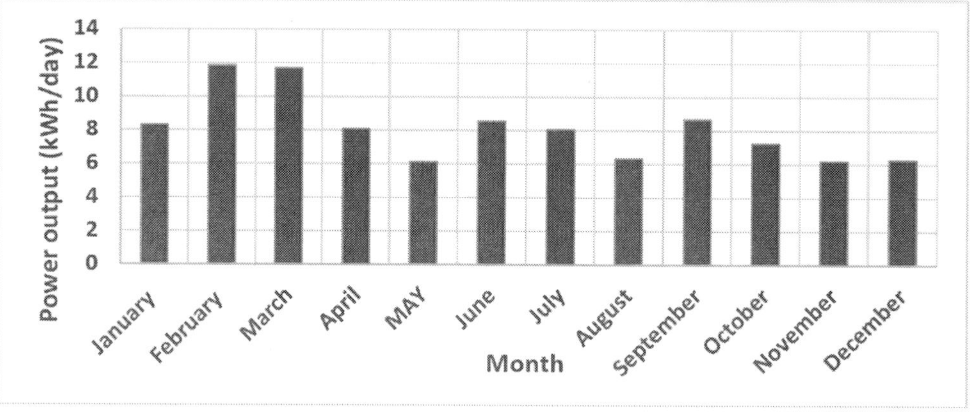

Fig 4. The daily power output for the Universiti Teknologi PETRONAS location in the year.

An analysis of the daily power output, DNI and the efficiency for the Universiti Teknologi PETRONAS location was carried out. Figure. 4 shows the annual gross electrical power output at various chosen days of every month during the year. It was clearly observed that the gross output of the plant was maximum at higher solar radiation. Furthermore, it was observed from results that the maximum net power is 11.84 kWh/day in February. The lowest of electrical power output is 6.11 kWh/day in May.

The fluctuations of solar radiation affect the collected thermal energy directly as well as the efficiency. Figure 5 shows the average efficiency of parabolic trough power plant. It was observed the average efficiency is 8.85 %, during low rain season in March. The lowest of efficiency of 6.23% was obtained in high rain season in November.

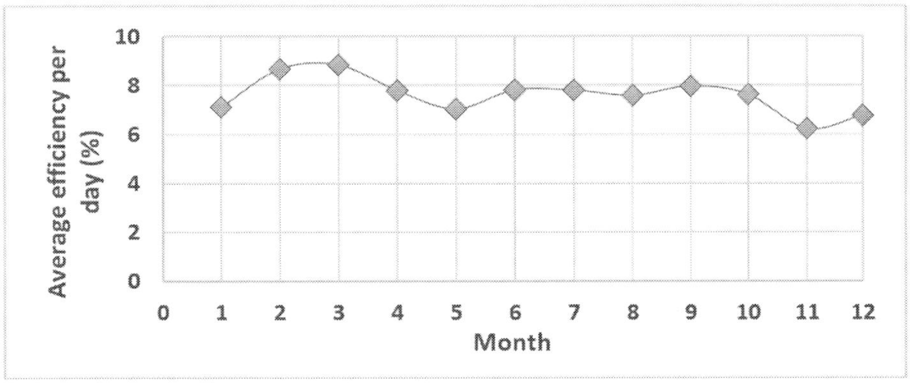

Fig. 5. The daily efficiency for the Universiti Teknologi PETRONAS location in the year.

4 Conclusion

This article presents a parabolic based thermal collector solar thermal power plant simulated mathematically for conceptual design calculations. The model was converted into a computer program within MATLAB environment to predict the required hydrothermal parameters of the system. The control ideology of plant consists the joint effect of the solar field under ambient conditions as well as mentioned solar radiation to account for the constant generation of power from turbine-generator unit in the hours of sunshine. As can be seen from the plotted results, the maximum useful heat gain 20701W was attained on a plane in March. The maximum net power is 11.84 kWh/day in February. These outcomes can be utilized to plan the operation and create a proper control strategy of power plant.

The authors acknowledge Universiti Teknologi PETRONAS for the financial support to conduct the research work. Also, appreciation are due to MOHE for the financial support under MyRA grant.

References

1. A. Poullikkas. RSER **13** 2474–2484 (2009).
2. S. D. Odeh, G. L. Morrison and M. Behnia. Solar Energy **62**, No. 6, pp. 395–406, (1998).
3. M. Sinclair. Renewable Energy Focus **9**, 64-70, (2008)
4. M. Eck and T. Hirsch. Solar Energy **81** pp.268–279, (2007).
5. L. Garcı´a-Rodrı´gueza, Blanco-Ga´ lvezb J. Desalination **212**:311, p.8, (2007).
6. M. Montes, A. Aba´nades and J. Martı´nez-Val. Solar Energy **83** pp. 679–689, (2009).
7. G. Kolb and R. Diver. Journal of Solar Energy Engineering, November **132**/041003-1, (2010).
8. N. B. Desai, S. B. Kedare and S. Bandyopadhyay. Solar Energy **107**, pp. 98–112, (2014).
9. D. Kumar and S. Kumar. Int. J Energy Environ, **6** (1):87–96, (2015).

CFD modeling of artificial vortex air generator for green electric power

Ali A. Ismaeel[1, 2], *Hussain H.* Al-Kayiem[1, *], *Aklilu T.* Baheta[1] and *Mohammed A.* Aurybi[1]

[1] Department of Mechanical Engineering, Universiti Teknologi PETRONAS, 32610 Bandar Seri Iskandar, Perak, Malaysia.
[2]Department of Electromechanical Engineering, University of Technology, Baghdad, Iraq

Abstract. This paper presents and discusses a Computational Fluid Dynamics (CFD) simulation of artificial vortex air generator as part of the preliminary of Solar Vortex Power Generator for an electrical power generation. A vortex air generator system was built, consisting of concentric cylinders. The inner cylinder was fitted with stationary air guide vanes and covered at the top by a transparent plate to capture the solar radiation and create swirling updraft flow which is able to rotate wind turbine and produces power. The influence of inlet air velocity and temperature on the swirling strength and mass flow generated has been evaluated by validated CFD simulation. ANSYS Fluent software was adopted to solve the 3-D, steady state of Navier-Stokes and energy equations in cylindrical coordinate system integrated with discrete ordinates (DO) radiation model. For the preliminary vortex generator design, the CFD results were validated first with previous experimental measurements. Then the variable operation parameters were carried out on the proposed model. The simulation result demonstrated that inflow velocity is a key parameter for enhancing the system performance. By increasing the inflow velocity from 0.4 m/s to 0.6 m/s and inflow temperature 323°k the enhancement rate of the mass air flow generated reached to 26% compared with 7% when increase the inflow temperature to 328°k and inflow velocity 0.4 m/s.

1. Introduction

The concept of harvesting energy from extreme atmosphere like dust devils and tornado phenomena to produce a clean electric power as a new renewable energy source have been evaluated by many researchers in the time past and still [1, 2]. This concept depends on a technique of generating an artificial updraft hot air like the vortex that transforms thermal energy into mechanical work during an upward heat convection motion of the air, which is similar to the idea of heat engine works [3]. To establish an artificial swirling air motion like a tornado, there are two considerable variables which must be implemented, i.e. circulation and updraft [4].

*Corresponding author: hussain_kayiem@utp.edu.my

The circulation is accomplished by getting the air tangentially in an inward direction using a mechanical swirling device such as (interior winding ribbing, screws, vortex tube, guide twirl vanes and others) [5]. Whereas, a approaches to build up the updraft vortex stream [6]. A heat source is utilized to upgrade the buoyancy forces of the hot air, such as the Solar Vortex Engine (SVE) was proposed by [7]. The SVE comprises of two main parts, solar air collector (SAC) and vortex engine (VE). The SVE has been developed by utilizing a new design for the guide vanes or deflectors inside the VE by using curvature deflectors to smooth and enhance the vortex generated by the engine [8]. For power generation, the axial flow turbine is located beneath the upstream of the air flow leaving the SAC at the base of the SVE. Although the SVE is a low-cost and simple configuration, but the drawback is the stack effect inside the vortex generator which is weak without the heat source to generate the initial vortex and maintain the updraft continuously. Thus, the enhancement and development concept of SVE using a simple construction design support was carried out in the Solar Research Site (SRS) at Universiti Teknologi PETRONAS – Malaysia. A new arrangement of artificial vortex generator (AVG) integrated with double solar absorbing units has been devised and hereby referred to as 'Solar Vortex Power Generator (SVPG)' model. The study objective is to investigate the effect of variable operating parameters on the swirling strength of the SVPG using CFD simulation. Also, to present brief description of the developed experimental model of the SVPG.

2. Preliminary model of plant

The preliminary model of SVPG comprises of main three parts. The greenhouse effect by the SAC converts the solar radiation to a kinetic energy in the air stream through a series of energy conversion processes and as a result, ambient air is heated and moves towards the artificial vortex generator (AVG) part.

The solar heat trap unit (SHT) consists of the transparent upper cover (TUC) with a transparent structure part, used as a source to establish the initial convective vortex by capturing the solar energy that is used to heat the air inside the AVG. Thus, creates the updraft flow while withdrawing hot air from the SAC to the AVG. Acceleration of the air flow through guide vanes imposes rotational swirling motion inside the AVG. In order to generate electricity, vertical – wind turbine with electric generator can be added in the path of vortex-induced flow, extracts kinetic energy and produces power. The detailed configurations of the preliminary design of SVPG are shown in the Figure 1. AVG has diameter and height of the inner cylinder were of 0.9 and 0.6 m, respectively. Diameter and height of the outer cylinder are 1 and 0.6 m, respectively. The guide vanes were set to 25° with a height of 0.3 m and length of 0.35 m. The entry slot width was 0.1 m with the height of 0.3 m. While the dimensions of TS was 0.6 m in height having a polygon shape and the diameter of the TUC was 0.1 m with the hole of 0.3 m in diameter at its center.

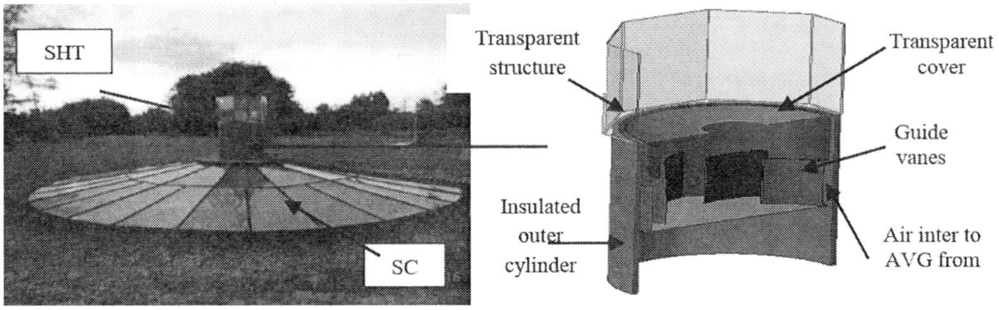

Fig. 1. Preliminary model of SVPG

3. Numerical investigation

3.1 Numerical Simulation Methods

Simulation using Fluent and the Boussinesq model for natural convection [9] was employed in this study. This model assumes that density remains a constant value in all solved equations, except for the buoyancy term in the momentum equation. Simulations were conducted based on $k-\varepsilon$ turbulence model with RNG. The Semi Implicit Method for Pressure Linked Equation – Consistent (SIMPLEC) algorithm was applied for second order discretization for pressure; momentum and energy obtain more converged solution. The Discrete Ordinates (DO) have been applied under the radiation model to represent the solar radiation load passing through the transparent Perspex to heat up the air underneath. The amount of radiation is highly dependent on the solar insolation available and the values were obtained from solar calculator pre-build in ANSYS Fluent 15. The (DO) method calculates the direct and diffuse intensity of the sun depending on the location of the plant related to the sun at a specified time and location which limited by the user, so there is a solar calculator to specify the directions and solar values.

3.2 Governing Equations

Various working principle and governing equations have been applied in numerical simulation done with ANSYS Fluent 15. Some of the examples are conservation of continuity equation and momentum, energy balance equations which have been applied to study the flow field within VAG as follows [10]:

Continuity equation:

$$\left(\frac{\partial u}{\partial x}+\frac{\partial v}{\partial y}+\frac{\partial w}{\partial z}\right)=0 \tag{1}$$

Momentum equation:

$$\rho\nabla.(uV) = -\frac{\partial p}{\partial x}+\frac{\partial \tau_{xx}}{\partial x}+\frac{\partial \tau_{yx}}{\partial y}+\frac{\partial \tau_{zx}}{\partial z}+S_x \tag{2a}$$

$$\rho\nabla.(vV) = -\frac{\partial p}{\partial y}+\frac{\partial \tau_{xy}}{\partial x}+\frac{\partial \tau_{yy}}{\partial y}+\frac{\partial \tau_{zy}}{\partial z}+S_y \tag{2b}$$

$$\rho\nabla.(wV) = -\frac{\partial p}{\partial z}+\frac{\partial \tau_{xz}}{\partial x}+\frac{\partial \tau_{yz}}{\partial y}+\frac{\partial \tau_{zz}}{\partial z}+S_z \tag{2c}$$

The source terms S_x,S_y, and S_z in (2a), (2b) and (2c) represent the body force.

An equation (3) the Boussinesq model is used by setting up the problem with fluid density as a function of temperature.

$$(\rho) = \rho_{\circ}(1-\beta\Delta T) \tag{3}$$

Equation (3) is valid for $(1-\beta\Delta T) \ll 1$

Turbulence model:

$$\rho\frac{\partial}{\partial x_i}(ku_i) = \frac{\partial}{\partial x_j}\left(\alpha_k\mu_{eff}\frac{\partial k}{\partial x_j}\right)+G_k+G_b-\rho\varepsilon-Y_M \tag{4}$$

$$\rho \frac{\partial}{\partial x_i}(\varepsilon u_i) = \frac{\partial}{\partial x_j}\left(\alpha_\varepsilon \mu_{eff} \frac{\partial \varepsilon}{\partial x_j}\right) + G_{1\varepsilon}\frac{\varepsilon}{k}(G_k + C_{3\varepsilon}G_b) - C_{2\varepsilon}\rho\frac{\varepsilon^2}{k} - R_\varepsilon \quad (5)$$

Energy equation:

$$\rho C_p\left(\frac{\partial(uT)}{\partial x} + \frac{\partial(vT)}{\partial y} + \frac{\partial(wT)}{\partial z}\right) = k\left(\frac{\partial^2 T}{\partial x^2} + \frac{\partial^2 T}{\partial y^2} + \frac{\partial^2 T}{\partial z^2}\right) + S_E \quad (6)$$

The radiation equation which is solved by the software can be written as:

$$\nabla.(I_\lambda(\vec{r},\vec{s})\vec{s}) + (a_\lambda + \sigma_s)I_\lambda(\vec{r},\vec{s}) = a_\lambda n^2 I_{b\lambda} + \frac{\sigma_s}{4\pi}\int_0^{4\pi} I_\lambda(\vec{r},\vec{s})\varphi(\vec{s}.\vec{s}')d\Omega' \quad (7)$$

The solar intensity in each direction $\vec{s}$ and position $\vec{r}$ can be written as follows [11]:

$$I_\lambda(\vec{r},\vec{s}) = \sum_k I_\lambda(\vec{r},\vec{s})\Delta\lambda_k \quad (8)$$

$\vec{r}$ = position vector, $\vec{s}$= direction vector, $\vec{s}$= scattering direction vector, α=absorption coefficient, n=refractive index, σ_s= scattering coefficient, σ= Stefan-Boltzmann constant (5.669×10^{-8} $W/m^2.K^4$), I= radiation intensity, which depends on $\vec{r}$ And $\vec{s}$, T=temperature, ϕ= face function, $\acute{\Omega}$= solid angle, λ=wave length, α_λ= spectral absorption coefficient, $I_{b\lambda}$= black body intensity, S_h = Heat source due to incident solar radiation, G_k = Production of turbulent kinetic energy, G_b = Generation of turbulence kinetic energy due to buoyancy, $C_{3\varepsilon}$ = constant to calculate the degree of ε effected by buoyancy.

3.3 Boundary Conditions and Simulation Parameter Cases

The values of the boundary conditions considered as an input parameter in the CFD – simulation were all adopted from the previous experimental measurements, as illustrated in Tables 1and 2.

Table1. Thermal properties of materials

Thermal property	Perspex	Aluminium
Properties	Solid, Semi-transparent	Solid ,Opaque
Thermal conductivity(w.m/k)	0.189	202.4
Density (kg/m^3)	1200	2719
Specific heat(J/kg.k)	1200	871

Table 2. Initial boundary conditions from experimental preliminary model

Surfaces components	No. of item	Boundary type	Boundary conditions
Transparent structure	1	Wall	Material= Perspex T= 322.7K°
Upper cover	1	Wall	Material= Perspex T= 322.7K°
Upper cover hole	1	Outflow	Opening hole , D= 0.3 m
Inner cylinder	1	Wall	Material= Aluminium T= 330K°
Guide vanes	8	Wall	Material= Aluminium T= 330K°
Air entry slots	8	Velocity-inlet	V=0.4 m/s T=323K°
Ambient condition	-	-	T= 309K° S=1000 W/m^2

3.4 Grid Independence

Grid independence is considered in the solution; five mesh sizes were used, tetrahedral cell sizes range between (0.05m-0.005m) is tested and logically the used of fine mesh with a minimum mesh size gave a more accurate solution, however, implies more computer cost and time. The mesh size (maximum 0.01m and minimum 0.001m) was used by percentage error (3.2%) for velocity and temperature (2.1%) for the preliminary model and they are acceptable.

4. Result and discussion

To validate the CFD assessment, the comparison of the CFD results with the preliminary experimental measurements was made. The outflow swirl updraft was considered as the main parameter for validation.The contours velocity magnitude distribution of the initial case is shown in Fig. 2a. From the figures, the average outlet velocity (1.85 m/s) was simulated at the center of the upper hole; the value obtained is in close agreement with the experimental measurement at the same inflow velocity condition (0.4 m/s) with percentage error of 3.2%. Whereas, the maximum velocity magnitude achieved was (2.3m/s) and occurred at 20 cm above the upper hole at the both ends of the vortex streams within the SHT. As the electric power generation is the main objective from SVPG, it is recommended for setting the turbine unit in this position to convert the peak velocity of vortices generated by the work.

4.1 Analyses of Vortex Flow Structure

Simulations the vortex flow behavior within the AVG & SHT are shown in Figure 2.b The Figure describes the outflow direction from the vanes as a slope vector along the guide vane height. Then, moves towards the upper hole with a small circulation motion like a convection currents as it begins to appear at the end of guide height in the bottom corner of the upper cover within the AVG. After the upper hole, the swirl flow structure diverges into two streams that are extended into SHT with the formation of convection currents around the swirl flow. The reason for the formation of the convection currents is due to the contact of the air inside the AVG and SHT with the hot surfaces in the SHT components.

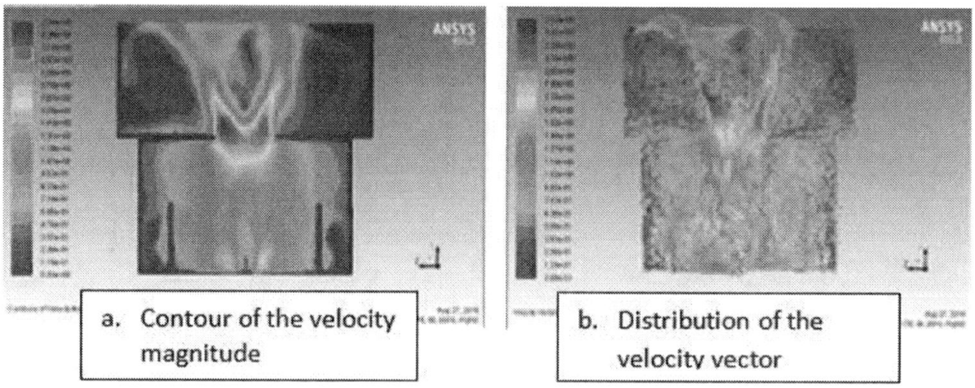

Fig. 2. Velocity distribution within AVG and SHT for the initial case.

4.2 Analyses of Initial Condition Parameters

The result of the mass flow rate and swirl updraft velocity generated in the different initial boundary conditions are listed in Table 3. It is clearly observed that there is a significant improvement in the swirl strength emerging from AVG when the air velocity is increased from the SC to AVG. Thus, swirl strength is enhanced; indicating an increase in the mass flow generated which translate to more electrical power extracted by the turbine. The CFD-results, obtained showed an enhancement in the buoyancy force into AVG because of the increased initial inflow temperature. Also, increasing the inner wall temperature of AVG cylinder contributes to the improvement in outlet velocity and the flow rate. However, this improvement is less compared with extended the inflow velocity state. On the other hand, the change of the initial parameters does not effect on the vortex structure within the VAG and SHT as shown in Figures.3. The flow behavior is observed to remain same in the initial case.

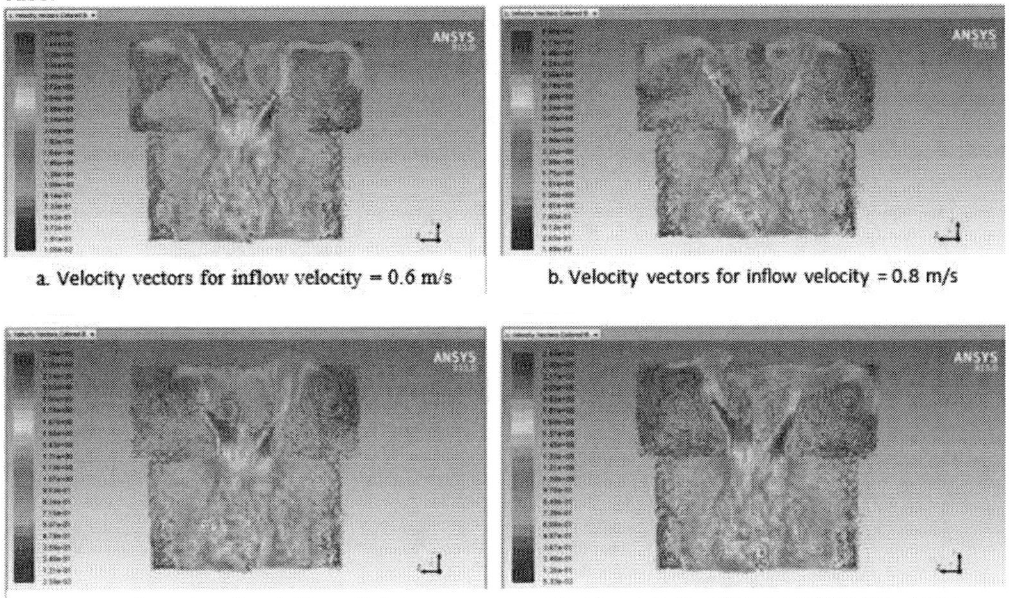

a. Velocity vectors for inflow velocity = 0.6 m/s
b. Velocity vectors for inflow velocity = 0.8 m/s
c. Velocity vectors for inflow Temp = 328 K
d. Velocity vectors for AVG wall Temp = 335 K

Fig. 3. Section of vortex structure within AVG and SHT in different boundary conditions.

Table 3. Result of the mass flow and swirl updraft velocity generated in the variables initial condition

Cases	Inflow Velocity (m/s)	Swirl updraft strength (m/s)	Mass flow rate (Kg/s)	Enhancement of mass flow (%) compared with initial case
Initial case	0.4	1.85	0.1595	-
Increase the inflow velocity	0.6	2.5	0.2155	26
	0.8	3.2	0.2759	42
Increase the inflow temperature from 323 K to 328 K	0.4	2	0.1724	7.4
Increase the inner wall temperature from 330K to 335K	0.4	1.97	0.1698	6

DOI: 10.1051/matecconf/201713102009

5. Conclusions

In this paper, modeling of artificial vortex generator (AVG) supported by transparent cover has been simulated computationally by using ANSYS FLUENT 15. The AVG is the main part of SVPG model that was proposed as a new system in the solar vortex updraft field for renewable power generation. The impact of the operating parameter on the swirl updraft flow characteristics was investigated and the results were compared with preliminary experimental measurements. For a given geometry of the AVG, The CFD results showed that increasing the inflow air velocity is more efficient to enhance the mass flow rate generated and extract power production compared to increase the air flow temperature.

The authors would like to acknowledge Universiti Teknologi PETRONAS for the technical and financial support to conduct this research. Also, the authors express their gratitude to The Ministry of Higher Education – Malaysia for permitting the utilization of the measuring instrumentations purchased under the FRGS/2015/1 grant under CS: 0153AB-K81.

References

1. S. Nizetic. Energy, **36** (2) pp 1236-1242, (2011).
2. A.A. Ismaeel, H.H. AI-Kayiem, A.T. Baheta and M.A. Aurybi. IET Renewable power generation, DOI:10.1049/ iet- rpg.2017-0058, (2017)
3. L.M. Michaud. Atmospheric Research, **39**, PP157-78, (1995)
4. A. Varaksin, M.E. Romash, V.N. Kopeitsev and M.A. Gorbachev. International Journal of Heat and Transfer, **55** (23-24), pp6567-6572, (2012).
5. A. Varaksin, M.E. Romash and V.N. Kopeitsev. 14th European Turbulence Conference, France, 1-4 September, (2013).
6. H.S. Takhar and O.A. Beg. International Journal of Fluid Mechanics Research, **32** (4), (2005).
7. H.H. AI-Kayiem and A.T. Mustafa, *Solar Vortex Engine*, Malaysia. Patent No PI2015702890, 2015.
8. H. H. Al-Kayiem, A. T. Mustafa & S. I. U. Gilani. WIT Transactions on Ecology and The Environment, **205**, 193-202. (2016) http://dx.doi.org/10.2495/EQ160181
9. H.F. Joe land P.E. Milovan. Third edition book Publications: Springer, (2002).
10. ANSYS Fluent Software 15 Help User Guideline.

DOI: 10.1051/matecconf/201713102010

CFD Modeling and Experimental Validation of a Solar Still

Tahir Mahmood[1], Muhammad Y. Naz[1], Shaharin A. Sulaiman[2], Yasir Jamil[1], Shazia Shukrullah[1], Muhammad Zahid[3], Muddasser Inayat[2]*

[1]Department of Physics, University of Agriculture, 38040 Faisalabad, Pakistan.
[2]Department of Mechanical Engineering, Universiti Teknologi Petronas, 32610 Bandar Seri Iskandar, Perak, Malaysia.
[3]Department of Chemistry, University of Agriculture, 38040 Faisalabad, Pakistan.

Abstract. Earth is the densest planet of the solar system with total area of 510.072 million square Km. Over 71.68% of this area is covered with water leaving a scant area of 28.32% for human to inhabit. The fresh water accounts for only 2.5% of the total volume and the rest is the brackish water. Presently, the world is facing chief problem of lack of potable water. This issue can be addressed by converting brackish water into potable through a solar distillation process and solar still is specially assigned for this purpose. Efficiency of a solar still explicitly depends on its design parameters, such as wall material, chamber depth, width and slope of the zcondensing surface. This study was aimed at investigating the solar still parameters using CFD modeling and experimental validation. The simulation data of ANSYS-FLUENT was compared with actual experimental data. A close agreement among the simulated and experimental results was seen in the presented work. It reveals that ANSYS-FLUENT is a potent tool to analyse the efficiency of the new designs of the solar distillation systems.

1 Introduction

Population of the world, especially in developing countries, is increasing exponentially and so does the consumption of potable water. The water is the most important in the contents of fundamental needs for the survival of living organisms [1]. Over 71.68% of the total area of earth is covered with water leaving a scant area of 28.32% for human to inhabit. Out of this total volume of water on the earth, the fresh water accounts for only 2.5% and the rest is the brackish water. Adding more, only 0.3% of the fresh water exists in liquid form. Since water is not used efficiently by the world, the sources of drinkable water are depleting with every passing moment. Other than the use for food and fiber by the human, many economic, technological and political factors are also intensifying the water shortage upto an alarming

* Corresponding author: shaharin@utp.edu.my

level. Therefore, the requirement of this age is to develop selfsustained systems to distill the available saline water into drinkable water for the better health of human being.

The countries like Pakistan, India, Bangladesh and African nations accommodate the notable part of the world population [2]. These countries also face high intensity of the solar radiations throughout the year. The issue of potable water shortage in such countries can be fairly reduced by processing brackish water with solar distillation systems [3, 4]. For converting saline water into potable water, solar still technology regarded as a low cost, simple and easily adaptable. Since solar still does not involve any specialty items in its fabrication, it is easy to maintenance and fabricate with locally available materials. A solar still works on the principle of evaporation and condensation due to a difference in temperatures of front glass-cover and the water in the basin. Solar still absorbs the solar radiations, falling on the front glass-cover, to raise the temperature of the impure water. The hot water evaporates and then condenses on the lower surface of the glass cover by leaving all the contaminants in the basin [4]. However, the evaporation-condensation in such systems is an essential assignment to raise the production rate of the fresh water. Therefore, to assess the efficiency of a solar still design, not only the experimental work but the development of suitable mathematical models investigating the different parameters of the solar still, are inevitably required.

Numerous research findings, addressing the effects of different factors on the solar still productivity, have been reported in the past literature. Tiwari et al. [3] considered inner glass temperature in performing the computer modeling of a passive/active solar still. They found that the solar still with cover-glass inclination angle of 30° is suitable for production of fresh water at reasonably good rates. Maximum heat transfer from the sun to the water was reported at this inclination angle. Work of Tiwari and Tripathi [4] showed that the semi-cylindrical shape of the solar distillation units is far better than the spherical geometry. It was also reported that the fresh water flows at higher rates during forced convection than that of natural convection. As a best factor that enhances the solar still productivity, Omar et al. [5] worked out at an active solar single slope at different functional parameters, such as variation in insulation thickness, effective transmissivity, effective absorptivity and solar intensity. Shakaib et al. [6] tested the effectiveness of a solar still and revealed that the efficiency of a solar still dependent on the fluid flow patterns during forced convection of the moist air. They carried out CFD analysis of a solar still and presented the results in terms of velocity and temperature profiles, distribution of shear stress and heat transfer coefficient. Their work revealed major flow recirculation within the solar still, which might be a useful parameter for increasing the rate of production of fresh water. Panchal et al. [7] investigated that all parameters of a solar still like chamber temperature, glass-cover temperature, vapor temperature, basin water temperature, etc. are directly proportional to the distillate output. The simulation results of their work showed a good agreement with experimental data.

This study is based on experimental and computational approaches to test a solar distillation system by comparing the experimental data and ANSYS-FLUENT simulation results. The properties of the fresh water, obtained with the single slope solar still design, are also tested to check the product quality and the process efficiency.

2 Methodology

Geometry of a single slope solar still design is shown schematically in Fig. 1. On experimental side, the solar distillation system was consisted of a wooden box with square base, rectangular top glass-cover and a metal tray sheet. The top surface of the solar still was covered with a 2 mm thick glass sheet. The metal try sheet of 3 mm thickness was used to raise the solar intensity and a wooden box to minimalize the heat losses from the system. The inclination angle of the glass-cover was fixed to 20°. Evaporation and condensation of brackish water taken place when the setup was exposed to direct sunlight. The experiments were performed in the main campus of the University of Agriculture Faisalabad, Pakistan during July 2017. Temperature of the glass cover, basin water and outlet water was measured after every hour starting from 9:00 am to 17:00 pm. The most favorable evaporation-condensation was attained at noon from 12:00 pm to 15:00 pm. The properties of the distilled water, obtained with the single slope solar still design, were tested to check its quality.

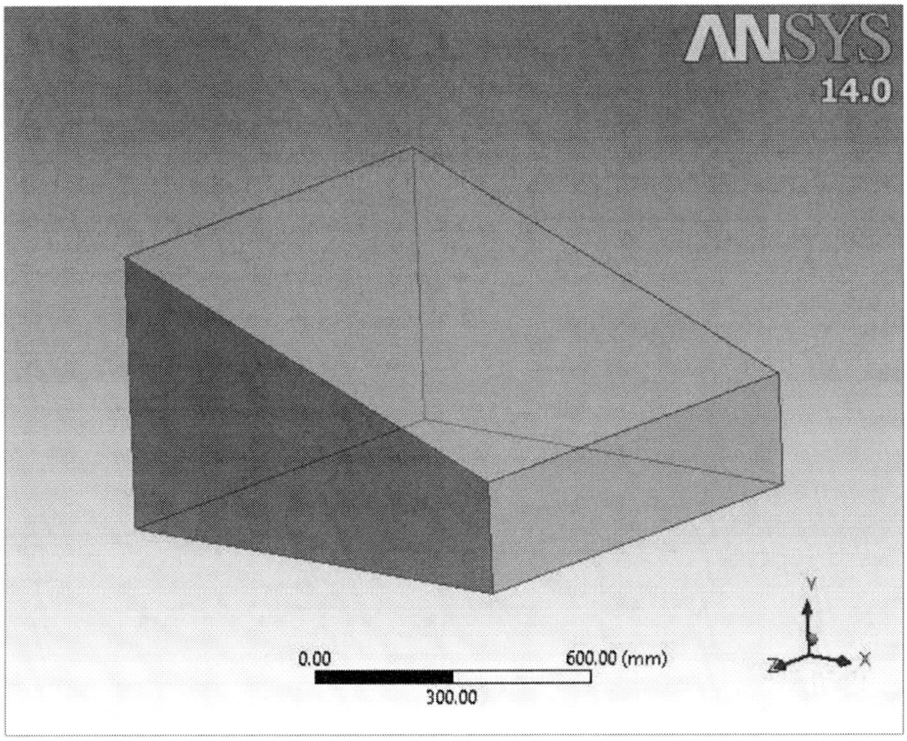

Fig. 1. Geometry of a single slope solar still design.

For ANSYS CFD modeling of the solar still, a 3-D model was made in ANSYS-FLUENT workbench design modeler which is an abelian software package in ANSYS CFD. The model had six faces, which were considered as waterproof walls. The temperature of the bottom and top surfaces was fixed at 293 K and 403 K, respectively. The remaining walls were assumed adiabatic. The height of the front and back walls of the unitary cubical solar still was 250 mm and 610 mm, respectively, which inclined the top surface slope (glass-cover) at 20°. The width and depth of the unit were fixed at 1 m. Fig. 2 shows the fine cell meshing of the solar still geometry. ANSYS FLUENT 14.0 was used to solve the equations of continuity, momentum and energy for the meshed geometry. Air is an ideal fluid because its thermal conductivity and viscosity varies with temperature. Therefore, the Rayleigh number for the simulation was set 1200. First order upwind scheme was used to discretize the convective terms while SIMPLE algorithm was used to couple the temperature and velocity profiles.

DOI: 10.1051/matecconf/201713102010

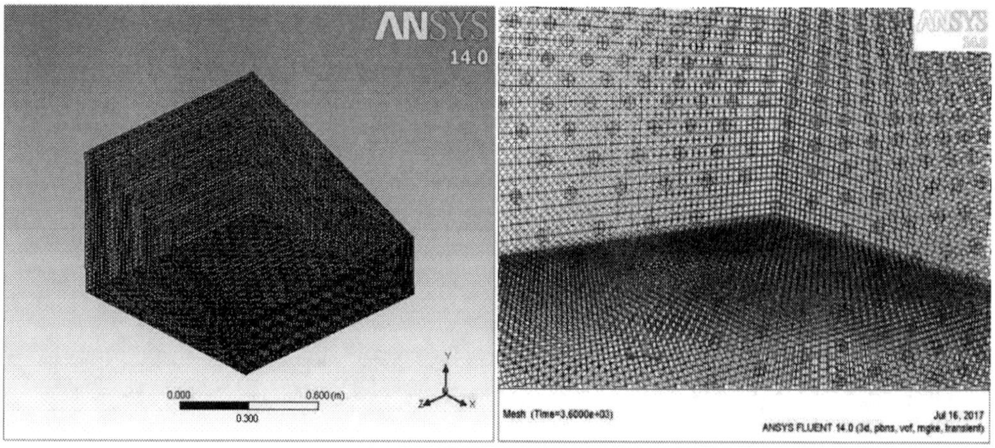

Fig. 2. Meshing of the solar still geometry for CFD simulation.

3 Result and Discussion

Temperature of the glass cover, basin water and outlet water was measured after every hour starting from 9:00 am to 17:00 pm, as shown in Table 1. The most favorable evaporation-condensation rates were attained at noon from 12:00 pm to 15:00 pm.

Table 1. Experimental data extracted from the solar still prototype.

Time	Collected water (ml)	Inlet temperature (°C)	Outlet temperature (°C)	Basin temperature (°C)	Glass cover temperature (°C)
09:00 to 10:00	35	34	36	40	33
10:00 to 11:00	50	34	37	45	36
11:00 to 12:00	95	34	40	45	36
12:00 to 13:00	185	34	41	48	38
13:00 to 14:00	200	34	42	52	40
14:00 to 15:00	180	34	41	49	38
15:00 to 16:00	145	34	40	49	37
16:00 to 17:00	115	34	36	45	36

The flow patterns in the solar still are shown in Fig. 3 with the help of velocity contour and path-lines on its all faces. On the front and back surface, the velocities were lower and this flow behavior can be considered suitable because the recirculating air would possibly drive the condensed water towards the distillate tray of the solar still.

The temperature contours in Fig. 4a shows variation of temperature in the vertical direction. Due to fixed temperatures of the top and bottom surfaces, the temperature variations were rapid near the front vertical wall of shorter height [7]. The enthalpy of the water was also higher at the front wall and in the center of the glass and absorber plate, as shown in Fig. 4b. The heat transfer coefficient was found to be higher near the vertical wall with shorter height due to higher temperature gradients, as shown in Fig. 5

DOI: 10.1051/matecconf/201713102010

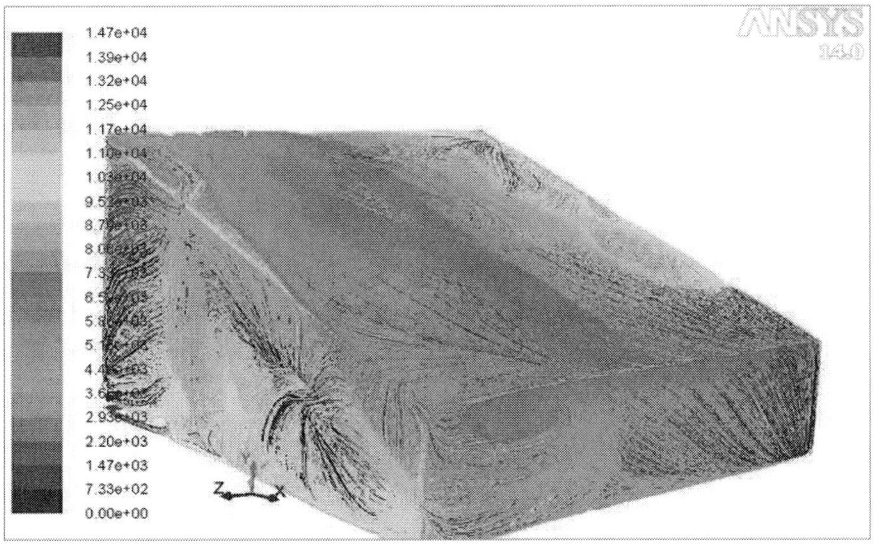

Fig. 3: Velocity path-line contour in the solar still.

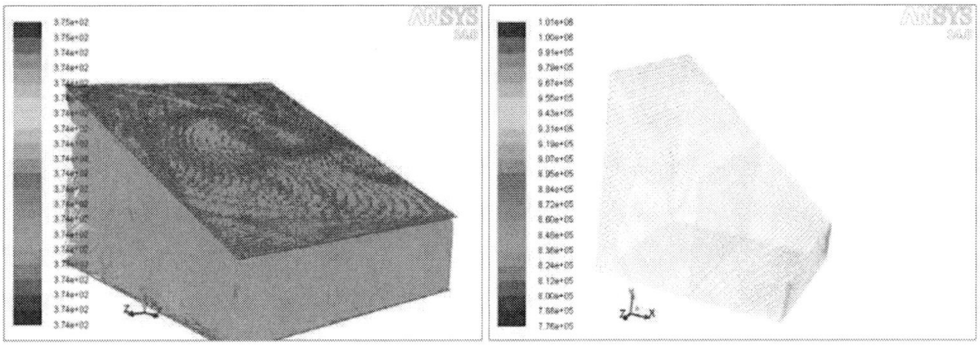

(4a) (4b)

Fig. 4. (4a) Temperature profiles at top and bottom surfaces, **(4b)** enthalpy of water (J/Kg).

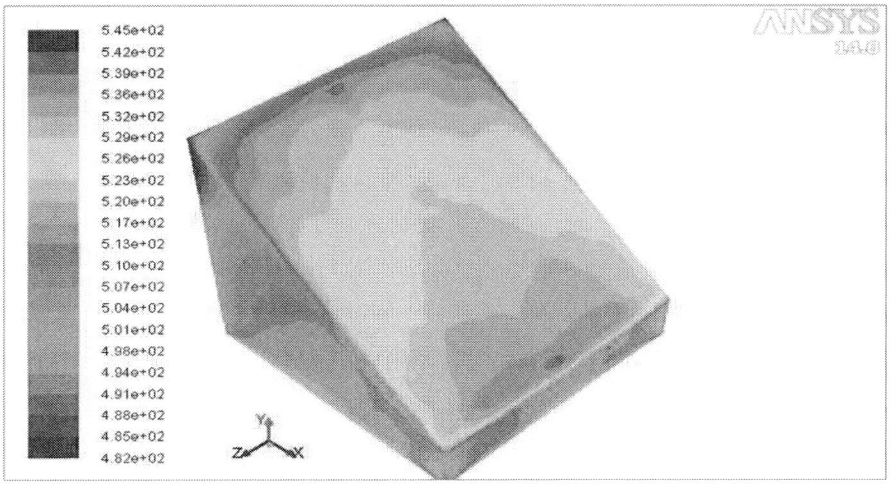

Fig. 5. Heat transfer coefficient.

Experimental and simulation results were compared by following the Dunkle correlation:

$$Nu = 0.075Ra^{1/3} \quad (1)$$

In this equation:

$$Nu = \frac{h_{av}L_c}{k} \quad (2)$$

$$Ra = \frac{g\beta(T_w - T_\infty)L_c^3}{v^2} P_r \quad (3)$$

Where Nu is the Nusselt number, Lc is the average height of the unit, Pr is the Prandtl number and Ra is the Rayleigh number. Fig. 6 provides a comparison of the experimental and ANSYS FLUENT data. A close agreement among the production rates of the potable water from the actual experiment and the simulation was predicted in the presented work.

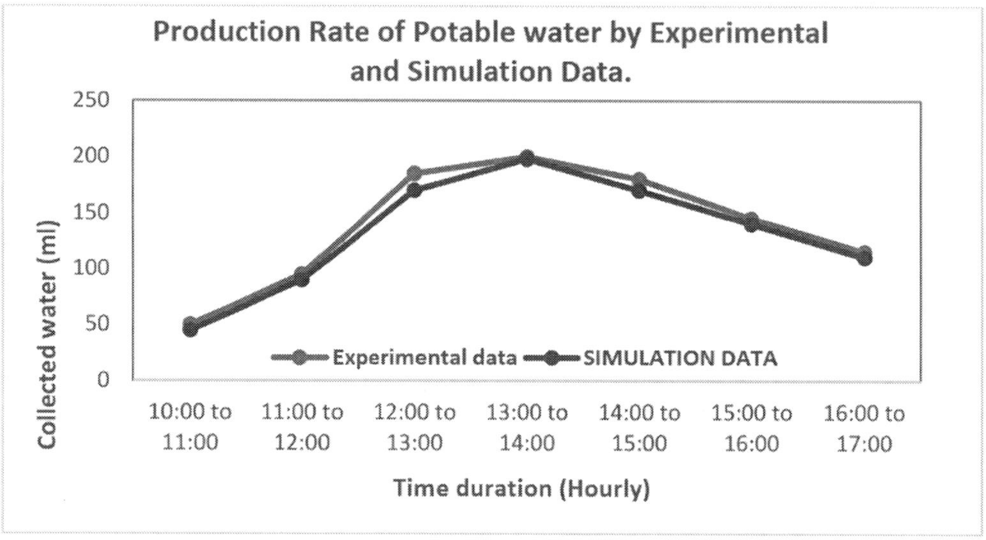

Fig. 6: Production rates of potable water from experiment and simulation.

4 Conclusion

To produce distillate output, the evaporation and condensation in a solar still need to happen at reasonably good rates. The simulation data showed lower velocities at the front and back surfaces and such flow behaviors are generally desired for high performance solar stills. The presented study revealed a close agreement among the simulation and experimental data, showing that the performance evaluation of a solar still is pretty simple with ANSYS-FLUENT.

References

1. Y. Arjen Hoekstra, M. M. Mekonnen. PAS, USA 109,3232–3237 (2012)
2. P. Micklin, AREPS 35, 47–72 (2007)
3. G. N. Tiwari., S.K. Shukla, I.P. Singh, JECM 154, 171 (2003)
4. G. N. Tiwari, R. Tripathi, JECM 154, 161-169 (2003)
5. M. Shakaib, M. A. Khan, ICMAEE-15, pp. 272-276 (2015)
6. H. N. Panchal, N. Patel, IJAE (2017)
7. N. S. Gokilavani, D. Prabhakaran, T. Kannadasan, IJIRSET 3, (2014)

MATEC Web of Conferences **131**, 02011 (2017) DOI: 10.1051/matecconf/201713102011
UTP-UMP SES 2017

On the Influence of Collector Size on the Solar Chimneys Performance

Sundus S. Al-Azawiey[2], *Hussain H.* Al-Kayiem[1,*], and *Suhaimi B* Hassan[1]

[1] *Mechanical Engineering Department, Universiti Teknologi PETRONAS, 32610 Bandar Seri Iskandar, Malaysia*
[2] *Electromechanical Engineering Department, University of Technology, Baghdad, Iraq*

Abstract. Performance of solar chimney power plant system is highly influenced by the design geometries. The collector size is logically enhances the solar chimney performance, but the trend of enhancement is not yet investigated. In the present work, experimental and numerical investigations have been carried out to ascertain, in terms of qualitative and quantitative evaluation, the effect of the collector diameter. Daily thermal efficiency has been determined at four different collector diameter. Two different collector diameters, 3.0 and 6.0 m, have been investigated experimentally, and then scaled up, to 9.0 and 12.0 m, by numerical simulation using ANSYS-FLUENT®15 software. Results demonstrated that collector diameter has effectively influenced the system performance. Larger collector diameter imposed increase in the velocity, temperature and the daily average thermal efficiency of the system. From the experimental results, increasing the collector diameter from 3.0 to 6.0 m has increased the daily average thermal efficiency of the collector from 9.81 to 12.8. Simulation results at 800 W/m^2 irradiation revealed that the velocity in the chimney have increased from 1.66 m/s at 3.0 m collector diameter to 2.34, 2.47 and 2.63 m/s for 6.0, 9.0 and 12.0 m collector diameters, respectively.

1 Introduction

The concept of solar energy induced updraft has been used for ventilation, power generation, drying and space heating. The use of solar updraft for electrical power generation using SCPP was first conceived in 1903 by a Spanish Artillery Colonel, Isadora Cabanyes in his "Proyecto de motor solar" (solar engine project) [1, 2]. In 1926, Prof. Bernard Dubos suggested the use of mountain slope as support to the chimney [3, 4]. SCPP gained attention in 1978 when Prof. Schlaich presented the SCPP technology in a congress [5]. Between 1980 and 1982, Prof Schlaich and his colleagues designed and constructed the first pilot SCPP on a site provided at Manzanares by the Spanish Utility Union Electrica Fenosa with funding from the German Ministry of Research and Technology [6-8].

SCPP collector component is the solar energy conversion component of the SCPP which converts solar energy to thermal energy and subsequently to kinetic energy in the air.

* Corresponding author: hussain_kayiem@utp.edu.my

Thus, the energy conversion process in the collector involves the absorption of solar radiation by the absorber/ground to thermal energy in the absorber, heat transfer from the absorber to adjacent air generating air in motion and the diving of the turbine by the buoyant air to generate electricity. The mass flow rate of the buoyant air at the collector exit/chimney base is a major determinant of the total power output expected from the plant. In his theses, [9] reported that the collector contributes about 50% of the total loss in a SCPP, which are mostly thermal losses to the atmosphere. To reduce the frictional drag, Bonnelle [10] proposed another design of air flow guides to direct the generated buoyant air towards the chimney, thus reduces the air circulation at the collector and reduces heat losses, as well.

Ref. [11] introduced the use of intermediate secondary cover/roof in the collector to reduce the losses. The intermediate roof divides the collector into top and bottom parts where the convective heat loss from the buoyant air generated from the bottom part of the collector is utilized to generate hot air at the top partition of the collector. To reduce air resident time in the collector and consequently reduce the convective loss from the air, [12] proposed a SCPP with inclined solar collector for high latitude location. Their claim is that the system had higher performances for locations at high latitudes than the use of traditional SCPP in same location. This claim might be as a result of increase radiation capturing and reduced frictional losses at the collector section for sloped collector, thus the thermal losses through the cover was reduced. Ref. [13] employed Constructal Design to evaluate analytically the effect the ratio between height of chimney and radius of collector over the power production for a reference plant. In spite of application of the same method for geometrical evaluation, the work brings different findings than [14] results.

It is obvious that most of the reported results and suggestions haven't included the influence of the size of the SAC on the performance of the SC. On the other hand, most of the works have been carried out numerically and mathematically. The literature suffers the lack for experimental data to be utilized for design of the SC, and also for validation of the numerical procedures that may be carried out by other researchers.

The present work intended to presents results on the SC performance under different collector sizes by experimental and numerical analyses. Experimental measurements have been repeated for two different collector diameters, 3.0 and 6.0 m. Numerical simulation has been carried out to extend the investigation to of 9.0 and 12.0 m collector diameter.

2 Experimental methodology

The experimental model of the SC was implemented in the Solar Research Site under the actual environmental condition at the location, Universiti Teknologi PETRONAS – Malaysia (latitude 4.39°N, 100.98°E).

Description of the experimental model

The experimental model of the solar chimney has a collector, chimney and power generation unit. The collector was 3.0 m diameter and then extended to 6.0 m with canopy made from Perspex, while the absorber was made of ground covered by black painted pebble. Water retention on the canopy surface has been considered, and hence it was designed with a slope. The chimney was 6.3 m height made of 152.5 mm (6.0") diameter PVC pipe installed in the center above the collector. The inlet of the canopy height above the ground was 0.05 m, for both 3.0 and 6.0 m collector diameter. The layout of the solar chimney model is shown in Figure1.

The collector thermal efficiency was calculated by equations (1).

 DOI: 10.1051/matecconf/201713102011

$$\eta_{coll} = \frac{q_u}{Q_{solar} + Q_{TES}} = \frac{q_u}{I \cdot A_{coll} + Q_{TES}} \quad (1)$$

Where: q_u useful energy gained

Q_{solar} is the energy gained from the solar radiation, which is the product of the global horizontal solar radiation, I, and the collector area, A_{coll}. Q_{TES} is the energy released from the TES to the air, expressed as equations (2)

$$Q_{TES} = m_{TES} \times c_{p.TES} \times (T_{TES,2} - T_{TES,1}) / 3600 \quad (2)$$

Where: $T_{TES.2}$ and $T_{TES.1}$ are the ground temperatures at the end and beginning of the measurement hours.

Measuring Instrumentations and Procedure

The instrumentation of the system was designed to capture all the components as a solar radiation, temperatures and velocities. The data generated from the experiments were used to determine the system performances under the various conditions following the experimental plans.

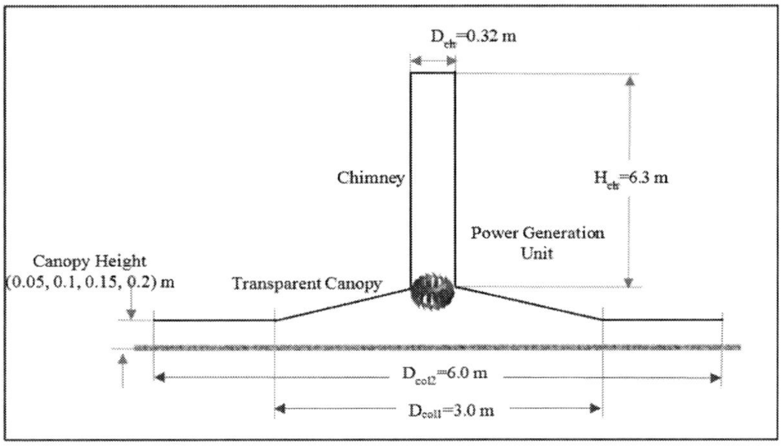

Fig. 1. Outlines of the experimental setup

The thermal performance of the system could only be determined through the measurement of the temperatures of all the components of the system and the temperature changes in the process. To get the temperature data at the different components of the system, the system was instrumented with thirteen thermocouples to collect the temperature data at the absorber surface, transparent cover material, air inside chimney, air in the collector and ambient air. All the thermocouples used for this study were Type J thermocouples, which included probe types used for air temperature and surfaces type for wall temperature measurements. To record the temperature values, all the thermocouples were connected to GRAPHTEC 820 data logger for instant temperature data collection and safe storage the record was every 5 mints and calculated the average of every one hours. For air velocities at the solar chimney model were measured at two different locations in the chimney using portable hotwire probe anemometer. The hotwire probe anemometer has accuracy of ±3% of reading or ±0.03 m/s. The ambient air velocity in the site was measured using portable thermo-anemometer. The global solar radiation was measured at the solar site in UTP using KIMO Solarimeter (SL200) which has measuring accuracy of ±5%.

3 Numerical methodology

The simulation procedure, using CFD technique, to analyze the fluid flow and thermal processes in the solar chimney is described in this section. Since the temperature is essential parameter in the process of solar energy conversion, software package ANSYS-FLUENT®15 was used.

Development of numerical model

Two different models have been created in this simulation. They are duplication of the experimental models. The first model was with 3.0 m diameter and the collector has 15° inclination, as shown in Figure 1. In the second model, the collector diameter was increased to 6.0 m. The extended canopy part has 5° inclination which is sloped towards the peripheral of the collector. The chimney has similar geometries of the experimental model. The ground, functioning as absorber, was simulated as TES made of black painted pebbles.

Mesh generation and mesh independency

The meshing criterion was structured of Tet/Hybrid cells type due to the sharp and curved edges in the flow passages. To study the independency of the results on the mesh, four different mesh sizes, coarse, medium, fine and finer were generated using the second model of 6.0 m diameter. The four different sizes of cells where used to predict the temperatures and air velocities in the chimney base. The mesh checking has been carried out with 0.05 m canopy height when the solar radiation was 800 W/m^2 and ambient temperature 33°C. Predicted temperatures and velocities by mesh with 988.499 cells have close results to the mesh results of 413571 cells with less than 1.0% error in the predicted both temperature and velocity. The results of the mesh with 413571 cells simulation was considered to continue the simulation at various boundary and design conditions. Table 1 shows the details for grid generation and grid Independency for SC model.

Table 1. Grid generation and grid independency for SC model.

Grid type	Grid sizes (No. of cells)	Grid dependent Parameters			
		Velocity at chimney base, $V_{ch.avg}$ (m/s)	Error (%)	Temperature at chimney base, T_{ch} (K)	Error (%)
Coarse Mesh	61165	1.38	38	317.8	2.93
Medium Mesh	195333	1.99	11.1	320.0	2.26
Fine Mesh	413571	2.22	0.89	327.0	0.12
Finer Mesh	988,499	2.24	-	327.4	-

Boundary conditions

For SC model, different boundary conditions were set for the different parts of the system and different materials. The ground material used in the entire simulation was black painted pebbles, collector roof material was Perspex, and the chimney wall material was

DOI: 10.1051/matecconf/201713102011

PVC. The materials were defined in the system with the different thermo-physical properties of the components and the operating fluid, air. The boundary conditions used in the simulation are presented in Table 2.

Table 2. Boundary conditions for the simulation of SC model.

Component	Boundary type	Value
Ground	Wall	Constant heat flux
Collector	Wall	Constant heat flux
Chimney	Wall	Adiabatic, $q = 0$ W/m^2
Collector inlet	Pressure inlet	$P_{inlet}= 0$ Pa, $T_{in}=T_{amb}$

4 Results and discussion

4.1 Experimental results

In this work, the measurements have been acquired over 24 hours daily for four days repeatability. The behavior of the collector with 3.0 and 6.0 m diameter is interesting and never being reported before. From figure 2, which shows the determined thermal efficiency of the collector over 24 hours, there is inconsistent behavior of the system between the day and in the night. During the sun hours, from 8.00AM to 8.00PM, the 3 m collector has efficiency higher than the 6.0 m collector. The reason is the larger area of the collector processes higher thermal losses to the environment, and also as frictional drag to the air flow inside the collector.

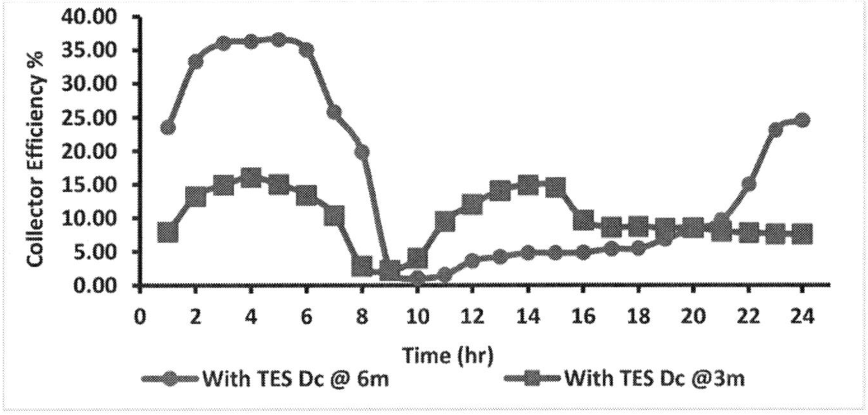

Fig. 2. Measured thermal efficiency at 3.0 and 6.0 m collector diameter over 24 hour (mean of four days repeatability)

In addition, the larger size of the TES absorbs larger amount of solar radiation and store it as heat and start to discharge the stored heat after sunset. Because the amount of stored heat in the case of 6.0 m collector, the heat supplied to the air inside the collector is higher compared to the amount of heat supplied to the air in 3.0 m case. This is why the 6.0 m collector diameter resulted in high collector thermal efficiency in the night.

DOI: 10.1051/matecconf/201713102011

4.2 Numerical results

The collector area effect on the SC performance in this work has been extended from the experimental investigation at 3.0 and 6.0 m to numerical investigation at 9.0 and 12.0 m collector diameter. The obtained results from the simulation have been presented in terms of air velocity, temperature and system thermal efficiency. The simulation considered 800 W/m^2 solar irradiance and prevalent ambient condition at the location when the solar intensity is between 700 W/m^2 and 900 W/m^2, which is at an average temperature of 33°C. The simulation considered optimum canopy height as captured in independent analysis from the experimental investigation.

The results showed that the velocity at optimum canopy height varied with collector diameter, as shown in Figure 3. This is because the air in the greenhouse gains the required thermal energy that can excite the internal energy and induce buoyancy.

By referring back to the experimental results, the measured velocity for the 6.0 m collector diameter is 2.25 m/s. The corresponding numerical result is 2.34. It could be seen that the percentage of error is 4%. This indicates the good agreement between the experimental and numerical analysis.

Figure 4 shows the simulation results of air velocity in vector contours, for 3.0 m collector diameter. Air velocity has a small radial component compared to the vertical component. The vertical component overrides the radial component because the distance from the ground to the canopy is relatively large as compared to the radial passage. The flow moves mostly up and reaches the canopy and return back with reversal manner, towards the chimney base. Once the outer narrow passage flow reaches the central part, it could gain larger vertical velocity component, due to the area enlargement of the configuration, in the vertical direction. However, the resultant of the flow direction was noted to be mainly towards the chimney base, due the stack effect.

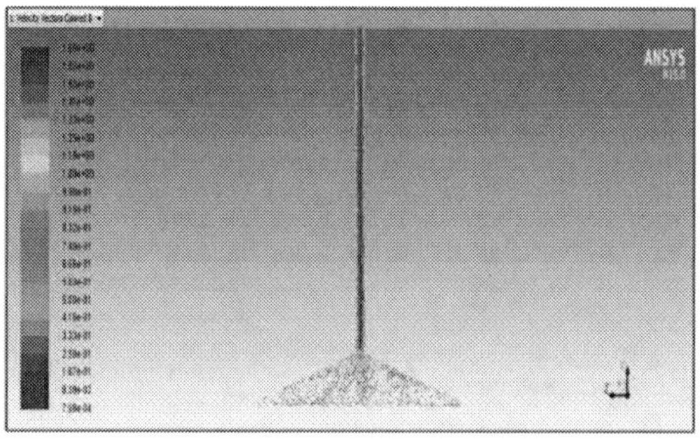

Fig. 4. Air velocity vector at 3.0 m collector diameters; 800 W/m^2 solar irradiance.

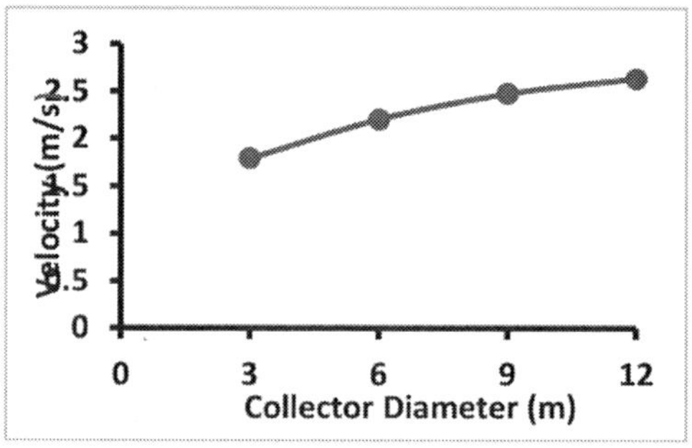

Fig. 3. Air velocity at different collector diameters

4.3 Validation of numerical simulation

Prior to carry forward on the numerical investigations, a level of confident has been established by validation process. The experimental results at 3.0 and 6.0 m collector diameter have been compared with the numerical results under similar conditions of ~800 W/m^2 irradiance and ~33°C ambient temperature.

In terms of velocity at the chimney, the comparison is shown in Table 3. Result of the simulation, in comparison with the experimental results showed small higher predicted velocity values. At 3.0 m collector diameter, the experimentally measured velocity at the chimney base was 1.56 m/s while the predicted one by the simulation was 1.6 m/s with difference of 2.5%. For the 6.0 m collector diameter, the experimental result was 2.25 m/s and the numerical result was 2.34 m/s, with difference of 2.22%. This margin of error represents acceptable accuracy of the numerical procedure.

Table 3: Comparison of Numerical and Experimental Velocity results for 3.0 m and 6.0 m Collector diameters.

D = 3.0 m				D = 6.0 m			
Solar irradiance (W/m^2)	Exp. $V_{ch.\ base}$ (m/s)	Num. $V_{ch.\ base}$ (m/s)	Error %	Solar irradiance (W/m^2)	Exp. $V_{ch.\ base}$ (m/s)	Num. $V_{ch.\ base}$ (m/s)	Error %
806	1.56	1.6	2.5	808	2.25	2.34	2.22

5 Conclusions

Experimental and numerical studies have been conducted on SC models with four different collector diameters. Results demonstrate that as the collector diameter increases, the system performance enhances. But, during the solar time, thermal efficiency of the collector is reduced as the collector diameter increased due to the high thermal losses to the environment.

The authors acknowledge Universiti Teknologi PETRONAS (UTP) in Malaysia for the logistic and technical support to produce this work. The first author appreciates the financial support from Iraqi government through sponsoring her PhD study under full scholarship.

References

1. E. Lorenzo, "De Los Archivos Históricos de la Energía Solar; Las chimeneas solares: De una propuesta española en 1903 a la Central de Manzanares,"(2002)
2. I. Cabanyes, Las chimeneas solares (Solar chimneys), La ernergia eléctrica, (1903)
3. M. A. dos Santos Bernardes. *Solar Energy,* A. Belmiloudi, Ed.: InTech, (2011)
4. C. Papageorgiou, Solar Energy, Ed.: R.D. Rugescu, 187-222 (2010)
5. N. Pasumarthi and S. Sherif, International Journal of Energy Research **22**, pp. 277-288, (1998)
6. J. Schlaich, R. Bergermann, W. Schiel, and G. Weinrebe, International: Journal of the International Association for Bridge and Structural Engineering (IABSE) **14**, pp. 225-229, (2004)
7. J. Schlaich, *The solar chimney: Electricity from the sun*: Edition Axel Menges, (1995)
8. J. Schlaich, R. Bergermann, W. Schiel, and G. Weinrebe, Journal of Solar Energy Engineering, Transactions of the ASME **127**, pp. 117-124, (2005).
9. M. A. d. S. Bernardes, MSc Theses published by Stuttgart: IER, Bibliothek, (2004) http://dx.doi.org/10.18419/opus-1630
10. D. Bonnelle. Doctoral thesis, University Claude Bernard-Lyon 1, France, 2004
11. J. P. Pretorius and D. G. Kröger, Solar Energy, **80**, 535-544, (2006)
12. E. Bilgen and J. Rheault, Solar Energy **79**, 449-458, (2005)
13. A. Koonsrisuk, S. Lorente, A. Bejan, Constructal solar chimney configuration, Int. J. Heat. Mass Transfer **53.** 327-333 (2010)
14. R. S. Vieira , A. P. Petry , L. A. O. Rocha , L.A. Isoldi and E. D. dos Santos. Renewable Energy **109.** 222-234 (2017)

DOI: 10.1051/matecconf/201713102012

Heating Performance Study on a Passive Solar Heater

Muhammad Y. Naz[1*], *Shaharin A. Sulaiman*[2], *Tahir Mahmood*[1], *Muhammad I. Hussain*[1], *Abdul Ghaffar*[1], *Yasir Jamil*[1], *Muddasser Inayat*[2]

[1]Department of Physics, University of Agriculture, 38040 Faisalabad, Pakistan.
[2]Department of Mechanical Engineering, Universiti Teknologi Petronas, 32610 Bandar Seri Iskandar, Perak, Malaysia

Abstract. A passive thermosyphon heating system was designed, fabricated and tested for its thermal performance in semi-arid and four-season climate of Pakistan. The heating system design was based on a two-stage storage and natural thermosyphon circulation of the water. The objective of the study was to enhance the heating performance of the thermosyphon systems by using a semicircular steel pot collector, water carrying copper coil cover, two step water storage, and side mirror reflectors. The experiments were conducted during April to July 2014 when ambient temperature was reported to vary between 32°C and 44°C. In continuous flow mode operation, the hot water temperature remained between 46°C and 78°C. Since water temperature in the range of 45°C to 50°C is considered suitable for the domestic use, the presented design can easily reach the temperatures even higher than those acceptable for the domestic use.

1 Introduction

Pakistan is one of those countries, which have abundant sunshine during the most days of a year. With 320 days of annual sunshine generally there are four seasons in the district Faisalabad of Pakistan. During this period, climate vary from very low to high temperatures. The residents of Faisalabad use flat plates (black coated solar collectors) to maximize the absorption of radiant energy [1-3]. However, such models have failed in least sunny days of winter season. The low performance of such collector models. So geometrical improvement are referred.

Based on construction, the solar water heaters are classified into active and passive systems. In active systems, there is a need of electric pump, valves and controllers to circulate the water, whereas the passive systems do not require external pumping for water circulation. Normally, direct systems are complicated, expensive, sensitive to freezing conditions and can only heat the water to moderate temperatures in the range of 50 to 60°C [4]. Nevertheless, few design alterations are possible to overcome the freezing issues. In passive solar heating systems, the heat transfer takes place by means of convection. Circulation of the water takes place through a natural process rather than forced circulation. Such systems are cheap, easy to handle and flexible, when compared to active systems.

Thermosyphon systems are generally more reliable type of passive heating systems due to their easy installation, low maintenance cost and long life [5]. The presented work is an example of thermosyphon heating system based on two stage storage and natural thermosyphon circulation of water. A passive thermosyphon heating system was designed and tested for its thermal performance. Effect of different methods of solar heat transformation were also investigated. Although some literature already exsists on such type of water heaters, the objective was to enhance the heating efficiency of thermosyphon system. For this purpose, a circular pot collector, double storage tanks, and side mirror reflectors were used to maxamine the heat transfer.

2 Materials and Methods

A passive thermosyphon heating system was designed and fabricated with locally available materials but not of specialty items. The detailed specifications of the system design are given in Table 1. A circular pot was housed in an air tight box and covered with spiral semi-circular copper pipes connected across the water tanks in such a way to form a thermosyphon system, as shown in Fig. 1.

Mirror reflectors were used at the back side of upper and lower storage tanks inside the box to focus the sun rays on water tanks. A steel reflector was used at the base of the box to reflect the sun rays. A circular pot of steel having diameter of 30 cm was used as a heat absorber. The top rough surface of the pot was painted black to maximize the absorption of radiations, while sides of the pot were polished with silver buffer for high reflection of the radiations. The copper pipe and storage unit were also painted black to maximize the heat absorption.

Table 1. Materials used in fabrication of a passive thermosyphon heating system.

Sr.	Item	Quantity	Specifications		
			Length	Width	Thickness
1	Steel sheet	1	150 cm	120 cm	26 gauge
2	Glass sheet	1	150 cm	120 cm	0.5 cm
3	Copper pipe	5	150 cm	1.27 cm dia.	-
4	Hard board sheet	2	150 cm	120 cm	-
5	Foam sheet	1	150 cm	120 cm	2.5 cm
6	Circular iron pot	1	-	Dia. = 30 cm	0.3 cm
7	Rubber pipe	1.25 cm	300 cm	-	Dia. = 1.27 cm
8	Mirror reflector	4			

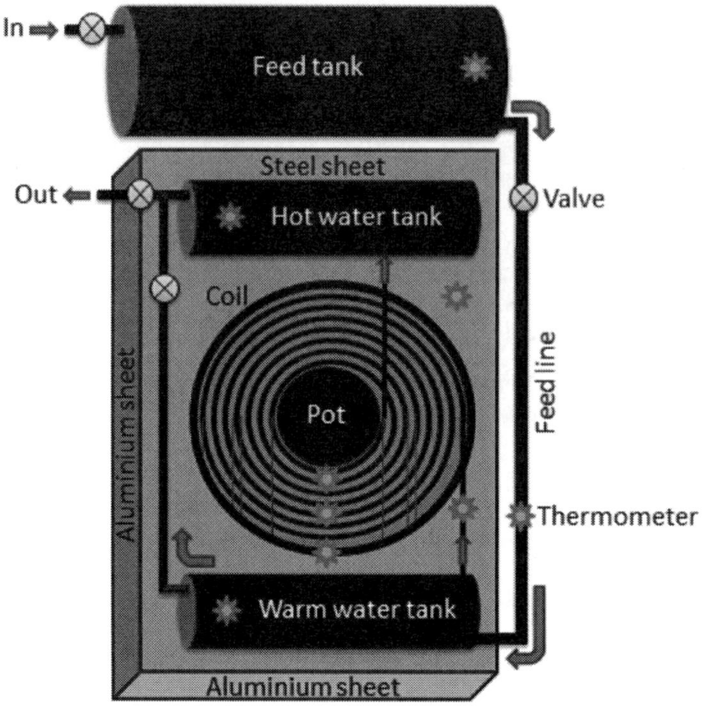

Fig. 1. Detailed drawing of the passive thermosyphon system with two-step water storage.

3 Results and Discussion

In this research work, a new configuration of passive solar water heater was tested with a group of thermosyphon effects, also known as gravity heat pipes. Totally solar based direct radiation energy was used to heat-up the water for domestic use. The water in the lower internal tank was heated through thermosyphon phenomenon. Being less dense, the hot water in the spiral copper pipe-line moved upward to the upper storage tank inside the box. The concerned parameters of the process were studied in the months of April, May, June and July of 2014, under different weather conditions of Faisalabad. The performance of the heater was monitored every day from 8:00 am to 6:00 pm.

The presented design was an example of thermosyphon heating system based on two-stage storage and natural thermosyphon circulation of water. The water pressure of the external tank at its full storage capacity was calculated as:

$$P_T = \rho g W H_T \tag{1}$$

where 'P_T' is the pressure of cold water storage tank, 'g' is the gravitational acceleration (m/s^2), 'ρ' is the density of the cold water (kg/m^3), 'W' is the weight and 'H_T' is the height of the water tank from ground. The equation (1) reveals that the water pressure increases linearly with height of the water tank. This increase in pressure inside the tank causes an increase in the flowrate of the water through flat-plate collector. It predicts an improvement in the system performance on account of faster heat transfer from collector to the water, thereby minimizing the convection and other losses from the collector.

The coefficient of cubic expansion of water was determined at different temperatures. The cylindrical water tanks, made-up of mild steel, were subjected to internal pressure of water at different temperatures. It was observed that the coefficient of cubic expansion

DOI: 10.1051/matecconf/201713102012

linearly increases with water temperature. The pressure is generally considered as uniformly distributed over the internal surface of the tank. If the wall thickness is equal to or less than 1/20 of the internal diameter, the vessel is said to be thin-walled; otherwise, it is considered as a thick-walled vessel. In this work, the water tanks were made of mild steel with yield strength of 176 MPa. The height of the cold-water tank from the ground was 0.9 m, height of the hot water tank was 0.2 m and the wall thickness of the tank was 0.5 mm. This information reveals that for a thin-walled cylinder, the wall thickness is related to the diameter via the following inequality '$t \leq \frac{1}{20} D_T$'. This relation holds for both cold water tank and hot water tank. Therefore a bursting stress also known as tangential or circumferential stress will be set-up on the wall of the cylindrical tank due to the water pressure. The tangential stress for cold water tank can be elaborated from the relation:

$$2\sigma_t H_T t - P_T D_T H_T \tag{2}$$

$$\sigma_t = P_T D_T / 2t$$

where 'σ_t' the tangential stress for cold water tank, 'H_T' is the height of the water tank from ground, 't' is the thickness and 'D_T' is the diameter. Same relation is applicable for tangential stress in hot water tank ($\sigma_t' = P_T' D_T' / 2t$), where P_T' is internal pressure of hot water tank and P_T is internal pressure of cold water tank. In the given heater design, P_T, P_T', σ_t and σ_t' were calculated about $8820\,\mathrm{N/m^2}$, $1910.6\ \mathrm{N/m^2}$, 7.9 MPa and 3.8 MPa, respectively. Since the water tanks were closed, the longitudinal stress σ_1 was also taken into account due to the water pressure on ends of the tanks. Assuming the uniform distribution of the stress, σ_1 was calculated about 9 MPa.

The efficiency on 1500 ml water of the solar water heater was checked, from 8:00 am to 6:00 pm after every 30 minutes during a day. In discontinu flow mode operation, the total water taken out during these ten hours was 30 litters. The cold-water temperature remained in the range of 18 to 28°C for the stated time period. In Fig. 2, the maximum water temperature remained between 48°C to 88°C when ambient temperature was in the range of 32 to 44°C [8].

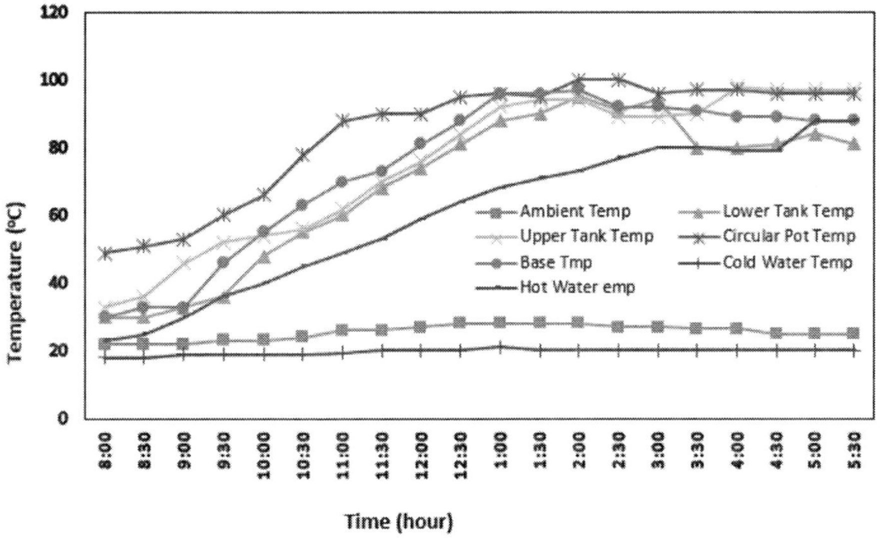

Fig. 2. Change in water temperature over time at different points on the solar water heater during intermittent water flow.

The water temperature in continuous flow mode operation of thermosyphon system was also measured, as reported in Fig. 3. The water discharge rate was fixed to 10 liters per hours. The maximum temperature of 78°C was achieved when ambient temperature was in the range of 36°C to 43°C. The high heating performance of the heater was attributed to the roughness and darkness of the pot surface. The copper coil, covering the pot, also contributed to the maximum energy conversion during the heating process [9, 10]. The reflected part of the radiative energy was trapped in the copper carrying the water. The airtight box with glass cover also minimized the heat and radiative losses from the wooden box.

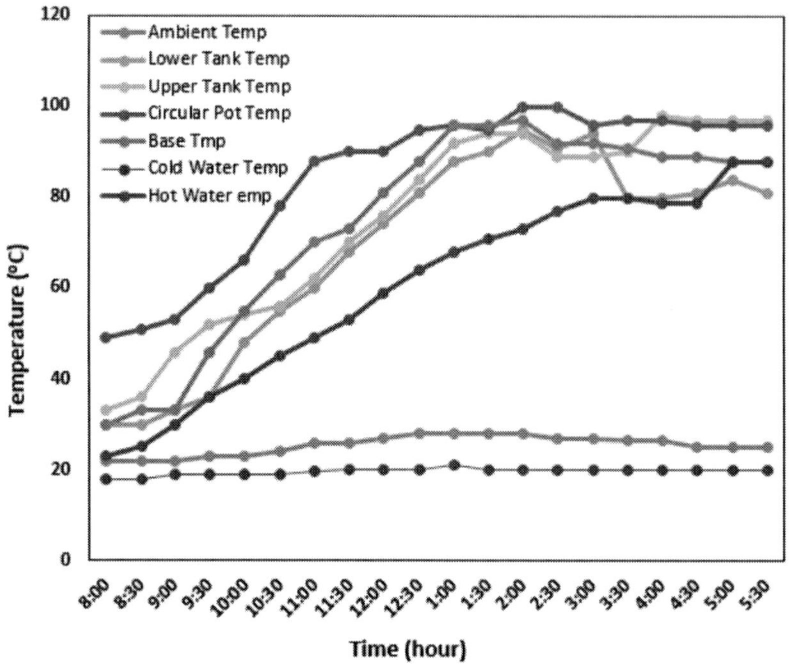

Fig. 3. Change in temperature over time at different points on the solar water heater during continuous flow.

The operating characteristics of the thermosyphon solar water heaters can also have significant impact on the efficiency of the system, because the performance of the system was based on the flowrate from the collector, absorber plate temperature and temperature rise of the water from inlet to outlet [9]. Ho and Chen [11] revealed that the sheet and tube based solar water heating is a convenient way of heating the water for domestic use. Considerable improvement in the collector performance was reported by Ho and Chen [11] by employing a water recycling operation. Similar findings are reported in the presented work.

4 Conclusion

This work reports the design, fabrication and performance study of a passive thermosyphon type solar water heater. The maximum water temperature, sustained by the system in intermittent flow mode operation, remained between 48°C to 88°C. In continuous flow mode operation, the hot water temperature reached to 46°C to 78°C. The water stored in the temperature range of 45°C to 50°C is considered suitable for domestic use. However, this study revealed that the presented design can easily reach the temperatures even higher than those acceptable for domestic use of water. The high performance of the presented heater design was attributed to the rough and dark surface of the pot and semicircular coil covering.

The copper coil also ensured the maximum energy conversion in the heating process. Most of the radiative energy incident on the pot was absorbed by its painted black surface. The reflected part of the radiative energy was trapped in the copper carrying the water. The airtight box with glass cover also helped minimizing the heat and radiative losses.

References

1. A. Hasan, ECM **38(9)**, 847-54 (1997)
2. R. Tang, Y. Cheng, M. Wu, Z. Li, Y. Yu, ECM **51**, 2548-2556 (2010)
3. A. Sakhrieh, A. A. L. Ghandoor, ECM **65**, 715-720 (2013)
4. R. Kumar, M. A. Rosen, ATE **30**, 1764 (2010)
5. S. A. Kalogirou, STCAPECS **30**, 231 (2004)
6. H. Zhai, Y. J. Dai, J. Y. Wu, R. Z. Wang, L. Y. Zhang, ECM **51**, 48-55 (2010)
7. O. Tiago, C. M. Joao. EP **30**, 1344-53 (2012)
8. B. J. Huang, C. T. Hsieh,SE **35**, 31-43 (1985)
9. S. Jaisankar, T. K. Radhakrishnan, K. N. Sheeba. SE **83**, 19431952 (2009)
10. S. Jaisankar, T. K. Radhakrishnan, K. N. Sheeba, S. Suresh, ECM **50**, 2638-2649 (2009)
11. C. D. Ho, T. C. Chen, RE **33**, 655-664 (2008)

Effect of Halogen Substitution on the Absorption and Emission Profile of Organometallic Perovskites

Ling Jin Kiong[1], and *Jose* Rajan[1]

[1]University Malaysia Pahang, Faculty of Industrial Sciences & Technology, Material Technology Department, Advance Material Laboratory, Lebuhraya Tun Razak, 26300 Gambang, Kuantan, Pahang Darul Makmur.

Abstract. Comparative study on the optical properties of methylamide lead mono- and hybrid-halide perovskite samples, from the perspective of its crystal structure, which were synthesized in ambient condition (temperature ~26.6 °C, humidity ~65%), a step toward industrial commercialization, were carried out. In this study, in-depth information on both crystal structure and optical properties of the methylamide lead halide perovskite was the main focus, including study of relationship between these two factors via real world experimentation. Increment in lattice stain can be observed when the bromide or chloride ions were substituted, resulting in higher lattice strain in hybrid-halide samples. The lattice strain provides kinetic energy to the electrons, facilitating the promotion of exciton which increased the number of photon emitted during recombination. However, increasing lattice strain results in the widening of the energy gap of the samples by shifting the energy level to a higher or lower level. The results provide some insight on the relationship between crystal structure and optical properties of perovskite which paces a route for future cross-factor experimentation under uncontrolled ambient environment, in an effort to unravel more underlying working mechanism of perovskite solar cells.

1 Introduction

In response to Paris Agreement, more than 120 countries had put in effort to counteract climate changes. One of their efforts is by implementing conversion of traditional industries to green industries, which result in advancement of photovoltaic industries all around the globe [1]. To ensure the further development of photovoltaic (PV) industry, improvement of PV efficiencies as well as performance and reduction in product's cost are both important. Perovskite solar cells (PSCs) were first introduced in 2009, with its efficiency leapfrogged from ~3.8% [2] to ~21% [3] in 2016, had attracted attention from the research community. Its capability of optimal operation even in low light intensity and solution-processability, which offer low cost fabrication, make PSCs a suitable candidate to replace non-economical silicon solar cells (SSCs) that monopolized nowadays photovoltaic industry. In addition, both nano-scale thickness and absence of liquid electrolyte opened up huge gateway for the application of PSCs technology, such as building integrated photovoltaic system (BIPV) or flexible photovoltaic device. Aside from pros, the biggest con of PSCs will be its stability. Efforts had been taken to improve the stability of PSCs, including introducing a polymeric framework to the perovskite crystal structure to prolong the decomposition of the perovskite film but results in poor fill factor (<40%) and large hysteresis (~50%) [4]. A study done by Fakharuddin et. al. had reported that not only the perovskite film itself but the morphology, porosity and chemical stability of the electron transport layer play a vital role as well in determining the stability of PSCs where the application of TiO_2 nanorod instead of conventional TiO_2 layer had demonstrated higher PSCs durability [5]. Though glance on some of the factors affecting the stability had been obtained, ways of improving the stability without affecting the PSCs performance had not yet been identified. Fabrication cost of PSCs, been the main factor determining PSCs marketability, had been revised to reduce the overall cost of PSCs manufacturing by remodeling the fabrication method, from one-step deposition method to two-step deposition method, as well as from solution deposition to evaporation techniques [6-9]. Studies on the underlying working mechanism of PSCs had also been carried out as an effort to understand the physics of PSCs and to gain new insight for further PSCs device improvement. Stranks et. al. had reported that the incorporation of chloride atoms into the $MAPbI_3$ structure will helps to improve the diffusion length of electrons and reduces the recombination rate with no focus on the optoelectronic properties [10]. The effect of non-stoichiometric precursors on $MAPbI_3$ PSCs had also been investigated, showing an improvement on the photovoltaic performance [11] as well as the fill factor [12] in step with the increasing PbI_2 amount. Instead of real-life experiment, several studies are done using computer simulation, mostly using Density Functional Theory (DFT), MD simulation, Group theory and GW approximation, to unravel the underlying working mechanism and physics of the ABX_3 perovskite material [13-16].

Computation simulation was able to estimate and explain the physics and working mechanism observed from PSCs devices under specific set of condition, which might be totally different in real-life ever-changing environmental condition. On the other hand, the stability of perovskite film cannot be studied using computer simulation method, indicating the importance of real-time experimentation in further perovskite investigation. With both real-time experimentations and computational simulations done by previous researchers, it is undeniable that adequate knowledge about perovskite had been revealed despite from some yet unraveled secret hidden within the physics of perovskite. Almost all the previous studies are carried out in line with 3 main directions, which are (i) improving perovskite stability, (ii) improving efficiency and performance of PSCs, and (iii) studying the physics of perovskite. Seldom a cross direction study is been carried out before, which might offer some recommendation on ways for PSCs improvement. One-directional study does provide some explanation on PSCs performance but seldom provide in-depth information regarding the physics of PSCs and the overall relationship between the crystal structure of perovskite and its optoelectronic properties. Therefore, a cross-directional study is crucial for PSCs to breakthrough from it efficiency bottleneck.

Herein, we studied the optical properties of perovskite from the perspective of crystal structure when the halide of $MAPbX_3$ (X = I, Br, or Cl) is substituted. A total of 9 samples are studied, including 5 mono-halide $MAPbI_3$ samples with different molar ratio and 4 hybrid-halide $MAPbI_{3-x}Y_x$ samples (where Y = Br or Cl; X = 1 or 2). One-step solution-decomposition technique is used to prepare all 9 samples under ambient environment (temperature ~26.6 °C, humidity ~65%), which is one of the crucial factor in future low cost industrial commercialization. It is well known that the optoelectronic properties of perovskite originated from its non-stoichiometry crystal structure. Our main focus is to study the effect of halide substitution on perovskite crystal structure and its effect on perovskite optical properties from the crystal structure point of view. The crystal structure of the perovskite samples were studied using X-ray Diffraction (XRD) and its optical properties were obtained using UV-Vis-NIR Absorption Spectroscopy as well as Photoluminescence Spectroscopy. We reported a detailed relationship between crystal structure and optical properties of perovskite samples when the halide, X of MAPbX3 is substituted.

2 Experimental Section

2.1 Perovskite Preparation

One-step solution deposition technique was adopted in this study to synthesis organometallic halide perovskite. All the chemicals used in this study were obtained from Aldrich-sigma and used without any further purification. Methylamine (CH_3NH_3, 33 wt% in absolute ethanol) was mixed with hydroiodic acid (HI, 57 wt% in H_2O) at 0 °C by stirring for 2 hours, followed by evaporation using rotatory evaporator at 70 °C for 1 hour. The precipitate was then cleaned using diethyl ether for three times until a white color precipitate was obtained followed by drying in oven at 70 °C overnight. The methylamide iodide, MAI powder synthesized was used for the following sample preparation without any further purification. Lead (II) iodide (PbI_2, 99%) powder and the prepared methylamide iodide powder were mixed, 40 wt% in dimethylformamide, DMF solution (EMSURE®) in the ratio of 1:3; 1:2; 1:1; 1:2; and 1:3. The resultant solutions were stirred at 70 °C overnight. Hybrid-halide perovskite samples were prepared by mixing the above prepared MAI powder with lead (II) bromide ($PbBr_2$, 99%) powder or lead (II) chloride ($PbCl_2$, 99%) powder, 40 wt% in DMF solution, followed by stirring at 70 °C for overnight. Ratio of (MAI:PbI2:PbX2) used in hybrid-halide perovskite preparation were (1:0.33:0.67) and (1:0.67:0.33). For perovskite film preparation, titanium dioxide, TiO_2 paste (Solaronix) was dissolved in ethanol with a weight ratio of 1 wt% of TiO_2 paste in 6 wt% of ethanol. The TiO_2 electron transport layer was then deposited on glass slide by spin-coating at 1000 rpm for 20 seconds. The substrates were heated at 500 °C, ~1 hour for complete drying. Once the substrate cooled down, perovskite solution was deposited via spin-coating at 1000 rpm for 15 seconds, followed by substrate heating at 70 °C for 1 hour. A dark colored perovskite film was obtained for all the sample.

2.2 Sample Characterizations

Absorption profile of the samples were recorded using UV-Vis-NIR absorption spectroscopy, UV-2600 of Shimadzu with integrated sphere 2600 plus. Single scan mode was applied, with medium scanning speed, scanning from 900 nm to 400 nm, peak's threshold value set to be 0.001 at points 4 and slit width of 5 nm. The light source changed wavelength at 323 nm whereas the detector change wavelength at 830 nm. The emission profiles of all the samples were recorded using photoluminescence spectroscopy, Edinburgh Instruments NIR 300/2. Source light path used is Xe900 with wavelength 300 nm whereas the emission detector light path used was Visible 500 nm with iris set at 95, scanning wavelength ranged from 620 nm to 900 nm with excitation wavelength set at 600 nm, bandwidth of 20 nm, and 0.5 s dwell time. To study the crystal structure of the samples, X-ray diffraction Bruker, D8 Advance model was employed. The XRD spectra were taken within the range of 5 to 60° 2θ at room temperature using CuKα radiation. Tube voltage was adjusted to be 40 kV, tube current of 40 mA, step-scan mode with a step size of 0.02° 2θ and counting step of 1-3s per step.

DOI: 10.1051/matecconf/201713103001

3 RESULTS AND DISCUSSION

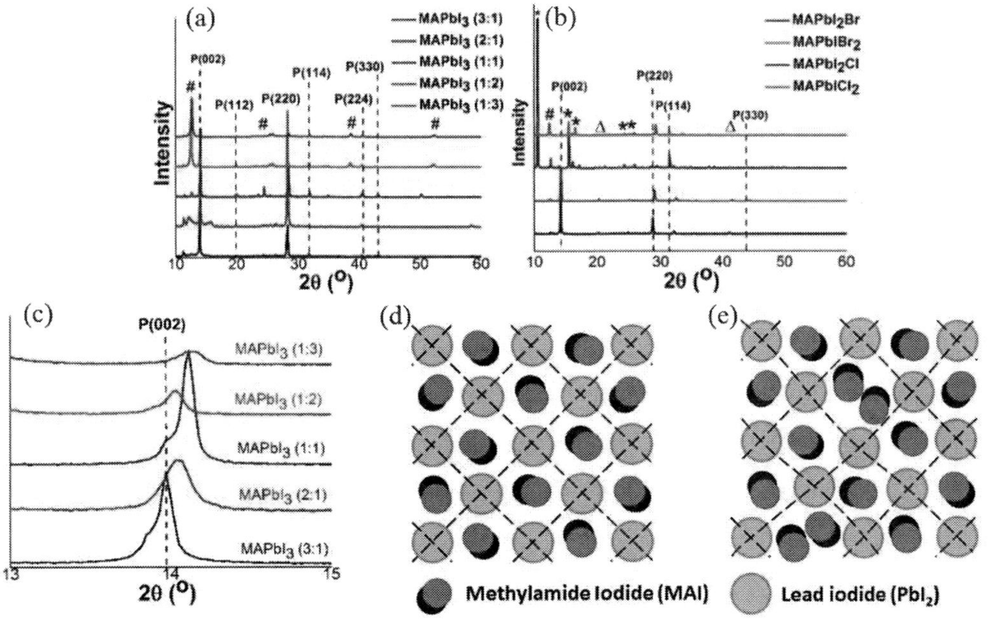

Figure 1: XRD spectra for mono-halide sample (a) and hybrid-halide samples (b), indicating peaks for perovskite (P), PbI_2 (#), $PbBr_2$ (Δ), and $PbCl_2$ (*). Graph (c) highlighted the left-skewed behavior of the perovskite (002) crystal plane when the MAI:PbI_2 ratio was altered. (d) and (e) illustrates the lattice strain experienced by the crystal structure when amount of MAI molecules increases.

Briefly, fig 1(a) and 1(b) show the XRD spectra of all the samples. All 9 samples show tetragonal crystal structure with lattice parameter a = b = ~8.9 Å, c = ~12.6 Å for $MAPbI_3$ as well as $MAPbI_{3-x}Cl_x$ samples, and a = b = ~8.7 Å, c = ~12.4 Å for $MAPbI_{3-x}Br_x$ samples respectively. Regardless to mono-halide or hybrid-halide perovskite samples, all the samples shown high peak intensity at ~14°, indicating plane (002) as the main crystal plane in all the perovskite samples, excluding mono-halide sample with ratio (2:1) where the main crystal plane was plane (220). By studying the position of the (002) crystal plane of mono-halide perovskite samples, the peak of the (002) planes were clearly left-skewed when the amount of MAI added was increased, as illustrated in fig 1(c). To study this left-skewed behavior of the (002) plane when MAI amount was altered, Williamson-Hall equation was applied to calculated the lattice strain and average crystallite size of the samples.

$$B_t \cos\theta = \frac{k\lambda}{L} + 4\varepsilon \cos\theta \quad (1)$$

$$B_t^2 = B_{measure}^2 + B_{Instrumental}^2 \quad (2)$$

B_t represent the sample line broadening, calculated using equation 2, k is the crystal constant, ~0.94, **λ** is the X-ray wavelength used in XRD testing, L is the average crystallite size, **ε** is the lattice strain, **θ** represent the angle, $B_{measure}$ represent the peak broadening at FWHD, and $B_{Instrumental}$ is the line broadening results from the measuring instrument which can be obtained from the peak of the standard sample.

DOI: 10.1051/matecconf/201713103001

Table 1: Lattice parameter, lattice strain, average crystallite size of the perovskite samples.

Sample	Lattice parameter (Å)		Lattice strain	Average crystallite size (nm)
	a = b	c		
$MAPbI_3$ (3:1)	8.9644	12.6764	0.0312	115.206
$MAPbI_3$ (2:1)	8.9097	12.6213	0.0091	448.697
$MAPbI_3$ (1:1)	8.8826	12.5484	0.0080	435.742
$MAPbI_3$ (1:2)	8.9187	12.6212	0.0103	335.776
$MAPbI_3$ (1:3)	8.8646	12.5122	0.0064	497.908
$MAPbI_2Br$	8.7758	12.4229	0.0103	337.370
$MAPbIBr_2$	8.7234	12.3875	0.0096	354.750
$MAPbI_2Cl$	8.9644	12.5665	0.0250	142.675
$MAPbICl_2$	8.9828	12.6212	0.0107	332.705

From table 1, it was noticeable that the lattice strain experienced by (3:1) ratio sample was higher compared to the equimolar ratio sample. The higher lattice strain experienced by (3:1) ratio sample originated from the vacancy occupation of excess MAI molecules, illustrated in figure 1(e). When excess MAI were mixed to the PbI2 powder, the cation A site will be occupied by more than one MAI molecules, stretching the crystal structure, creating some degree of strain on the crystal structure. The strain increased when more number of MAI molecules occupied the cation A site, stretching the crystal lattice. Besides, the peak for PbI2 hexagonal crystal structure can be observed on perovskite sample with ratio (1:2) and (1:3). The insufficient amount of MAI molecules to occupy the cation A site resulted in empty cation A site, which increased the tendency for the PbI2 tetragonal cage to form pure PbI2 hexagonal crystal structure. The peak intensity for PbI2 plane, located at ~12.6°, increased when the amount of PbI2 in the perovskite sample increased, as shown in the XRD spectra for both sample with ratio (1:2) and (1:3) respectively. No PbI2 hexagonal crystal structure peak can be observed in the (3:1) perovskite sample where all the cation A sites were occupied by the over sufficient MAI molecules, and therefore reducing the convertion tendency from PbI2 tetragonal cage to hexagonal crystal structure. The PbI2 hexagonal crystal structure peak can also be detected by XRD from hydrated perovskite structure upon longer exposure duration to the ambient environment. In this study, the XRD spectra were obtained instantaneously after the samples were prepared. Therefore, the possibility that the PbI2 hexagonal crystal structure peak originated from a hydrated perovskite sample was very insignificant and can be ruled out. Comparison in between hybrid-halide samples highlighted higher lattice strain experienced by the chloride-substituted sample than the bromide-substituted sample. Chloride ions, with a smaller ionic radii compared to the bromide, created more vacancy when substituting iodide ions, causing the crystal space to shrink in order to reduce the vacancy space, generating more strain on the crystal lattice compared to bromide ions. However, when the amount of substitution ions, such as bromide and chloride, increases, the lattice strain experienced by the crystal structure reduced. In terms of lattice parameters, the bromide-substituted samples had a smaller value of lattice parameters, due to the bromide's smaller ionic radii. However, that was not the case of chloride –substituted samples. In spite of smaller ionic radii, the lattice parameter of the chloride-substituted samples were almost the same as the mono-halide samples. The radius of chloride ions might be too small to have any effect on the overall perovskite crystal parameter. With the study done on the effect of halogen substitution on the crystal structure of the perovskite, changed on the optical profile of the perovskite can be studied in the perspective of perovskites' crystal structure.

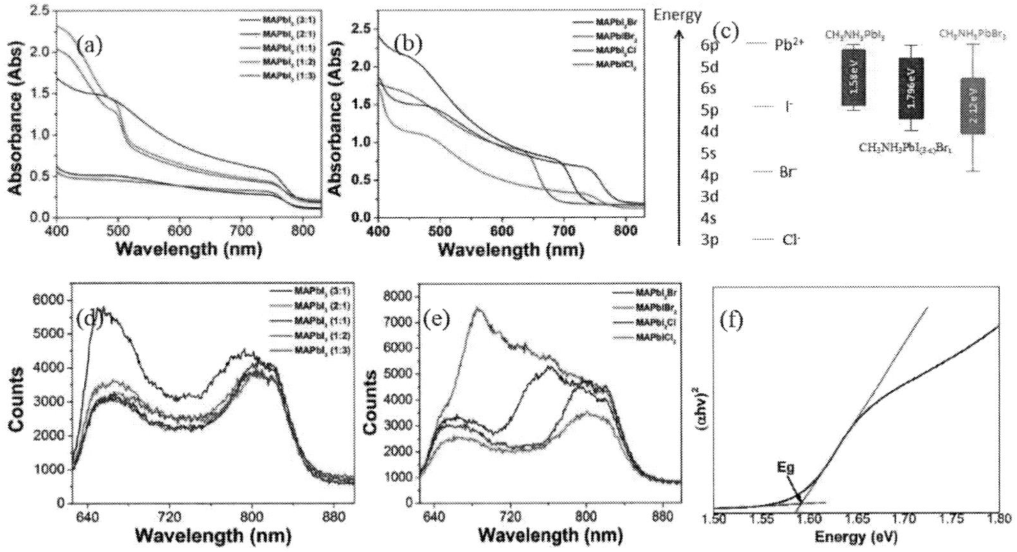

Figure 2: (a) and (b) illustrated the absorption spectra for mono-halide and hybrid-halide samples respectively, (c) illustrated the energy diagram of the mono-halide and hybrid-halide samples, (d) and (e) illustrated the photoemission spectra for mono-halide and hybrid-halide samples respectively, lastly (f) energy gap investigation using Tauc's plot technique.

Table 2: The calculated tolerance factor and energy gap calculated using Tauc's plot technique.

No.	Sample	Tolerance Factor (t)	Energy gap (eV)	Absorption cross section	Photoemission cross section (10^5)
1.	$MAPbI_3$ (3:1)	0.9175	1.598	166.1	9.03
2.	$MAPbI_3$ (2:1)	0.9175	1.582	162.2	6.52
3.	$MAPbI_3$ (1:1)	0.9175	1.575	405.5	6.74
4.	$MAPbI_3$ (1:2)	0.9175	1.566	378.9	7.05
5.	$MAPbI_3$ (1:3)	0.9175	1.557	352.9	6.94
6.	$MAPbI_2Br$	0.9233	1.710	470.5	8.56
7.	$MAPbIBr_2$	0.9294	1.849	373.6	11.60
8.	$MAPbI_2Cl$	0.9271	1.597	421.9	7.20
9.	$MAPbICl_2$	0.9376	1.590	268.7	5.97

Figure 2 (a), (b), (d) and (e) illustrated the absorption and photoemission spectra for mono-halide and hybrid halide perovskite respectively. The energy gap of the perovskite materials were obtained using the Tauc's plot method where the absorption coefficient, α of the materials were calculated using equation 3.

$$\alpha = \left(\frac{1}{t}\right)\left(\frac{A}{\log e}\right) \tag{3}$$

Where t represent the thickness of the sample, A is the absorbance of the material, and e is the electron charge ($e = 1.6\times10^{-19}$ C). Graph of $(\alpha hv)^2$ against hv was plotted, as shown in fig 2(f), where the intersection point of the two tangent point was the energy gap of the studied materials. The calculated energy gaps were recorded in table 2 where clear observation can be made where the energy gap widened when the amount of cation A added increases. Additional amount of cation A might played an important part in altering the total number of charge within the perovskite crystal structure, where different amount of free charges will altered the level of the Fermi energy. However, the electric field in perovskite samples were not generated by the free electrons, but from the overall charge of the perovskite structure due to the position orientation of the cation A. The lattice strain experienced by the crystal lattice do have some degree of influence on the Fermi energy level, but the small lattice strain experienced by the crystal structure had no significant effect on the Fermi energy level in this case. From Schrondinger equation, illustrated in equation 4, the different in charges might affect the conduction level as well.

$$n = 2\left(\frac{2\pi m_e^* kT}{h^2}\right)^{3/2} exp\left(\frac{E_F - E_C}{kT}\right) \quad (4)$$

$$E_{F1} - E_{C1} = \log\left(\frac{n_1}{n_3}\right)(E_{F2} - E_{C2}) \quad (5)$$

Equation 4 can be further derived into eq 5, explaining the relationship between different number of charges on the energy level of Fermi energy and conduction band energy level of two different samples. In this study, since the different in Fermi energy level due to the lattice strain was very insignificant, conduction energy level will be the main focus. The additional MAI molecules in cation A site might reduce the overall charge of the perovskite crystal by forming secondary bonding with the surrounding PbI_2 tetragonal cage, which were previously left unbounded. From eq 5, less charge will eventually widened, or increased, the energy level of the conduction band, results in the widening of the energy gap for the mono-halide samples. Another interesting fact regarding to the mono-halide samples was the absorption cross section of the (3:1) and (2:1) molar ratio samples. Samples with additional MAI molecules in cation A site showed low absorptivity compared to the other samples where samples with less MAI molecules absorbed more photons. The fact that additional MAI in cation A site formed more secondary bonding with the PbI_2 tetragonal cage might eventually (i) reduced the number of charges available to absorb photons and get excited, and (ii) reducing the internal electric field and increased the binding energy of the charges where more energy was required to excited the charges. Samples with depletion of MAI molecules show considerably high absorptivity and multi-bandgap can be observed with the second energy gap located around 2.2 eV, which was the energy gap of the PbI_2 hexagonal crystal structure.

Instead of altering energy level, $MAPbI_{3-x}Br_x$ samples show yet another behavior in response to the halogen substitution, which was band-bending. The energy gap of the bromide-substituted samples lied right in between the energy gap value of mono-halide $MAPbI_3$ and $MAPbBr_3$ perovskites, illustrated in fig 2(c). When both PbI_2 and $PbBr_2$ powder were mixed with the MAI powder, both $MAPbI_3$ and $MAPbBr_3$ perovskite crystal phases were formed with their own respective electronic band structure. Both electronic band structure, especially valence and conduction band, underwent relocation of energy band, which was also known as band-bending, and results in an intermediate energy value. However, this was not the case for the chloride-substituted samples. Two crystalline phase, the $MAPbI_3$ phase and $MAPbCl_3$ phase, might also co-exist in the hybrid-halide chloride-substituted samples. However, the energy level of the valence band for $MAPbCl_3$ might be too low to post any effect on the valence band of the $MAPbI_3$ phase. The valence band of the $MAPbI_3$ perovskite was formed by the I 5p orbital whereas the valence band of the $MAPbCl_3$ perovskite was formed by the Cl 3p orbital, where the conduction band of both perovskites were formed by the Pb 6p orbital. The difference in energy between I 5p and Cl 3p orbitals might be too great that most of the charges were excited from the I 5p orbital and not from the Cl 3p orbitals. Though, charges can still be excited from Cl 3p orbital at a longer wavelength, but the absorption can only took place at the forbidden zone between the I 5p prbital and Pb 6p orbital.

No sharp peaks were observed in the photoemission spectra illustrated in fig 2(d) and 2(e). This broadening peaks can be a results of degraded samples. However, sharp peaks were obtained from the XRD analysis, indicating that the samples were not degraded. Therefore, the broadening of the photoemission peaks were the only results of spin-orbit coupling effect. The spin-orbit coupling effect results in the splitting of orbitals into several small sub-orbital. With the splitted sub-orbital, electrons were relaxed to several different energy level, emitting photons with different wavelength depending to the energy level they ended up in. This phenomenon results in the broadened photoemission peak. The irregular crystal surface, results from the addition of MAI ions or substitution of ions with different ionic radii, gave rise to even more energy trap step, trapping the exciton before recombination on the valence band.

4 Conclusion

In conclusion, we studied the crystal structure as well as optical properties of the perovskite samples upon halide substitution and the relationship between the crystal structure and optical properties of this sample. Hybrid-halide experienced higher lattice strain compared to the mono-halide samples due to the difference in size between the substituted halide ions and the existing halide ions where substituting chloride ions generated more lattice strain compared to the bromide ions. In term of optical properties, band bending phenomenon can be observed in MAPbI3-xBrx samples when iodide ions were substituted by bromide ions. However, the energy gap of MAPbI3-xClx bends slightly when iodide ions were substituted by chloride ions, mainly due to the huge different between the Cl 3p orbital and I 5p orbital. Both spin-orbit coupling and irregular surface or impurities due to the uncontrolled ambient environment gave rise to the formation of sub-orbital energy level and energy trap steps, resulting in energy loss during exciton recombination. This study aimed to provide some knowledge on providing new perovskite materials by studying the effect of ionic radii of different ions on the crystal structure and optical profile of the perovskite. The effect of halogen substitution on the full PSC device's performance need to be carried out in order to gain full understand on the halogen substitution effect.

Acknowledgments

Million thank Professor Dr. Jose Rajan for his financial and knowledge support throughout the study. This study is funded by the Material Technology Department, Faculty of Industrial Sciences & Technology (FIST) of Universiti Malaysia Pahang. I would like to also thank Dr. Azhar Fakharuddin for all the useful discussions and assistances.

References

1. Movellan, J., *The 2016 Global PV Outlook: US, Asian Markets Strengthened by Policies to Reduce CO2*. 2016.
2. Kojima, A., et al., *Organometal Halide Perovskites as Visible-Light Sensitizers for Photovoltaic Cells*. Journal of the American Chemical Society, 2009. **131**(17): p. 6050-6051.
3. Saliba, M., et al., *Cesium-containing triple cation perovskite solar cells: improved stability, reproducibility and high efficiency*. Energy & Environmental Science, 2016. **9**(6): p. 1989-1997.

4. Manshor, N.A., et al., *Humidity versus photo-stability of metal halide perovskite films in a polymer matrix.* Physical Chemistry Chemical Physics, 2016. **18**(31): p. 21629-21639.
5. Fakharuddin, A., et al., *Vertical TiO2 Nanorods as a Medium for Stable and High-Efficiency Perovskite Solar Modules.* ACS Nano, 2015. **9**(8): p. 8420-8429.
6. Liu, D., J. Yang, and T.L. Kelly, *Compact Layer Free Perovskite Solar Cells with 13.5% Efficiency.* Journal of the American Chemical Society, 2014. **136**(49): p. 17116-17122.
7. Liu, M., M.B. Johnston, and H.J. Snaith, *Efficient planar heterojunction perovskite solar cells by vapour deposition.* Nature, 2013. **501**(7467): p. 395-398.
8. Chen, Q., et al., *Planar Heterojunction Perovskite Solar Cells via Vapor-Assisted Solution Process.* Journal of the American Chemical Society, 2014. **136**(2): p. 622-625.
9. Zhou, H., et al., *Interface engineering of highly efficient perovskite solar cells.* Science, 2014. **345**(6196): p. 542.
10. Stranks, S.D., et al., *Electron-Hole Diffusion Lengths Exceeding 1 Micrometer in an Organometal Trihalide Perovskite Absorber.* Science, 2013. **342**(6156): p. 341.
11. Wang, Q., et al., *Large fill-factor bilayer iodine perovskite solar cells fabricated by a low-temperature solution-process.* Energy & Environmental Science, 2014. **7**(7): p. 2359-2365.
12. Roldan-Carmona, C., et al., *High efficiency methylammonium lead triiodide perovskite solar cells: the relevance of non-stoichiometric precursors.* Energy & Environmental Science, 2015. **8**(12): p. 3550-3556.
13. Kim, J., C.-H. Chung, and K.-H. Hong, *Understanding of the formation of shallow level defects from the intrinsic defects of lead tri-halide perovskites.* Physical Chemistry Chemical Physics, 2016. **18**(39): p. 27143-27147.
14. Mosconi, E., P. Umari, and F. De Angelis, *Electronic and optical properties of MAPbX3 perovskites (X = I, Br, Cl): a unified DFT and GW theoretical analysis.* Physical Chemistry Chemical Physics, 2016. **18**(39): p. 27158-27164.
15. Carignano, M.A., et al., *A close examination of the structure and dynamics of HC(NH2)2PbI3 by MD simulations and group theory.* Physical Chemistry Chemical Physics, 2016. **18**(39): p. 27109-27118.
16. He, Y. and G. Galli, *Instability and Efficiency of Mixed Halide Perovskites CH3NH3AI3–xClx (A = Pb and Sn): A First-Principles, Computational Study.* Chemistry of Materials, 2017. **29**(2): p. 682-689.

Development and implementation of bottom ash crushing system in Submerged Scrapper Conveyor (SSC) for Coal-fired Power Plant

Firas Basim Ismail[1*], *Elaine* Why Siew Kuan[1], *Mohammad* Shakir Nasif[2]

[1]Power Generation Unit, Institute of Power Engineering (IPE), Universiti Tenaga Nasional, 43000 Kajang, Selangor, Malaysia.
[2]Mechanical Engineering Department, Universiti Teknologi Petronas, 32610 Tronoh, Malaysia.

Abstract. The existence of Submerged Scrapper Conveyor (SSC) in coal-fired power plant is to handle the by-product of bottom ash. However, soot-blowing will be performed sometimes, in order to remove slag formed at the boiler furnace wall. Thence, this lead to a sudden loading of large amount of slags and bottom ash at SSC after soot-blowing, causing SSC conveying system to jam and conveying chain breakage. In this paper, a new SSC design with additional crushing system is proposed. By implementing the new design proposed, it is expected to improve the overall current performances, and to reduce the trip issue of SSC in coal-fired power plant. The new 3D model of SSC is designed, and stress-strain simulation of the model is analysed by using software of PTC Creo Parametric. Final cost and safety factor analysis of model is made to prove its validation.

1 Introduction and background

The Submerged Scrapper Conveyor (SSC) is known as to convey and remove the by-product of combusted pulverized coal, bottom ash and slag. It is normally installed at the bottom of boiler. The SSC conveyor drag chain consists of flights, which are located between two parallel length of chain that scrape along the lower surface of the water impounded trough. Further discharge of the bottom ash is transported by belt conveyor, from SSC to ash silo or ash ponds [1]. The conveyor is submerged in water trough below the boiler ash reject hopper. Water contain inside the conveyor box quenches hot bottom ash which falls from the boiler. The fallen bottom ash is then dewatered, as it being scrapped up the inclined section [2].

During the incomplete combustion of pulverized coal at boiler furnace wall, the process will cause accumulation of ash residue on boiler furnace wall surface. Hence, the layers of ash residue formed may seriously inhibit heat transfer to the steam cycle [3]. Therefore, soot blowing process is required, in order to remove these residues. In this soot blowing process, soot blowers are used by using high pressure steam, which impacts on a residue, and leads to its fracture [4]. In this case, residue removal occurs due to mechanical and thermal stresses, induced in the residue [5, 6]. The fallen slag will then pass through the ash reject hopper, and then submerged in the water content of SSC [7, 8]. Further ash discharging process after SSC, is settled by belt conveyors to the bottom ash silo, or to the ash pond.

In this paper, it is aimed to enhance the current SSC design by adding extra feature of crushing system, to reduce the falling rate and size of slag into SSC after soot blowing process. By doing this, SSC tripping issue can be prevented, and improves its durability. It is also expected that the modification of SSC proposed will contribute to plant performance improvement, by reducing the numbers of unplanned outage on SSC. The project discussed in this paper will focus on PTC Creo Software for designing and stress-strain analysis simulation.

2 Methodology

In order to design a SSC with new additional crushing system, methodology of five phases are required. The phases are: Studies of problem root cause phase, SSC 3D model designing phase, auxiliaries and material selection for design phase, Simulation analysis phase, and model analysis phase

2.1 Studies of problem root cause

Before designing a new implementation of crushing system to the present SSC, it is important to have a review on slag formation and soot blowing process at boiler furnace wall. The important properties of both bottom ash and boiler slag are studied, and listed in Table 1.

* Corresponding author: Firas@uniten.edu.my

 DOI: 10.1051/matecconf/201713103002

Table 1. Physical properties of bottom ash & boiler slag [9].

Property	Bottom Ash	Boiler Slag
Specific Gravity	2.1 – 2.7	2.3 – 2.9
Dry Unit Weight	720 – 1600 kg/m³ (45 – 100 lb/ft³)	960 – 1440 kg/m³ (60 – 90 lb/ft³)
Shear Strength (Friction Angle)	38 – 42° 32 - 45° (<9.5mm size)	38 – 42° 36 - 46° (<9.5mm size)
Maximum Dry Density kg/m³ (lb/ft³)	1210 – 1620 (75 – 100)	1330 – 1650 (82 - 102)

The properties provided are used in solving for feed rate of crusher by

$$Q=\pi 60DW\omega L\rho_{\beta} \tag{1}$$

where Q is feed rate of crusher (t/hr), D is diameter of roller (m), W is width of roller (m), ω is roller speed (rpm), L is gap between two rollers (m), and ρ_{β} is the bulk density of material (t/m³) [10].

2.2 3D model design

Firstly, the desired output size of the crushed slag load is set, referring to the calculation of crusher distance, Nip angle formula by

$$\cos\theta = \frac{(R+L)}{(R+d)} \tag{2}$$

where θ is the nip angle (°), R is the radius of roll crusher (m), L is the gap between two rollers (m), and d is the input material size (m) [10]. The optimum nip angle is set to be 35° as constant variable, and expected output size of the slag infeed is to range between 2mm to 3mm.

Next, the dimension (diameter) of input shaft at the driving motor and the driving pulley is calculated, using the DE-Goodman criterion by

$$d_{in} = \left(\frac{16n}{\pi}\left\{\frac{1}{S_e}\left[4(K_fM_a)^2 + 3(K_{fs}T_a)^2\right]^{1/2} + \frac{1}{S_{ut}}\left[4(K_fM_m)^2 + 3(K_fT_m)^2\right]^{1/2}\right\}\right)^{1/3} \tag{3}$$

where d_{in} is the input shaft diameter, n is the safety factor which is set to be 2.5, S_e is the endurance limit at the critical location of a machine part in the geometry and condition of use, S_{ut} is the ultimate tensile stress of the shaft material, K_f and K_{fs} are the fatigue stress-concentration factors for bending and torsion of shaft, M_a and M_m are the alternating and midrange bending moments, and T_a and T_m are the alternating and midrange torques [10].

3D model drawing is done by using software of PTC Creo Parametric. The model dimension unit used in software drawing is set to be millimetre (mm), and it is drawn in scale 1:1 according to general dimension. The SSC model parts are firstly drawn in separate individual, and then followed by model assembly. The drawing of SSC model only focused on the crushing system, instead of the conveying system inside.

2.3 Auxiliaries and material selection for design

The material for each parts of the SSC model are selected, based on current existing crushing system technology. The material selected for each SSC individual parts, are assigned to the model drawn, by referring to the material dictionary in PTC Creo Parametric.

The suitable crusher duty motor for the new SSC model crushing system developed is selected after the designing process. The rated output power (kW and hp) of the motor is chosen according to direct calculation of the crusher power by

$$Design\ Hp = Transmitted\ Hp\ x\ N_{sf} \tag{4}$$

where Hp is horse power, and N_{sf} is service factor. After the rated output power of the motor is determined, the total number of model to operate is selected according to current existing technical data.

 DOI: 10.1051/matecconf/201713103002

2.4 Stress-strain simulation

Once the selected type of material is assigned to the new SSC model, simulation of stress-strain analysis procedure is then carried on. The use of software PTC Creo Parametric is remained for this procedure.

Firstly, the constraint is set at the bottom surface of the SSC model, as a mean of the model is fixed on ground. Then, the estimated value of slag load is applied perpendicularly to each unit of the roll crusher in SSC model as shown in Figure 1. The setting of the load is done on the roll crushers, in order to illustrate the falling process of slag after soot blowing process performed at boiler furnace wall. The stress-strain simulation of full SSC model is first being analyzed, then followed by individual parts. Three types of simulation result are obtained, such as von Mises stress, maximum shear stress, and displacement of parts deformation.

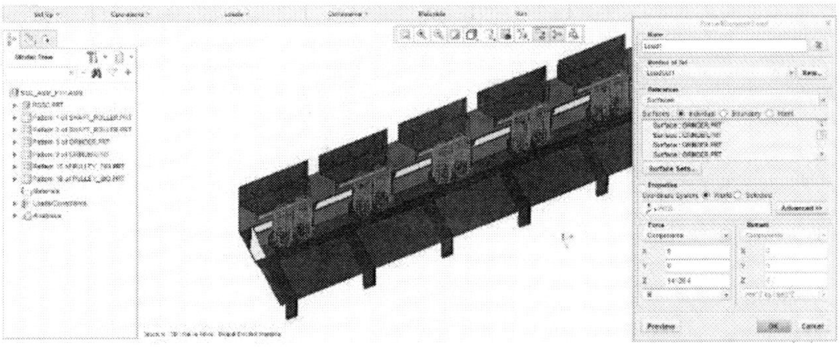

Figure 1. Stress application on model parts for stress strain simulation.

2.5 Model analysis

The calculation of the SSC model safety factor is done, right after the results of stress-strain simulation are generated. Only the critical part of the SSC model which has experienced the highest stress and strain, will be analyzed for its safety factor by

$$SF = \frac{Ultimate\ Tensile\ Strength}{Allowable\ Stress} \tag{5}$$

The optimum safety factor of SSC model is set to be 2. Any of the safety factor figure calculated less than 2 is considered to be unsafe.

3 Design overview

The crusher shown in Figure 2 consists of two sets of running roll crushers, attached with a pulley belt and driving power model at each side respectively. The power model is arranged as such, so that the two roll crushers are able to run at opposite direction facing each other. Hence, crushing action of infeed material will be done. After the crushing process take place at the roll crushers, the slag will be crushed into smaller pieces, and then continue by scrapping and conveying of the SSC below the roll crushers.

Figure 2. The close-up view of double roller crushing system.

DOI: 10.1051/matecconf/201713103002

4 Stress-strain and deformation displacement analysis

Based on simulation made, the forces are applied perpendicularly on the surface of each roll crusher, with estimation of 14,126.4 N (1.6 ton). The simulation results presented in this section includes the criteria of model deflection, von Mises stress, and also maximum shear stress.

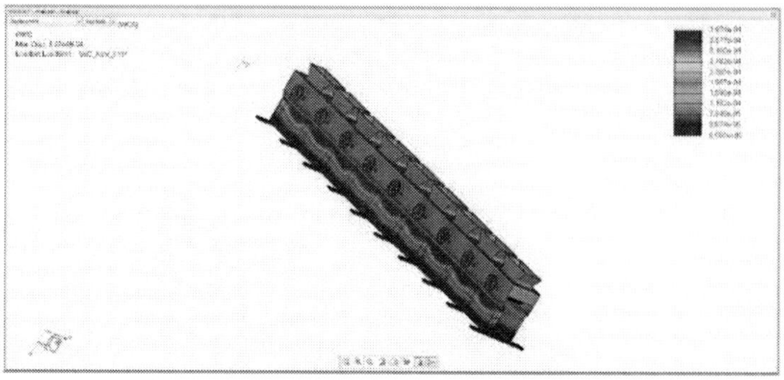

Figure 3. Deflection of SSC model.

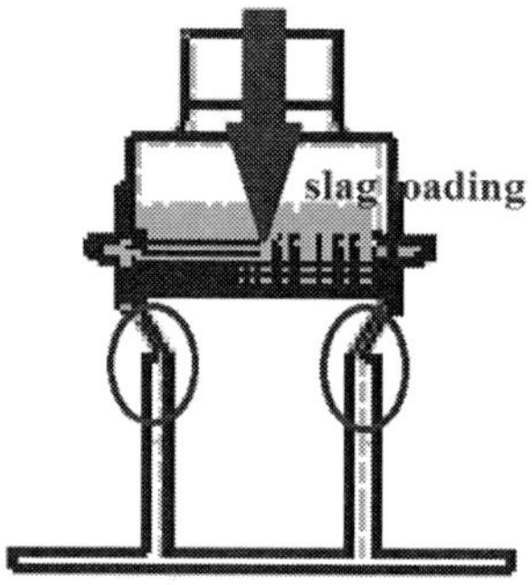

Figure 4. Side view of SSC with slag loading illustration.

The highest deflection of 0.0003974 mm occurred at SSC model shown in Figure 3 is highlighted in red area. This is due to the geometry design as shown in Figure 4, where loads are exerted in maximum at the bending feature of the design. The support of the model at the highlighted location is said to be the weakest. However, the value of deflection occurred at SSC is considered very low and will not cause failure to the model.

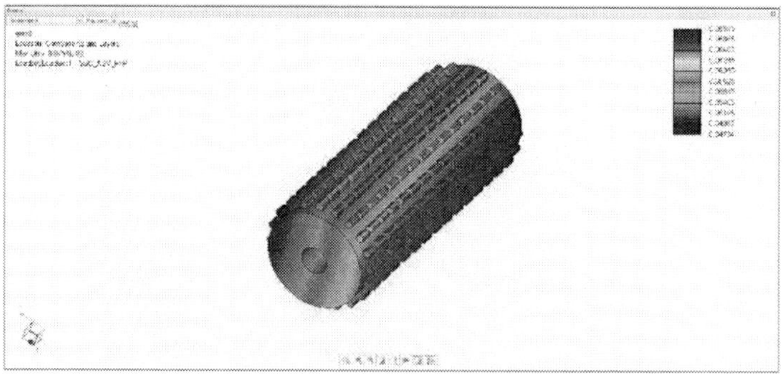

Figure 5. Deflection of roll crusher.

Based on the roll crusher deflection simulation result shown in Figure 5, roll crusher experienced the highest deflection of 0.06875 mm. This is shown in the highlighted red area of the roll crusher, which is facing upward during the simulation process. The high deflection is due to the direct force is applied perpendicularly on the highlighted surface of the roll crusher, with estimation of 14,126.4 N (1.6 ton). Roll crusher experienced the lowest deflection of 0.04774 mm, as represented in the highlighted blue area, which is facing downward during the simulation process. No force is applied on this area of roll crusher.

DOI: 10.1051/matecconf/201713103002

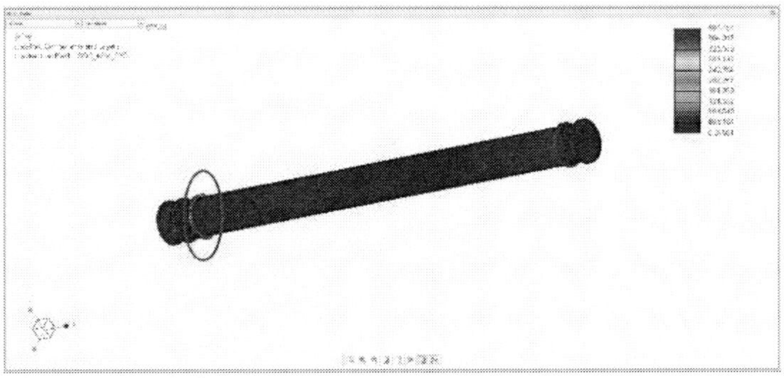

Figure 6. von Mises stress for roll crusher shaft.

A part is said to start yielding when its von Mises stress reaches a critical value known as the yield strength. The von Mises stress is used to predict yielding of part under any loading condition from results of simple uniaxial tensile tests [10]. In this simulation, model part of crusher shaft had experienced the highest value of von Mises stress, which is 404.454 kPa as shown in the highlighted edge in Figure 6. However, by referring to the endurance limit at critical location of crusher shaft and the material ultimate tensile strength, with the value of 262.2 MPa and 690 MPa respectively, the safety factor gained for the roller shaft is to be 2.6316. The roll crusher shaft experienced the lowest von Mises stress of 0.07891 kPa, which is represented by the highlighted blue area of the part.

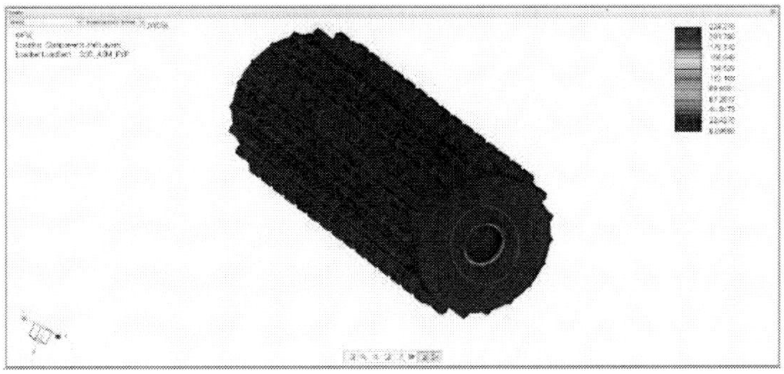

Figure 7. Maximum shear stress for roll crusher.

Shear stress is kwon as a stress state where the stress is parallel to the surface of the material, as opposed to normal stress when the stress is perpendicular to the surface. Shear stress is relevant to the motion of fluids upon surfaces, which result in the generation of shear stress [10]. Based on the roll crusher stress simulation result shown in Figure 7, roll crusher experienced the highest shear stress of 224.210 kPa. This phenomenon occurred at the highlighted area of the roll crusher. Roll crusher experienced the lowest shear stress of 0.00660 kPa, as represented in the highlighted blue area.

5 Conclusion and future work

The input-output relation of the crushing system is highlighted, where the constant outcome of crushed particle in size of 2.313mm is achieved. Next, PTC Creo Parametric engineering drawing software plays the main role in completing this project. The software is used for the SSC model 3D drawing, and also to run for stress-strain simulation. Based on the simulation result, roll crusher shaft experienced the highest von Mises stress value of 404.454 kPa. Hence, safety factor calculation is done on the roll crusher shaft. The optimum safety factor of SSC model is set to be 2. The final safety factor value of SSC model is calculated to be 2.6316, which means the model is safe to be implemented.

Future works on the new invention of SSC can be made based on the issue faced in coal-fired power plant from time to time. New invention of SSC can be either done by adding extra auxiliary system, or modification on the present technology. Other engineering drawing software such as Solidworks and CATIA are also recommended. By using the recommended software, complete mechanical product including functional tolerances, 3D annotations and kinematics of any designed model can be defined by their modelling tools. Process oriented tools of the recommended software also capture the manufacturing process intent in the early stages of design.

The authors would like to express our gratitude to the power generation unit, institute of power engineering Universiti Tenaga Nasional (UNITEN) TNB Janamanjung and Ministry of Higher Education (MOHE), Malaysia for the opportunity to sponsorship this corresponding research.

DOI: 10.1051/matecconf/201713103002

References

1. Clyde Bergemann Power Group Americas Inc. (2013). Submerged Scraper Coveyor – Wet Bottom Ash Handling [Online].Available: http://www.cbpg.com/sites/default/files/public%3A/products_solutions/materials_handling/SSC_low%20res.pdf
2. B. Michelbrink. "Wet Bin for Collection and Quenching of Ashes from a Pulverized Coal Combustion Chamber." U.S. Patent 3 915 105, (Oct. 28, 1975).
3. Zbogar et al. (2008, Nov.). "Shedding of ash deposit.". Progress in Energy and Combustion Science. 38(1), pp. 1-26.
4. D. H. Archer and M. M. Ahmed. "Method for soot blowing automation/optimization in boiler operation." U.S. Patent 4 996 951 A, (Mar 5, 1991).
5. P. Kittipongpattana and T. Fongsamootr. "Stress Analysis of Water Wall Tubes of a Coal-fired Boiler during Soot Blowing Operation," World Academy of Science. vol. 10, No:3, pp. 451-455, (2016).
6. F. Graube et al. "Optimisation of water-cannon cleaning for deposit removal on water walls inside waste incinerators" Waste Management & Research. vol. 34, No. 2, pp. 139-147, (2015).
7. Lei Deng et al. "A study on water-quenching waste heat recovery from molten slag of slag-tap boilers". Applied Thermal Engineering. Vol. 108, pp. 538-545, (2016).
8. P. Regucki et al. "Study of thermal-flow processes in ash cooler cooperating with CFB boiler," Perspectives in Science. Vol. 7, pp. 353–356, (March 2016).
9. "User Guidelines for Waste and Byproduct Materials in Pavement Construction," U.S. Department of Federal Transportation, New Jersey, FHWA-RD-97-148, (August, 2016).
10. Budynas and Nisbett, "Design of Mechanical Element," in Shigley's Mechanical Engineering Design, 10th Ed. New York, (2015).

 DOI: 10.1051/matecconf/201713103003

Hybrid Intelligent Warning System for Boiler tube Leak Trips

Deshvin Singh[1] , *Firas* B. Ismail[1*] , and *Mohammad* Shakir Nasif[2]

[1]Power Generation, Institute of Power Engineering (IPE), Universiti Tenaga Nasional, 43000 Kajang, Selangor, Malaysia
[2]Mechanical Engineering Department, Universiti Teknologi Petronas, 32610 Tronoh, Malaysia

Abstract. Repeated boiler tube leak trips in coal fired power plants can increase operating cost significantly. An early detection and diagnosis of boiler trips is essential for continuous safe operations in the plant. In this study two artificial intelligent monitoring systems specialized in boiler tube leak trips have been proposed. The first intelligent warning system (IWS-1) represents the use of pure artificial neural network system whereas the second intelligent warning system (IWS-2) represents merging of genetic algorithms and artificial neural networks as a hybrid intelligent system. The Extreme Learning Machine (ELM) methodology was also adopted in IWS-1 and compared with traditional training algorithms. Genetic algorithm (GA) was adopted in IWS-2 to optimize the ANN topology and the boiler parameters. An integrated data preparation framework was established for 3 real cases of boiler tube leak trip based on a thermal power plant in Malaysia. Both the IWSs were developed using MATLAB coding for training and validation. The hybrid IWS-2 performed better than IWS-1.The developed system was validated to be able to predict trips before the plant monitoring system. The proposed artificial intelligent system could be adopted as a reliable monitoring system of the thermal power plant boilers.

1 Introduction

Steam boilers represent one of the main component in the power plant. Boiler tubes continuously pass fluids, withstanding extremely high temperatures and pressures of the steam generator [1]. This eventually leads to leakages in boiler tubes. Malfunction of boiler tube causes frequent boiler trips which eventually increase the operating costs of a plant [2]. Conventional detection methods of detecting tube leaks by measuring the make-up water or inspecting is ineffective as it gives information on the existing tube leak only when serious damage is already done. Consequently prediction of boiler tube leak trips is crucial to maintain normal and safe operational conditions of the plant [3]. Benefits of an early detection of boiler tube leak trip are: to increase operating profit by reducing repair costs and secondary damage, increase safety of the plant, increase availability and tube life, and avoid unplanned outages [4]. Prediction can be achieved by developing intelligent warning system which monitors the operating conditions. Using information provided by measurements from

* Corresponding author: firas@uniten.edu.my

several sensors and actuators that are abundantly stored throughout operations, the intelligent warning system is able to predict a failure [5].

One of the most commonly used intelligent systems for early fault detection is the Artificial Neural Network. A study on ANN had adopted the methodology of feed forward backpropagation to diagnosis a steam boiler based on superheater monitoring, in which the minimisation algorithms were applied [6]. ANN as a stand-alone system is effective however there are several other intelligent systems that have been applied together with ANN for early fault detection. Genetic algorithm is an optimization tool which reduces the possibilities of getting into local minimum in the process of finding an optimum solution [7]. Momeni et al. have applied hybrid genetic algorithm-based ANN for the prediction of pile bearing capacity. From this study it was concluded that using the optimum Genetic Algorithm (GA) parameters, the hybrid model outperformed the conventional ANN model in the prediction process [8].

In all proposed solutions, there has not been any intelligent monitoring system established particularly for boiler tube leak trip in thermal power plants. Several researches use mathematically simulated data instead of real plant data for training and validation, which is inaccurate for decision making. It can be noted from previous applications that the raw data are used directly for the training process. However the data are not always complete and the quality is limited. Therefore a data preparation framework is important for the complex data analysis. Previously, trial and error approach has been applied for the selection of best NN topology combination and the optimization of the boiler operation parameters. The selection of suitable boiler operation variables and the design of the optimal neural network topology can be achieved using additional artificial intelligent system technique such as GA. Therefore the objective of this research is to develop two Intelligent Warning Systems (IWSs) for boiler trip prediction by adopting pure artificial neural network system and hybrid intelligent system. IWS-1 represents the use of the pure artificial neural network system for boiler trip prediction. The ELM training algorithm was also adopted for IWS-1 and the results were compared with the traditional training algorithms. IWS-2 involves the hybrid intelligent system using the application of artificial neural networks with genetic algorithms. In order to ease the selection of suitable boiler operation variables and to achieve the optimal neural network topology, encoding and optimization process is done using genetic algorithms.

2 Description of power plant and trip cases

This study was focused on three real cases of boiler tube leak trip that had occurred in a thermal power plant in Malaysia. This particular power plant has six operational units with capacity of generating up to 2420MWe. The units have triple fuel firing capability using gas, oil and coal. The steam generating capacity for the unit 1 is 169 bar whereas the steam outlet has a temperature of 560 °C. The tubes used in the boilers are made up of stainless alloy heater tubes of various grades. The ratio of unplanned outages against total annual hours was at 12.5% in the year 2005. Over the years the number of unplanned outages have declined significantly for all the units, however the operating company is targeting for lower number of planned outages. One of the main reason has been boiler tube leak trips which eventually causes shutdown and outage.

The scope of this work focusses on the working fluid of the boiler components including drum, superheater, boiler feed pump and water tube. These components are chosen based on the frequency of the trip occurrence within a set period of time. The real cases of boiler tube leak trips that had occurred unit 1 have been considered for this study. The first case was a boiler tube leak trip caused due to sudden overheating of boiler tubes. The second trip was caused by steam temperature fluctuation whereas the third case was due to extremely high superheater steam temperature.

3 Data acquisition and preparation

The on-line plant control system consists of 1250 observations of actuators and sensor signals which are identified as the boiler process observations. This large number of observation is reduced to 67 based on the work scope which focuses on the diagnosis of working fluid components only in the boiler. Since many of the observations were measured by multiple sensors, eventually only 17 influential operation boiler variables were identified to be used as input. The enormously large data collected from the power plant containing thousands of values is filtered and passed through three important steps. They are visual cleaning of data, missing data treatment and data normalisation. Data normalisation was used as a data transformation technique to scale the data between 0 and 1 (0 represents running mode and 1 represents faulty). The Min-Max data normalisation was applied. For each variable, the behavioural analysis was done based on its time interval to reach high alarm indicator at "1".

4 Design and implementation of IWSs

In this study the feed-forward methodology was adopted using the back-propagation neural network. Activation functions are applied to the weighted sum of its input. The main NN topologies are training algorithms, learning rate, momentum coefficient, activation functions, number of hidden layers and the number of neurons in each hidden layers. In this study up to 2 hidden layers have been used. The performances of these systems are indicated by Root Mean Square Error (RMSE) where output error between the actual output and the expected output is evaluated. It is noted that lower error value indicates better performance. Based on the best performance of the IWS, training and validation is done. The 17 variables presented before are used as inputs in the IWS and the outputs are either '0' or '1'.The ANN training is done in two phases: the preliminary training phase and the basic training phase.

4.1 Design Implementation of IWS-1

The main objective of this preliminary training is to determine with the optimal NN topology combination. The IWS was input with faulty data set which makes up 60% of the data set. The network input includes all 17 operation variables. The optimal topology combination obtained from the preliminary training results is used for the basic training. Real boiler data set were presented as basic training sets for the IWS-I. Based on the proposed system performance indicator, the basic training results were analysed. The calculated RMSE values of the optimal NN topology combination is compared with the RMSE value of the preliminary training process. The aim of this process is to get smaller or equal calculated RMSE values than the preliminary training process. In the validation and testing process the last 40% of the data set was used as a validation real data set. Applying the optimal NN topology combination obtained from the preliminary training process, validation of real data set was used to determine the rapidness of the proposed IWS in predicting the specific trip. The performance of the system during normal boiler operation is also validated.

4.2 Design Implementation of IWS-2

In the development of IWS-2, GA is used for automated design and NN parameter optimisation. In this work, IWS-2 combined ANN with GA in order to determine the best ANN structure and to optimise the ANN input parameters. The NN topologies were encoded together with the NN training parameters into the genes of the GA using Weak Specification Representation (WSR). The proposed GA individual has genotypes which represent every phenotype of ANN topologies together with training parameters. Binary representations were established for activation function, neuron number and input parameters. At first, an initial population is generated randomly in which the parents are selected. Crossover and mutation operators are applied on the parents to generate new generation on population. In this study, random initialisation is adopted for population initialisation because heuristic initialisation can result in the population having similar solutions. The population size is set by the user in which a bigger population will consume longer computational time. Fitness function is a function which determines how effective the solution is with respect to the considered problem. In this study the fitness function was represented by the RMSE value obtained from the ANN training. The selection operator used in this study is the fitness based selection. Using this selection operator, the new offspring replaces the least fit individuals in the population. The number of generation is set to be the termination condition for GA. GA is run for maximum number of iterations until there is no improvement in the population.

5 Results analysis

The performance of each system is analysed and discussed in this section based on the best NN topology combination together with the most influential boiler variables. In the training process of IWS-1 for case 1, the resilient back-propagation algorithm performed best among the traditional algorithms. However ELM was able to produce better results in terms of RMSE value. It can be concluded that for all cases, the minimum RMSE values were obtained using ELM. It can be concluded that the ELM algorithm performed better than the traditional ANN algorithm. Therefore the ELM algorithm is applied for the next step which is validation. Using the optimal results achieved from IWS-2, the validation for each case was done with the same validation data sets used for IWS-1 validation. Similar approach of IWS-1 was adopted in indicating the fault in IWS-2. Figure 1 represents IWS-2 output results for all cases. In case 1, the trip was introduced at the 15220th interval. IWS-2 was able to detect the fault at the 15210th interval with an output of 0.7. This output is considered above average warning indication. For case 2, the total data sampling interval is at the point of 15063th minute before the shutdown. The introduction of fault in the operation occurred at the 15051st interval. The intelligent system was able to predict the fault within the 15040th interval. The fault indication is considered a strong alarm warning indication (close to one) with system output of 0.91. In case 3, the total data sampling interval is at the point of 16600th minute before the shutdown. The introduction of fault in the operation occurred at the 16570th interval. The intelligent system was able to predict the fault within the 16550th interval. The fault indication is considered a very strong alarm warning indication (close to one) with system output of 0.93.

5.1 Comparison between IWS-1 and IWS-2 performance results

Both the systems were able to detect specific boiler tube leak trips based on the trip cases with varied performances. For IWS-1, it was established that the ELM training algorithm outperformed the conventional ANN training algorithms. Therefore ELM was adopted in the hybrid system IWS-2. Table 1 summarises the results of IWS-1 and IWS-2 for all cases. In case 1, IWS-1 detected the trip 9 minutes earlier than IWS-2. Similar comparison can be done for case 2 and 3 in which the detection time of IWS-1 outperformed IWS-2. However in all the trip cases, IWS-2 resulted in lower RMSE values compared to IWS-1. This indicates that IWS-2 performed better than IWS-1 in all cases. The output value is referred to categorise the alarm warning level based on the thresholds set. The closer the output value is to '1', the stronger the alarm warning indication. Therefore a direct comparison of results using RMSE or detection time cannot determine the better system. Despite being able to achieve the objective of detecting the boiler tube leak trips, IWS-1 required longer time in developing the structure of ANN and selecting the input variables. This is due to the trial and error approach which is not efficient when it comes to applying the same IWS method for other boiler units or trip cases. Optimisation of the effective variables and NN architecture is essential, considering the large problem space of topology combinations and input variables available. IWS-2 merged optimisation techniques with NN techniques in a hybrid system, to select the best structure combination and input variables. This way, the trial and error process can be avoided all together saving time in developing the IWS. Using IWS-2 the structure development and input variables selection was simplified, thus making the application of the system for other boiler units and trip cases less tedious.

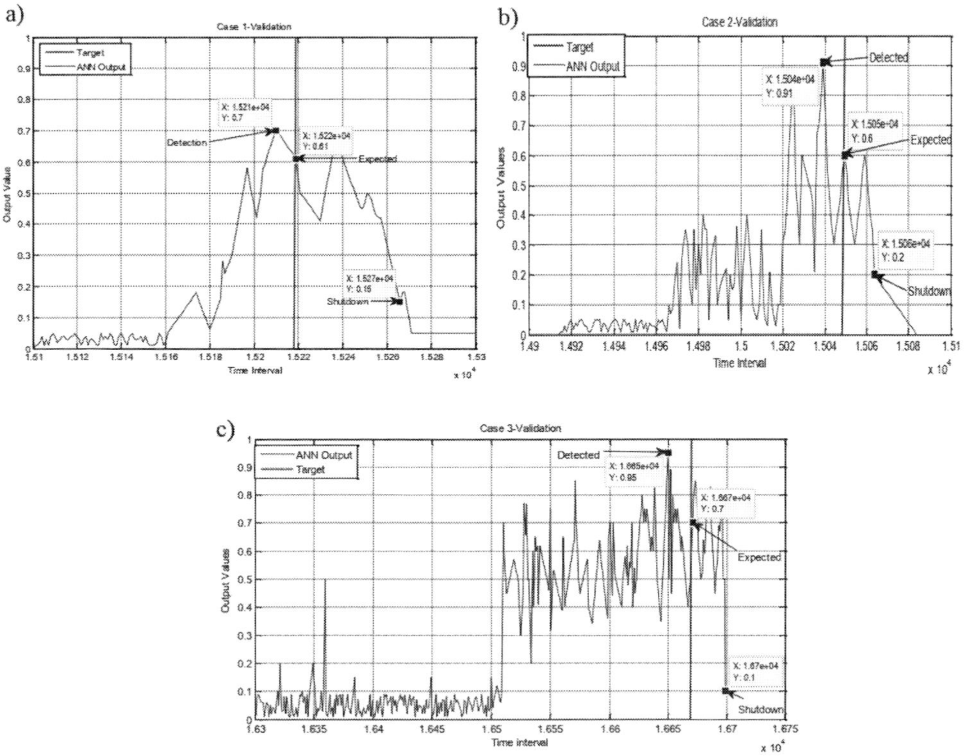

Fig. 1. IWS-2 output for: a) Case 1; b) Case 2; c) Case 3

Table 1. Summary of IWS-1 and IWS-2 results for all cases.

Trip cases	Case 1		Case 2		Case 3	
Systems	IWS-1	IWS-2	IWS-1	IWS-2	IWS-1	IWS-2
Architecture	9HL1	7HL1	5HL1	7HL1	8HL1	6HL1
Activation function	L+P	L+P	L+L	L+P	L+P	L+L
RMSE	0.1421	0.1419	0.1618	0.1312	0.0931	0.0701
Detection	19	10	22	11	30	20
Output	0.75	0.7	0.9	0.91	0.95	0.93
Trip probability	High	High	High	Very high	Very high	Very high

6 Conclusion

Early detection of boiler tube leak trip was achieved for all three cases. Two IWS were developed using pure ANN and combination of ANN with GA. Using real faulty data, the proposed systems were validated. An integrated data preparation framework has been proposed to process the large amount of data. The single layer feed-forward neural network with ELM training algorithm performed better than the traditional training algorithms. The best ANN topology combination for each trip case was determined. IWS-1 was able to detect the trips earlier than IWS-2. However taking into consideration the optimization capability of GA, IWS-2 was selected as the reliable system for this boiler unit. The optimization of ANN topology and input boiler parameters has made the development of IWS simpler and faster. Hybrid intelligent warning system can be a reliable system for warning and analyzing faults in power plants. Boiler tube leak trips were successfully detected earlier by the IWS in order to allow the plant operator to take appropriate action in maintaining continuous operation of boilers.

Acknowledgement to Ministry of Higher Education Malaysia (MOHE) for sponsoring this study in Universiti Tenaga Nasional through the MyBrain15 Scholarship Program.

References

1. P.K. Nag, Power Plant Engineering (Tata McGreaw-Hill, 2002).
2. T. Zhong, A.T. Alouani, R.A. Smoak, System Theory (1997).
3. Jianping Ma and Jin Jiang, Progress in Nuclear Energy (2011) pp.255-266.
4. D. Flynn, Thermal power plant simulation and control (The Institution of Electrical Engineers, 2003).
5. L. Fortuna, S. Graziani, A. Rizzo, Xibilia M.G., Advances in Industrial Control (2007).
6. B. Firas and H. Hussain, Journal of Applied Sciences (2011).
7. T. Jingwen and G. Meijuan, "Network intrusion detection method based on high speed and precise genetic algorithm neural network", Conference on Network Security, Wireless Communications and Trusted Computing (2009).
8. E. Momeni, R. Nazir, D. Jahed and H. Maizir, Measurement, 57 (2014) pp.122-131.

Comparison of nanofluid heat transfer properties with theory using generalized property relations for EG-water mixture

Seshu Kumar Vandrangi[1,a)], *Suhaimi bin* Hassan[1,b)] *Sharma* K.V. [2,c)], and *Prasad* Reddy[3,d)]

[1]Department of Mechanical Engineering, Universiti Teknologi PETRONAS 32610 Tronoh, Malaysia

[2]Department of Mechanical Engineering, JNTU Hyderabad, India

[3]Department of Chemical Engineering, Universiti Teknologi BRUNEI

Abstract. A numerical analysis for the determination for turbulent characteristics of fluid flow and heat transfer have been developed by employing the eddy diffusivity equation of Van Driest. The properties of Silicon dioxide (SiO_2) nanofluid with spherical particles in base liquid ethylene glycol (EG) -water (W) mixture of 60:40 ratio is employed for a wide range of concentrations and bulk temperature. A good agreement of the numerical results with the experimental data for properties and heat transfer is observed. A comparison of Copper oxide (CuO), Aluminum dioxide (Al_2O_3) and Silicon dioxide (SiO_2) nanofluids revealed that SiO_2 attain higher temperature gradients in comparison to CuO nanofluid at the same concentration and temperature.

1 Introduction

Studies using conventional heat transfer fluids such as water, ethylene glycol (EG) and engine oil are subjected to system alterations to meet the desired performance requirements. The methods of augmentation include the use of additives with micron size metal particles dispersed in the liquid [1, 2] are quite outdated. Moreover, the size of these particles for heat transfer enhancement proved to be undesirable, due to problems associated with clogging, sedimentation, erosion of pipe lines and large pressure drop. In order to meet the requirements of greater heat transfer rates, the application of nanotechnology has been employed successfully.

Nanofluids are formed by the dispersion of small quantities of nanometer size metal and metal oxide particles in base liquids such as water or ethylene glycol proved to possess higher values of thermal conductivity compared to base liquid. Recent trends suggest that using nanofluids dispersed in EG and water mixture as base liquid has proved to be beneficial for low temperature applications. Experiments undertaken with nanoparticles dispersed in EG-water mixture as base fluid in different ratios seems to be encouraging.

* Corresponding author: suhaimiha@utp.edu.my

Vajjha et al. [3-5] have conducted various experiments to determine the properties of Al_2O_3, CuO, SiO_2 and ZnO nanofluids. Praveen et al. [6] determined the rheological properties of CuO nanofluids. Sundar et al. [7] reported enhanced thermal properties of nano-diamond nanofluids. Sahoo et al. [8, 9] carried out experiments with SiO_2 and Al_2O_3 nanofluids to determine the thermal conductivity and viscosity. Kulkarni et al. [10] performed experiments for the estimation of viscosity with SiO_2 nanofluid.

Experiments for the determination of nanofluid forced convection heat transfer coefficients in the turbulent range are undertaken with Al_2O_3 (45nm), CuO (29nm), and SiO_2 (20, 50, 100nm) for a maximum concentration of 10.0% for temperature varying from 20-90°C by Vajjha and Das [11] in base liquid EG-water mixture in 60:40 ratio. They have reported an enhancement of 81.74% in heat transfer for Al_2O_3 nanofluid at a concentration of 10%. The pressure loss is reported to increase with volume concentration of the nanofluid. The increase of pressure loss for a 10% Al_2O_3 nanofluid at a Reynolds number of 6700 is about 4.7 times that of the base fluid. The authors developed equations for the nanofluid properties and the Nusselt number. Kulkarni et al. [10] undertook experiments to estimate the convective heat transfer and pressure loss of SiO_2 (20, 50, 100nm) nanofluids dispersed in EG-water mixture for volume concentration in the range of 0-10% with temperatures varying between 20-90°C. They have reported an enhancement of 16% in heat transfer coefficient at a concentration of 10.0% with 20nm particle size at Re = 10000.

Theoretical analysis for turbulent flow has been modeled by Sarma et al. [12] who introduced a correction factor for the mixing length. Expressions for eddy diffusivity of momentum $(\varepsilon_{nf}/\vartheta_{nf})$ and heat $(\varepsilon_H/\alpha_{nf})$ have been proposed, where ε_{nf} is momentum eddy diffusivity (m^2/s) of nanofluid, ε_H is thermal eddy diffusivity (m^2/s), α_{nf} is thermal diffusivity (m^2/s) of nanofluid and ϑ_{nf} is kinematic viscosity (m^2/s) of nanofluids. They compared the theoretical values of Nusselt number with experimental data of water based Al_2O_3 nanofluid for a maximum volume concentration of 0.5%. The analysis has been extended to 4% vol. concentration by Sharma et al. [13], . A good agreement of the numerical results with the experimental data have been reported. The numerical values of temperature gradients at the wall determined are reported to be inversely proportional to nanoparticle density for water based nanofluids.

The determination of nanofluid heat transfer coefficient in base liquid EG-water mixture of 60:40 ratio is mostly undertaken through experiments. The heat transfer enhancement reported by Vajjha et al. [11] is 82% with Al_2O_3 nanofluid while Kulkarni et al. [10] reported only 16% with SiO_2 nanofluid at the same volume concentration of 10%. The reason for the deviation in heat transfer enhancement with Al_2O_3 and SiO_2 nanofluids has not been explained by the authors in the turbulent range and hence numerical analysis is undertaken in the turbulent range of Reynolds number for EG-water mixtures of 60:40 ratio. The characteristics of SiO_2 nanofluid flow and heat transfer employing the van Driest eddy diffusivity equation is undertaken to determine the influence of properties on heat transfer coefficient. Only Salient result from the numerical analysis is presented.

2 Base fluid and nanofluid properties

The properties such as density, specific heat, viscosity and thermal conductivity of EG-W mixture were developed through regression equations employing experimental data available in the literature [14]. The properties of the base liquid are estimated with the regression equations in the numerical analysis and are given in Eqs. (4)- (7) by Sharma et al. [15]. Nanofluid properties are required for the estimation of heat transfer coefficients and friction factor. The nanofluid properties such as density and specific heat are well known conventional mixture relation and are given by Pak and Cho [16]. The thermal conductivity and viscosity of EG-water based CuO, Al_2O_3 and SiO_2 nanofluids are developed for different particle sizes, temperatures and concentration that are available in the literature [4]. The data available from experimental results of literature are used in the development of Eqs. for viscosity and thermal conductivity through regression analysis using numerical program and are given by Sharma et al. [15] in eqs (14) and (15). The experimental heat transfer coefficients of nanoparticles dispersed in base liquid EG-water mixture in 60:40 ratio by Vajjha and Das [11] is used to develop a regression equation resulting in Eq. (1)

$$\mathrm{Nu} = 0.0257\mathrm{Re}^{0.8}\mathrm{Pr_{bf}}^{0.4}(1+\mathrm{Pr_{nf}})^{-0.04297}(1+\frac{\emptyset}{100})^{5.205} \tag{1}$$

Where Nu, Re are Nusselt number and Reynolds number respectively and $\mathrm{Pr_{bf}}$ and $\mathrm{Pr_{nf}}$ are prandtl number of base fluid and nanofluid respectively.

The Eq. (1) is developed with an average deviation of 7.8% and a standard deviation of 9.3% with the exception of a few data points deviating from the correlation equation by 18%. The equation is valid in the temperature range of 20-90°C for a maximum concentration of 4% for particle diameters lower than 53nm. It can be observed that the substitution of $\emptyset = 0$ and $Pr_{nf} = 0$ in Eq. (1) simplifies to

$$\mathrm{Nu} = 0.0257\mathrm{Re}^{0.8}\mathrm{Pr_{bf}}^{0.4} \tag{2}$$

which is the conventional Dittus-Boelter equation applicable for pure fluids with a deviation of 11%.

3 Results and discussions

The base liquid and nanofluid properties, heat transfer coefficients are validated with the experimental data. The experimental data of base liquid and SiO_2/EG-W nanofluid thermal conductivity are compared with Eqs. (6) and (15) of Sharma et al. [15] respectively. The thermal conductivity data is plotted in Fig.1 for two concentrations. The results show a satisfactory agreement with a maximum deviation of 7% among all the concentrations.

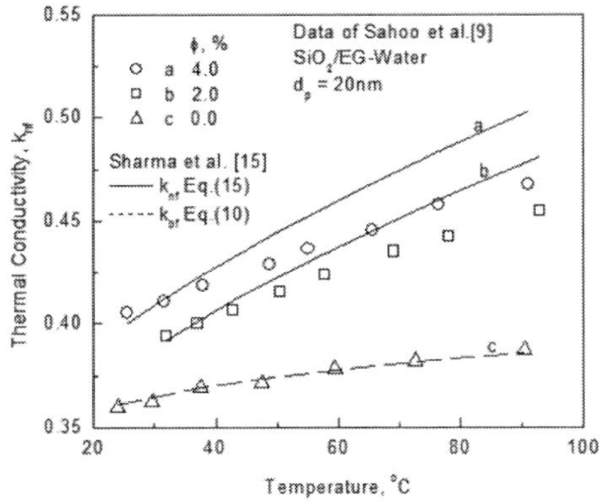

Fig.1 Comparison of thermal conductivity data with theory

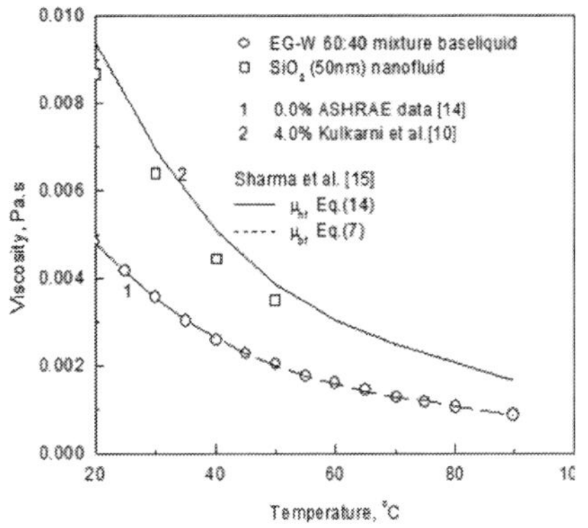

Fig.2 Comparison of viscosity data with theory

The values of viscosity estimated with Eqs. (7) and (14) given by Sharma [15] for base liquid and nanofluid respectively are in good agreement with the experimental data with a maximum deviation of 14% among all concentrations, shown plotted in Fig.2. The experimental determination of heat transfer coefficients with base liquid EG-water 60:40 mixture were performed by Kulkarni et al.[10] and Vajjha et al. [11] in the turbulent range of Reynolds Number at unspecified temperatures. The temperatures are predicted using the theoretical results which is observed to be in agreement with the base liquid experimental ata at 30°C and 70°C for Kulkarni et al.[10] and Vajjha et al. [11] with a maximum deviation of 20% and 15% respectively as shown plotted in Fig.3.

Thus, the base liquid temperatures determined, the nanofluid heat transfer coefficient at different Reynolds number for 2 and 4% concentration are shown plotted in Fig.4. The numerical results employing the property relations indicate that the experiments could have been undertaken at 30°C and 70°C by Kulkarni et al. [10] and Vajjha et al. [11] respectively.

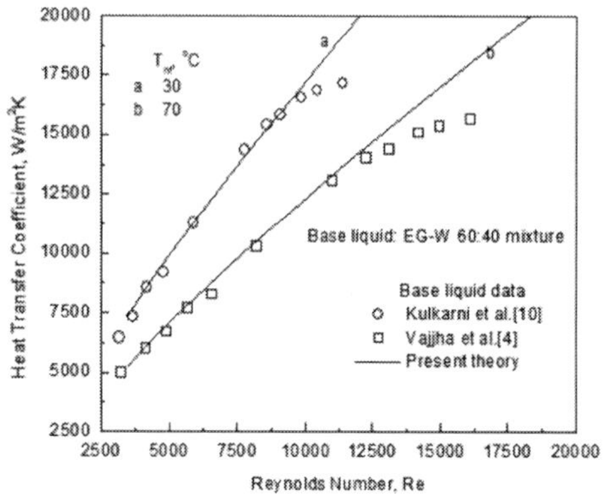

Fig.3 Comparison of base fluid heat transfer coefficient with theoretical results

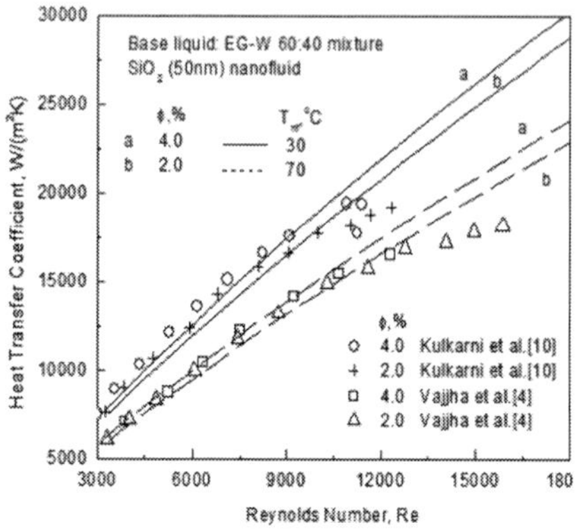

Fig.4 The variation of heat transfer coefficients with Reynolds number

It can be observed that the theoretical results are in close agreement with the experimental data with Eqs. (4)- (7) and (12)-(15) given by sharma et al. [15] are employed in the theoretical model. The concentration at which maximum heat transfer enhancement is possible can be determined with Enhancement Ratio (ER). The ER is defined as the ratio of enhancement in viscosity to that of thermal conductivity given in Eq. (21) of Sharma et a. [15].

The variation of nanofluid concentration with temperature for maximum heat transfer enhancement at various operating temperatures are shown plotted in Fig.5 for laminar and turbulent flow conditions. From Fig.5, it is evident that at 30°C, heat transfer enhancement for SiO_2 nanofluid is not attainable for concentrations greater than 1%. For experiments undertaken at 70°C, Fig.5 indicates heat transfer enhancements can be obtained up to 5% volume concentration under turbulent flow conditions.

The influence of particle properties on heat transfer coefficients is undertaken by a comparison of the temperature gradients rawn for SiO_2, Al_2O_3 and CuO nanofluids shown plotted in Fig.6 with Reynolds number**.** It can be observed that the temperature gradients decrease with increase in operating temperature which is reflected in lower values of heat transfer coefficients obtained at higher temperatures. Further, the thermal conductivity of CuO is greater than SiO_2 nanoparticle. It is generally expected that CuO nanofluid to predict higher heat transfer coefficients compared to SiO_2 nanofluids. However, the temperature gradient decrease with increase in nanoparticle density of SiO_2 (2220kg/m^3) to CuO (6900kg/m^3). Similar observation was reported for water based nanofluids [17]. Further studies are undertaken in this regard to determine the flow and heat transfe characteristics.

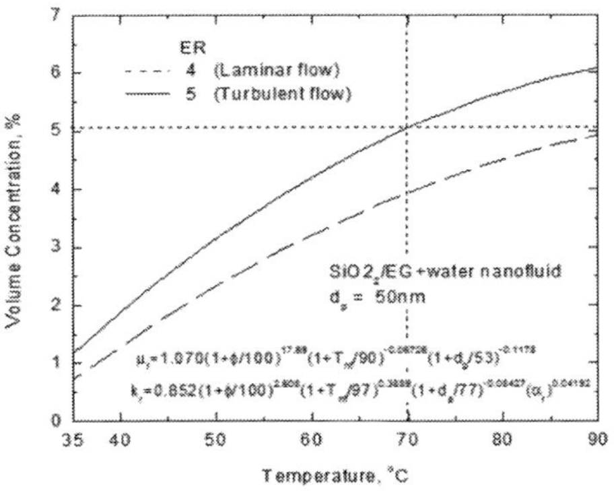

Fig.5 Evaluation of concentration for optimum heat transfer coefficient of SiO_2 nanofluid at different temperatures

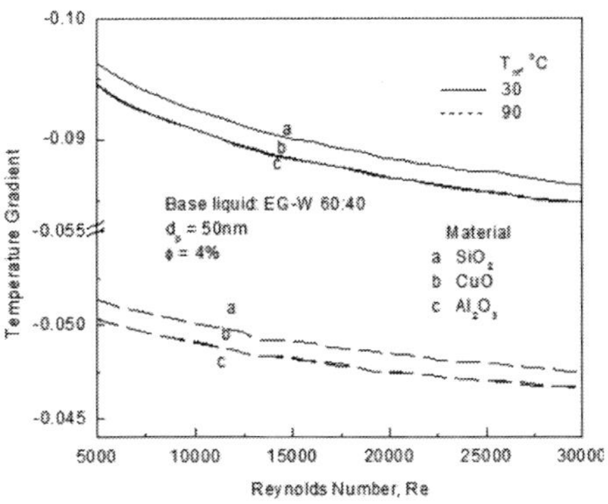

Fig. 6 Variation of temperature gradient with Reynolds number for different nanofluids

4 Conclusions

The salient observations from the numerical results on the nanofluid properties and heat transfer coefficients can be listed as:

(i) The proposed correlations of nanofluid viscosity and thermal conductivity can predict heat transfer coefficients satisfactorily with a maximum deviation of 7 and 14% respectively.
(ii) The Enhancement Ratio predicts an increase in heat transfer coefficient for nanofluid concentration up to 5% under turbulent flow and 4% for laminar region at an operating temperature of 70°C.
(iii) The increase in Nusselt number of SiO_2 nanofluid with concentration is not significant.
(iv) Higher heat transfer coefficients can be expected with SiO_2 in comparison with CuO nanofluid, based on the temperature gradients obtained.
(v) SiO_2 nanofluids predicts a higher temperature gradient when compared to Al_2O_3 and CuO based nanolfuids.

The authors acknowledge the financial assistance from Department of Mechanical Engineering, Universiti Teknologi PETRONAS, Malaysia.

5 References

1. A. S Ahuja, J of App Phy, **46**(8): p. 3408-3416 (1975).
2. A. E. Bergles, A. E., *Techniques to augment heat transfer*, in *Handbook of heat transfer applications*, p. 31-380 (1985).
3. R. Vajjha, D. Das, and B. Mahagaonkar, PETROL SCI TECHNOL, **27**(6): p. 612–624 (2009).
4. R. S Vajjha, and D. K. Das, Int J Heat Mass Transf,. **52**(21): p. 4675–4682 (2009).
5. R. S. Vajjha, and D. K. Das, J. Heat Transfer, **131**(7): p. 071601–071601 (2009).
6. P. K. Namburu, D. P. Kulkarni, D. Misra, and D. K. Das, Exp. Therm Fluid Sci. **32**(2): p. 397–402 (2007).
7. L. S. Sundar, L. S., M. K. Singh, E. V. Ramana, B. Singh, J. Grácio, and A. C. M. Sousa,. Sci. Rep. **4** (2014).

8. B. C. Sahoo, D. K. Das, R. S. Vajjha, and J. R. Satti, J Nanotechnol Eng Med. **3**(4): p. 041006 (2012).
9. B. C Sahoo, R. S. Vajjha, R. Ganguli, G. A. Chukwu, and D. K. Das, PETROL SCI TECHNOL,. **27**(15): p. 1757–1770 (2009).
10. D. P. Kulkarni, P. K. Namburu, H. Ed Bargar, and D. K. Das, HTrEn,. **29**(12): p. 1027-1035 (2008).
11. R. S. Vajjha, D. K. Das, and D. P. Kulkarni,. Int J Heat Mass Transf,. **53**(21–22): p. 4607–4618 (2010).
12. P. K. Sarma, Chada, K. V. Sharma, L. S. Sundar, P. S. Kishore, and V. Srinivas, IJHT,. **28**. (2010).
13. K. V. Sharma, W. H. Azmi, S. Kamal, and S. Hassan. Appl Mech Mater.. Trans Tech Publ (2016).
14. ASHRAE, *Handbook of fundamentals.* American Society of Heating Refrigerating and Air Conditioning Engineers, Atlanta, GA, (2005).
15. K. V. Sharma, S. K. Vandrangi, K. Habib, and S. Kamal, IJSER..7(6) (2016)
16. B. C. Pak and Y. I. Cho, EXP HEAT TRANSFER,. **11**(2): p. 151–170(1998).
17. K. V. Sharma, W. Azmi, S. Kamal, P. Sarma, and B. Vijayalakshmi, Can. J. Chem. Eng., (2015).

Effect of ventilation on thermal comfort in campus hostel bedrooms after sunset

Shaharin Anwar Sulaiman[1,*], *Ahmad Faridzuan* Zakeria[1], *Muhammad Fahmi* Ramely[1], *Mohamad Nazmi* Zaidi Moni[1] and *Mohamad Firdaus* Basrawi[2]

[1]*Mechanical Engineering Department, Universiti Teknologi PETRONAS, 32610 Bandar Seri Iskandar, Perak, Malaysia*
[2]*Faculty of Mechanical Engineering, Universiti Malaysia Pahang, 26600 Pekan, Pahang*

Abstract. After sunset, the indoor of buildings would be relatively warmer as compared to that of the outdoor due to thermal was associated with the heat stored in the building structures and furniture. In tropical countries, doors and windows are usually shut after sunset in order to avoid mosquitoes. Thus, the condition of the indoor air is worsened as the hot air is trapped inside the bedroom and takes long time to be fully dissipated to the surrounding. This causes discomfort to the occupants as bedrooms where they are expecting to have a good rest to rejuvenate from hard day works. The objective of this study was to analyze the nature of air temperature profile in bedrooms of a hostel at night and the effect of ventilation systems in reducing the indoor air temperature. The study was conducted by measuring temperature of air in the bedrooms throughout a night. It was found that the exhaust fan gave the best results by reducing the air temperature by 1.5°C to 2.5°C.

1 Introduction

During the day, buildings experience heat gains from several sources. The heat radiated from the sun during the day is absorbed and stored by equipment, furniture and structure of the building. After the sun is set, due to temperature different the stored heat is emitted to the surrounding causing the indoor temperature to increase and becomes relatively higher than that of the outdoor air[1]. This situation creates uncomfortable environment to the bedroom occupants during night time because the thermal comfort condition in the bedroom cannot be maintained. The concept of thermal comfort is closely related to thermal stress. People do not perform well under thermal stress. People's performance under thermal stress is about 11% lower than their performance at normal thermal conditions[2]. In this context, student's life should not be excluded. In university's residential buildings, the residents which are the students spend a lot of their time in the bedroom during night time as the building is designed with bedrooms only. Hence, most of student activities especially studying is done in the respective bedroom. The increase in temperature during night time in the bedroom space could affect the overall academic performance of the students if it is not solved properly. Besides that, prolonged heat can give consequences to the psychological aspect especially to those that have history in psychological problems. People that exposed to prolonged elevated temperature may behave differentially and aggressively in respond to the surrounding thermal[3]. Thus, it is very important to keep the bedroom environment in thermal comfort condition to avoid any commotion in student. Malaysia has hot and humid climate with temperatures ranging between 20°C and 32°C during daytime, and 21°C to 27°C during the night with relative humidity of around 75%[4]. The occupants of houses are usually open their doors and windows for natural ventilation during daytime. Thus, the indoor air temperature is increased due to higher outdoor air temperature and the natural ventilation will induce air flow into buildings to increase heat transfer from indoor space and human body[5].

On the other hand, windows and doors are usually shut after sunset in order to prevent mosquitoes from coming into the rooms and also for safety reasons. There is a need to conduct a study on residential building rise university hostels because it is rarely done. Issues that often arise, and often raised by the hostel residents are feeling less comfortable with the environment of their bedroom because of high indoor air temperature. The problem arises mainly during the night time. To fix this, the indoor air of the hostel bedroom is vented to the surrounding using the help of mechanical ventilator as the use of air conditioner is not preferable.

Corresponding author: shaharin@utp.edu.my

Therefore, it is necessary to know the thermal condition of the bedroom that use mechanical ventilation system and which is not, in order to determine whether the system can help reducing the indoor temperature rise or not[6-8]. The thermal condition in hostel bedrooms during the night time has to be considered carefully because of the high occupancy at that time on campus. This is because the hostel residents which consisting of pre-graduate, undergraduate and postgraduate students favour a comfortable bedroom for them to rejuvenate after long exhausting daytime activities. The objective of this work was to study the thermal indoor conditions of hostel rooms in a university after sunset and the effectiveness of an innovative portable mechanical ventilation system. The effectiveness of ventilation to help in achieving thermal comfort environment will be measured based on the thermal conditions of bedroom before and after the ventilation is installed and perception by the occupants.

2 Methodology

The experimental site located in a hot and humid tropical climate zone and therefore treated according to the ventilation requirement for such an area based on published works by other researchers[9-14]. The mechanical ventilation fan must be capable to discharge the minimum air changes required per hour (ACH) for that particular room. To calculate the value of the ACH, the following Equation (1.0) must be satisfied. The volumetric flow rate or ACH required in a bedroom is depending on the volume of bedroom space and the number of air changes rate for typical bedroom space. The equation is very beneficial in selecting the proper fan for ventilation purposes:

$$Q = \frac{VN}{60} \tag{1.0}$$

where Q is the volumetric flow rate of air in cubic feet per minute (cfm), V is the volume of space in cubic feet, and N is the number of air changes rate per hour (ACH)[4]. In order to select the correct ventilation fan for the bedroom, it is important to know how much air required to be moved. Table 1 below shows the suggested number of air changes rate for typical bedroom from various ventilation fan manufacturer around the world.

Table 1. Recommended air changes rate for typical bedroom by various ventilation fan manufacturer

Manufacturer	Air Changes Rate	Room Rate Used
A	2 - 4	4
B	5	5
C	2 – 4	4

2.1 Equipment

For the purpose of this experiment, OM – DaqPRO 5300 data logger is used to collect various data of temperature in the bedrooms. The data logger will store temperature data and can be extract using computer through DaqLAB software. A typical data logger of this type can support maximum eight points of thermocouple and therefore suitable for temperature measurement activities in this study, at which six points are used. Each of the bedroom composed of the same dimension which is 4mx4mx3m. This makes the volume of the bedroom to be 48m, or approximately 1700ft. The number of air discharge is taken as 5 (based on Table 1). Hence, the volumetric flow rate can be determined which is 142cfm. The capability of a ventilation fan used in this project is 90cfm. However, with two identical fans, a total of 180cfm of volumetric flow rate of air can be discharged. Hence, the air exchanges rate can be achieved, which is more than minimum 142 cfm required from the calculation.

2.2 Building and bedrooms selection

In this study, Village 5 Block F (V5F) multi-storey hostel building in UTP has been selected as the location to carry out the temperature measurement experiment. The building composed of 5 floors with a total built area of 1246m. Each floor comprised of 4 identical houses with 6 bedrooms per house. It is selected because it receives hundred percent of solar intensity and does not blocked by other building surround it. The bedrooms are selected based on the difference of floor which is ground floor, third or middle floor and the top floor, and the windows orientation which facing towards Northeast, Northwest, Southeast and Southwest. A total of nine bedrooms are involved in this study. Fig. 1 shows the floor plan with the indication of the bedrooms selected to be monitored.

 DOI: 10.1051/matecconf/201713103005

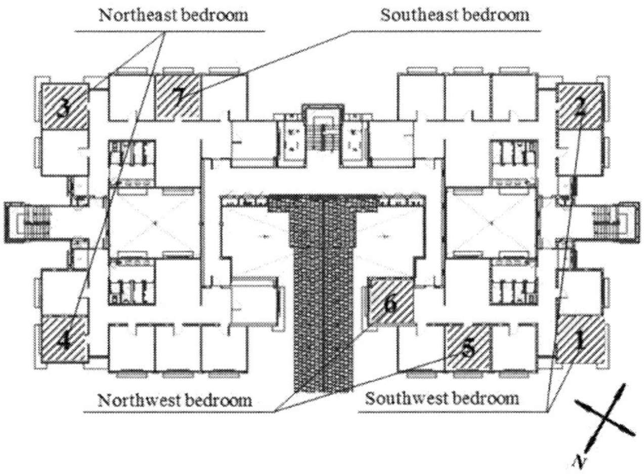

Fig 1. Top floor plan with indication of the monitored bedroom

3 Experimental setup

The indoor and outdoor air temperatures of the bedrooms are collected by DaqPRO ¯portable data logger. The data logger is connected to several thermocouple wire probe which placed inside and outside the bedroom. The data logger is programmed using the computer software DaqLAB. The same programme is used to download the data to a computer. The data were taken every 10 minutes throughout the period of temperature measurement process. A total of six thermocouples type-J wire probes are installed during the measurement where five are installed in various location inside the bedroom and the other is outside. Thermocouple probes for indoor temperature 1, 2 and 3 are mounted at approximately 1.5 m from the floor level except for indoor temperature 4 and indoor temperature 5 which installed at 1.0 m and 2.0 m height. Fig. 2 shows the location of thermocouple probes of indoor and outdoor temperatures taken.

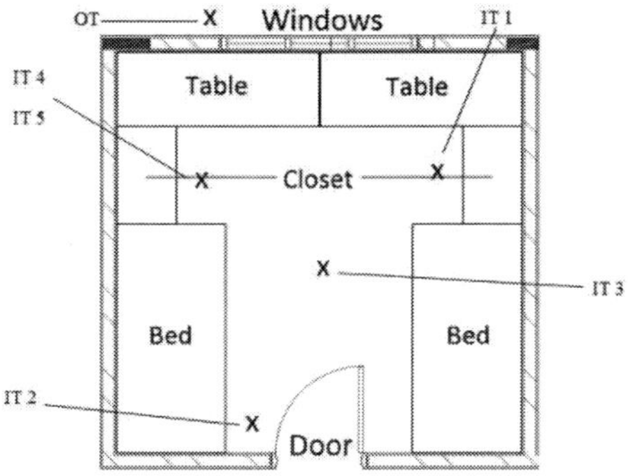

Fig. 2. Indoor and outdoor points of temperatures taken in a bedroom.

The nine bedrooms are monitored during the night time over 12 hours per day from 5 pm to 5 am the next morning. The measurement of the bedroom temperature is taken during sunny days only in order to get an even data collection. During the process, the bedrooms are unoccupied and the windows and the doors are kept closed. The indoor lighting also turned on throughout the period. The bedroom indoor air temperature is measured at various location and height to investigate the existence of air temperature different inside a bedroom. There are three different conditions to take the indoor air temperature measurement in order to study the effect of mechanical ventilation system towards the indoor air temperature with the interest of ceiling fan operation. Table 2 shows the condition for temperature measurement.

Table 2. Experimental condition of bedrooms for temperature measurement.

Case	Condition
1	Only ceiling fan is operating
2	Only ventilation fans are operating
3	Both ceiling fan and ventilation fans operate at the same time throughout the experiment

4 Results and discussions

4.1 Case 1 (Normal Condition)

Measurements were conducted in October and November 2012, in which, different condition of bedrooms were applied. Based on the temperature measurements on the top floor bedrooms, it is shown that bedrooms having orientation towards Southwest has the highest indoor temperature while bedrooms having orientation towards Southeast has the least indoor temperature during the night time. Hence, Southwest bedrooms are relatively hot compare to other bedrooms on the same floor. Thus, the same orientation of bedrooms is selected on the first and third floor in order to investigate the different in level of the floor towards the indoor thermal. The same experimental condition is applied to both bedrooms. The indoor temperature of Southwest orientation bedrooms are still high regardless the floor level they are. However, among the three Southwest orientation bedrooms experimented in this study, the bedroom on the top floor has slightly higher indoor temperature compare to the same orientation bedrooms in the other floors. Hence, the Southwest bedroom on the top floor has been selected to be the critical bedroom for further experiments in Case 2 and Case 3 of the experimental conditions. Fig. 3 shows the temperatures for Southwest bedroom on the top floor.

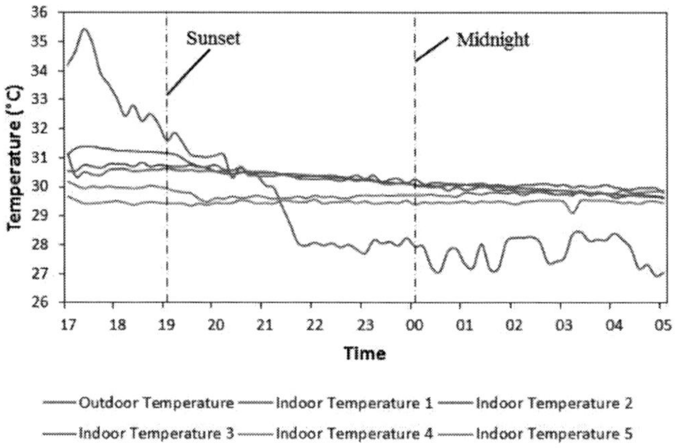

Fig. 3. Hourly variation of indoor and outdoor temperature of top floor Southwest bedroom.

The temperature of the bedroom decreases since the beginning of the experiment approximately at 5 pm up until midnight. Then the temperature maintains and decreases slightly until the end of the experiment at 5 am. In this study, thermocouple probes to measure indoor temperature 1, 2 and 3 are located at the same height which is 1.5 m from the floor. However, the Indoor Temperature 1 (IT 1) has the highest temperature between them, followed by Indoor Temperature 3 (IT 3) in the middle of the room and Indoor Temperature 2 (IT 2) placed near the door. This is because the emission of heat stored by the furniture and heat emission from the light effect the IT 1. The IT 3 was less affected by the heat emission from the furniture because it is located away from the furniture and plus it is situated below the ceiling fan. IT 3 experienced the lowest indoor temperature because it is situated far away from the sources of heat emission. Between Indoor Temperature 4 and 5 (IT 4& IT 5), IT 4 is higher than IT 5 even though it is located nearer to the floor compare to IT 5. IT 5 should be higher than IT 4 as the fact that hot air rises. However, because of the location of IT 5 that near to the ceiling fan, the wind produced from the fan decreases the temperature and thus make IT 5 lower than IT 4. This is proven in Case 2 when the ceiling fan is not operating; IT 5 is higher than IT 4 as the hot air rises without being pushed back down by the fan.

DOI: 10.1051/matecconf/201713103005

4.2 Case 2

Fig. 4 shows the result of thermal measurement in top floor Southwest bedroom under the condition of only ventilation fans operating without the operational of the ceiling fan. The windows and door remain closed and the bedroom's light still turned on throughout the experiment.

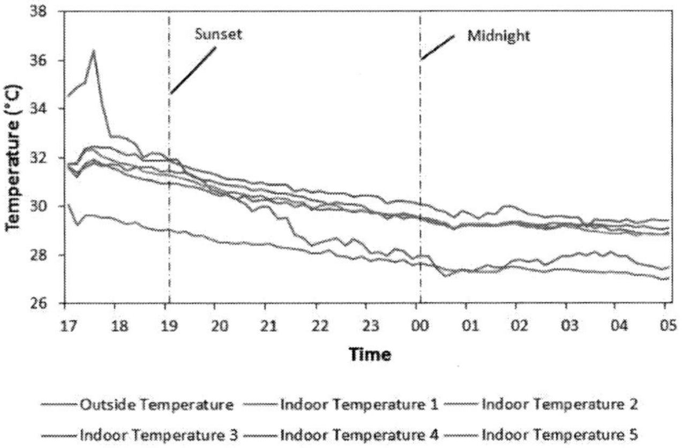

Fig. 4. Hourly variation of indoor and outdoor temperature of top floor Southwest bedroom with only ventilation fans operating.

The indoor temperature is dropping faster compare to the indoor temperature in normal condition. This shows that the hot air inside the bedroom is successfully discharged to the surrounding by the ventilation fans during early after sunset time. As night grows, the temperature drop rate becomes slower as heat is dissipated from the furniture and the light to the bedroom space affect the indoor thermal. The average of indoor temperature is still around 30°C and above until midnight. During this period, the bedroom is cooler compare to the normal condition (case 1). However, it is still considered warm and near to hot condition as the temperature is still fluctuating around 30°C until midnight time. After midnight, the indoor temperature begins to drop at a faster rate for two hours because of large temperature different between the indoor and the outdoor thermal. Then, indoor temperature decreases slowly at almost steady rate until 5 in the morning.

4.3 Case 3

In this case, both ventilation fans and ceiling fan were operated at the same time during the course of the temperature measurement period. Other conditions remained the same. Fig. 5 shows the result of the thermal measurement of the bedroom in this kind of condition. The indoor temperature at the beginning of the experiment is lower than indoor temperature in Case 2 with average of 30°C despite high outdoor temperature around 33.5°C. As the time approach to sunset, the indoor temperature drops almost equal to the outdoor temperature. By 10 pm, the indoor temperature already beyond 29°C and continue to drop even though the outdoor temperature drops at a slow rate. The indoor temperature maintained around 29°C until midnight and dropping to 28°C at the end of the experiment with the outdoor temperature dropping to minimum 27°C. This means that the indoor air is circulating and at the same time the hot air is removed from the bedroom even at a slow rate. This helps the bedroom temperature to drop continuously and prevent the thermal condition to become high.

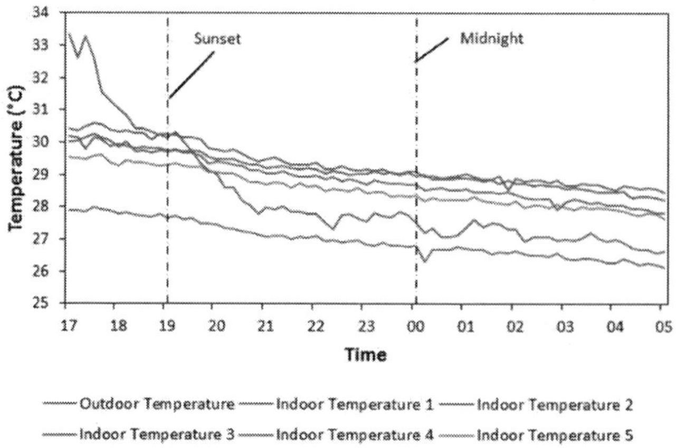

Fig. 5. Hourly variation of indoor and outdoor temperature of top floor Southwest bedroom when both ventilation fans and ceiling fan operating.

4.4 Temperature difference in critical bedroom

The temperature difference between the indoor and outdoor thermal of the critical bedroom is determined in all experimental condition to give a better picture on the temperature results. Fig. 6 shows the temperature difference between the indoor and outdoor thermal of the critical bedroom.

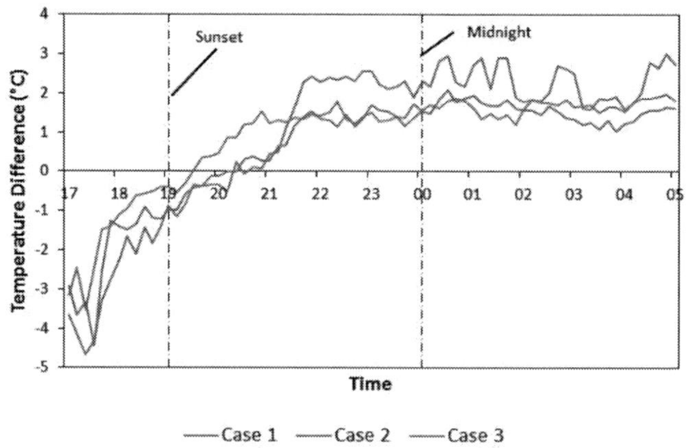

Fig. 6. Temperature difference between the indoor and outdoor thermal of critical bedroom

Negative temperature difference means that the indoor temperature is lower than the outdoor temperature. In this situation, the indoor temperature is selected at Indoor Temperature 1 for comparison because of the occupants tend to spend their night time at this location (table area) to do their homework, surfing the internet, watch movies and other activities in order to get some rest from their tiring days' activities. At the beginning of the experiment until sunset time, the indoor temperature is lower compare to the outdoor temperature. In this case, the temperature difference is highest compare to case 2 and 3 with average of -2.5°C, followed by Case 2 and Case 3. This means that the indoor thermal of Southwest bedroom before sunset is the lowest compare when it is in Case 2 and Case 3. However, after sunset, the temperature different from the indoor and the outdoor rises rapidly at time between 9 to 10 pm which is around 2.5°C and kept steady until midnight. The same rising pattern can be observed in Case 2 where only ventilation fan is operating. The only different is the maximum temperature difference before midnight is lower than the normal condition which is 1.5°C. The same maximum temperature difference occurs in Case 3. However, in Case 3 with both ventilation fans and ceiling fan operating, the temperature starts to differ early, at 7.30 pm and gradually rises and kept at 1.5°C of difference starting at 9 pm onwards. This is due to the cloudy weather on the night time when the experiment was carried on compare to the other cases which are the weather is fine during the night time. This situation causes the occupants to feel hot at this time during normal condition compare when ventilation fans are operating. After midnight and onwards, same pattern of temperature difference is experienced in all three cases where maximum 0.5°C of rises are seen from 12 am to 2 am. After that, it drops again 0.5°C until 4 am before starts to rise gradually until the end of the experiment.

5 Conclusion

From the study, it is found that the bedrooms facing towards the sunset has the highest possibility to become hot during the night time regardless which level of floor the bedrooms are in. However, bedrooms in the top floor are warmer than the bedrooms at a lower level due to fact that hot air rises. Temperature at the table area in the bedroom which the area the occupants spend most of the night time is the highest compare to the temperature under the ceiling fan and near the door. Temperature near the ceiling is higher than the temperature near the floor if there are no cooling sources near it such as the operational of ceiling fan. Otherwise, the temperature near the ceiling would be lower than the temperature near the floor area. In practicing ventilation, the indoor temperature is lower compare with normal condition which ventilation just relies on the ceiling fan operation during the night time. The reduction is about 1.5- 2.5°C. Supporting by the temperature difference between indoor and outdoor, which is 1.0-1.5°C, the hot indoor thermal condition in a bedroom can be reduced.

Ventilation during the night time is importance especially in a bedroom that facing the sunset because the indoor thermal is high. Using both mechanical ventilation fans and ceiling fan are the best way to reduce the hot indoor condition as the hot air can be discharged from the bedroom and at the same time circulation of air can be promoted in the bedroom by the ceiling fan. Even though the power consumption would be higher when both types of fans are used, the occupants' perception is that it is better than having just ventilation fans to operate in the bedroom because no circulation of air or no movement of air when the ceiling fan is not operating.

References

1. W.J. Younger, A. Thumann, *Handbook of Energy Audits* 7th edition (New York: Fairmont Press, Inc. CRC Press, 2007)
2. L.R. Leon, Toxicology, Applied Pharmacology **1**(223), pp. 146 (2008)
3. P. Gut, D. Ackerknecht, Climate Responsive Building, pp. 16 (1993).
4. V.I. Soebarto, S. Handjarinto, *Proceedings of the Eleventh Symposium on Improving Building Systems in Hot, Humid Climates*, pp. 482-491 (Fort Worth, TX, June 1998)
5. ASHRAE Standard 55-1992. *'Thermal Environmental for Human Occupancy'* (American Society of Heating, Refrigerating, Air-Conditioning Engineers, Atlanta, 1992)
6. ASHRAE Standard 62.2-2010. *'Ventilation, Acceptable Indoor Air Quality in Low-Rise Residential Buildings'*, (American Society of Heating Refrigerating, Air-Conditioning Engineers, Atlanta, 2010)
7. D.W. Bearg, 'Indoor Air Quality, HVAC Systems' *CRC Press,* pp. 64 (1993)
8. H. Feriadi, N.H. Wong, Energy, Buildings **36**, pp. 614-626 (2004).
9. T.H. Karyono, Energy, Buildings **23**, pp. 161-167 (2003)
10. A.M. Nugroho, Economics, Engineering **2** (1), (2011)
11. F.H. Mallick, Energy, Buildings, pp. 628-637 (2004)
12. Z. Lin, S. Deng, Building, Environment **39**, pp.1101-1114 (2004)
13. S.N. Ekasiswi, Architecture & Environment **6** (1), pp. 60-72, (2007)
14. R. Ruzaimah, H.A. Abdul., Z.A.W Abdul, International Conference on Environment Science, Engineering (ICESE), pp. 167-170 (2011)

Trends in Supersonic Separator design development

Rami Ali Altam [1], *Tamiru Alemu* Lemma [1,*], and *Shiferaw Regassa* Jufar [1]

[1]Universiti Teknologi PETRONAS, 32610 Seri Iskandar, Perak Darul Ridzuan, Malaysia.

Abstract. Supersonic separator is a new technology with applications in hydrocarbon dew pointing and gas dehydration which can be used to condensate and separate water and heavy hydrocarbons from natural gas. Many researchers have studied the design, performance and efficiency, economic viability, and industrial applications of these separators. The purpose of this paper is to succinctly review recent progress in the design and application of supersonic separators and their limitations. This review has found that while several aspects of this study are well studied, considerable gaps within the published literature still exists in the areas such as turndown flexibility which is a critical requirement to cater for variation of mass flow and since almost all the available designs have a fixed geometry and therefore cannot be considered suitable for variable mass flow rate, which is a common situation in actual site. Hence, the focus needs to be more on designing a flexible geometry that can maintain a high separation efficiency regardless of inlet conditions and mass flow variations. This review is focusing only on the design and application of the supersonic separators without going through the experimental facilities, industrial platform, pilot plants as well as theoretical, analytical, and numerical modelling.

1 Introduction

The presence of CO_2 in a natural gas mixture has a negative effect. As such, (i) it reduces heating value of the natural gas, (ii) it causes corrosion of the pipes and hence corrosion-resistant materials have to be used which leads to a costly facility, (iii) high CO_2 production requires separation process and huge storage facilities. However, the high capital cost for the Acid Gas Removal System (AGRS) considered being the major factor obstructing the monetization of a high CO_2 gas field. Therefore, several technologies and processes have been used and evaluated by the industry which have the potential to remove CO_2 from the natural gas. Each of these methods has its landscape of performance, energy requirement and system footprint [1], namely; amine and membrane which however were economically impossible. Besides, these technologies were technically non-viable as CO_2 proportion gets higher [2]. Extensive studies then introduced a new technology called 'vortex tube'. This technology has been developed on the pressure difference principle. This principle is employed to a nozzle pressure that has hot and cold ends, these ends create a differential

[1] Corresponding author: tamiru.lemma@utp.edu.my

pressure state, which is used in the supersonic flow. The role of supersonic flow is to create a low-temperature area inside the nozzle. This process can separate CO_2, sulfur and other impurities from the natural gas that consist of heavier hydrocarbons. The credit for this goes to a physical principle called 'centrifugal force' which is caused by the swirling flow [3].

The supersonic separation unit consists of (i) Casing, (ii) swirling device, (iii) supersonic nozzle, (iv) liquid extraction and (v) diffuser. As shown in Fig. 1.

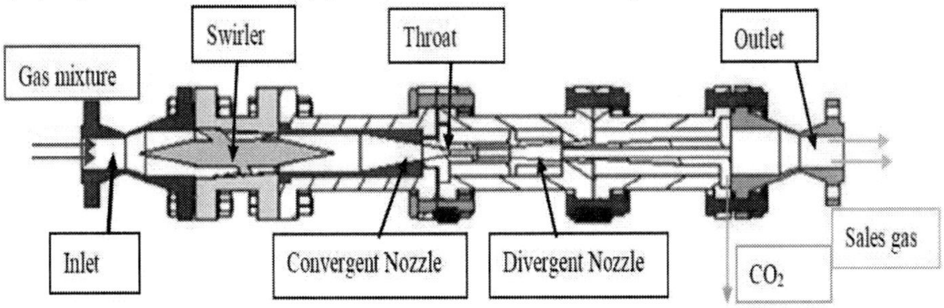

Fig. 1. Current supersonic separator UTP patent design.

One of the supersonic separation advantages is that it is more efficient in extracting dissolved impurities in natural gas. Moreover, supersonic separation unit has no rotating parts. Hence it needs less compression power which simplifies its scheme, footprint, and maintenance. Also, supersonic separation unit is cost-efficient due to its small and compact design, it also requires less operating cost.

There are several technologies by which CO_2 can be separated. Two technologies are considered the most recent and efficient CO_2 separation devices; namely, the 3S supersonic separator and the second one is the Twister separator. A detailed review of the literature is presented in the sections that follow.

2 Geometry of nozzle

The Laval Nozzle is a tube of an hourglass shape that is pinched in the middle. It uses the principle of the Joule-Thompson effect where expanding the pressurized gas to low pressure at constant enthalpy results in transforming the potential energy (temperature and pressure) into kinetic energy (velocity) that eventually leads to accelerating the gas to a supersonic speed [4]. The Laval Nozzle is composed of three sections: expander, cyclone separator, and compressor as in Fig. 2.

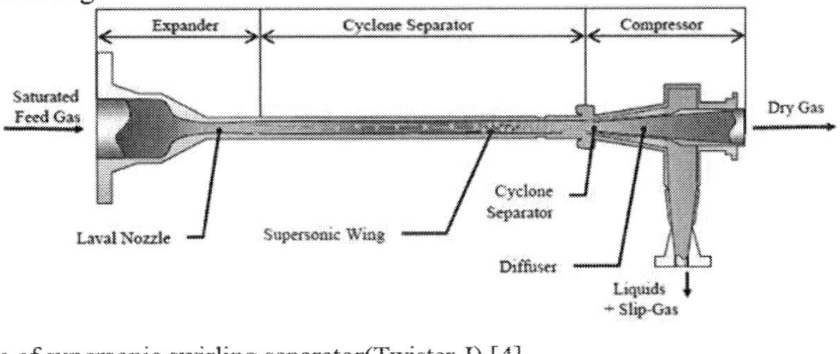

Fig. 2. Scheme of supersonic swirling separator(Twister-I) [4].

1. The expander (convergent section)

The expander (subsonic zone) is located at the inlet section. Where the swirling generator located in the expander is composed of several blades tangent to the nozzle. The gas flow passes through these blades leading the gas flow to accelerate and enter the nozzle with the tangential velocity of a particular value. The pressure and temperature are decreasing due to the supersonic speed of the wet gas in the expander. Where the drop of the temperature results in condensation of the water vapors existing in the gas[5, 6].

2. The Cyclone Separator (throat section)

After the expansion, in the cyclone separator (critical zone) where the already formed droplets are being separated to the walls due to the centrifugal force that can reach values up to 500000g causing the cyclonic separation [5]. The droplets separated on the walls are then forming a thin film which is moving in the direction of the flow.

3. The Compressor (divergent section)

The Laval Nozzle ends with the compressor (supersonic zone). A separation channel is located in the compressor to separate the thin water film on the walls from the dry gas stream. Compressor part leads to the diffuser in order to recover the initial pressure of the gas where a shock wave is produced. The shock wave is generated by changing the velocity from supersonic to sonic speed. From here, the gas is slowed down in the diffuser that results in recovering about (65-80%) of the initial pressure [7].

In the Laval nozzle, the inlet diameter must be greater than $\sqrt{5}$ of the throat diameter, in order to obtain the sonic speed at the throat [8]. On the other hand, the throat diameter should be equal to or smaller than the convergent length based on Wen et al. [5, 9-14] calculations where the Foelsch's analytical calculation [15] was used to compute the dimensions of the convergent section of the Laval nozzle in order to obtain the sound speed in the throat area and generate the stable supersonic flows. The recommended geometry is governed by:

$$\begin{cases} \dfrac{D-D_{cr}}{D_1-D_{cr}} = 1-\dfrac{1}{X_m^2}\left(\dfrac{x}{L}\right)^3 & \text{for} \quad \left(\dfrac{x}{L} \le X_m\right) \\ \dfrac{D-D_{cr}}{D_1-D_{cr}} = \dfrac{1}{(1-X_m)^2}\left(1-\dfrac{x}{L}\right)^3 & \text{for} \quad \left(\dfrac{x}{L} > X_m\right) \end{cases} \tag{1}$$

For converging-diverging nozzle, the optimum design proposed by Betting [16] dictates that $50 \le L_2\,(D_2 - D_{cr}) \le 220$. Many of the designs use straighteners in the substantially gaseous region. However, the decision on the geometry and location of this component is still not clear.

3 Twister BV

Prast et al. [17] who was associated with Twister BV, studied the nucleation and growth of droplets in a 2D supersonic Laval nozzle. Twister separator which is based on air drying device was originally developed by Stork Product Engineering BV in 1989 [18]. Twister BV company is a joint venture between the Beacon Group and Shell launched in the year 2000 for developing the first generation of supersonic separators, trademarked Twister [19]. This model used a deflecting blade in the supersonic section of the nozzle to generate the swirl required for condensed particles separation and is now referred to as Twister-I as illustrated in Fig. 2. However, Twister design is reported to have a temperature drop as low as (-40°C) with a corresponding pressure drop of up to 30%. But then, a significant portion of the gas leaves with the mist, forcing the design to have a secondary separator called degassing vessel. Erosion of the wing or deltoid plate paneled to the inside wall downstream to the throat is a further drawback that is making the swirl action prone to extinction in a short period of time. The residence time for the droplets is also short rendering the separation performance very low. Therefore, the design of the nozzle was optimized by including a central body and an annular nozzle configuration [20]. The swirl generation was also moved upstream, in the subsonic part of the nozzle inlet (Twister-II) as showed in Fig. 3.

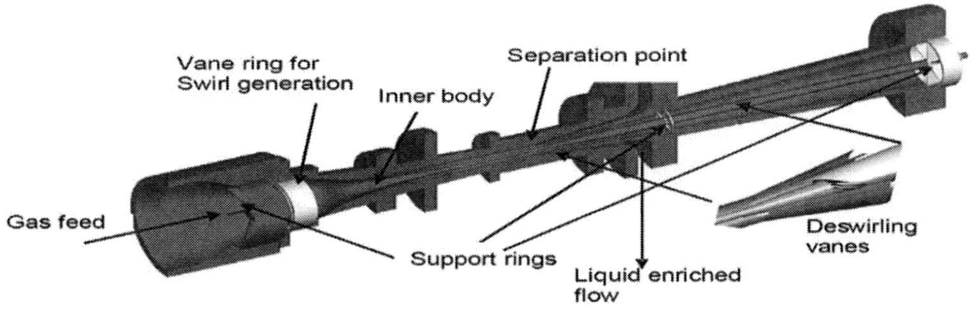

Fig. 3. Scheme of supersonic swirling separator (Twister-II) [21].

The improved design is being used by PETRONAS and Sarawak Shell Berhad (SSB). Nevertheless, inability to handle variable mass flow rates and high-pressure loss (30%) are issues yet to be resolved.

4 3S supersonic separator

3S separator is supersonic separator designed by TransLang Technologies Limited which is an engineering services company. The 3S supersonic separator is a hybrid of the two Twister devices. This device has a swirl generator in the subsonic entrance to the nozzle and has no central body as in Fig. 4.

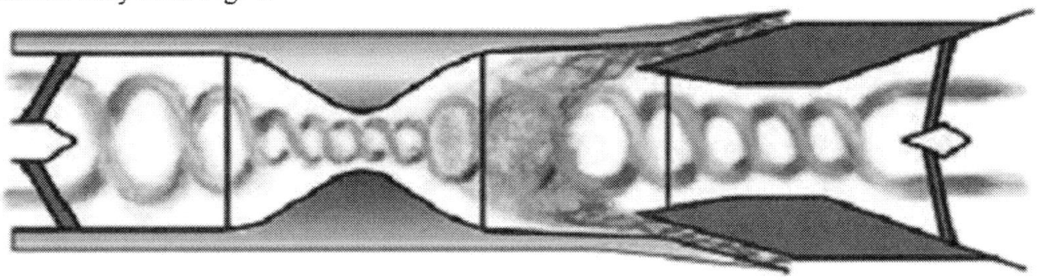

Fig. 4. 3-S separation [21].

3S Separator uses the fundamental principle of adiabatic expansion through a Laval nozzle. Additionally, however, a swirler is included to increase the centrifugal force in the separation zone. A de-swirler or directing vanes at the outlet is another feature that makes it different from other designs. As reported, a temperature drop of up to (-70°C) is possible with a corresponding pressure ratio of about 2.5 and Mach numbers in the range of 1.5 to 2. Because of the low temperature, the potential use of 3S extends to LNG applications, CO_2 extraction, and ethane recovery and the cryogenic temperature for natural gas is within the capacity limit of the separator. 3S is patented in many countries including US, Canada, and Netherlands. Additionally, the 3S separator was capable of dew pointing lean gases where Joule Thompson systems were ineffective. Some of these accomplishments were published in an article in the Oil and Gas Journal [22]. However, it seems at this point that there have been no further contributions from this group to the technology available from Twister.

5 Garrett design [23]

The separator is featured by supersonic expansion of a multi-component gas in a planar bend. The bend is shaped to have converging and diverging sections with an intervening minimum throat area. The condensed fluid is collected through a permeable wall as shown in Fig. 5.

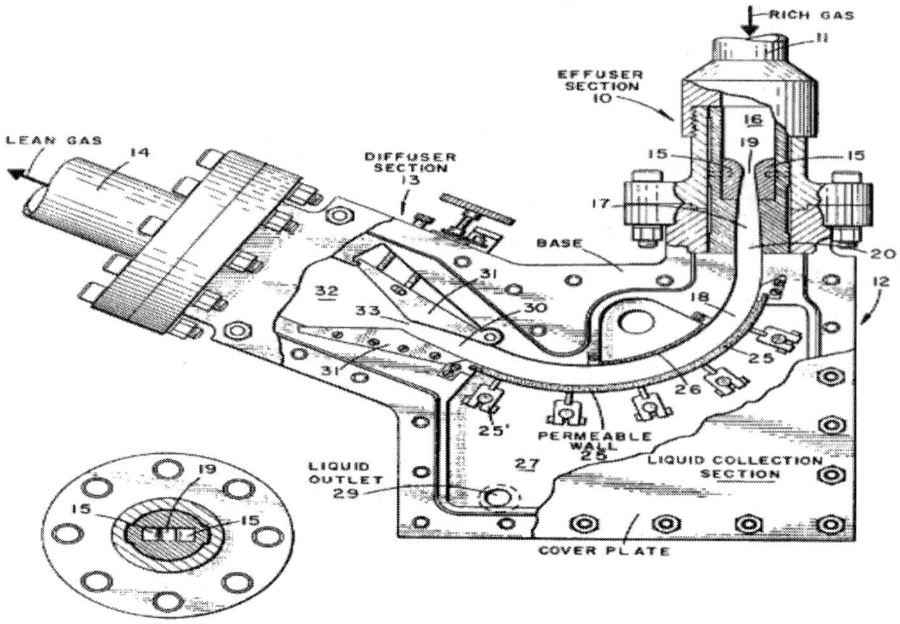

Fig. 5. Supersonic flow separator by Garrett [23].

The design is unique in the sense that it is equipped with (i) a mechanism to adjust size of the flow region and fully control the starting of supersonic flow and also the location of shockwave for maximum separation efficiency, and (ii) a permeable wall with electrostatic field to enhance particle ejection and droplet coalescence. One drawback of this design, however, is that the separation performance depends on the radius of curvature of the duct, which in turn is affected by the presence of unwanted shockwaves.

6 Keisuke design [24]

Designed to separate two or more gaseous mixtures, it has a converging-diverging nozzle with a swirler at the inlet, and two vacuum pumps connected to the exits. The oblique shock in the expansion zone is reported to have used for vaporizing the condensate back to the gaseous state. As documented by other authors, however, the design is featured by high-pressure drop due to the long axial distance. It also needs sufficient energy to sustain strong swirl throughout the separation zone.

7 Van Holten design [25]

Unlike Garret Design, the separator by Val Holten has a cylindrical shape that converges to a nozzle and diverges into a swirl nozzle. The gas that enters to the separator at subsonic speed first flows through a converging section and then expands to supersonic speed in the diverging section of the separator. Plate-like elements attached to the inside wall and protruding radially into the flow imparts a swirl to the supersonic flow. The lean gas at supersonic speed is discharged through a centrally arranged pipe. Condensate reach mixture, on the other hand, leaves the separation zone through a diverging section where a suction fan is installed to draw out the fluid. This design suffers from erosion of the deltoid plates and the energy it needs to run the fan. Furthermore, there is no mechanism to deal with variable mass flow rate. According to the discussion in Betting et al. [16], Van Holten's separator has low efficiency due to the presence of shock waves in the first and second outlets. Which in other words is that substantial amount of the condensed gas leaves through the secondary outlet.

8 Borissov design [26]

Two version of Borissov designs have been reported. The first design called SUSTOR-I entails a plug or central body and inlet manifold where the swirl is initiated using guide vanes, as described in Fig. 6 (a). Some of the features are: (i) temperature drop up to (-73°C), (ii) tangential inlet and tangential exit, (iii) reduced pressure drop for an equivalent temperature reduction, (iv) liquid droplets attached to the sidewalls by centrifugal action are removed through slits, (v) dry gas that is spirally expanding in the separator leaves the vortex chamber through a tangential outlet. The second design called SUSTOR-II involves a chamber consists of two disc-like parts (marked by the A-A and B-B cross-sections in Fig. 6 (b)) and the converging-diverging nozzle located in between. SUSTOR-II advantage is that the tangential exit is thought to have mitigated the high-pressure loss in SUSTOR-I. Nevertheless, the cast geometry is relatively complex.

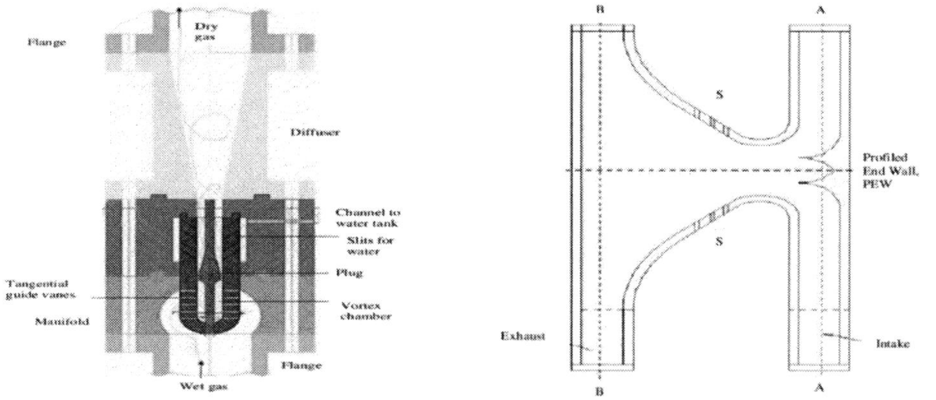

(a) (b)

Fig. 6. Schematics of the (a) SUSTOR-I and (b) SUSTOR-II Separators [26].

9 Wen et al. design [27]

Wen and his colleagues have worked on the analysis and modelling of the supersonic separators over the last few years. This design has a cylindrical casing starting from inlet flange all the way to the expanding pipe. Akin to Twister-II and SUSTOR-I, it also has a central body and swirl generating blades. However, the shape of the central body is different, and the longitudinal cross-section has ellipsoid shape at the inlet while the rest of the edges are straight lines. The presence of a swirler at the inlet and resistive vortex vanes at the central outlet also making the design similar to 3S separator. In terms of structure, is reported simple. However, the claim that it runs stably and has high efficiency is hard to verify for, and there is limited information on its performance and working range. Additionally, the design is fixed geometry and therefore cannot be considered suitable for variable mass flow rate, which is a common situation in actual site.

10 Beijing university of technology [28]

A supersonic separator test apparatus with a throat diameter of 10 mm was built with stainless steel. The separator has no central body, in which a swirl generator was located downstream of the nozzle and in the supersonic flow region. The device also has a total of seven thermocouple probes along the nozzle, humidity and temperature sensors, pressure gauges, flow meters at the inlet and dry gas outlet of the separator, and also an extra pressure gauge at the wet outlet.

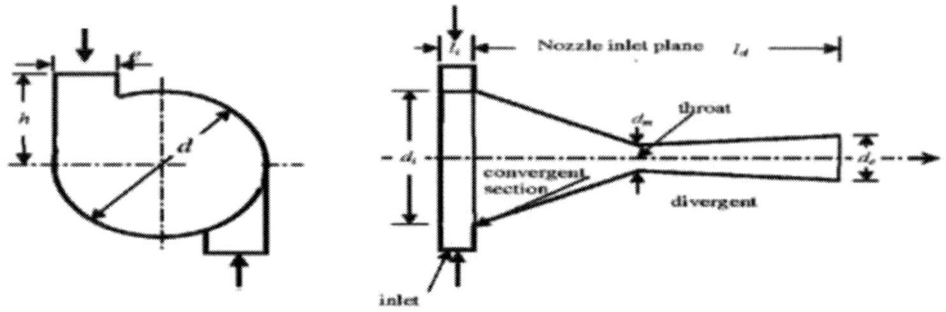

Fig. 7. Tangential inlet openings that generate the swirl [29].

As showed in Fig. 7, the flow is introduced as a tangential flow by passing through the two symmetrical inlet openings in which the swirl generator was relocated upstream to the subsonic section and also the blade system to generate swirl is not needed [29]. The dimensions of the two openings can be adjusted in order to increase or decrease the inlet velocity. However, changing the size of the inlet openings had no effect on the flow rate. Another factor to consider in this design was the strong gradients in tangential velocity near the axis of the nozzle.

11 Discussion

Table 1 shows the comparisons between the available designs of the supersonic separators. In terms of the main features, it is clear that almost all the separators contain a swirler generator installed except for Garrett design. On the other hand, Twister, SUSTOR-I and Wen et al. designs have a central body inside the Laval nozzle in order to increase the separation performance. Another main feature is the temperature drop which found to be the lowest of (-70°C) in the 3S separator and (-73°C) in the SUSTOR separator. Garrett separator has a unique design with electrostatic field to enhance the separation and adjustable flow region that might be able to handle the flow rate variations. SUSTOR separator can also be scaled down for any rate of the gas flow. Regarding the limitation, almost all the designs suffer from the high-pressure losses and the turndown flexibility problem. SUSTOR-II considered being the latest and best design available recently. SUSTOR-II was able to minimize the pressure losses and solve the turndown flexibility. However, the cast geometry is relatively complex.

 DOI: 10.1051/matecconf/201713103006

Table 1. Comparison between the available supersonic separators designs.

Design	Main features	Limitations
TWISTER BV [7]	• Deflecting blade to generate the swirl. • Central body in order to increase the separation performance. • A temperature drops as low as (-40°C).	• Inability to handle variable mass flow rates and high-pressure loss up to 30%.
3S Supersonic Separator [22]	• Hybrid of the two Twister devices. • A swirler is included. • A de-swirler or directing vanes at the outlet. • Temperature drop of up to (-70°C).	• Fixed geometry that lacks the ability to handle the variable mass flow rates.
Garrett Design [23]	• Featured with a planar bend contains the converging and diverging sections with an intervening minimum throat area. • A permeable wall to collect the condensed fluid. • Adjustable flow region. • Electrostatic field to enhance the separation.	• The separation performance depends on the radius of curvature of the duct, which in turn is affected by the presence of unwanted shock waves.
Keisuke Design [24]	• A converging-diverging nozzle with a swirler at the inlet and two vacuum pumps connected to the exits.	• The design is featured by high-pressure drop due to the long axial distance. • It needs sufficient energy to sustain strong swirl throughout the separation zone.
Van Holten Design [25]	• A cylindrical shape that converges to a nozzle and diverges into a swirl nozzle. • Plate-like elements attached to the inside wall and protruding radially into the flow imparts a swirl to the supersonic flow. • The gas at supersonic speed is discharged through a centrally arranged pipe. • A suction fan is installed in the diverging section to draw out the fluid.	• The design suffers from erosion of the deltoid plates and the energy it needs to run the fan. • There is no mechanism to deal with variable mass flow rate. • Has low efficiency due to the presence of shock waves in the first and second outlets.
SUSTOR-I&II [26]	• SUSTOR-I entails a plug or central body and inlet manifold where the swirl is initiated using guide vanes. • A temperature drops up to (-73°C). • Reduced pressure drop for an equivalent temperature reduction.	• The cast geometry is relatively complex.

 DOI: 10.1051/matecconf/201713103006

	• The liquid is removed through slits. • Dry gas leaves through a tangential outlet. • The tangential exit in SUSTOR-II is thought to have mitigated the high-pressure loss in the SUSTOR-I design. • Can be scaled down for any rate of the gas flow.	
Wen et al. Design [27]	• A cylindrical casing is starting from inlet flange all the way to the expanding pipe. • A central body and swirl generating blades. • Simple structure.	• The claim that it runs stably and has high efficiency is hard to verify for, and there is limited information on its performance and working range. • Fixed geometry which is unable to handle variable mass flow rates.
Design by Beijing University of Technology [28]	• No central body. • Swirl generator located downstream of the nozzle. • Contains total of seven thermocouple probes along the nozzle. • Tangential flow by passing through two symmetrical inlet openings that can be adjusted to increase or decrease the inlet velocity.	• Changing the size of the inlet openings had no effect on the flow rate. • Strong gradients in tangential velocity near the axis of the nozzle due to the lack of the central body.

12 Conclusion

- The application of supersonic separators in natural gas treatment has been studied and practiced for over two decades now. Several aspects of the technology including physical design, sensitivity to flow conditions, and various design features such as swirl generation and diffuser properties have been studied using analytical, numerical, and experimental techniques.
- Using the conventional method of separation, it is impossible to initiate condensation at pressures higher than 100atm. Besides, the condensation process might not be amenable to full control over a wider operating region. Therefore, it is necessary to rely on a separator that partly or entirely uses supersonic flow.
- Turndown flexibility is a critical requirement to cater for variation of mass flow rate. The reported designs fail to have this feature.
- Swirl generation device is also an important part of the design of these separators, and further research can potentially increase the knowledge base to improve the swirl generation effectiveness significantly.
- Many of the designs discussed above use straighteners in the substantially gaseous region. However, the decision on the geometry and location of this component is still not clear.
- The research on mechanical gas separators for the last few decades had studied the sizing of the supersonic zone, design of the swirler, design of the separation section, prediction of the gas property, condensation enhancement, and CFD simulation.

- This paper reviewed the available supersonic separator designs and limitations in a very simple way, as well as making a brief comparison between them in terms of the main features and limitations.

Abbreviation

AGRS	Acid Gas Removal System
°C	Celsius
CFD	Computational Fluid Dynamics
LNG	Liquefied natural gas
SSB	Sarawak Shell Berhad

Symbols

D_1	Inlet diameter
D_{cr}	Throat diameter
D	Convergent diameter
D_2	Outlet diameter
L	Length of the convergent nozzle
L_2	Length of the diverging nozzle
X_m	Convergent curve relative coordinate
x	Distance from the inlet to arbitrary cross section

The authors would like to acknowledge Universiti Teknologi PETRONAS (UTP) financial assistance under Graduate Assistance (GA) Scheme.

References

1 M. S. Onn, A. A. Seman, Z. Kassim, and M. A. Esa, "CO_2 Separation from Natural Gas Through Hydrate Formation," (2014).
2 R. A. Samawe, K. Rostani, A. Mohd Jalil, M. Esa, and N. Othman, "Concept Proofing of upersonic Nozzle Separator for CO_2 Separation from Natural Gas using a Flow Loop," (2014).
3 S. Kong, J. Liu, L. Zhao, P. Cai, and Y. Huang, "Study of CO_2 Separation Characteristic in Supersonic High-Pressure Vortex Tube," in *Materials for Renewable Energy & Environment (ICMREE), 2011 International Conference on*, (2011), pp. 936-940.
4 T. González, M. Netušil, and P. Ditl, "Raw Gas Dehydration on Supersonic Swirling Separator," *Czasopismo Tech. Mech. J.*, vol. **109**, (2012).
5 C. Wen, X. Cao, J. Zhang, and L. Wu, "Three-dimensional Numerical Simulation of the Supersonic Swirling Separator," in *The Twentieth International Offshore and Polar Engineering Conference*, (2010).
6 C. Wen, X. Cao, Y. Yang, and J. Zhang, "Effects of Swirls on Natural Gas Flow in Supersonic Separators," in *The Twenty-first International Offshore and Polar Engineering Conference*, (2011).
7 F. T. Okimoto and J. M. Brouwer, "Supersonic Gas Conditioning," vol. **223** (2002).
8 H. C. Man, J. Duan, and T. M. Yue, "Design and Characteristic Analysis of Supersonic Nozzles for High Gas Pressure Laser Cutting," *J. Mater. Process. Technol.*, vol. **63**, pp. 217-222, (1997).

9 C. Wen, X. Cao, and Y. Yang, "Swirling Flow of Natural Gas in Supersonic Separators," *Chem. Eng. Process.*, vol. **50**, pp. 644-649, (2011).

10 C. Wen, X. Cao, Y. Yang, and Y. Feng, "Prediction of Mass Flow Rate in Supersonic Natural Gas Processing," *Oil Gas Sci. Technol.- Rev. IFP. J.*, vol. **70**, pp. 1101-1109, (2013).

11 C. Wen, X. Cao, Y. Yang, and J. Zhang, "Supersonic Swirling Characteristics of Natural Gas in Convergent-Divergent Nozzles," *Pet. Sci. J.*, vol. **8**, pp. 114-119, (2011).

12 C. Wen, Y. Feng, X. Cao, Y. Yang, and P. Witt, "Effects of Operating Parameters on Flow Characteristics of Natural Gas in Supersonic Separators," in *Offshore Technology Conference*, (2013).

13 C. Wen, Y. Feng, P. Witt, X. Cao, and Y. Yang, "CFD Simulation of Supersonic Swirling Separation of Natural Gas Using a Delta Wing," in *Proceedings of the Ninth International Conference on CFD in the Minerals and Process Industries*, (2012).

14 Y. Yang, C. Wen, S. Wang, Y. Feng, and P. Witt, "The Swirling Flow Structure in Supersonic Separators for Natural Gas Dehydration," *RSC ADV J.*, vol. **4**, pp. 52967-52972, (2014).

15 K. Foelsch, "The Analytical Design of an Axially Symmetric Laval Nozzle for a Parallel and Uniform Jet," *J. Aeronaut. Sci.*, (2012).

16 M. Betting, T. V. Holten, and J. M. H. M. V. Veen, "Supersonic Separator Apparatus and Method," US Patent US 20020194988A1, (2002).

17 B. Prast, R. A. van Dam, J. F. H. Willems, and M. E. H. van Dongen, "Formation of Nano-Sized Water Droplets in a Supersonic Expansion Flow," *J. Aerosol Sci*, vol. **27**, pp. S147-S148, (1996).

18 P. Schinkelshoek and H. Epsom, "Supersonic Gas Conditioning-Low Pressure Drop Twister for NGL Recovery," (2006).

19 F. T. Okimoto, S. Sibani, and M. Lander, "Twister Supersonic Gas Conditioning Process," (2002).

20 B. Prast, P. Schinkelshoek, B. Lammers, and M. Betting, "CFD for Supersonic Gas Processing," in *NEL Multiphase Separation and Multiphase Pumping Technologies Conference*, (2005), pp. 53-58.

21 M. Haghighi, K. A. Hawboldt, and M. A. Abdi, "Supersonic Gas Separators: Review of Latest Developments," *J NAT GAS SCI ENG*, vol. **27**, pp. 109-121, (2015).

22 V. Alfyorov, L. Bagirov, L. Dmitriev, V. Feygin, S. Imayev, and J. R. Lacey, "Supersonic Nozzle Efficiently Separates Natural Gas Components," *Oil Gas J.*, vol. **103**, pp. 53-58, (2005).

23 R. L. Garrett, "Supersonic Flow Seprator," US Patent 3559373, (1971).

24 S. Keisuk, "Separation of Gaseous Mixtures," Japan Patent JPH0217-921A, (1990).

25 V. Holten, "Method and Device for Separating a Gas from a Gas Mixture," (1992).

26 A. Borissov, G. Mirzoev, and V. Shtern, "Supersonic Swirling Separator 2 (Sustor2)," ed: Google Patents, (2014).

27 C. Wen, Y. Yang, and W. Jiang, "Supersonic Expansion Refrigeration and Cyclone Separation Device for Natural Gas," (2012).

28 L. Hengwei, L. Zhonggliang, F. Yongxun, G. Keyu, and Y. Tingmin, "Characteristic of a Supersonic Swirling Dehydration System of Natural Gas," *Chin. J. Chem. Eng.*, vol. **13**, pp. 9-12, (2005).

29 Z. Liu, J. Ding, W. Jiang, J. Zhang, and Y. Feng, "Numerical Simulation of Highly-Swirling Supersonic Flow Inside a Laval Nozzle," *PROG COMPUT FLUID DY*, vol. **8**, pp. 536-540, (2008).

The effect of milling duration on silicon dioxide characterization

Mohd Aizat Hafiz Mohd Sohor[1]*, *Mazli* Mustapha[1], and *Othman* Mamat[1]

[1]Mechanical Engineering Department, Universiti Teknologi PETRONAS (UTP), 32610 Seri Iskandar, Perak, Malaysia

Abstract. This paper reports the study of milling duration on silicon dioxide characterization. The silicon dioxide has undergone mechanical milling treatment from 0 to 100 hours. The product after milling process was characterized by using scanning electron microscopy and X-ray diffraction. Williamson-Hall plot technique was used in this study to analyse the silicon dioxide crystallite size and lattice strain. The longer milling duration resulted in the reduction of crystallite size and increase in lattice strain.

1 Introduction

Silicon carbide (SiC) is an interesting ceramic material that has unique properties in mechanical, optical and electrical aspects. These special properties and characteristics have made SiC to be widely applied in industrial and engineering fields. Carbothermal reduction process involved long hours of heating raw materials at extremely high temperatures which was around 2200 - 3300 °C. Quartz sand and petroleum coke were used as precursor materials for this technique [1]. Large particles powder was produced by this process because of high temperature and long heating hours. Therefore, extensive mechanical milling (MM) of raw materials was needed to produce smaller diameter size and will aid the reaction to reduce sintering temperature [2]. This paper aims to present the effect of milling duration on silicon dioxide characterization.

2 Experimental procedure

The precursor materials of silicon and carbon were prepared from amorphous silicon dioxide, SiO_2 (Johore silica, 99.5% purity) and carbon black powder from petroleum coke was originally from Petronas refinery. Silicon dioxide was performed mechanical treatment in a stainless-steel jar by using iron as ball milling media. The ball to powder ratio was set to 20:1 with 200 rpm grinding speed. The mechanically milled silica and carbon were mixed with ≈10% silica and ≈60% carbon beyond the stoichiometric ratio C/SiO_2 of 3:1 and were placed in the rotary mixture.

* Corresponding author: mohd.aizat_g03388@utp.edu.my

The morphology of samples was analyzed under scanning electron microscopy (SEM) (VPFESEM, Zeiss Supra55 VP) and chemical composition was analysed by energy dispersive spectroscopy (EDS). The phase composition was determined by Bruker model D8 Advanced X-ray Diffractometer (Cu K_α radiation, $\lambda = 0.15406$ nm) from Braggs angle (2θ) of 10 to 90° with steps of 0.040°. Phase identification of the powders were carried out using EVA software. Positions of the peaks in a diffraction profile are integral in X-ray diffraction (XRD) when conducting phase identification. The relative intensities of a peak in a diffraction profile as well as relative intensities of these peaks also influence it. Additional vital information can be gleaned from the shapes of the peaks. In particular, the width of a peak indicates the amplitude of thermal oscillations of the atoms at their regular lattice sites. Furthermore, it is an indication of vacancy and element concentrations impurities and even plastic deformation; any factor which affects the distribution of d-spacings.

The broadening of a peak represents the crystallite size. Scherrer equation satisfies the Bragg condition for non-adjacent diffraction planes as it explains peak broadening in terms of incident beam divergence [3]. The peak width (described as full width at half maximum peak intensity (FWHM)), peak position and wavelength are used in a function to calculate crystallite size once instrument effects have been excluded.

The Williamson Hall plot method's is based on a Fourier deconvolution of the measured peaks and the instrument broadening to obtain the true diffraction profile [3]–[6]. This method can yield both the crystallite size distribution and lattice macrostrain. Uniform residual strain will lead to a parallel shift in the x-ray diffraction peaks for the specimens that experienced large plastic deformation. Conversely, the broadening of x-ray diffraction peaks are caused by varying residual strain. Since the broadening parameter of the x-ray peaks is due to strain effect as well as size effect, it could then be written as following:

$$B_s = \frac{0.9\lambda}{d\cos\theta} + \eta\tan\theta \tag{1}$$

By multiplying *cos* θ to (1), the formula becomes:

$$B_s\cos\theta = \frac{k\lambda}{d} + \eta\sin\theta \tag{2}$$

where B_s refers to full width at half maximum (FWHM), θ the Bragg angle, d is the crystallite size, η the lattice strain, λ the x-ray wavelength, k is Scherrer constant (=0.9). The instrumental broadening (B_i) should be corrected by the following equation:

$$B_s^2 = B_e^2 - B_2^2 \tag{3}$$

where B_e is the FWHM of the measured XRD peak. Thus, when B *cos* θ is plotted against *sin* θ, a straight line will be obtained with the slope as η (lattice strain) and the intercept being $k\lambda/d$ where crystallite size, d, can be calculated by using this method.

3 Results and discussions

3.1 Influence of mechanical milling treatment on silicon dioxide

The XRD patterns shown in Figure 1 had different duration of grinding which ranged from 0 hour to 100 hours while Fig. 2 (a) – (f) shows the surface morphology of the silica sand which was subjected to various milling time. For reference, the highest peak for Un-MM silica sand is used and it is at 26.2° (2θ angle). After 20 hours of grinding, it has shown that the intensity of the peak had reduced up to 89.5% whereas 98.3% was achieved after 100 hours of grinding. The increment of broadening effect at the peak base of the spectrum is caused by longer mechanical milling time in order to reduce particle size (MM) silica sand. Similar finding by Suryanarayana where talc sample is used instead has shown the same XRD pattern [3]. Szewezak et al have reported that the maximum size reduction limit for a particle is equivalent to the size of tough-brittle transition which lead to plastic deformation rather than breakage of particle and it is accompanied by the growth of structure distortions [6].

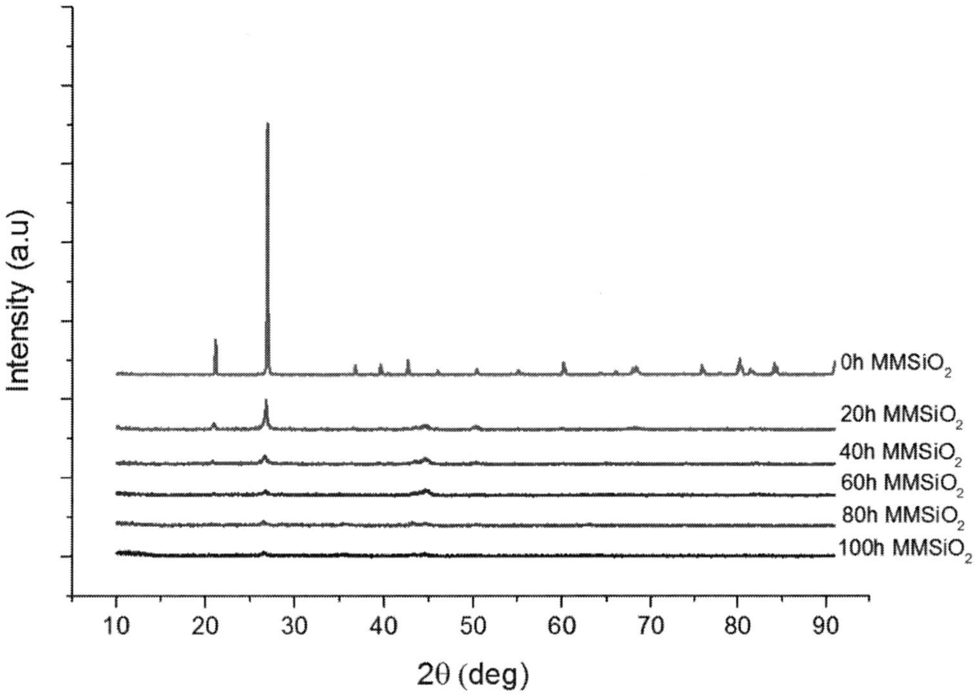

Fig. 1. X-ray diffraction pattern of SiO_2 at various milling duration

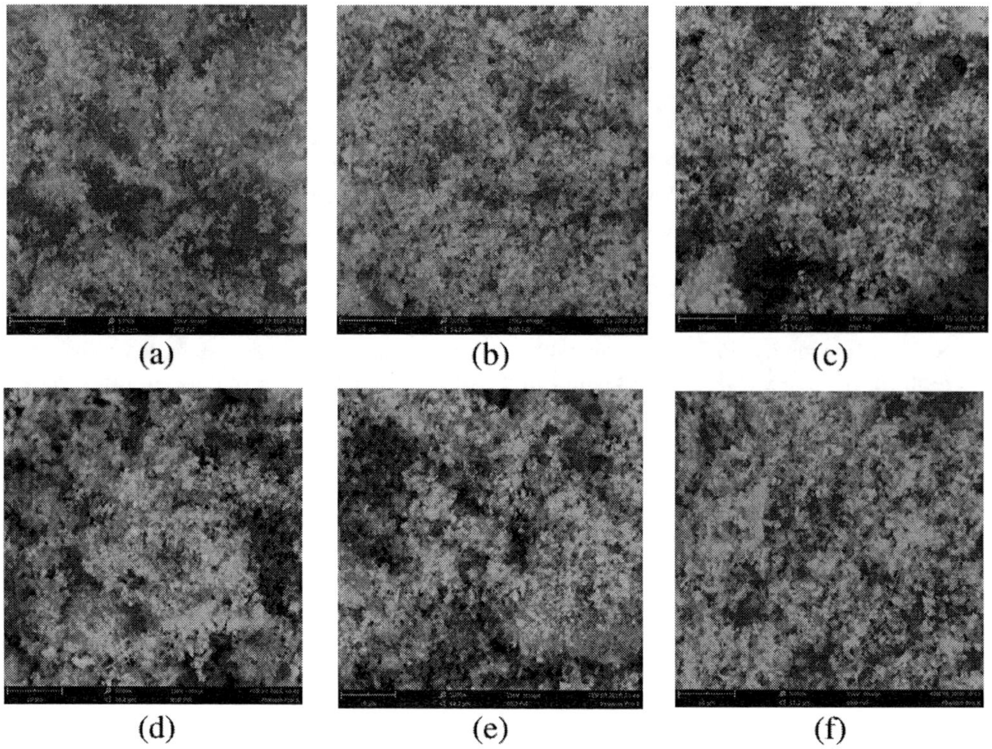

Fig. 2. The SEM images of SiO_2 sand after undergoing mechanical treatment at various milling time: (a) 0 h (b) 20 h (c) 40 h (d) 60 h (e) 80 h (f) 100 h

Table 1 shows the intensity value at milling time ranging from 0 hour to 100 hours. It can be seen that the degree of random disorder in (MM) silica sand increases as milling time increases. At 100 hours of milling time, the degree of random disorder is about 98.3%. This indicates that the mechanical milling process which is based on impact and friction among particles not only causes size reduction but also produces important changes in the physio chemical properties of solid. This leads to structural alteration because of loss of regularity in the crystalline network. The observation of line broadening in the X-ray diffraction spectra is an important method that was used for distortion investigation. The degree of random disorder and lattice distortion of the particles are indicated by the intensity and half width of X-ray lines in different patterns.

Table 1. Intensity value at different milling time

Milling Time (h)	Intensity (a.u)	Random Disorder (%)
0	1818	-
20	191	89.5
40	60	96.7
60	39	97.9
80	33	98.2
100	30	98.3

3.2 Crystallite size and microstrain

The crystallite size, D and the microstrain, η for the fabricated foam at various SiO_2 after milling process were calculated from the Williamson-Hall plot (Fig. 3) and tabulated in Table 2. The table reveals that as milling time increases, the crystallite size decreases. Abdoli et al. also reported a similar finding whereby a ductile Al has a higher chance to plastically deform which can lead to grain refinement and changes in lattice distortion [7]. They further added that low stiffness or high work hardening capacity powders will densify more effectively. The increment of milling time leads to the increase of microstrain which can be attributed to SiO_2 milling. During the process, they were in a complicated stress strain state after pressure was being applied by the iron ball which acted as milling media inside the stainless-steel jar.

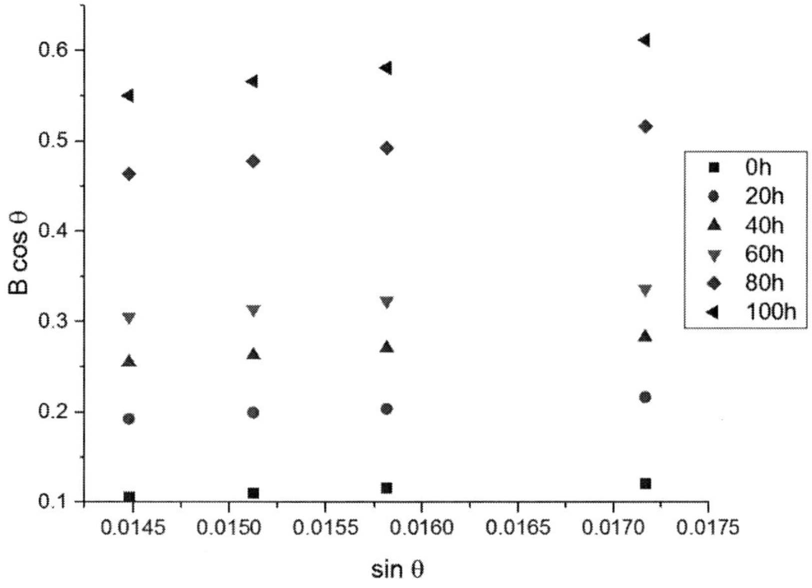

Fig. 3. Williamson-Hall plot for SiO_2 with different milling duration

Table 2. Lattice strain and crystallite size mechanically milled SiO_2 calculated from Williamson-Hall plot method

MMSiO_2 (h)	MMSiO_2-Crystalline Size (μm)	Lattice Strain (%)
0	4.95	5.47
20	2.04	8.62
40	1.31	10.4
60	0.96	11.23
80	0.76	19.84
100	0.63	22.74

4 Conclusion

In summary, the preparation of precursor material namely SiO_2 has been done successfully through mechanically milled treatment at various times. XRD pattern showed that the intensity peak lowered with longer duration of SiO_2 milling time. Similarly, the increase of mechanical milling time caused the size reduction of SiO_2 crystalline and increase in SiO_2 lattice strain.

The authors want to thank Universiti Teknologi PETRONAS for necessary support throughout the project. This project is financially funded by the Malaysian Ministry of Higher Education (MOHE) under the Fundamental Research Grant Scheme (FRGS) no. 0153AB-K82.

References

1. S. Cetinkaya and S. Eroglu, *J. Eur. Ceram. Soc.*, **31**, no. 5, pp. 869–876, (2011)
2. B. M. Moshtaghioun, A. Monshi, M. H. Abbasi, and F. Karimzadeh, *Int. J. Refract. Met. Hard Mater.*, **29**, no. 6, pp. 645–650, (2011)
3. C. Suryanarayana, *Progress in Materials Science*, **46**, no. 1–2. pp. 1–184, (2001)
4. J. S. Lee, Y. K. Byeun, S. H. Lee, and S. C. Choi, *J. Alloys Compd.*, **456**, no. 1–2, pp. 257–263, (2008)
5. F. Charlot, E. Gaffet, B. Zeghmati, F. Bernard, and J. C. Niepce, *Mater. Sci. Eng. A*, **262**, no. 1–2, pp. 279–288, (1999)
6. E. Szewczak and J. W. Wyrzykowski, *Nanostructured Mater.*,**12**, no. 1, pp. 171–174, (1999)
7. H. Abdoli, H. Farnoush, E. Salahi, and K. Pourazrang, *Mater. Sci. Eng. A*, **486**, no. 1–2, pp. 580–584 (2008)

 DOI: 10.1051/matecconf/201713103008

Torrefied Biomass Gasification: A Simulation Study by Using Empty Fruit Bunch

Muhammad Bilal Muslim, *Suriyati* Saleh, and *Noor Asma Fazli* Abdul Samad*,

Faculty of Chemical & Natural Resources Engineering, Universiti Malaysia Pahang, Lebuhraya Tun Razak, 26300 Gambang, Pahang, Malaysia

Abstract. A fixed bed gasification model has been developed by using Aspen Plus as the simulation and modelling tool. The feedstock chosen consisted of raw empty fruit bunch (EFB) and torrefied EFB at 300°C. The hydrogen gas composition obtained using fixed bed gasification model from all feedstock were then analysed together with lower heating value (LHV) and cold gas efficiency (CGE). The simulation results show that as the gasifier temperature is increased, the amount of hydrogen produced is also increased and an increment of air-to-biomass ratio (ABR) reduces the amount of hydrogen produced. Besides, the torrefied EFB produced a higher amount of hydrogen gas (6.52%) at gasifier temperature of 800°C and air-to-biomass ratio (ABR) of 0.2 compared to raw EFB (5.09%). This indicates that torrefied biomass at higher temperature improves hydrogen production. In addition, the use of torrefied biomass also contributes to higher LHV and increases the efficiency of the gasification process.

1 Introduction

Biomass is an important contributor to the world economy. Today, various forms of biomass energy are consumed all over the world. Biomass provides a clean, renewable energy source that could dramatically improve the environment which can help reduce greenhouse gas (GHG) emission and can help nature climate. In particular, conversion of non-edible biomass such as agriculture residues, wood chips, fruit bunches, stalks, and industrial and municipal wastes into fuels and useful chemicals would solve waste disposal and energy issues [1]. Palm oil wastes are the main biomass resources in ASEAN countries. In Malaysia alone, there are 9.66, 5.20, and 17.08 million tonnes of fibres, shell, and empty fruit bunches, respectively [2]. Thus, to treat this tremendous amount of wastes, novel technologies with improved efficiencies and reduced environmental impacts such as gasification are necessary.

Biomass gasification is a process of incomplete combustion of biomass which produces combustible gases consisting of carbon monoxide (CO), hydrogen (H_2), and traces of methane (CH_4). One type of reactor that is utilised for gasification process is fixed bed reactor. Fixed bed gasifiers can be divided into updraft (counter-current) and downdraft (co-current) gasifiers based on the direction of the flow of the gasifying agent. For updraft

* Corresponding author: asmafazli@ump.edu.my

gasifiers, the biomass moves downwards through a bed, while reacting with the gasifying agent that moves in the opposite direction. Meanwhile, for downdraft gasifiers, the gasifying agent moves downwards through the bed in the same direction as the biomass feedstock. The advantages of this gasifier are its ability to produce syngas with low tar content and no requirement of extensive clean-up [3].

To use biomass more efficiently, a variety of pre-treatment methods for improving biomass have been developed. Torrefaction is one of the most noteworthy routes. Torrefaction removes moisture from raw biomass by decomposing the biomass in the absence of oxygen at temperatures ranging from 200°C to 300°C. The advantages of torrefaction of biomass include intensifying calorific value and energy density, reducing O/C and H/C ratios, producing hygroscopic materials, and making the grindability and storage of biomass easier. Moreover, the properties of torrefied biomass become more uniform compared to those of raw biomass [4]. Therefore, the objective of this work is to develop and simulate gasification process using fixed bed reactor in the Aspen Plus software. The fixed bed reactor is chosen because it is simple, reliable, and proven for certain fuels; provides high carbon conversion efficiency; and has high thermal efficiency. Here, the biomass used is EFB. Two different types of EFB are used for the feed of gasification process which consists of raw EFB and torrefied EFB at 300°C. The synthesis gas produced is then evaluated. In addition, the effects of temperature and air-to-biomass ratio on low heating values (LHV) and cold gas efficiency (CGE) are also discussed.

2 Simulation Model Developments

2.1 Model Description and Properties

The main processes in this gasification are classified into three stages which consist of the drying of the feed, the decomposition of the feed, and char gasification and combustion. The gasification flow sheet for fixed bed reactor is shown in Figure 1. The first stage is the drying of the feed in order to remove water content in the feed. Next is the RSTOIC reactor which is used to simulate the decomposition of the feed. Here, the extent of the reaction (*Feed*: 0.0555084H_2O) is used in the RSTOIC reactor to transform some part of the inlet feed to water. When the dry feed enters the RYIELD reactor, the decomposition of the feed takes place and the feed is then converted into atoms of carbon (C), hydrogen (H), oxygen (O), sulphur (S), nitrogen (N), and ash. The decomposed feed and air enter the RGIBBS block for combustion and gasification reaction to produce the desired syngas. The feed used consists of raw and torrefied EFB at 300°C. The proximate and ultimate analysis of all the feedstock is shown in Table 1.

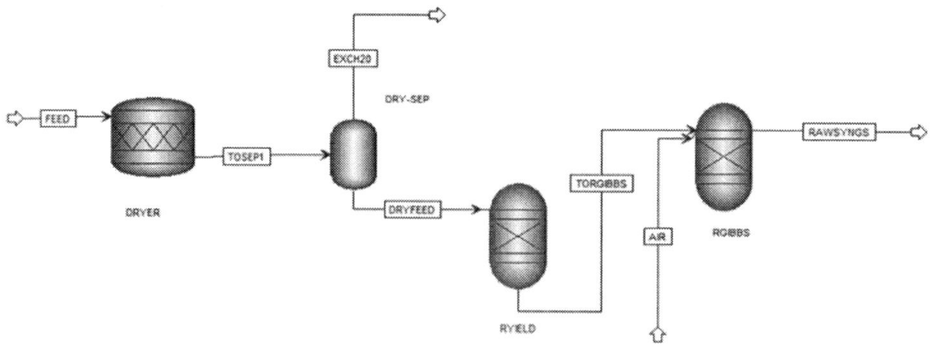

Fig. 1. Fixed bed gasification process flow sheet

The model is simulated as a standalone model using the feed flow rate of 10 kg/h, temperature of 700°C, and pressure of 1 bar. For gasifying agent, the air flow rate of 10 kg/h is used. The Redlich-Kwong-Soave (RKS) cubic equation of state with Boston-Mathias alpha function (RKS-BM) is used as a thermo-physical property method in the gasification process. The assumptions used for this gasification model are (i) the process is in the isothermal and steady-state conditions; (ii) the composition of char consists only carbon, void, and ash; (iii) there is no pressure loss; (iv) all the particles are assumed to be in the spherical shape; and (v) the syngas produced are in ideal gas state, including hydrogen (H_2), carbon monoxide (CO), carbon dioxide (CO_2), and methane (CH_4).

Table 1. Proximate and ultimate analysis for raw and torrefied empty fruit bunch [5]

Proximate and Ultimate Analysis	Raw EFB	Torrefied EFB at 300°C
Volatile Matter	65.01	48.44
Fixed Carbon	15.37	43.86
Ash	3.85	7.70
C	43.53	54.63
H	7.20	5.63
O	47.09	36.04
N	1.73	6.37
S	0.46	0.21
HHV (MJ/kg)	15.49	19.60

The dry product gas lower heating value (LHV) and cold gas efficiency (CGE) are defined as follows:

$$LHV_{Bio} = HHV - 0.212 \times H - 0.0245 \times M - 0.008 \times O \tag{1}$$

$$\eta_{cg} = \frac{\dot{m}_{PG}\, LHV_{PG}}{\dot{m}_{Bio}\, LHV_{Bio}} \times 100\ \% \tag{2}$$

where *HHV* is the high heating value of the biomass (MJ/kg), and *H*, *M*, and *O* are the weight percentages of hydrogen, moisture content, and oxygen obtained from ultimate analysis. Meanwhile, cold gas efficiency (CGE) is defined as the fraction of the chemically-bound energy in the biomass that is converted into chemically-bound energy in the product gas from the gasification process where $\dot{m}_{PG}$ and $\dot{m}_{Bio}$ are the mass flow of product gas and biomass in kg/h, respectively, and LHV_{PG} and LHV_{Bio} are the lower heating value of the product gas and the biomass in MJ/kg, respectively [6].

3 Results and Discussion

3.1 Model Validation

The data of syngas component in the literature is compared with the predicted syngas produced in this work as shown in Table 2. By comparing both of the results, it can be seen that the syngas contents are quite similar in both studies. The final results for CO_2, CO, CH_4, and H_2 are very comparable between the simulation and experimental results. The root mean square error (RMSE) obtained for every syngas components are relatively low and generally in good agreement with the literature data. This indicates that the developed model is validated and indeed reliable.

Table 2. Synthesis gas production comparison between this work and literature data [7]

Components (%)	CO_2	CO	CH_4	H_2
Simulation result	15.88	10.12	0.00	7.26
Experimental results from Moni & Sulaiman [7]	14.00	10.39	0.01	8.20
RMSE	0.118	0.026	0.097	0.115

3.2 Effects of Gasifier Temperature and Air-to-Biomass Ratio (ABR)

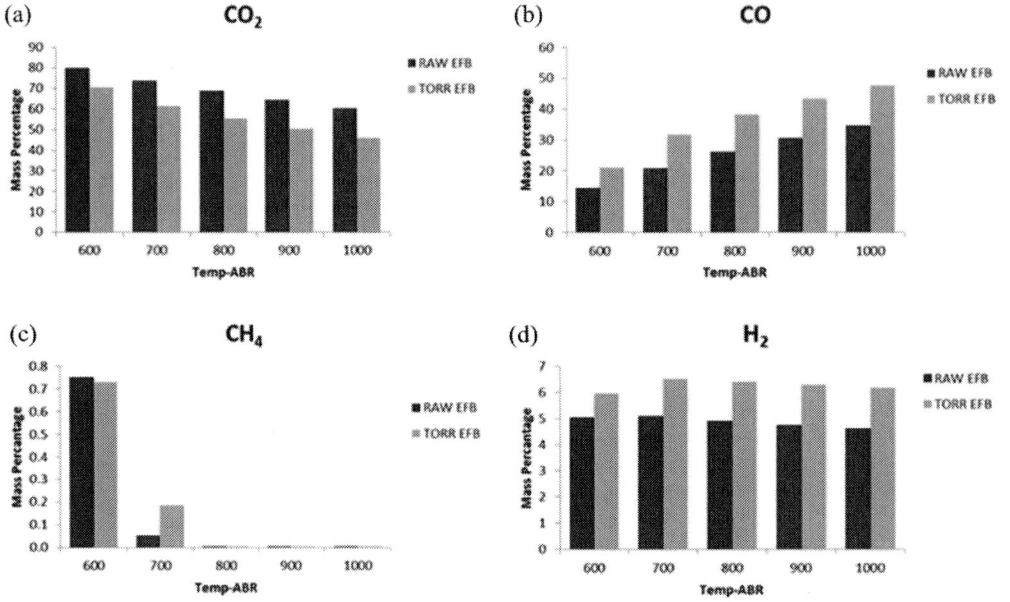

Fig. 2. Mass percentage on dry basis for temperature cases for syngas (a) carbon dioxide, (b) carbon monoxide, (c) methane, and (d) hydrogen

The effects of gasification temperature on the syngas production are shown in Figure 2. The amount of hydrogen gas shows a slight increase as the gasifier temperature is increased from 600°C to 700°C. This is due to the water-gas shift reaction ($C + H_2O \leftrightarrow CO + H_2$) and Boudouard reaction ($C + CO_2 \leftrightarrow 2CO$) which cause the increasing amount of hydrogen and carbon monoxide and the decreasing amount of CO_2. However, at 800°C, the amount of hydrogen gas is lower than the previous gasifier temperature due to the dominant Boudouard reaction which contributes to the increasing of CO gas and decreasing of CO_2 gas. In addition, some of the hydrogen is also converted into methane ($C + 2H_2 \rightarrow CH_4$), which explains the low amount of hydrogen gas produced.

Figure 3 shows the syngas production based on different values of ABR. The effect of air-to-biomass ratio (ABR) is investigated by varying the ABR from 0.2 to 1. When the value of ABR is increased, the H_2 production is decreased but the CO_2 shows the opposite trend. This is due to the exothermic reaction of combustion reactions ($C + O_2 \leftrightarrow CO_2$ and $C + 0.5O_2 \leftrightarrow CO$) where the oxygen gas supplied to the gasifier enhances the combustion process and subsequently the production of CO, CO_2, and H_2O are increased. More oxygen supply to the reactor will ultimately increase the ABR ratio which promotes the oxidation reaction and thus, lowering the amount of CO content but increasing the CO_2 content. The amounts of H_2 also decrease as the oxidation of hydrogen occurred to become water ($H_2 + 0.5O_2 \leftrightarrow H_2O$). Hence, the amount of H_2O will increase as the value of ABR increased. In terms of hydrogen production, torrefied EFB provides higher amount of hydrogen percentage compared to raw EFB. This is due to the fact that higher carbon content in the ultimate analysis of torrefied EFB enables more carbon conversion rate to produce more hydrogen gas and carbon monoxide gas [8].

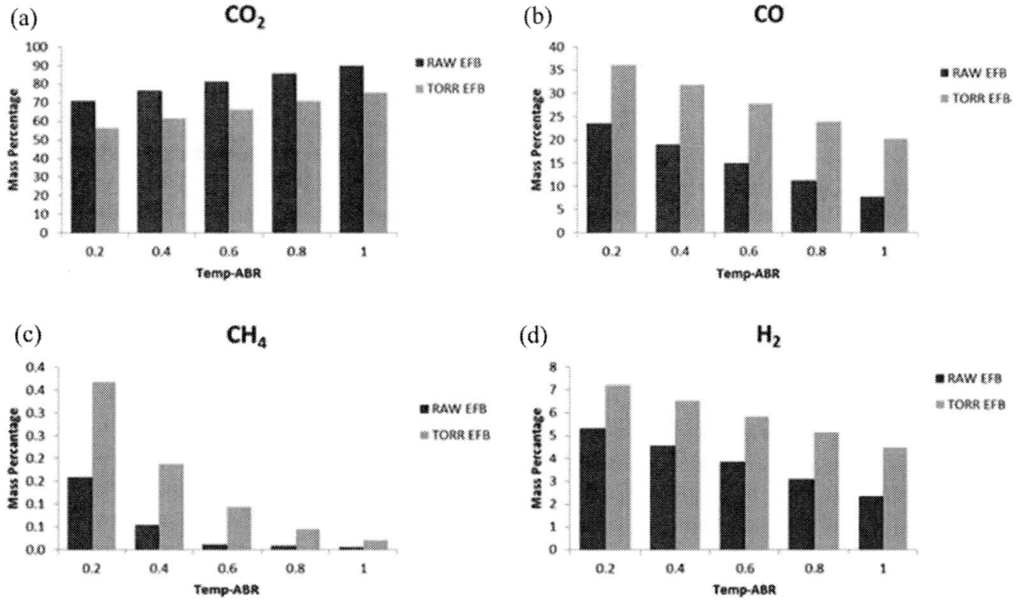

Fig. 3. Mass percentage on dry basis for air-to-biomass ratio cases for syngas (a) carbon dioxide, (b) carbon monoxide, (c) methane, and (d) hydrogen

3.3 Lower heating value (LHV) and cold gas efficiency (CGE)

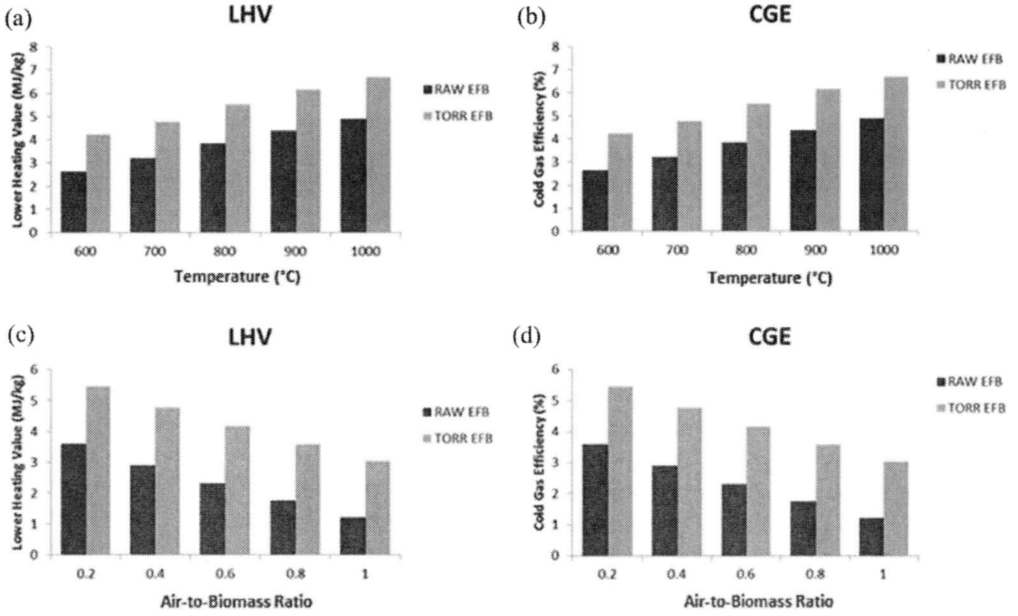

Fig. 4. Lower heating value and cold gas efficiency for temperature and air-to-biomass ratio cases

Figure 4 shows the effect of temperature and air-to-biomass ratio on the efficiency of gasification and net calorific value. The torrefied EFB provides more net calorific value and gas efficiency compared to raw EFB. As shown in Equation (2), the LHV depends on the HHV of feedstock and its ultimate analysis. The amount of HHV for torrefied EFB is higher than raw EFB, which indicates better conversion of solid biomass to gas production, and thus, ultimately increases the carbon conversion rate. Therefore, the CGE of the feedstock will improve significantly. As for ABR, the amount of LHV and CGE value shows a decreasing trend as the ratio of air to biomass continues to increase. The consumption of combustion gas causes the lower or higher air equivalence ratios, which results in lower gasification temperatures or deterioration of heating values of syngas [9]. Thus, increasing the ABR will lower the CGE of both raw and torrefied EFB.

4 Conclusions

The fixed bed gasification process has been developed in the Aspen Plus and validated for evaluating the hydrogen production. Two different types of EFB are employed as input for the gasification process, namely, raw EFB and torrefied EFB at 300°C. Subsequently, the sensitivity analysis of gasification temperature and air-to-biomass ratio (ABR) has been performed and their effects on the synthesis gas production, LHV, and CGE have been discussed for each feedstock. The increment of the gasification temperature in the sensitivity analysis improves synthesis gas production, particularly H_2 and CO concentrations, which ultimately lead to higher LHV and CGE. In addition, the torrefied EFB at 300°C provides higher amount of hydrogen production (6.52%) at the gasification temperature of 800°C and ABR ratio of 0.2 compared to the other feedstock. Also, by increasing the ABR ratio, both CGE and LHV show sharp decreasing trends; on the other hand, higher LHV and CGE were obtained when the gasification temperature is increased.

References

1. R. Warnecke, Biomass and bioenergy **18**, 489-497 (2000).
2. R. E. Konda, S. A. Sulaiman, and A. Bambang, Journal of Applied Sciences **12**, 2555-2561 (2012).
3. P. Basu. "Combustion and gasification in fluidized beds". CRC press (2006).
4. N.H.H.M. Harun, F.R.A.A. Wahid, S. Saleh, and N.A.F.A. Samad, Chemical Engineering Transactions **56**, 1195-1200 (2017).
5. F.R.A.A. Wahid, N.H.H.M. Harun, S.R.M. Rashid, N.A.F.A. Samad, and S. Saleh, Chemical Engineering Transactions **56**, 199-204 (2017).
6. Arena, Umberto, Waste management **32**, 625-639 (2012).
7. M. N. Z. Moni, and S. A. Sulaiman, Asian Journal of Scientific Research **6**, 197-206 (2013).
8. F.R.A.A. Wahid, M.B. Muslim, S. Saleh, and N.A.F.A. Samad, ARPN Journal of Engineering and Applied Sciences **11**, 2673-2680 (2016).
9. M.B. Muslim, S. Saleh, and N.A.F.A. Samad, Chemical Engineering Transactions **56**, 1495-1500 (2017).

MATEC Web of Conferences **131**, 03009 (2017) DOI: 10.1051/matecconf/201713103009
UTP-UMP SES 2017

Analyzing The Behavior of Classical Functionally Graded Coated Beam

A. Toudehdehghan[1, 2, a)] and M. M. Rahman[2, b)] and F. Tarlochan[3, c)]

[1] Department of Mechanical Engineering, INTI International University, Jalan BBN 12/1, Bandar Baru Nilai, 71800 Nilai, Negeri Sembilan.
[2] *Department of Mechanical Engineering, Universiti Tenaga Nasional Malaysia, km7, Jalan Kajang-Puchong, 43009 Kajang, Selangor, Malaysia*
[3] *Mechanical and Industrial Engineering, College of Engineering, Qatar University, P.O. Box 2713, Doha, Qatar.*

[a)]Corresponding author: Abdolreza.toudehdehghan@newinti.edu.my
[b)] mujibur@uniten.edu.my
[c)] faris.tarlochan@qu.edu.qa

Abstract. The governing equation of a classical rectangular coated beam made of two homogeneous layers at top ceramic coated layer and bottom metal layer and also single Functionally Graded Material (FGM) as a sub coated layer subjected to uniform distributed mechanical load are derived by using principle of virtual displacements and based on Euler-Bernoulli deformation beam theory (EBT). In FGM layer the material properties are presumed differ as an exponential function form in thickness coordinate. Hence, the aim of this paper is analyzed the static behavior of clamped-clamped thin coated beam under mechanical load.

NOMENCLATURE

a = Length of beam
E_c = Modulus elasticity of ceramic
E_m = Modulus elasticity of metal
b = Width of beam
h = Total thickness of beam
M_x = Total moment
u = Displacement in x direction
v = Displacement in y direction
w = Displacement in z direction
u_0, w_0 = Displacement mid-plane
I = Second moment of area
δW = Total virtual work
δW_I = Internal virtual work
δW_E = External virtual work
D_{xx} = Flexural rigidity of beam
ε_{xx} = Strain
R = Matrix
σ_{xx} = Stress
m =Series member
γ = Row number
F_γ = Force column
A_i = Fourier coefficient column

INTRODUCTION

Generally, the surface of structures or elements is weaker than the inside of them. Therefore, the coating has been used for decades to increase the resistance of the surface of structures, reduce the stress concentration on the surface and stop cracks which have been generated. In the conventional surface coating method, one layer as a coated will be covered the surface of the body and the thickness range is from a micrometer to several millimeters. The coating thickness usually is depended on the type of material which is used as a substrate layer [1]. For example, in Thermal

Barrier Coating (TBC) one-layer ceramic is bonded to the substrate layer. The disadvantage of this type of coating is that due to the use of two completely different materials (ceramic-metal), the concentration of stress between the two layers' increases, and thus the two layers are separated [2]. To improve the weakness, Functionally Graded Materials (FGMs) have lately been suggested to modify the conventional coating [3]. Due to the characteristic of FGM composites, the properties of materials in the composite change very gradually between two different materials. This property makes the separation between the composite layers not present. Mainly, under a high-temperature atmosphere like a nozzle of the shuttle, nuclear fusion reactor, internal combustion engine and so on at the interface of two layers due to different thermal expansion, the mismatch will be happened [4]. Coating structures with FGM are called FGC. FGC beams under mechanical, thermal and thermos-mechanical loads are studied by several researchers [5-11]

The present article, a static analysis of FGC beam is studied by using a series displacement as a linear combination of known function which satisfies the boundary conditions and unknown parameters. The governing equation of FGC beam based on classical beam displacement theory and principle of virtual is derived. The material properties of FG layer through the thickness is considered an exponential function. In order to demonstrate the behavioral difference of a conventional composite with a modern composite which is consisting of an FGM layer. Two composite models are considered. Lastly, the effect of the rectangular clamped-clamped beam under distributed load is investigated.

EXTRACTING GOVERNING EQUATIONS

The displacement field components in Timoshenko's beam are considered in the form below [12]:

$$u(x,y,z)=u_0(x,y)-z\frac{dw_0(x)}{dx}$$
$$v(x,y,z)=0 \tag{1}$$
$$w(x,y,z)=w_0(x)$$

Thus, the axial strain component in mid-plane direction of beam is

$$\varepsilon_{xx}=\frac{du}{dx}=\frac{du_0}{dx}-z\frac{d^2w_0}{dx^2} \tag{2}$$

Based on EBT strains are small in the case of the following terms are negligible compared to equation (2)

$$\frac{du}{dz},\left(\frac{du}{dx}\right)^2,\varepsilon_{zz},\varepsilon_{xy} \tag{3}$$

FGM material properties

The continuous composition of FGM can be described by using several mathematical models to show the material properties of the composite. The material properties of FGM are inhomogeneous microscopically, therefore the material properties of FGMs depend on position. One of the mathematical models of material properties was used in a lot of research in order to study fracture mechanics, crack propagation, vibration, and bending, is exponential function [4, 5, 13-20].

Position dependent material properties with exponential function of FGM are

$$E_{fe}(z)=E_c e^{\left(\frac{1}{h}\ln\left(\frac{E_m}{E_c}\right)\left(z+\frac{h}{2}\right)\right)} \tag{4}$$

DOI: 10.1051/matecconf/201713103009

Governing equation of beam

To derive the governing equations of the beam, we use the principle of virtual work defined as [21]:

$$\delta W = \delta W_I + \delta W_E = 0 \tag{5}$$

Where δW_I and δW_E are virtual work cause of internal forces and virtual work cause of external forces respectively. The boundary conditions of clamped-clamped beam are

$$w(0) = w(a) = \frac{dw}{dx}(0) = \frac{dw}{dx}(a) = 0 \tag{6}$$

Also, the variation form of the beam should satisfy the equation (6):

$$\begin{aligned} &\delta w(0) = \delta w(a) = 0 \\ &\frac{d\delta w}{dx}(0) = \frac{d\delta w}{dx}(a) = 0 \end{aligned} \tag{7}$$

The stress-strain relation in the mid-plane direction is

$$\sigma_{xx} = E(z)\varepsilon_{xx} \tag{8}$$

Substitution equation (2) into equation (8), the resulting moment per unit length of the beam is

$$M_x = \int_{-\frac{h}{2}}^{\frac{h}{2}} E(z) b \left[z\frac{du_0}{dx} - z^2 \frac{d^2 w}{dx^2} \right] dz \tag{9}$$

The total internal virtual work done is

$$\delta W_I = \int_0^a \int_A \sigma_{xx} \delta\varepsilon_{xx} dA dx \tag{10}$$

The virtual work done by external distributed load by using the virtual displacement is

$$\delta W_E = -\int_0^a bP_0 . \delta w dx \tag{11}$$

By using equations (10) and (11) into equation (5) and with consideration the boundary conditions (equations (6) and (7)), it can be obtained the governing equation of beam, taking into account Euler-Bernoulli theory, according to the following:

$$\int_0^a D_{xx} \frac{d^2 w}{dx^2} \frac{d^2 \delta w}{dx^2} dx - \int_0^a bP_0 \delta w dx = 0 \tag{12}$$

Where $D_{xx} = IE_{fe}(z)$

SOLUTION TO RECTANGULAR COATED BEAM

An approximate solution, the deflection of the beam will be obtained based on boundary conditions (equation (7)) and geometry which is occupying the space defined by

$$0 \le x \le a \qquad -\frac{h}{2} \le z \le \frac{h}{2} \tag{13}$$

Where a and h are the length of the beam and the total thickness of beam included the three layers respectively.

To find the answer to equation (12), based on boundary condition and geometry of beam, an approximate solution is obtained as:

$$w = \sum_{i=1}^{m} A_i \sin^2\left(\frac{i\pi x}{a}\right) \tag{14}$$

By using the solution equation (14) and variation form of solution in equation (12) we get

$$R_{(\gamma)(i)} A_i - F_\gamma = 0 \tag{15}$$

NUMERICAL CALCULATIONS

In order to analyze the behavior of coated beam, two cases are considered. In case one, the coated beam has two layers (coating ceramic layer with (4 mm) thickness on top and substrate metal layer with (6 mm) thickness on bottom). Case two, the coated beam has three different layers. The homogeneous ceramic layer with (2 mm) thickness as a coated, FGM layer with (2 mm) thickness as an under coated (bonded) of coated layer and substrate as a homogeneous metal layer with (6 mm) thickness which are located from top to bottom of the model in thickness direction respectively. In the whole layers of the coated beam, the Poisson's ratio was presumed to be constant $(\nu = 0.3)$ [22, 23]. In the exponential function which defined in Equation (4), the ratio of Young's modulus $\left(E_m/E_c\right)$ was assumed constant and equal to $\left(1/10\right)$, ($E_m = 10$ GPa and $E_c = 100$ GPa).

Static analysis

To illustrate the numerical approach, the dimensions of the coated beam in both cases were taken as a = 50 cm, b = 5cm and h = 1cm. The coordinate axes were located in the middle of the volume of the coated beam. The uniformly distributed load is assumed to be $P_0 = 22kPa$ on top of the beam with clamped-clamped boundary conditions (Equation (6)). With consideration $i = \gamma = 2$ in equation (15), the series of the solution will be achieved with two members.

In Figure (1), the deflection of the coated beam is shown for two cases. The behavior of beam in both case slightly was almost identical, yet, in FGM beam (case two) it was apparent that the deflection at the middle of the beam was observed to be less than 2mm bigger in contrast with the conventional coated beam (case one). Figure (2) shows the variation of the stress (σ_{xx}) throughout the thickness of the coated beam for both cases. In case one, it clearly shows that the stress singularity at the interface between the ceramic layer and substrate layer. Because of this phenomenon, de-bonding will appear at the interface between two layers. For FGM beam graph or case two in Figure (2), because of FGM layer at the interface between the ceramic coated layer and substrate layer, it was observed that the stress gradually changed. Thus, stress concentration and singularity will be decreased. It means, in case two, there is the possibility to create separation at interface two layers will be much less compared to case one.

DOI: 10.1051/matecconf/201713103009

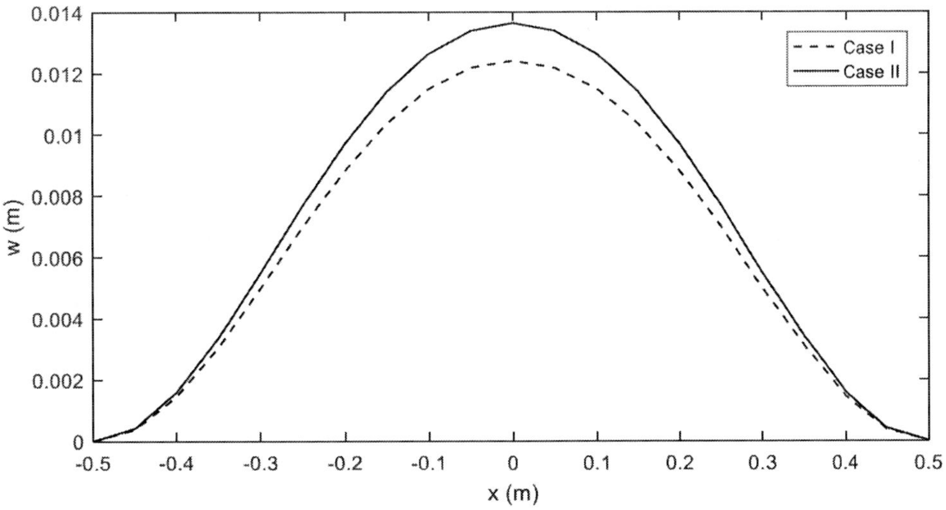

Fig. 1. The deflection (w) of coated beam in case I and case II along x axis $\left(-\frac{a}{2} \leq x \leq \frac{a}{2}\right)$

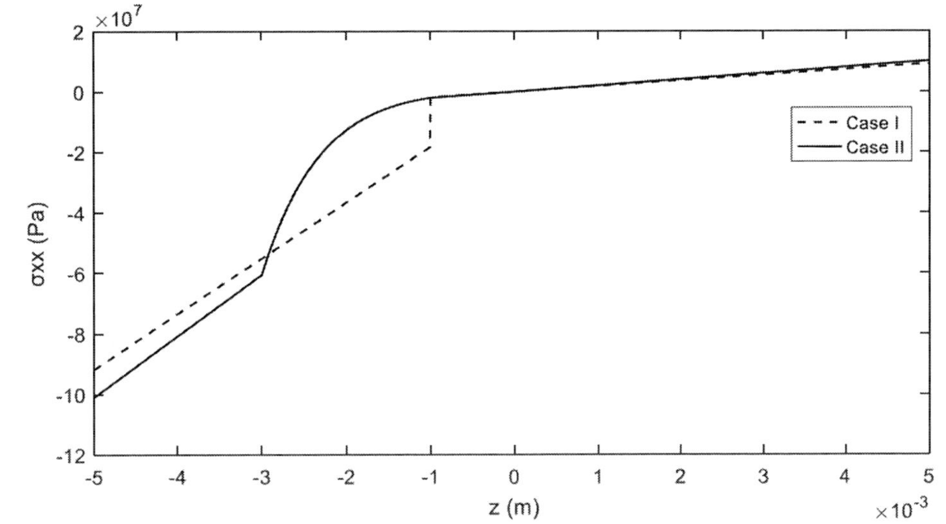

Fig. 2. The stress (σ_{xx}) throughout the thickness of the coated beam in case I and case II

CONCLUSION

The static analysis of conventional coated and FGC under coated rectangular beam under the transverse distributed load was performed. The exponential function in demonstrating the material's properties behavior was considered. Henceforth, the governing equation was then derived based on the Euler-Bernoulli theory and exponential function for FGM layer by using the virtual work principle with only the transverse displacement and stress throughout the thickness of the beam being considered. In conclusion, the obtained approximation function for the displacement or

DOI: 10.1051/matecconf/201713103009

behavior of structure under mechanical load was found. It should be noted that this approximation function is supported by the clamped-clamped boundary conditions. Clearly, from this study, the superiority of FGC composite against conventional coated composite has been demonstrated.

REFERENCES

[1] J. Mencik, *Mechanics of components with treated or coated surfaces* vol. 42: Springer Science & Business Media, 2013.

[2] Y.-L. Chung and C.-F. Pon, "Boundary element analysis of cracked film–substrate media," *International journal of solids and structures,* vol. 38, pp. 75-90, 2001.

[3] M. Kashtalyan and M. Menshykova, "Three-dimensional elastic deformation of a functionally graded coating/substrate system," *International Journal of Solids and Structures,* vol. 44, pp. 5272-5288, 2007.

[4] S.-H. Chi and Y.-L. Chung, "Mechanical behavior of functionally graded material plates under transverse load—Part I: Analysis," *International Journal of Solids and Structures,* vol. 43, pp. 3657-3674, 2006.

[5] B. Sankar, "An elasticity solution for functionally graded beams," *Composites Science and Technology,* vol. 61, pp. 689-696, 2001.

[6] R. Shahsiah, K. Nikbin, and M. Eslami, "Thermal buckling of functionally graded beams," *Iranian Journal of Mechanical Engineering,* vol. 10, pp. 64-80, 2009.

[7] A. Davoodinik and G. Rahimi, "Large deflection of flexible tapered functionally graded beam," *Acta Mechanica Sinica,* vol. 27, pp. 767-777, 2011.

[8] G. Rahimi and A. Davoodinik, "Thermal behavior analysis of the functionally graded Timoshenko's beam," *IUST International Journal of Engineering Science,* vol. 19, pp. 105-113, 2008.

[9] S. Esfahani, Y. Kiani, and M. Eslami, "Non-linear thermal stability analysis of temperature dependent FGM beams supported on non-linear hardening elastic foundations," *International Journal of Mechanical Sciences,* vol. 69, pp. 10-20, 2013.

[10] B. V. Sankar and J. T. Tzeng, "Thermal stresses in functionally graded beams," *AIAA journal,* vol. 40, pp. 1228-1243, 2002.

[11] L. Ma and D. Lee, "A further discussion of nonlinear mechanical behavior for FGM beams under in-plane thermal loading," *Composite Structures,* vol. 93, pp. 831-842, 2011.

[12] C. Wang, J. N. Reddy, and K. Lee, *Shear deformable beams and plates: Relationships with classical solutions*: Elsevier, 2000.

[13] F. Erdogan, "Fracture mechanics of functionally graded materials," *Composites Engineering,* vol. 5, pp. 753-770, 1995.

[14] J. Yang and Y. Chen, "Free vibration and buckling analyses of functionally graded beams with edge cracks," *Composite Structures,* vol. 83, pp. 48-60, 2008.

[15] A. Kaya and P. Joseph, "The crack problem in bonded nonhomogeneous materials," *J. Appl. Mech.,* vol. 58, pp. 410-418, 1991.

[16] W. B H, "The surface crack problem for a plate with functionally graded properties," *Journal of applied Mechanics,* vol. 64, p. 449, 1997.

[17] H. J. Choi, "An analysis of cracking in a layered medium with a functionally graded nonhomogeneous interface," *Journal of Applied Mechanics,* vol. 63, pp. 479-486, 1996.

[18] Z.-H. Jin and N. Noda, "Crack-tip singular fields in nonhomogeneous materials," *Journal of Applied Mechanics,* vol. 61, pp. 738-740, 1994.

[19] N. Noda and Z.-H. Jin, "Crack-tip singularity fields in nonhomogeneous body under thermal stress fields," *JSME international journal. Ser. A, Mechanics and material engineering,* vol. 38, pp. 364-369, 1995.

[20] B. Wang, J. Han, and S. Du, "Thermoelastic fracture mechanics for nonhomogeneous material subjected to unsteady thermal load," *Journal of applied mechanics,* vol. 67, pp. 87-95, 2000.

[21] J. N. Reddy, *Energy principles and variational methods in applied mechanics*: John Wiley & Sons, 2017.

[22] Y. Hao, W. Zhang, J. Yang, and S. Li, "Nonlinear dynamic response of a simply supported rectangular functionally graded material plate under the time-dependent thermalmechanical loads," *Journal of Mechanical Science and Technology,* vol. 25, pp. 1637-1646, 2011.

[23] Y.-L. Chung and W.-T. Chen, "Bending behavior of FGM-coated and FGM-undercoated plates with two simply supported opposite edges and two free edges," *Composite Structures,* vol. 81, pp. 157-167, 2007.

A brief review of functionally graded materials

Abdolreza Toudehdehghan[1,2,*], *J. W.* Lim[1] , *K. E.* Foo[1], *M.I.N* Ma'arof[1], and *J.* Mathews[1]

[1] Department of Mechanical Engineering, INTI International University, Jalan BBN 12/1, Bandar Baru Nilai, 71800 Nilai, Negeri Sembilan, Malaysia
[2] Department of Mechanical Engineering, Universiti Tenaga Nasional , km7, Jalan Kajang-Puchong, 43009 Kajang, Selangor, Malaysia.

Abstract. Material engineering has become one of the corner stones in various engineering applications. The scientific alteration of base materials into various inorganic and organic compounds has resulted in the development of Functionally Graded Material (FGM). FGM can be described as a cutting edge material in which the particular material's properties could be made varied as the dimension changes. The aims of this study was to review on FGM's applications and its manufacturing processes. In addition, the mathematical idealization of the FGM was also reviewed. Based on the review, since FGMs are highly heterogeneous, it can be idealized as their mechanical properties changes smoothly with respect to their spatial coordinates. Nonetheless, in order to analyze FGMS in an efficient manner, the homogenization schemes are needed to be simplified from their complicated heterogeneous microstructures. Closed-form solutions of some fundamental solid mechanics' problems can be obtained by this idealization. Conclusively, FGM is unique in relation to any of the individual materials that structures it. Suggestively, this provide a limitless possibility for future engineering applications.

1 Introduction

Recently, distinct materials are often integrated together and used as a part of structural components. This is commonly associated with complex or demanding engineering applications. Hence, in reducing the local stress concentrations initiated by sudden changes in material properties, the idea of functionally graded materials (FGMs) was proposed in Japan during the mid-1980s. Via the used of FGM, the results had shown significant decrease in the severity of interfacial stresses whilst showing the conservation of structural quality and ductility. This key advantage given by FGM is highly beneficial to various present day endeavours. This is especially true in developing new materials for applications related to biomedical gadgets /prosthetics, solar panels, aviation transport, sensors and ceramics. It should be noted however, for FGM, the production process requires a metric as a method for controlling performance specifications. Spatial deformities in the compound structure or physical properties of a material volume will affects the FGM's performance. Therefore, in overcoming this issue, modern coatings are produced to create a homogeneous surface with

* Corresponding author: abdolreza.toudehdehghan@newinti.edu.my

the objective of limiting potential failure areas. This concept is the reasoning behind systematic efforts to define and characterize determinations or specifications for FGM. Conventionally, the FGM are manufactured in using a blend of metals and ceramics and are characterized in a way that composition of each and the volume division of materials are changed gradually. Therefore, with the gradual fluctuation of the volume fraction on the constituent materials, the particular material properties show a smooth and nonstop change starting with one surface then onto the next.

2 Production methods of functionally graded coating

Conventionally in the industry, the surface thin layer FGMs is manufactured via coating methods. In this section, several coating methods are reviewed.

2.1 Power Metallurgy (PM)

Powder metallurgy (PM) is commonly utilized in the industry in order to deliver practically graded material [1] via the following three (3) steps: (i) weighing and blending of powder as indicated by the pre-composed spatial distribution as dictated by the functional prerequisite, (ii) stacking and ramming of the premixed-powders, (iii) sintering. In short, the PM strategy gives ascend to a stage wiser structure. Even so, if a nonstop structure is required to be achieved or desired, the centrifugal technique is utilized instead.

2.2 Vapor deposition technique

There are multiple versions of the vapour deposition techniques. Examples of these techniques are Chemical Vapour Deposition (CVD), Physical Vapour deposition (PVD) and sputter deposition, among others. These vapour deposition techniques are utilized to ensure slim surface covering and high-quality microstructure finish. Yet, these techniques produce harmful gases by-products [2]. Hence, precaution and safety measures need to be undertaken if the Vapor Deposition Technique is chosen as the manufacturing methods.

2.3 Centrifugal methods

The centrifugal method is similar to centrifugal casting where the gravity force is used through turning of the mold to frame mass functionally graded material [3]. The graded material is created via the distinction with respect to the materials' densities and the spinning of the mold. It should be noted that consistent reviewing in grading can be accomplished via the centrifugal method. Nevertheless, this method is limited only to circular and hollow shapes. Another issue in relation to the centrifugal method is that there is limiting point to specification of gradient produced [4] since the gradient is formed through common process (centrifugal force and thickness difference). Therefore, in understanding and overcoming these problems, researchers are practicing alternative manufacturing techniques such as the Solid Free Form (SFF).

2.4 Solid Free Form (SFM) fabrication method

Solid free form is an added substance fabricating process that offers loads of points of interest. Examples of these characteristics of interest are the capacity to create complex shapes, lesser energy, high material usage, higher speed of generation, and design freedom as parts are delivered straightforwardly from CAD (e.g. AutoCAD) data [5]. There are various types of SFF innovations, even so, the laser based forms are mostly utilized in creating functional graded material [6]. Laser based SFF process for FGM [7] include: (i) laser cladding based strategy [8, 9], and (ii) Selective Laser Sintering (SLS) [10, 11]. SFF includes five essential steps [12]. The five steps are: (i) the generating of CAD information from the product like AutoCAD, Solid edge etc., (ii) the transformation of the CAD information to Standard Triangulation Language (STL) file, (iii) the slicing of the STL into two dimensional cross-area profiles, (iv) developing of the part layer by layer, and (v) expulsion and finishing. In short, the Solid Free Form (SFF) fabrication method is a great method in producing the FGMs.

2.5 Other methods in FGMs fabrication

Aside from the aforementioned methods, there are several other methods which are commonly practiced in the industry in order to produce the FGMs. Examples of these techniques are the electro-deposition, plasma spraying, electrophoretic, Ion Beam Assisted Deposition (IBAD), Self-Propagating, High-temperature Synthesis (SHS) [3]. Nonetheless, these methods cannot be used to create mass FGM since these methods requires longer manufacturing duration and energy intensive. Henceforth, these methods are uneconomical to be utilized as a part of delivering mass FGM.

3 Field of application of FGM

These days, in view to the capacity to produce composite materials which in accordance to functional stability and application requirement, the scope of FGM utilization is exceptionally wide. The distinctive ranges of FGM utilization is summarized in Figure 1.

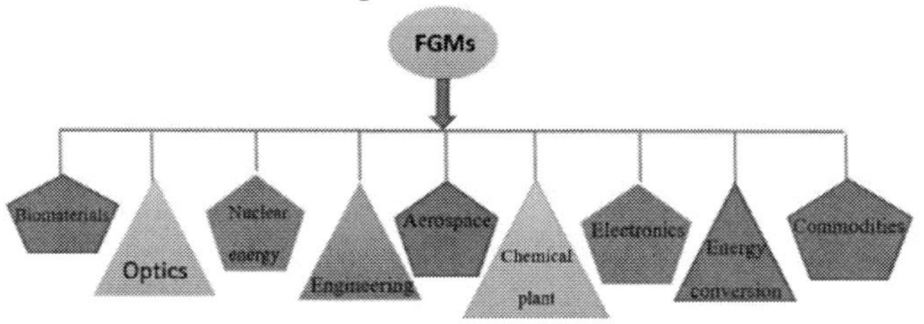

Fig.1 Schematic field of FGMs applications

For medical application, FGMs are used to supplant for naturally occurring FGM such as bones and teeth [13]. Hence, FGMs are extensively used in dental and orthopedic applications for teeth and bone replacement [14]. Whereas in nuclear projects, FGMs are widely used where great thermal opposing material is required. This is usually with respect to having a certain structure being sealed in order to maintain a strategic distance from modern mishaps which happens because of combination reaction process. For similar purposes, the FGM is utilized as defensive covering on turbine edges in gas turbine motor. Since FGMs can withstand high thermal gradient, this makes it appropriate for use in rocket engine components, structures and space plane body [15]. For optoelectronic applications, FGMs are used as graded refractive list material and in audio –videodiscs having magnetic capacity as storage media. Closely related to this field is the utilization of FGMs are in the communication field. The FGMs are used for optical fiber wires, which require adequate electrical resistor at external cover and adequate electric transistor material property on inward side. In regards to the defense industry, the FGMs are utilized in safeguarding application of penetration resistant material used for shield plates and bullet – proof vest [16] due to its capacity to hinder crack propagation. Other miscellaneous used of the FGMs are as lenses points and in the semiconductors industry. In short, the FGMs could be utilized in various fields and industries. With more extensive research on this particular material, the scope of usage could be further expanded.

4 Mathematical idealizations of FGMs

Since FGMs are highly heterogeneous, it can be idealized as continua with their mechanical properties changing smoothly with respect to the spatial coordinates. To analyze FGMS in an efficient manner, the homogenization schemes are needed to simplify their complicated heterogeneous microstructures. Closed-form solutions of some fundamental solid mechanics' problems can be obtained by this idealization. It will help in evolving and developing related numerical models of the structures. A typical FGM often represents a particulate composite with a prescribed distribution of volume fractions of constituent phases [17]. The FGM properties are generally assumed by following gradation through the thickness in continuous manner.

4.1 Exponential Law

This particular idealization for FGM modeling is very common in the fracture mechanics studies. The uniform thickness h, the typical material properties 'P(z)' at any point located at a distance 'z' from the reference surface is being modelled to this equation:

$$P(z) = P_t \exp(-\lambda\left(1 - \frac{2z}{h}\right)) \tag{1}$$

Where $\lambda = \frac{1}{2}\ln(\frac{P_t}{P_b})$.

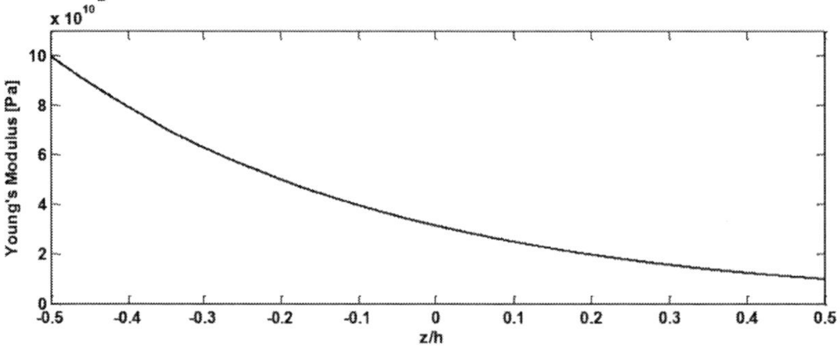

Fig.2 Exponential position dependence of elastic modulus of FGM layer

From figure 2, it is observable that the Young's modulus shows decrement as the z/h value increases. Hence, adequate assessment is warranted to address this changes in material's mechanical properties. This is especially vital if the particular properties is related to design and/or safety issues.

4.2 Exponential Law

This is more common in the stress analysis of FGM [17, 18] and given by:

$$P_Z = (P_t - P_b)\left(\frac{z}{h} + \frac{1}{2}\right)^k + P_b \tag{2}$$

From the terminology, 'P(z)' denotes a typical material property, viz., Young's modulus of elasticity (E), shear modulus of elasticity (G), Poisson's ratio (t), material density (q), etc. of the structures made of FGM [33]. 'h' is the total thickness of structure. 'Pt' and 'Pb' are the material properties at the top-most (z = +h/2) and bottom-most (z = -h/2) surfaces. 'k' in the exponential model, and 'k' in the power model are the material grading indexes respectively which depend upon the design requirements.

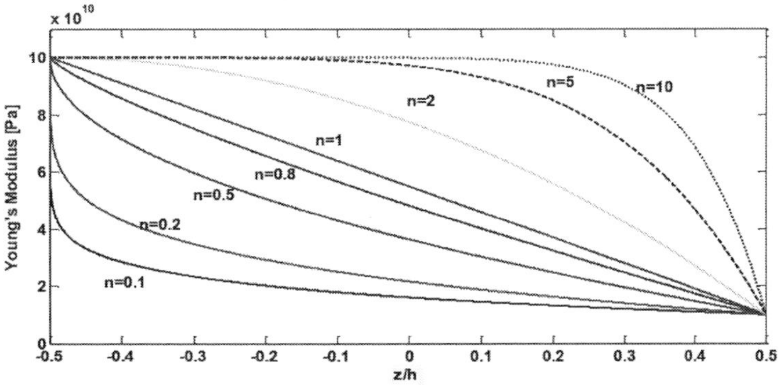

Fig. 3 Power-law position dependence of thermal expansion coefficient of FGM layer for several material parameters

From Fig. 3 it is observable as the z/h value increases, the Young's Modulus characteristics changes with respect to the varying value of n. The n=1 line creates an inversing mirror image for the power-law position dependence of thermal expansion coefficient of the FGM layer.

4.3 Sigmoid Function

The volume fraction of the sigmoid function of FGM structures which is consisting of two power-law functions [19] can be written as below:

$$\mathrm{P(z)} = (P_t - P_b)(\frac{1}{2}\left(\frac{z}{h} - \frac{1}{2}\right)^k) + P_b \qquad 0 \le z \le \frac{h}{2} \tag{3}$$

$$\mathrm{P(z)} = (P_t - P_b)(1 - \frac{1}{2}\left(\frac{z}{h} + \frac{1}{2}\right)^k) + P_b \quad -\frac{h}{2} \le z \le 0 \tag{4}$$

Where h, n are the thickness of the FGM layer and the material parameter respectively.

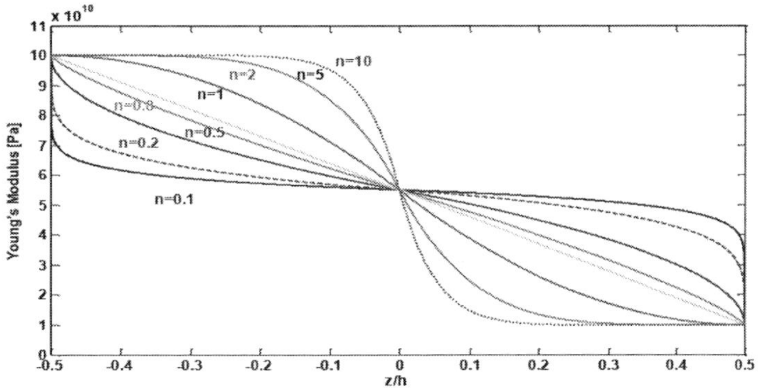

Fig.4 Sigmoid position dependence of Young's Modulus of FGM layer for several material parameters

As shown in Figure 4, at the top of layer is typically Pb and the bottom layer is typically Pt. Through the thickness of the FGM layer, there is a mixture of the two materials composition. From figure 4 it is observable that point 0 (center of thickness) is the converging and diverging point for the sigmoid position dependence of Young's Modulus of FGM layer for several material parameters.

5 Conclusion

Conclusively, FGM are one of a kind and unique in relation to any of the individual material that structures it. Suggestively, this provide a limitless possibility for future engineering applications. In order to analyze FGM in an efficient manner, the homogenization schemes are needed to simplify their complicated heterogeneous microstructures. Closed-form solutions of some fundamental solid mechanics' problems can be obtained by this idealization. It will help in evolving and developing related numerical models of the structures.

References

[1] M. M. Nemat-Alla, M. H. Ata, M. R. Bayoumi, and W. Khair-Eldeen, "Powder metallurgical fabrication and microstructural investigations of aluminum/steel functionally graded material," *Materials Sciences and Applications,* vol. 2, p. 1708, 2011.

[2] J. Groves and H. Wadley, "Functionally graded materials synthesis via low vacuum directed vapor deposition," *Composites Part B: Engineering,* vol. 28, pp. 57-69, 1997.

[3] Y. Watanabe, Y. Inaguma, H. Sato, and E. Miura-Fujiwara, "A novel fabrication method for functionally graded materials under centrifugal force: the centrifugal mixed-powder method," *Materials,* vol. 2, pp. 2510-2525, 2009.

[4] B. Kieback, A. Neubrand, and H. Riedel, "Processing techniques for functionally graded materials," *Materials Science and Engineering: A,* vol. 362, pp. 81-106, 2003.

[5] X. Lin and T. Yue, "Phase formation and microstructure evolution in laser rapid forming of graded SS316L/Rene88DT alloy," *Materials Science and Engineering: A,* vol. 402, pp. 294-306, 2005.

[6] D. W. Hutmacher, M. Sittinger, and M. V. Risbud, "Scaffold-based tissue engineering: rationale for computer-aided design and solid free-form fabrication systems," *TRENDS in Biotechnology,* vol. 22, pp. 354-362, 2004.

[7] Y. T. Pei and J. T. M. De Hosson, "Functionally graded materials produced by laser cladding," *Acta materialia,* vol. 48, pp. 2617-2624, 2000.

[8] M. Ivosevic, R. Knight, S. Kalidindi, G. Palmese, and J. Sutter, "Solid particle erosion resistance of thermally sprayed functionally graded coatings for polymer matrix composites," *Surface and Coatings Technology,* vol. 200, pp. 5145-5151, 2006.

[9] K. A. Mumtaz and N. Hopkinson, "Laser melting functionally graded composition of Waspaloy® and Zirconia powders," *Journal of materials science,* vol. 42, pp. 7647-7656, 2007.

[10] M. Erdal, S. Dag, Y. Jande, and C. Tekin, "Manufacturing of functionally graded porous products by selective laser sintering," in *Materials Science Forum*, 2010, pp. 253-258.

[11] M. M. Gasik, "Functionally graded materials: bulk processing techniques," *International Journal of Materials and Product Technology,* vol. 39, pp. 20-29, 2010.

[12] M. A. Boboulos, *CAD-CAM & rapid prototyping application evaluation*: Bookboon, 2010.

[13] W. Pompe, H. Worch, M. Epple, W. Friess, M. Gelinsky, P. Greil, *et al.*, "Functionally graded materials for biomedical applications," *Materials Science and Engineering: A,* vol. 362, pp. 40-60, 2003.

[14] F. Watari, A. Yokoyama, M. Omori, T. Hirai, H. Kondo, M. Uo, *et al.*, "Biocompatibility of materials and development to functionally graded implant for bio-medical application," *Composites Science and Technology,* vol. 64, pp. 893-908, 2004.

[15] L. Marin, "Numerical solution of the Cauchy problem for steady-state heat transfer in two-dimensional functionally graded materials," *International Journal of Solids and Structures,* vol. 42, pp. 4338-4351, 2005.

[16] M. Malinina, T. Sammi, and M. M. Gasik, "Corrosion resistance of homogeneous and FGM coatings," in *Materials Science Forum*, 2005, pp. 305-310.

[17] H.-S. Shen, *Functionally graded materials: nonlinear analysis of plates and shells*: CRC press, 2016.

[18] R. Javaheri and M. Eslami, "Thermal buckling of functionally graded plates based on higher order theory," *Journal of thermal stresses,* vol. 25, pp. 603-625, 2002.

[19] Y.-L. Chung and W.-T. Chen, "Bending behavior of FGM-coated and FGM-undercoated plates with two simply supported opposite edges and two free edges," *Composite Structures,* vol. 81, pp. 157-167, 2007.

PDC Bit Hydraulic & Mud Rheological simulation to model pressure drop across Bit

Ong Kai Sheng[1], *Tamiru* Alemu Lemma[1,*], and *Shazaib* Ahsan[1]

[1]Universiti Teknologi PETRONAS, 32610 Seri Iskandar, Perak Darul Ridzuan, Malaysia.

Abstract. During fluid flow from larger to smaller diameter pipes, a drop in pressure is experienced. High pressure drop across bit indicated high energy loss in the hydraulic system and also a setback to ROP performance. This is inefficient and pressure pumps would have to be of bigger sizing to make up for the losses. Present form of pressure drop models is in terms of mud density, flow rate, and total flow area. The objective of this paper is focused on the analysis of CFD simulation and to propose optimized parameters for improved ROP. Single phase flow study of Yield Power Law mud rheology was simulated at bottom hole of horizontal section. Parametric study on mud rheology was carried using DOE. Design points of DOE were sampled mostly using Latin Hypercube Sampling and a few by Central Composite Design. It is found that Kriging Response Surface method generated the best regression model where the predicted values are closest to the observed values and has the lowest Maximum Relative Residual (0.000336%). Inlet velocity and Power Index have significant effect on pressure drop. Consistency Index showed moderate effect while Yield Stress showed small effect to pressure drop. This research has proven that pressure loss model should take into account of mud rheology. Further research can be done with PDC bit rotation and its effect on mud behaviour.

1 Introduction

In the oil and gas industry, drilling cost and time are of major concerns. Operators of this field have their focuses centred on minimizing the overall drilling cost while maintaining safe practices and environmental friendly operations. Rate of Penetration (ROP) is a measure of drilling speed. Based on the relationship between drilling cost and ROP, it had been shown that maximizing the ROP will result in minimizing the drilling cost [1]. Studies have been done on factors affecting ROP. These factors are categorized into bit design parameters and operational parameters. Bit design parameters that significantly affect ROP are Junk Slot Area (JSA) and Bit Size (BS). Operational parameters are Weight on Bit (WOB), Rotation of drill bit (RPM), Hydraulic Horsepower (HHP), Flow Rate, Nozzle Size, and Mud Weight (MW) [2]. High ROP would generate high rate of cuttings and vice versa. The removal of cuttings is undeniably necessary so that the bit can be in direct contact with bottom hole formation and drill deeper and faster. Cuttings are removed as mud circulates to the top and

* Corresponding author: tamiru.lemma@utp.edu.my

carries the cuttings along. When cuttings accumulate at bottom hole, Equivalent Circulating Density (ECD) increases and ROP decreases. Besides that, bit hydraulics plays an important role during drilling. Good bit hydraulics help jet through the formation, keep the PDC cutters cool and clean, and prevents the JSA and nozzles from clogging up and balling. Moreover, mud rheology plays a huge role in drilling as well. Two main mud properties that have direct impact to removal of cuttings are viscosity and gel strength. Mud viscosity and gel strength primarily suspense cuttings and effectively sweeps the cuttings out of hole.

There are many issues during operations like bit hydraulics is crucial for the removal of cuttings and cutting the PDC cutters. With poor bit hydraulics, PDC bit may face problems like bit balling and plugged nozzle. These phenomena are the obstruction of JSA and nozzles which are caused by poor cuttings removal away from the PDC bit. Another issue is high pressure drop across bit nozzles is an indication of energy loss in the hydraulic system. If it is inefficient, then the pressure pumps have to be of bigger sizing to make up for the losses. How does mud rheology affect pressure drop across bit nozzles? From the issues discussed, the objectives of this investigation comprises of (i) to develop a CAD and CFD model for a typical PDC bit, (ii) to develop regression models for the pressure loss around the PDC bit, and (iii) to analyze the CFD simulation result and propose optimized parameters for improved ROP. The scopes of study based on the objectives can be simplified as (i) PDC bit size of 8.5 inches with length of CAD model 5 times the diameter of PDC bit, (ii) Single phase flow, (iii) Flow rate: 100 to 1000gpm and (iv) Mud rheological parameters (Yield Stress, Consistency Index, and Power Index).

1.1 Bit Hydraulics

Bit Hydraulics plays important role in cuttings removal away from PDC bit. Poor hydraulics may cause bad cuttings removal away from PDC bit and results in many problems as mentioned earlier. Optimization of bit hydraulics is through maximizing bit horsepower or nozzle-jet impact force [3]. This brings about effective cuttings removal as the cuttings are removed as fast as they are generated [4]. Flow rate of mud has significant positive effect on cuttings removal away from PDC bit [5-9]. Increasing the annular velocity by increasing the flow rate decreases the cuttings bed height significantly. At constant mud flow rate, smaller-sized nozzles increase cutting transport velocity as they provide higher jet velocity at bottom hole. Two nozzles showed higher cuttings-transport ratio as compared to three nozzles.

This is because the two nozzles generated asymmetrical flow which result in higher jet velocity and improved cutting transport [10]. For a similar total flow area (TFA) of nozzles, higher number of nozzles improves cuttings removal away from PDC bit as more nozzles provide a more-uniform fluid low distribution [11]. Bigger face volume of bit is at higher risk of bit balling when drilling at low ROP. And lower face volume achieved maximum ROP without balling. There is no correlation between face volume, JSA, and cuttings removal away from PDC bit efficiency [12].

1.2 Pressure drop across bit

Previous studies on pressure drop models are summarized in Table 1. Pressure drop across a bit happens when mud flows through the bit nozzles. The mud experiences this drop in pressure simply because it moves from large diameter drill pipes into the small diameter bit nozzles. This is highly important for the optimization of drilling hydraulic with the objective of maximizing hydraulic horse power or impact force without neglecting effectiveness of cutting removal. This sudden loss of pressure can be calculated from mud weight along with various parameters and can be derived from potential or kinetic energy.

 DOI: 10.1051/matecconf/201713103011

Table 1. Previous studies on Pressure Drop Models. Where ΔP_b is bit nozzle jets pressure loss (lb/in^2); *HHP* is horse power at bit (hp); *MW* is mud weight (lb/gal); V_n is velocity of mud (ft/s); *Q* is flow rate of mud (gpm); *TFA* is total flow area of bit nozzles (in^2); C_d is coefficient of bit nozzles (0.95 or 1.00 or 1.03 unitless); *ROP* is rate of penetration (ft/hr); *RPM* is drill bit rotation per minute (rpm).

Source	Equations	Remark
[13]	$\Delta P_b = \frac{(MW)(Q^2)}{11884(C_d{}^2)(TFA^2)}$	C_d = 0.8 (Conventional Bit) C_d = 0.95 (Jet Bit)
[14-16]	$\Delta P_b = \frac{(MW)(Q^2)}{12032(C_d{}^2)(TFA^2)}$	C_d = 1.03
[17]	$\Delta P_b = \frac{(MW)(Q^2)}{8795[(TFA)(e^{-0.832})(\frac{ROP}{RPM})]^2}$	-

1.3 Mud Rheology

Mud type has small to moderate positive effect on cuttings removal away from PDC bit [18]. Different mud types lead to different bed consolidation. Conventionally, there are two types of mud which is either oil based or water based mud. The two mud properties that have direct impact on cuttings removal away from PDC bit are viscosity and density. Viscosity plays function of the suspension of cuttings which is crucial for cuttings removal away from PDC bit. Rheology is defined as the science of deformation and flow of matter [19]. To date, all fluids are classified as either Newtonian or Non-Newtonian. Several rheological models have been developed based on research over time. The Herschel-Bulkley model is also called the Yield Power Law (YPL) model, since it takes both a yield point (YP) and a power law development into account. Effectively, it is a combination of the Bingham and power law fluid models. The Herschel-Bulkley model is often used to describe oil-well drilling fluids, since it considers both a yield point and power law development with increasing shear rate. The yield point factor is due to gelling.

$$\tau = \tau_y + k\dot{\gamma}^n \tag{1}$$

The rheological characteristics of drilling mud such as plastic viscosity (PV) and YP and the flow behavior indicators such as k and n, of drilling mud play an important role in cleaning of drill cuttings. These fluid properties, especially the low shear rate rheological properties that prevail at annular section between the drill pipe and borehole wall have a major impact on the cuttings removal efficiency of drilling mud.

2 Methodology

2.1 CAD Model and Mesh independency study simulations

To simulate flow around a drill bit at bottom hole, it is required to prepare CAD drawings. A typical drill bit of length and diameter of 1.12 m and diameter of 220 mm, respectively, is drawn in ANSYS Modeler. This is followed by a drawing of a borehole of the same length and diameter. Both CAD drawings are then aligned together with the same axial axis. After that, the drill bit is subtracted from the borehole. This leaves an annulus through which the mud will flow from drill pipe passing through bit nozzles into the annulus and then flow away from the bit. After the CAD model is prepared, it is necessary to lay mesh on the model. The smaller the mesh size, the more accurate the results but longer simulation time. This calls for mesh independence study which is the optimization of simulations on various mesh sizes ranging from 0.01 to 0.004 element size. For the study, water is selected with default values of properties for faster simulation time. Once the solver parameters are settled, hybrid initialization method is initialized and followed by the run of calculation. This optimization aims to reduce needless simulation time and produce consistent results. Validity of the results can be justified with small percentage error of less than five percent amongst all convergence criteria. In ANSYS Fluent, the solver used is pressure-based with absolute velocity formulation running in steady state of time. Since this research only revolves around single phase flow, the multiphase model is turned off. Viscous model of realizable k-epsilon is selected with standard wall functions and default values for constants. Density of all muds are fixed to 1198 kg/m^3. Viscosity of the muds are governed by YPL. Inlet velocity is varied according to flow rate over a fixed TFA as in Table 2.

2.2 Parametric Study Simulations

In the parametric study simulations, various muds rheology which are governed by yield power law are collected from past studies and compiled into upper and lower bounds as tabulated in Table 3. These upper and lower bounds were inputted into ANSYS' Design of Experiments program. Latin Hypercube Sampling design is chosen as it brings about no two points of equal value. Central Composite Design is used as backup when some of LHS' design points do not show expected results.

Table 2. Varying flow rate and inlet velocity over fixed TFA for preliminary simulations.

Flow Rate, Q (gpm)	300	400	500	600	700	800
Inlet Velocity, V_{in} (ft/s)	35.30	47.03	58.79	70.55	82.31	94.07

Table 3. Upper and lower bound mud rheology variables govern by Yield Power Law for parametric study simulations.

Mud Rheology	Unit	Min.	Max	Average
Volume Flow Rate, q	gpm	100	1000	550
Inlet Velocity, V_{in}	m/s	2.5	40	21.25
Yield Stress, τ_0	Pa	0.35	12	6.175
Consistency Index, K	Pa.sn	0.031	9	4.5155
Power Low Exponent, n	-	0.2	0.9	0.55

3 Results and Discussions

3.1 Preliminary Simulations

Based on mesh independence study, mesh size of 0.06 provides consistent results within optimal time. Hence, further simulations are conducted using this particular mesh size. Flow rate of muds ranged from 300 to 800 gpm. Inlet velocities were calculated over the fixed TFA of CAD PDC bit model. Inlet pressures were recorded while values for outlet pressures were offset away from outlet boundary. This is because direct collection of data from the outlet boundary may give inaccurate data.

When mud flow rate increases, the pressure drop across bit increases. However, both simulations on water based mud (WBM) rheology show smaller pressure drop across bit nozzles as flow rate increases as shown in Fig. 1(a). The smaller the shear stress (τ) of WBM, the smaller the pressure drop. Rheological properties have significant positive effect on pressure drop; up to 50% reduction in losses. The pressure drop points for the present study are shown in Fig. 2(a).

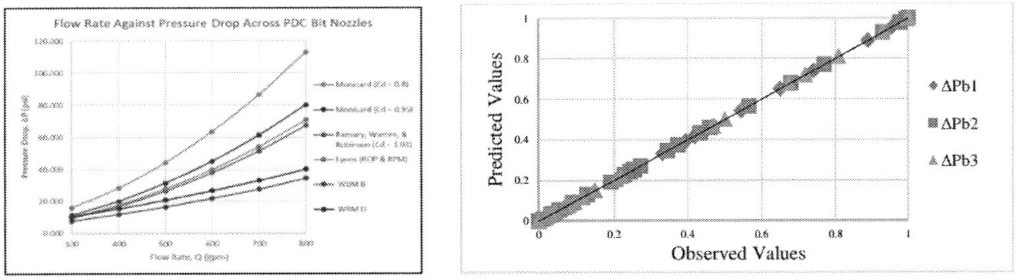

(a) (b)

Fig. 1. (a) Graph of Flow Rate against Pressure Drop across PDC Bit Nozzles for WBM B & D. (b) Goodness of Fit generated over Kriging Response Surface Types.

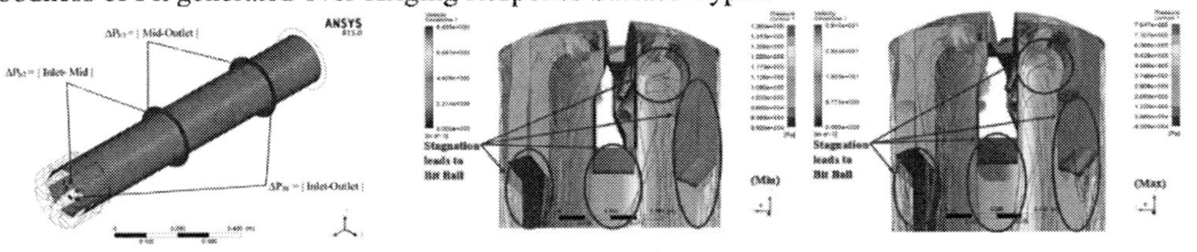

(a) (b)

Fig. 2. (a) Pressure drop data points for this study. The outlet plane is offset away from the outlet to avoid boundary conditions and to obtain more accurate data. (b) Top and bottom represents overall input parameters at minimum and maximum each respectively.

3.2 Parametric Study

Type of DOE used to generate design points are mainly by Latin Hypercube Sampling with a few by Central Composite Design. This is because some of design points from LHS did not simulate expected result where the pressure drop is too small means invalid result.

To overcome this problem, CCD was used to sample additional design points and simulated results substituted those invalid results. Stagnation points are indications of bit balling as shown in Fig. 3(b). These points have very little or no velocity vector paths which means mud would be not circulated at these points. No proper mud circulation would definitely lead to bit ball despite high or low input parameters (emphasis on flow velocity). This can be concluded with bad geometry of the model and an optimized tweak to the geometry would eliminate the probability of bit balling.

3.3 Regression Analysis

There are a total of 4 types of Response Surfaces which are Standard Response Surface, Kriging, non-parametric Regression and Neural Network. The design points in all Goodness of Fit graphs are generally inline linearly. The blue and orange dots as shown in Fig. 1(b), which represents ΔP_{b1} (Inlet-Outlet) & ΔP_{b2} (Inlet-Mid) respectively, are noticeably inline lineally and consistently over all 4 types of response surfaces.

However, the grey dots, which represents ΔP_{b3} (Mid-Outlet), tend to be disoriented and dispersed in the Neutral Network. And, they appear to be closer to the linear line with Standard Response Surface Visually, Kriging and Non-Parametric Regression Response Surfaces would give the best Goodness of Fit. Between Kriging and Non-Parametric Regressions, Kriging yields the better results as shown in Table 4.

All response charts are plotted in terms of XYZ axes as depicted in Fig. 3 in which V_{in}, ΔP_b, and either τ_o, K, or n each, respectively. This is done as such to study the correlation between the parameters. Based on the generated charts, there are unnoticeable or insignificant differences between ΔP_{b1} & ΔP_{b2}. This is because majority of pressure drop (Max = 265759 Pa) occurs nearer to the PDC bit while minority of pressure drop (Max = 13344 Pa) occurs away from the bit. Minor pressure drop is approximately 5% of the total pressure drop across the bit.

V_{in} & ΔP_b have the strongest correlation in all cases. Around the bit, ΔP_{b1} & ΔP_{b2} increases as V_{in} increases. However, away from the bit, ΔP_{b3} increases as V_{in} increases until 34 m/s and then ΔP_{b3} decreases. This shows that V_{in} has very strong positive effect on ΔP_b around the bit and slightly less effective away from the bit. To avoid large pressure drop across bit, flow velocity should be to a minimum 2.5 m/s.

τ_o & ΔP_b show the good correlation in all cases. ΔP_{b1}, ΔP_{b2}, & ΔP_{b3} increase as τ_o increases. ΔP_{b1}, ΔP_{b2}, & ΔP_{b3} peak when τ_o is 5.6 Pa and then ΔP_{b1}, ΔP_{b2}, & ΔP_{b3} decrease as τ_o increases. τ_o has positive and negative effect on ΔP_{b1}, ΔP_{b2}, & ΔP_{b3}. The value of τ_o should be away from 5.6 Pa; either very small or large value. τ_o is better off at 11.25 Pa; the higher the better.

K & ΔP_b also show the good correlation in all cases. ΔP_{b1}, ΔP_{b2}, & ΔP_{b3} decrease slightly as K increases until 2.5 Pa.s^n and then ΔP_{b1}, ΔP_{b2}, & ΔP_{b3} increase as τ_o increases. To minimize pressure drop across bit, K value should be kept low; approximately 2.5 Pa.s^n.

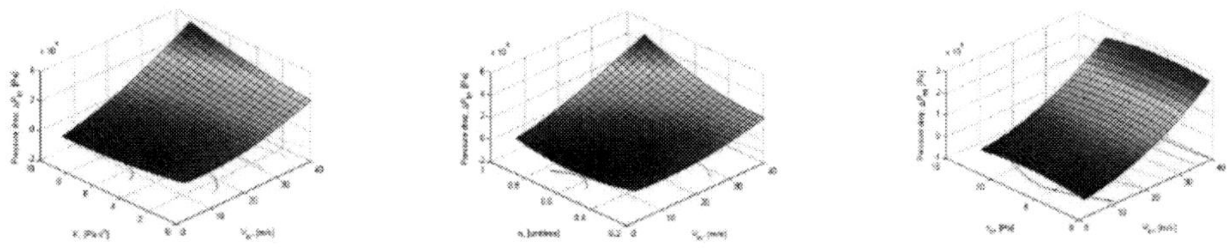

(a) (b) (c)

Fig. 3. (a) 3D Response Charts of ΔP_b against K. (b) 3D Response Charts of ΔP_b against K. (c) 3D Response Charts of ΔP_b against K.

Table 4. Goodness of Fit details on various types of Response Surfaces.

Kriging	ΔP_{b1}	ΔP_{b2}	ΔP_{b3}
Coefficient of Determination (Best Value = 1)	1	1	1
Adjusted Coefficient of Determination (Best Value = 1)	0	0	0
Maximum Relative Residual (Best Value = 0%)	0.000336	0.000302	5.73E-05

4 Conclusion and Recommendations

This research is about pressure drop across bit with respect to mud rheological parameters. Objectives were achieved with successful development of regression model of pressure drop across bit and optimization of parameters for improved ROP. Some important conclusions that can be drawn are: (i) Mud flow rate has significant effect to pressure drop across the bit. The lower the inlet velocity, the lower the pressure drop, (ii) Mud rheology parameters have been proven to show moderate effects to pressure drop across the bit. Power Index showed significant effect to pressure drop. Consistency Index showed moderate effect while Yield Stress showed small effect to pressure drop. This research has 4 input parameters and optimization analysis were done individually where the other 3 input parameters are kept at average values. Optimized parameters for reduction in pressure drop across bit and improvement in ROP: (Inlet Velocity V_{in} = 2.5 m/s, Yield Stress τ_0 = 11.25 Pa, Consistency Index K = 2.5 Pa.sn, Power Index n = 0.425) and (iv) Simulation on mud rheology showed a reduction of about 50% in pressure drop across the bit as compared to existing models. Existing pressure drop models should be revised to include mud rheology parameters.

The recommendations from this research is further research on the effect of drill bit rotation (RPM) and its effect on the mud behavior should be considered as it was not considered in previous study.

The Authors are thankful to Universiti Teknologi PETRONAS for facilities provided for the research.

References

1. A.T. Bourgoyne, *Applied drilling engineering*. (SPE (U.S.), Richardson, TX, 1991).
2. D. Hankins, S. Salehi, and F.K. Saleh, J. PETROL ENG **2015**, 1 (2015).
3. H.A. Kendall and W.C. Goins, Jr., AIME SPE **219**, 238 (1960).
4. B. Guo and G. Liu, *Applied Drilling Circulation Systems: Hydraulics, Calculations and Models*. (GPP., Burlington, MA., 2011).
5. S. M. Hussaini and J.J. Azar, SPE J. **23(01)** ,11 (1983)
6. P.H. Tomren, A.W. Iyoho, and J.J. Azar, SPE DRILL ENG **1(01)**, 43 (1986).
7. J. Li and S. Walker, SPE J. **6(04)**, 356 (1991).
8. H. Cho, S.N. Shah, and S.O. Osisanya, SPE ANTEC (2001).
9. K. Ravi and T. Hemphill, IADC SPE (2006).
10. K.M. Lim, and G.A. Chukwu, SPE WRM (1996).
11. G. Mensa-Wilmot and M. Fear, IADC SPE (2002).
12. M. Wells, T. Marvel, and C. Beuershausen, IADC SPE (2008).
13. R. Monicard, *Drilling Mud and Cement Slurry Rheology Manual*. (Springer, Netherlands, 1982).
14. L. H. Robinson, M. S Ramsey, AADE **01-NC-HO-31** (2001).
15. T. M. Warren, SPE DRILL ENG **4(04)**, 335 (1989)..
16. L. Robinson, AADE **10-DF-HO-26**, 1 (2010).
17. W.C. Lyons, G.J. Plisga, M.D. Lorenz, *Standard Handbook of Petroleum and Natural Gas Engineering (Third Edition)* (GPP., Boston, 2016).
18. T. Nazari, G. Hareland, and J.J. Azar, SPE WRM (2010).
19. R. Darby, *Viscoelastic Fluids. An Introduction to Their Properties and Behavior*. (Marcel Dekker, New York, 1976).

Analysis of contributing factors for synthesizing silicon carbide using Taguchi's approach

Mohd Aizat Hafiz Mohd Sohor[1]*, *Mazli* Mustapha[1], and *Othman* Mamat[1]

[1]Mechanical Engineering Department, Universiti Teknologi PETRONAS (UTP), 32610 Seri Iskandar, Perak, Malaysia

Abstract. This study concerns the carbothermal reduction of milled amorphous silicon dioxide (SiO_2) to produce silicon carbide (SiC) using Taguchi's approach. The $L_9(3_4)$ orthogonal design matrix was selected involving four operation specifications; temperature, mechanical milling time, heating rate and time on amorphous SiO_2 under carbothermal reduction. The responses were then analysed and evaluated by Analysis of Variance (ANOVA) technique. SiC yield was optimized to the highest when synthesized using amorphous SiO_2 with the highest-level setting for temperature (1450°C), milling duration (100 minutes), time (180 minutes) together with the minimum heating rate of 5°C/min and only SiC formation was observed at $T \geq 1400$°C.

1 Introduction

Silicon carbide (SiC) is considered a highly valued ceramic because of its properties such as high abrasiveness, corrosion resistance and high stiffness with low bulk density. The main issue faced by manufactures during the production of SiC materials is that there is no precise configuration of operating specification to synthesis maximum yield of SiC. Another main disadvantage of the current manufacturing method which largely focuses on the usage of furnace with extremely high temperature of about 2400°C and consumes about 23J/kg of electricity shows the inefficiency and high energy wastage during the SiC production [1]. This process leads to a SiC product with huge grain size because of the employed synthesizing method, which is the Acheson method. It also requires further milling and purification process to produce a quality component from the SiC precursor.

The strategy for selecting a predetermined number of examinations which creates the most data is known as partial fraction experiment [2]. Even though this technique is notable, there are no broad rules for its application or the investigation of the outcomes acquired by the tests. Dr. Genechi Taguchi built a unique arrangement of general plan rules for factorial trials that covers various engineering applications which is known as Taguchi's Method. Taguchi's method enables reduction of experiment runs while being able to investigate effects of all processing parameters. In this research, there will be four factors selected which are temperature, mechanical milling time, heating rate and time with three levels of variations for each factor which substantially reduce the number of experiment to nine times. The

* Corresponding author: mohd.aizat_g03388@utp.edu.my

objective of this study is to have a clear set of operating specifications for carbothermal reduction of amorphous SiO_2 by analysing SiC yield to study the significance of each operating specification using Taguchi's Design matrix together with Analysis of Variance (ANOVA) technique.

2 Methodology

The $L_9(3_4)$ orthogonal design matrix was selected for this research as it involved four processing parameters while running three levels for each factor and using the degree of freedom approach, nine experiments are needed to understand the effect of all four parameters. The lowest and the highest range each of operating parameters from the literatures were identified and tabulated in Table 1.

Table 1. Operation Specifications of Experiment

Parameter	Operating Specification	Level 1	Level 2	Level 3
A (min)	Mechanical Milling Duration	20	60	100
B (min)	Time	60	120	180
C (°C/min)	Heating Rate	5	10	15
D (°C)	Temperature	1350	1400	1450

Nine sets of experiments using carbothermal reduction process were conducted according to designated setting of $L_9(3_4)$ orthogonal design matrix involving four operating specifications. The responses in terms of SiC yield were recorded two times for each of the experiment.

Larger amount of SiC yield will be the most favourable feature, and thus, was categorized in the 'higher the better' quality characteristic. The mean square deviation (MSD) related to the 'higher the better' feature was calculated from the equation (1):

$$MSD = \frac{\left(\frac{1}{R_1^2} + \frac{1}{R_2^2}\right)}{2} \tag{1}$$

These values were then converted into their equivalent signal to noise ratios (S/N) as in equation (2):

$$S / N = -10 \log(MSD) \tag{2}$$

The four-operating specifications affect the SiC yield with contrasting amplitude. Analysis of variance (ANOVA) was conducted to analyse the proportionate contribution of various operating specifications through decomposition of the total variation into appropriate components.

3 Results and discussions

Considering the contributing factors studied in this research as mentioned in the introduction with their respective levels chosen as portrayed in Table 2 to produce silicon carbide using carbothermal reduction method with amorphous silicon dioxide and carbon as the precursor material, statistical evaluation utilizing the Taguchi's Model were executed largely to forecast and anticipate the most favourable operation specification for this study. Experimental response indicated that the most noteworthy operating specification influencing the production of SiC was temperature in which the amorphous SiO_2 and carbon mixture was burned under carbothermal reductin. It is then followed by milling duration, reaction time and heating rate being the least significant factor. From these experiments, it can be observed that Run 8 resulted with the highest SiC yield as compared with Run 5 which showed the least SiC yield. With these results, no conclusive statement can be made as a more detailed analysis was required to study the effects and significance of each operation specifications on SiC yield.

Table 2. The results based on $L_9(3_4)$ orthogonal design matrix

No.	Factor				Operation Specification				SiC weight		S/N
	A	B	C	D	Milling Time (min)	Time (min)	Heating rate (°C/min)	Temp. (°C)	R1 (g)	R2 (g)	
1	1	1	1	1	20	60	5	1350	6.26	6.12	15.83
2	1	2	2	2	20	120	10	1400	7.77	7.45	17.62
3	1	3	3	3	20	180	15	1450	9.96	9.73	19.86
4	2	1	2	3	60	60	10	1450	10.62	10.81	20.60
5	2	2	3	1	60	120	15	1350	6.11	5.90	15.57
6	2	3	1	2	60	180	5	1400	10.80	10.50	20.54
7	3	1	3	2	100	60	15	1400	8.78	8.77	18.86
8	3	2	1	3	100	120	5	1450	12.16	12.36	21.77
9	3	3	2	1	100	180	10	1350	9.50	9.44	19.53

As part of Taguchi's optimization technique, mean effects of each operating specification for respective levels were calculated and plotted on the graph as shown in Figure 1(a)-(d). From these data, it can be observed that there is a steady escalation for the effects from the first to the third level of the three-operating specifications: mechanical milling duration, time and temperature. From these graphs, it can be analysed that the SiC response rate will step-up if these three-operating specifications are levelled up. On the contrary, when the heating rate is increased, SiC yield of the resultant reaction decreases.

Based on the first parameter that was studied in this research which was mechanical milling duration, it can be seen from Fig 1(a) that the SiC yield increases as mechanical milling rate increases from level setting of 20 minutes to 100 minutes [3]. This means that, the random degree of disorder for the precursor amorphous structure of silicon dioxide sand still increases after mechanical milling. The second parameter that was studied to be optimized was the time of carbothermal reduction process. As portrayed in Fig. 1(b), as the time increases from 60 to 180 minutes, the SiC yield also increases. This phenomenon was explained by Seo et. al [4] whereby the increase in reaction rate increases the SiC whiskers thickness. It can also be explained that for reaction time that is longer than 120 minutes, the generation of SiO gas is more homogenous; thus, promoting the reduction process which leads to thick whiskers formation and a higher SiC yield with low aspect ratio [5].

Contrarily with the rest of the operating specifications studied, heating rate had a negative effect for the SiC yield as portrayed in Fig. 1(c). This finding can be co-related with past literature analysis which stated that the increase in heating rate reduces the surface area for reaction to take place, which in turn, reduces the SiC formation [6]. High heating rate will be able to transfer more heat for the reaction but it will reduce the time feasible for the reactant to react. As the transport of reactants and chemical reaction being time-dependent phenomena, inadequacy of time is thus expected to hinder the transformation of amorphous SiO_2 to SiC [7]. It was found that the increase in temperature increases the SiC yield which can be seen in Figure 1(d). This temperature factor can be explained from the thermodynamics point of view which prefers SiC formation starting at critical temperature of 1350 °C. This temperature effect on SiC yield is in accord with Chen et al. [8] who showed similar low SiC formation at temperature of below 1400 °C.

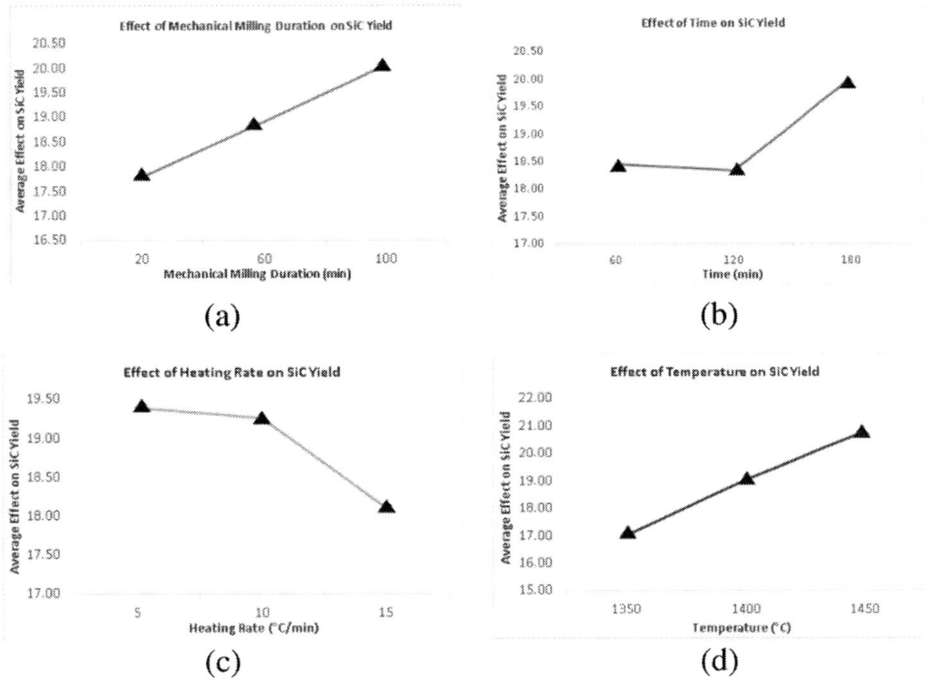

Fig. 1. Effect of each operating parameters on SiC yield (a) milling duration (b) time (c) heating rate and (d) temperature

3.1 Analysis of variance (ANOVA) technique

Analysis based on the "Percent" column as of Table 3 showed the operating specification of temperature as the highest contributing percentage (57.229%), followed by mechanical milling duration (20.931%), time (13.820%) and heating rate (8.0200%). The heating rate parameter was observed as the least significant specification towards the response data because of its low effect which can be seen to be lower than 10%. Hence, it was treated as an inconsequential parameter to SiC formation and is thus termed together with the error value. Pooling, as described by Ibrahim et al. [9], is the process of ignoring the contribution of a selected factor and subsequently adjusting the contribution of the other factors. Hence, the new ANOVA results after pooling are as tabulated in Table 4.

Table 3. Analysis of variance (ANOVA)

No.	Operating Specification	DOF	Sum of Squares	Variance	F	Percent (%)
1	Mechanical Milling Duration	2	7.807	3.904		20.931
2	Time	2	5.155	2.577		13.820
3	Heating Rate	2	2.991	1.496		8.0200
4	Temperature	2	21.346	10.673		57.229
All other/ Errors		0	-	-		-
Total		8	37.300			**100.000**

Table 4. Pooled ANOVA Table

No.	Factors	DOF	Sum of squares	Variance	F	Percent (%)
1	Mechanical Milling Duration	2	7.807	3.904	2.610	12.911
2	Time	2	5.155	2.577	1.723	5.799
3	Heating Rate	{2}	2.991	-		
4	Temperature	2	21.346	10.673	7.136	49.209
All other/Errors		2	2.991	1.496		32.081
Total		8	37.300			**100.000**

After conducting the analysis with heating rate being termed together with the error value, the percentage for each operating specification reduced marginally, without changing the ranking sequence of the rest of the operating specifications. The mean impact of the four-operating specification as shown in Table 5 gives a clear indication that the best combination to achieve maximum SiC yield would be when the third level of temperature, milling duration and time are selected. As shown in Table 5, for best SiC yield, the approximated enhancement is 4.046 against the existing mean achievement of 18.910. Based on these results, the highest SiC yield could be attained at the combination setting of A3, B3, C1 and D3, i.e. mechanical milling duration of 100 min, carbothermal reduction time of 180 min, processing heating rate of 5 °C/min^{-1} and temperature of 1450 °C. This combination of settings was not used in the experiments. Hence, the orthogonal array can be used to identify optimum factors in a multidimensional factor space.

Table 5. Contribution of each factors

Operation Specification	Level Setting	Level	Attribution
Mechanical Milling Duration	100 min	3	1.144
Time	180 min	3	1.068
Temperature	1450 °C	3	1.834
Contributions acquired from sum of all parameters			4.0464
Existing performance value			18.910
Predicted response with optimum conditions			22.956

4 CONCLUSION

Based on the three levels of setting of the four operating specifications investigated; the increase of the mechanical milling duration (min), temperature (C) and time (min) to the highest-level setting respectively to 100 minutes, 1450C and 180 minutes lead to maximum SiC response. In contrast, when the heating rate (C/min) was increased to the maximum setting of 15C /min, SiC yield reduced significantly. This trend leads to the conclusive statement of milling duration, temperature and time of reaction increment having a positive impact for SiC enhancement whereas increased heating rate having an adverse effect on the SiC yield.

Finally, from the Analysis of Variance (ANOVA) statistical calculation, it exhibited that temperature (57%) was the most significant factor for SiC formation; followed by milling duration (21%), time and heating rate (8%) which has the least significant parameter for SiC formation. This information acquired from this research are crucial for manufacturers as they can now target to optimize factors based on their significance to obtain an optimal SiC yield for a higher production rate in a more economical manner.

The authors acknowledge the support provided by Universiti Teknologi PETRONAS and Ministry of Higher Education (MOHE) of Malaysia through Fundamental Research Grant Scheme (FRGS) no. 0152AB-K82.

References

1. S. Cetinkaya and S. Eroglu, *J. Eur. Ceram. Soc.*, **31**, no. 5, pp. 869–876, (2011)
2. E. Szewczak and J. Wyrzykowski, *Nanostructured Materials,* **12**, pp. 171-174, (1999)
3. L. C. P. Ajoy Kumar Ray, S.K. Das, *Mater. Lett.*, **57**, no. January, pp. 1120–1123, (2003)
4. M. Alcala, J. Criado, and C. Real, *Solid State Ionics,* **141**, pp. 657- 661, (2001)
5. K. Komeya and H. Inoue, *Journal of Materials Science,* **4**, pp. 1045-1050, (1969)
6. A. W. Weimer, D. F. Carroll, D. W. Susnitzky, and D. R. Beaman, *Journal of the American Ceramic Society,* **82**, pp. 1635-1638, (1999)
7. W. Duan, X. Yin, Q. Li, X. Liu, L. Cheng, and L. Zhang, *Journal of the European Ceramic Society,* **34**, pp. 257-266, (2014)
8. D. Kata, J. Lis, and R. Pampuch, *Solid State Ionics,* **101**, pp. 65-70, (1997)
9. M. Ibrahim, N. Muhamad, A. B. Sulong, K. Jamaludin, S. Ahmad, and N. Nor, *International Journal of Mechanical and Materials Engineering,* **5**, pp. 282-289, (2010)

A new cooling technique for stingless bees hive

Ahmad Syazwan Ramli[1], *A. H.* Luqman[1], *Firdaus* Basrawi[1,*], *Ahmed N.* Oumer[1], *Azizuddin Abd* Aziz[1], and *Zulkifli* Mustafa[2]

[1]Meliponini Engineering Laboratory (MePEL), Energy Sustainability Focus Group (ESFG), Faculty of Mechanical Engineering, Universiti Malaysia Pahang (UMP), 26600 Pekan, Malaysia
[2]Department of Neurosciences, School of Medical Sciences, Health Campus, Universiti Sains Malaysia (USM), 16150 Kubang Kerian, Malaysia

Abstract. Stingless bees are a type of insect that are very sensitive to the changes of their surroundings, especially to severe heat wave. A report stated that at temperature as high as 38 °C can cause death of bees especially to the pupae. Therefore, the objective of this research is to evaluate a new method in regulating the temperature in the hive. Greenroof, a type of roof which contains green vegetation and soil, was used as the cooling method in this study. Two units of MUSTAFA-hives were exposed under sunlight, one is without temperature control and another one was fitted greenroof. The temperatures inside each hive was measured at two points and was compared with the hive without temperature control. It was found that, for the hive integrated with greenroof, the average hive temperature was about 3 °C and 6 °C lower in the honey cassette and brood-cells compartment, respectively. Therefore, it can be concluded that the implementation of greenroof could solve the problem of stingless bee hive overheating, and the greenroof has an impressive cooling performance besides being an economic and simple solution.

1 Introduction

Stingless bees are fascinating creatures. They predate the stinging honey bee, roaming the tropical earth for over 65 million years. They are diverse and although they share some similarities with Apis they are biologically distinct. Stingless bees species are common to tropical and subtropical areas such as the South-east asia region. Diversity of stingless bees makes classifying them a difficult task, especially in Indo-Malayan region [1]. Like honeybees, stingless bees produce honey in nest containing sterile workers and a queen. However, stingless bees produce less honey, and do not use water to cool their nest. [2]. Their hives usually consist of sections of broodcell area, as well as honey pot area. In tropical region, their nests are made in trees.

Hive temperature of 38 °C and above can cause death in pupae development [3]. This is detrimental towards the growth of colony. In natural hives, thermostatic effect helps thermoregulation, where the wood insulation along with shades, are able to keep the temperature of the hive relatively constant [4].

* Corresponding author: mfirdausb@ump.edu.my

Typically, trees are cut down to obtain log containing the stingless bees nests. The log is then attached with a topping compartment to facilitate the honey production. However, such practice can be damaging towards the natural ecosystem especially for tropical jungles and are not sustainable in long term. As stingless bees are natural pollinator for tropical forests, the human disturbance to stingless bees community could potentially disrupt and alter the forest composition [5]. Hence, once a colony is obtained from a log, it can be transferred into artificial hives such as MUSTAFA-hives. Then, it can be split into two colonies. Once the colonies settled and become stable it can be divided again and again. The honey can be easily extracted as MUSTAFA-hives has separate honey cassette compartment. Thus, this minimizes the disturbance and intrusion into the colony.

Colonies in hives are often placed under trees or other suitable shaded areas. This is to prevent the hives from overheating especially in tropical hot and humid climate. Placing the hives under direct sunlight can cause temperature to build up and can be dangerous to the colony as pupae in broodcell could die when subjected to extreme heat. Hence, suitable cooling method for hives may be developed to prevent overheating of hives when subjected to outdoor conditions.

A possible cooling method is by using greenroof, a natural cooling technique that have been used for centuries [6]. Although the technique is centuries old, due to scarcity of planted surface in urban areas, greenroof become stabilising and efficient choice [7]. Essentially, a green roof is a type of roof made up of vegetation and soil as its outermost surface [8]. As vegetation have potential to provide cooling benefits [9], buildings using greenroofs could reduce their inside temperatures as much as 3.4 °C [10]. Therefore, greenroof have potential to be adapted to cool stingless bees hive.

From literature reviewed, there is no study reported on the greenroof based cooling method for stingless bees hive. Thus, the main objectives of this study are:

- to investigate the temperature profile of a hive exposed to outdoor conditions, and
- to evaluate the temperature regulation in the hive using greenroof as a cooling method.

The study was carried out by exposing two MUSTAFA-hives to outdoor condition. One of the hive was fitted with greenroof and another hive was left as control. The temperature profile of the hives was recorded.

2 Materials and method

An experiment was carried out at Universiti Malaysia Pahang, Pekan campus (coordinates: 3.5437° N, 103.4288° E, GMT + 8). Two identical MUSTAFA hives were exposed to outdoor conditions; without any shade. One of the hive was fitted with greenroof as the cooling technique. Temperature readings were taken inside the broodcell compartment and honey cassette for each hives every five minutes during the day using thermocouples. The measurement was recorded by a data logger (Lutron BTM-4208SD), which was then transferred into computer for further analysis. Figure 1 shows the typical construction of a MUSTAFA-hive, where the hive is divided into three main components, which are the broodcell compartment, the honey cassette and the roof. In the brood cell compartment, a 25 mm diameter hole acts as the only entry and exit for stingless bees. The broodcell compartment is connected to the honey cassette with a 35 mm diameter hole. This provided access for the bees to store honey. The hives used in the experiment are all empty hives without colony, to ensure no thermal exchange occur with a living colony as well as to no disrupt them.

 DOI: 10.1051/matecconf/201713103013

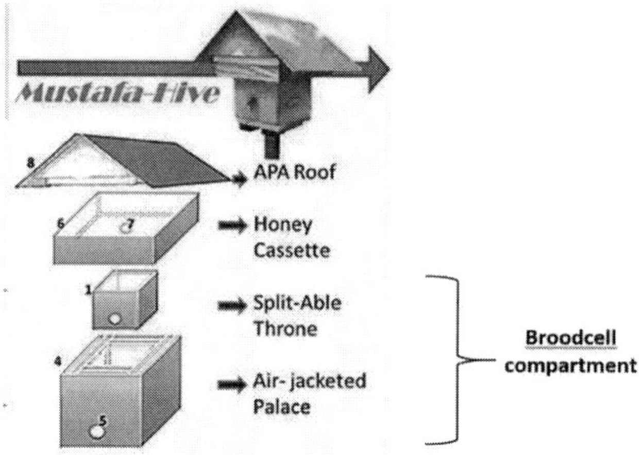

Figure 1: Typical construction of a Meliponiculture Using a Split-able Throne within Air-jacketed palace For Amplification-hive (MUSTAFA-hive)

Error! Reference source not found. shows the overall greenroof construction used in the experiment. The greenroof used was made up of four main components, which are the vegetation or the plant, some soil and sand mixture in semi dry condition, a polyethylene layer as root barrier, and wooden planks as roofing material and container.

Figure 2: Greenroof used in experiment, made up of vegetation, soil and sand mixture, polyethylene layer, and wooden planks

Figure 3 shows the overall experimental setup. Two hives were exposed to outdoor conditions without any shades. The hive on the right was fitted with a greenroof, and the hive on the left was not fitted with any cooling technique and acted as a control hive.

Figure 3: Overall setup, with thermocouples measuring temperatures inside honey cassette and brood cell compartment of each hives, connected to a temperature data logger

3 Results and discussion

Table 1 shows the maximum, minimum, and average temperature inside the honey cassette and broodcell compartment of each hives. The control hive showed that, without cooling technique, the maximum temperature in the broodcell compartment could rise as high as 39.8 °C while in the honey cassette the maximum temperature was at 34.7 °C. This showed the need for cooling especially in the broodcell compartment.

Table 1: Minimum, maximum, and average temperatures in the honey cassette and broodcell compartment of the hives

	Honey cassette		Broodcell	
Temperature (°C)	Control	Greenroof	Control	Greenroof
Minimum	30.1	28.3	30.6	27.9
Maximum	34.7	30.8	39.8	31.6
Average	33.0	29.7	36.7	30.2

Greenroof showed significant cooling effect, where the average temperatures recorded are, 29.7 °C and 30.2 °C in the honey cassette and broodcell compartment respectively. This translated to an average of 3.3 °C and 6.5 °C lower than the control hive. Apart from that, greenroof recorded 3.9 °C and 8.2 °C lower maximum temperatures for the honey cassette and broodcell compartment, respectively.

Figure 4 shows the temperature profile inside the honey cassette, represented by solid lines and broodcell compartment, represented by the pecked lines. The circle, and triangle represented the control hive, and hive with greenroof, respectively.

 DOI: 10.1051/matecconf/201713103013

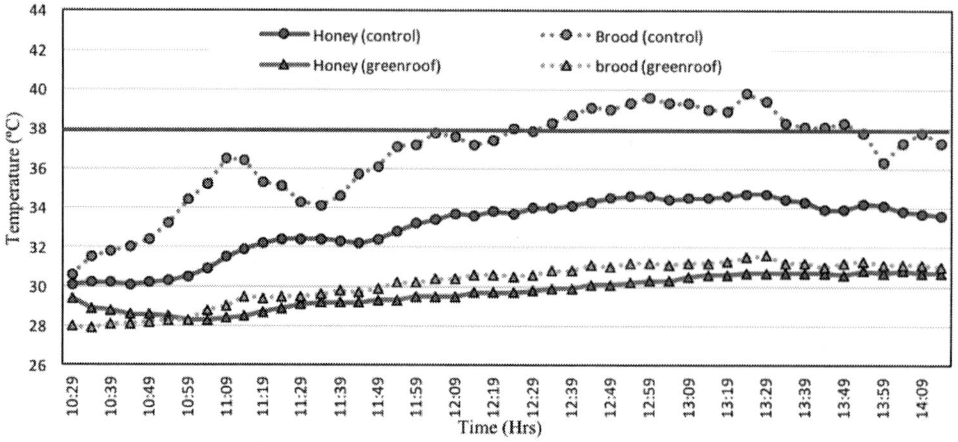

Figure 4: Temperatures inside broodcell compartment and honey compartment during the day for the control hive and hive fitted with greenroof

In the control hive, the broodcell compartment experienced a prolonged temperature rise above 38 °C, more than an hour, recorded around noon. For a colony, such extreme temperature are lethal to pupae in the broodcell compartment [3]. This can cause the colony to collapse and die out. The broodcell compartment also recorded significantly higher temperature compared to the honey cassette. This could partly be due to the roof covering most of the honey cassette from sun radiation while less covering of the broodcell compartment. As the hive was a closed system, the hot air may be trapped inside the broodcell compartment with only one exit, disrupting natural hot air circulation upwards.

The hive with greenroof recorded stable temperatures, with the broodcell compartment recorded just slightly higher temperatures than the honey cassette. When compared to the control hive, the temperatures were consistenly lower than the control hive, showing significant cooling effect around noon. The temperatures recorded were close to optimum temperature for stingless bees species such as *Trigona denoiti*, which the broodcell area optimum temperatures are between 31 °C and 32 °C [11]. These results were impressive considering the greenroof is relatively easy and simple to set up.

4 Conclusion

Two hives were exposed to outdoor conditions, with one hive without cooling technique and another hive was fitted with greenroof, acted as a control hive. The broodcell compartment in both hives recorded higher temperatures compared to the honey cassette. This could be explained by low air convection inside the hive, which caused most of heat transfer to occur inside the broodcell compartment. In the control hive, the broodcell compartment experienced prolonged temperature rise above 38 °C of more around noon. This could lead to death of pupae in broodcell. On the other hand, it was found that the hive fitted with greenroof has an impressive cooling performance. The temperatures inside the hive fitted with greenroof were significantly lower than the control hive when exposed to outdoor conditions without any shades. The average temperature reduction from the experiment were 3.3 °C and 6.5 °C in the honey cassette and broodcell compartment respectively. The temperatures recorded were relatively stable and were very close to optimum temperature of broodcell area in natural hives. Thus, it can be concluded that the greenroof is a feasible and economical solution towards cooling stingless bees hives.

The authors would like to thank Universiti Malaysia Pahang for providing Research Grant under RDU140119.

References

[1] F. Schwarz, "The Indo-Malayan Species of Trigona," Am. Museum Nat. Hist., vol. **76**, no. 3, pp. 83–141, (1939).
[2] D. W. Roubik, "Stingless bee nesting biology," Apidologie, vol. **37**, no. November, pp. 124–143, (2006).
[3] A. Vollet-Neto, C. Menezes, and V. L. Imperatriz-Fonseca, "Behavioural and developmental responses of a stingless bee (Scaptotrigona depilis) to nest overheating," Apidologie, vol. **46**, no. 4, pp. 455–464, (2015).
[4] I. H. Sung, S. Yamane, and S. Hozumi, "Thermal characteristics of nests of the Taiwanese stingless bee Trigona ventralis hoozana (Hymenoptera: Apidae)," Zool. Stud., vol. **47**, no. 4, pp. 417–428, (2008).
[5] H. Samejima, M. Marzuki, T. Nagamitsu, and T. Nakasizuka, "The effects of human disturbance on a stingless bee community in a tropical rainforest," Biol. Conserv., vol. **120**, no. 4, pp. 577–587, (2004).
[6] E. P. Del Barrio, "Analysis of the green roofs cooling potential in buildings," Energy Build., vol. **27**, no. 2, pp. 179–193, (1998).
[7] A. Niachou, K. Papakonstantinou, M. Santamouris, A. Tsangrassoulis, and G. Mihalakakou, "Analysis of the green roof thermal properties and investigation of its energy performance," Energy Build., vol. **33**, no. 7, pp. 719–729, (2001).
[8] D. J. Sailor, "A green roof model for building energy simulation programs," Energy Build., vol. **40**, no. 8, pp. 1466–1478, (2008).
[9] S. A. Shiflett, L. L. Liang, S. M. Crum, G. L. Feyisa, J. Wang, and G. D. Jenerette, "Variation in the urban vegetation, surface temperature, air temperature nexus," Sci. Total Environ., vol. **579**, pp. 495–505, (2017).
[10] V. W. Y. Tam, J. Wang, and K. N. Le, "Thermal insulation and cost effectiveness of green-roof systems: An empirical study in Hong Kong," Build. Environ., vol. **110**, pp. 46–54, (2016).
[11] D. J. C. Fletcher and R. M. Crewe, "Nest structure and thermoregulation in the stingless bee, Trigona denoiti, Hymenoptera: Apidae," J. Entomol. Soc. South Africa, vol. **44**, no. 2, pp. 183–196, (1981).

A new dewatering technique for stingless bees honey

Ahmad Syazwan Ramli[1], *Firdaus* Basrawi[1,*], *Daing Mohamad Nafiz* Daing Idris[1], *Mohd Hazwan* bin Yusof[1], *Thamir* Khalil Ibrahim[1], *Zulkifli* Mustafa[2], and *Shaharin* A. Sulaiman[3]

[1]Meliponini Engineering Laboratory (MepEL), Energy Sustainability Focus Group (ESFG), Faculty of Mechanical Engineering, Universiti Malaysia Pahang
[2]Universiti Sains Malaysia, Kubang Kerian, Kelantan, Malaysia
3Department of Mechanical Engineering, Universiti Teknologi Petronas, Bandar Seri Iskandar, 31750, Tronoh, Perak, Malaysia

Abstract. One of the problems faced in stingless bee honey storage is spoilage by the fermentation process occurs in honey due to its high water content. There are a few techniques available currently, but they are time consuming and there is excessive heat involved in the process. The temperature of the process must be kept low because excessive heat can deteriorate nutrition value and biochemical content in honey. Hence, a new method of honey dewatering was developed using a Low Temperature Vacuum Drying (LTVD) with induced nucleation technique.The objective of this research is to investigate the performance of a LTVD with induced nucleation to reduce the water content in honey. First, the honey was placed in a pressure vessel, and then air was removed. Then, the honey was slightly heated at 30°C and the water content before and after the experiment was measured by a refractometer. The steps were repeated until the water content reached below 20%. It was found that the LTVD method improved the water removal rate significantly with an average of 0.15% of water content per minute. That is 3 times much faster than the conventional method of low temperature heating by Tabouret. Higher temperature during dewatering process improved the dewatering rate significantly. It can be concluded that LTVD is a promising option in tackling the high water content in stingless bee honey issue.

1 Introduction

Honey is a very sweet flavour natural liquid made by bees from nectar which is taken from flowers or from secretions of living parts from plants [1]. It can consist up to 65% sugar which is primarily fructose and glucose [2]. The taste and aroma of the honey are ranging in color from pale yellow to dark amber which varies according to geographical and seasonal conditions [3]. In addition, the water content of honey in humid area is normally much higher than the low humid area [4]. According to Marinus (2006), honey from stingless bee in Trinidad for example has a very high water content which can reach up to 42% [5]. The water content in honey especially in stingless bee honey in general has become a major issue since the high water content promotes microbial activities that causes fermentation process to occur in honey as reported by Stephen (1946)[6]. On the other hand, honey is considered to be a heat sensitive material in general with most of its content can altered due to certain amount of excessive heat especially the antimicrobial antioxidant in it as explained by Fauzi et al. 2014[7]. Turhan et al. (2008) stated that excessive heating may also cause crystallization which will decrease the quality of the honey as well [8]. It was also found by Kretavičius et al. (2010) that heating the honey lower than 50°C will not damage the enzyme activity in the honey [9].

Therefore, in recent years there has been a lot of effort to implement low temperature drying for heat sensitive material in order to preserve the quality, nutrients and other qualities that might have change or damage by high temperature heating. In order to lower the temperature for water removing purpose, the process is subjected to low pressure or vacuum condition so that the boiling point of water can be lowered down as low as possible. Hence this combination is commonly known as low temperature vacuum drying (LTVD). In general, vacuum drying always involves with low temperature since by introducing vacuum condition, the boiling point of water is definitely lower than 100°C. Hence the term low temperature for vacuum drying is often referring that only a little heat is supplied so that the temperature for the drying process is lower than 100°C. LTVD is a common method used to dehydrate and remove the water content from heat sensitive and high quality mediums such as food biomedical and pharmaceuticals products for preservation purpose before they reach consumers in the market. Since only low temperature is used in the drying process, there is no excessive heat that could damage the heat sensitive material such as stingless bee honey in this case.

But there are currently many variants of LTVD available. For instance, there is also superheated steam drying developed by Devahastin et al. (2004) which also utilizes the LTVD concept. However, this method takes too much time for dewatering viscous liquid such as honey [10]. There is also freeze drying LTVD which the medium is first frozen before it is slightly heated to sublimate the water content but the freezing process consumes high amount of energy and not very efficient as stated by Nam et. al. (2007)[11]. There is also spray drying combined with LTVD but this method can only convert the liquid medium directly to solid and the dewatering level cannot be controlled [12]. There was also an attempt to assist LTVD with microwave to enhance the vaporization of water. However, this method will damage the composition and nutrient of the medium due the high radiation energy emitted from the microwave [13]. It is proposed that LTVD should be combined with nucleation boiling to enhance the rate of honey dewatering. A LTVD with induced nucleation boiling has been developed to meet this requirement of maintaining the original content of nutrients, quality, aroma and taste of the honey throughout the dewatering process. From literature reviewed, there is no study reported on the LTVD with nucleation boiling especially on the effect of temperature on dewatering process. Thus the objective of this research is to investigate the performance of a LTVD with induced nucleation to reduce the water content in honey.

2 Materials and Method

For the first phase of this research, an experimental test rig was prepared to properly measure the effectiveness of the honey dewatering process. Fig. 1 depicts the schematic diagram of the experimental setup.

DOI: 10.1051/matecconf/201713103014

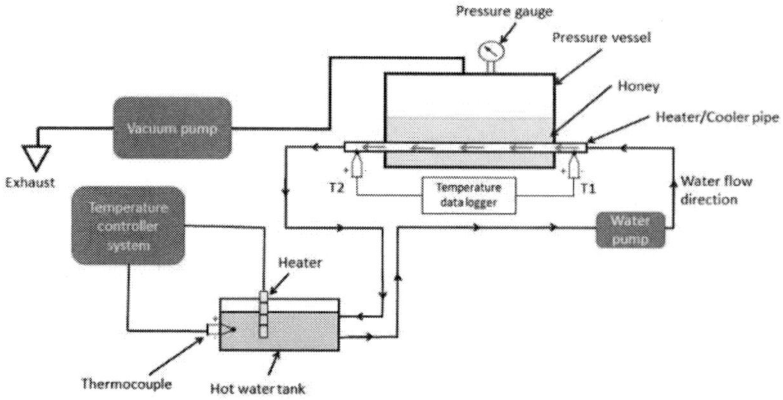

Fig. 1. Schematic diagram of the experimental setup

The main part of this setup was the pressure vessel where it the sample for LTVD is placed. A vacuum pump was connected to the vessel which created a vacuum condition inside the vessel. For this experiment, a cylindrical container type of vessel was chosen. A pressure gauge attached to the vessel is used to monitor the pressure inside the vessel. The right vacuum pressure and temperature are crucial to induce the nucleation in honey. There was also a stainless steel pipe inside the vessel which acts as a heating element for the sample. A thermocouple was attached at the water inlet and another one at the outlet of the stainless steel pipe. Both of these thermocouples are attached to a temperature data logger where the water inlet temperature, T_1 and the water outlet temperature, T_2 were used to monitor the heat transfer from the hot water to the honey.

For heating up the sample, a heater in a water tank heated up the water up to the determined temperature and the water temperature of the tank was held constant by a temperature controlling system with a thermocouple. For this experiment, 30°C was set for the dewatering. The honey samples used in this experiment was obtained from a farm in the state of Perak, Malaysia. Fig. 2 shows the sample obtained from a farm in Perak. The water content of the honey sample was measured by a refractometer. Fig. 3 is an example of measurement taken by a refractometer. The initial reading for the sample was around 26% of water content.

Fig. 2. Stingless bee honey samples from Perak in liquid container.

DOI: 10.1051/matecconf/201713103014

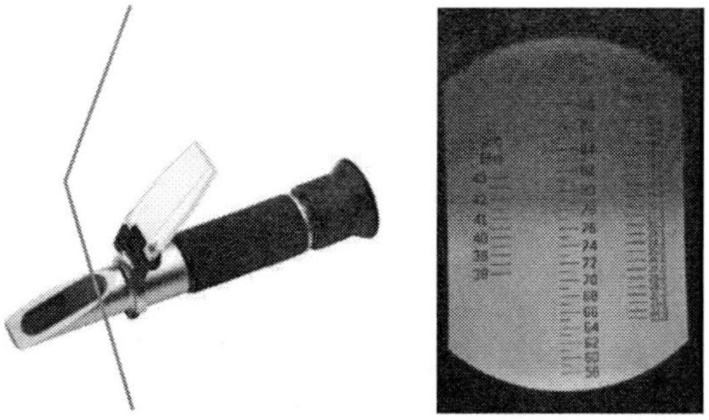

Fig. 3. Measurement sample by a refractometer.

The experiment was done by introducing the honey sample inside the pressure vessel and closes the lid tightly. 200g of the honey sample was weighted first as shown in Fig. 4. Then the reading of the initial water content of the sample was taken by using a refractometer. After 10 minutes, the experiment was stopped to measure the final water content. The process was repeated until the water content of the sample reaches 20% level or below.

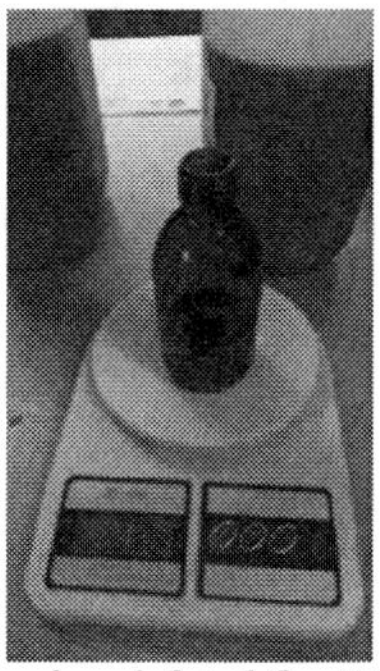

Fig. 4. Honey sample was put in a bottle and weighted for 200g.

The data that needs to be collected for this research is the water content of the sample against 10 minutes time interval. In this case, the temperature of heater, heat transfer rate in honey and the pressure inside the vessel were assumed to be constant. By having the water content data against time, the rate of water removal can be determined for each parameter by using equation (1) below:

$$\eta_D = \frac{\Delta W_h}{t} \quad (1)$$

where η_D is the dewatering [%/min], ΔW_h is the change of water content [%], and t is the time taken for the dewatering process [min].

3 Results and Discussion

During the experiment, nucleation bubbling was observed in the honey sample. The nucleation occurred in the sample was observed and the view of the nucleation density is recorded via a camera. Fig. 5 is an example of the nucleation bubbling that occurred during the dewatering. For this experiment, the size, density and shape of the nucleation was considered to be constant.

 DOI: 10.1051/matecconf/201713103014

Fig. 5. Nucleation bubbling occurred during honey dewatering.

Fig. 6 shows water content changes by time for each experiment. As shown in Fig. 6, the increase of temperature improved the rate of dewatering significantly. At temperature of 30°C, the dewatering reached the desired level of water content within 60 minutes. The average dewatering rate, η_D with the temperature of 30°C was 0.1%/min.

If the dewatering rate is compared to the traditional method of using hot air stream for example, this dewatering method is much better because the hot air stream method takes about 2 hour to reduce approximately 2% of water content. So the dewatering rate is about $\eta_D = 0.017$. On top of that, the hot air is very difficult to control and may damage the content of the honey [14]. A large dehumidifier built by Paysen for example can dewater 57 tonnes of honey but the dewatering rate is only 0.6% of water content within 30 hour [15].The dewatering rate for this method is approximately $\eta_D = 0.00033$. With the help of low pressure vacuum, Tabouret managed to remove the water content at the rate of η_D = 0.048% per minute but at a temperature of 63°C which is considered to be too high for dewatering process[16].This means that induced nucleation LTVD performed 3 times much faster than the result achieved by Tabouret. By using LTVD with the assistance of microwave on the other hand, the dewatering rate is much higher which is around η_D = 0.25% per minute but the radiation emitted by the microwave have the potential to damaged and alter the original quality, aroma and taste of the honey. Fig. 7 compares dewatering rate of present result with other techniques

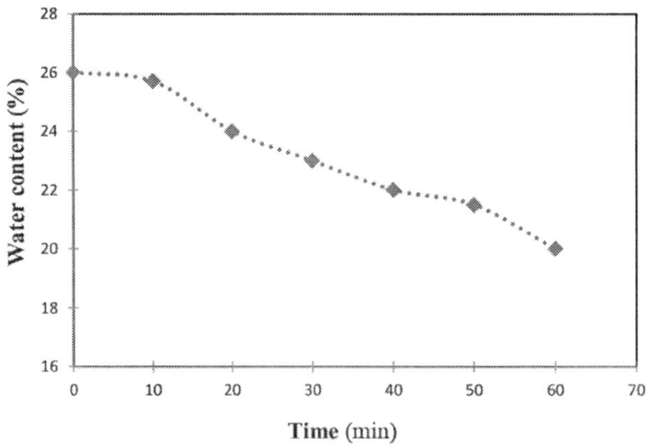

Fig. 6. Dewatering result for honey sample at 30°C operating temperatures.

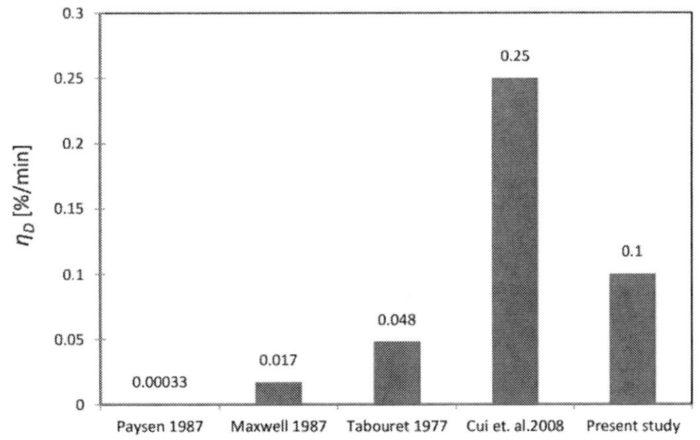

Fig. 7. Comparison of the experiment's dewatering rate with other techniques.

4 Conclusion

The effectiveness of LTVD with induced nucleation at 30°C for dewatering stingless bee honey was investigated, and the performance was compared with other dewatering methods. It was found that by introducing the nucleation boiling in LTVD, the performance and efficiency of the dewatering rate can be further improved for up to an average of 3 times faster than the traditional methods without the need to increase any energy and operational cost. The η_D for induced nucleation LTVD was $\eta_D = 0.1$ at temperature 30°C. This means that the method is much better compared to dehumidifier method which only had $\eta_D = 0.00033$, hot air drying with $\eta_D = 0.017$ and low pressure heating with $\eta_D = 0.048$. The rate of microwave LTVD is much better which is $\eta_D = 0.25$ but the radiation generated by the microwave damaged the nutritional content in honey. However, further testing to check the content of the honey before and after the proposed LTVD with induced nucleation should be done to verify that this method does not give any side effect or any damage to the original content of the honey.

The authors would like to thank the Ministry of Education Malaysia for providing Research Grant under FRGS. The authors gratefully acknowledge Prof. Ir. Dr. Hassan Ibrahim of the Faculty of Mechanical Engineering, University Malaysia Pahang for his kind advice and the help in editing this manuscript.

References

1. Mendes, E., Brojo Proença, E., Ferreira, I. M. P. L. V. ., and Ferreira, M. ., Quality evaluation of Portuguese honey, *Carbohydr. Polym.* **3** (1998) 219–223.
2. Nurhadi, B., and Roos, Y. H., Water Sorption and Water Plasticization Behavior of Vacuum Dried Honey, *Int. J. Food Prop.* **19** (2016) 1370–1380.
3. Aparna, A. R., and Rajalakshmi, D., Honey—its characteristics, sensory aspects, and applications, *Food Rev. Int.* **15** (1991) 455–471.
4. Chen, G., Sun, X., Huang, Y., and Chen, K., Tracking the dehydration process of raw honey by synchronous two-dimensional near infrared correlation spectroscopy, *J. Mol. Struct.* **1076** (2014) 42–48.
5. Marinus, J. S., Water content of stingless bee honeys (Apidae, Meliponini): interspecific variation and comparison with honey of Apis mellifera, *Apidologie* **37**

(2006) 480–486.
6. Stephen, W. A., The Relationship of Moisture Content and Yeast Count in Honey Fermentation 1, *Sci. Agric.* **6** (1946) 258–264.
7. Silva, F., Fauzi, N. A., and Farid, M. M., High-Pressure Processing of Manuka Honey- Improvement of Antioxidant Activity, Preservation of Colour and Flow Behaviour, *Food Bioprocess Technol.* **7** (2014) 2299–2307.
8. Turhan, I., Tetik, N., Karhan, M., Gurel, F., and Tavukcuoglu, H. R., Quality of honeys influenced by thermal treatment, *LWT-Food Sci. Technol.* **41** (2008) 1396–1399.
9. Kretavičius, J., Kurtinaitienė, B., Racys, J., and Ceksteryte, V., Inactivation of glucose oxidase during heat-treatment de-crystallization of honey, *Žemdirbystė (Agriculture)* **4** (2010) 115–122.
10. Devahastin, S., Suvarnakuta, P., Soponronnarit, S., and Mujumdar, A. S., A Comparative Study of Low-Pressure Superheated Steam and Vacuum Drying of a Heat-Sensitive Material, *Dry. Technol.* **22** (2004) 1845–1867.
11. Nam, J. H., and Song, C. S., Numerical simulation of conjugate heat and mass transfer during multi-dimensional freeze drying of slab-shaped food products, *Int. J. Heat Mass Transf.* **23** (2007) 4891–4900.
12. Cuevas-Glory, L. F. P., A., J., Sosa-Moguel, O., Sauri-Duch, E., and Bringas-Lantigua, M., Optimization of the Spray-Drying Process for Developing Stingless Bee Honey Powder, *Int. J. Food Eng.* **13** (2017).
13. Cui, Z., Sun, L., Chen, W., and Sun, D., Preparation of dry honey by microwave–vacuum drying, *J. Food Eng.* **84** (2008) 582–590.
14. Crane, E., The removal of water from honey, *Bee World* **77** (1996) 120–129.
15. Tabouret, T., Vacuum drying of honey, *Apiacta* **12** (1977) I57-164.

 DOI: 10.1051/matecconf/201713103015

Investigation on the effect of blending ratio and airflow rate on syngas profile produced from co-gasification of blended feedstock

Muddasser Inayat[1,*], *Shaharin Anwar Sulaiman*[1], *and Jundika Candra Kurnia*[1]

[1]Department of Mechanical Engineering, Universiti Teknologi PETRONAS, 32610 Bandar Seri Iskandar, Perak Darul Ridzuan, Malaysia

Abstract. Shortages of feedstock supply due to seasonal availability, high transportation costs, and lack of biomass market are creating serious problems in continues operation of bioenergy industry. Aiming at this problem, utilization of blended feedstock is proposed. In this work blends of two different biomasses (wood and coconut shells) were co-gasified using externally heated downdraft gasifier. The effects of varying biomass blending ratio and airflow rate on gaseous components of syngas and its heating value were investigated. The results obtained from the experiments revealed that W20:CS80 blend yielded higher values for H_2 (20 Vol.%) and HHV (18 MJ/Nm^3) as compared to the other blends. The higher airflow rate has a negative effect on syngas profile and heating value. The CO and CH_4 were observed higher at the start of the process, however, CO was observed decreasing afterward, and the CH_4 dropped to 5.0 Vol.%. The maximum H_2 and CH_4 were obtained at 2.5 LPM airflow rate. The process was noticed more stable at low air flow rates. The HHV was observed higher at the start of process at low airflow rate. It is concluded that low airflow rate and a higher ratio of coconut shells can improve the syngas quality during co-gasification.

1 Introduction

Malaysia is a tropical country blessed with plenty of biomasses by the nature especially forestry and palm oil industry. The biomasses from these sources could be potential feedstock for bioenergy industry via gasification [1]. It is projected that exploitation of forestry wastes and agricultural wastes will contribute significantly as the substitution of fossil fuel in the future [1]. Malaysian woodland and timber processing factories generate a huge amount of wood residue. These timber factories and woodland are considered as one of the major unexploited biomass sources in Malaysia. Malaysia has five wood mills, which are using the wood residue as fuel and each wood mill has potential to generate 900 kW to 10 MW of energy. In 2010, 12 million m^3 of timber logs were produced from the Malaysian forest area [2]. The wood waste generated during the logging operation was 5.1 million m^3 in form of stumps, branches, bark, tops, broken logs, defective logs and injured standing trees. All these wastes are equal to 43 % of total volume of the trees [3]. Beside this, coconut is also potentially available in Malaysia as a source of edible oil and coconut milk, and it is one of the main ingredient of most Malaysian foods. Coconut shells are the by-product of coconut milk processing industries, which can be utilized as potential feedstock for bioenergy [2].

Dependency on only a single biomass feedstock could be problematic for continuous gasification operation. It was reported that gasification of biomass often experienced the problems of unsteady supply of feedstock throughout the year due to various reasons such as; seasonal harvesting, high transportation cost, and climate change; for instance, flood, heavy rain and drought [1, 4]. In addition, some biomass materials are available in small quantity and therefore, gasification process could not be sustained for longer periods. However, with co-gasification, the small amount of biomass could be meaningful as it can be mixed with other biomass materials to produce syngas for a prolonged period [5].

Although co-gasification of coal and biomass is extensively studied in the published literature. However, information about co-gasification is very rare [6, 7]. Therefore, residual of lignocellulosic biomasses such as wood and coconut shells are considered as a potential source of energy via co-gasification. The biomass feedstock mixtures have a strong influence on the resulting syngas composition [4]. In this study, co-gasification of wood and coconut shells was carried out using an externally heated downdraft gasifier. The effects of varying blending ratios and airflow rate on the gas profile of syngas components and its heating value were investigated.

2 Materials and methodPreparation and characterization of feedstock

Blends of wood and coconut shells (W/CS) at various ratios were used as feedstock for this study. The wood was collected from the landscape area of Universiti Teknologi PETRONAS Malaysia and coconut shells were collected from the local market at Batu Gajah, Perak, Malaysia. A lab-scale granulator machine was used for size reduction of feedstock in the range of 3 - 5 mm [8]. Characterization of wood and coconut shells includes ultimate analysis, proximate analysis, and heating value. The ultimate analysis was carried out as per ASTM D3176-09 in elementar vario MICRO CHNS analyzer. Proximate analysis was carried out on LABsys Evo TGA analyzer as per ASTM E1755-01. The moisture content feedstock was determined as per ASTM E871-82. According to ASTM D4809-00 standard procedure higher heating value of the feedstock was determined [9]. Characterization results are presented in Table 1.

Table 1. Characterization Results of Wood and Coconut Shells

Feedstock	Ultimate Analysis (%)					H/C	O/C	Proximate Analysis (%)				HHV(MJ/kg)
	C	H	N	S	O*			MC	VM	FC*	Ash	
Wood	43.54	3.59	0.31	0.08	51.70	0.99	0.89	4.25	88.58	6.02	1.16	17.53
Coconut Shells	46.93	3.96	1.33	0.25	48.21	1.01	0.77	2.29	82.09	14.79	0.83	19.43

*On difference basis

2.2 Experimental setup and experiment design

The experiments were conducted using a batch type externally heated downdraft gasifier. The detail of experimental setup is presented in Fig. 1. Gasifier consisted of a cylindrical body surrounded by the ceramic electric furnace. The desired temperature of gasifier was controlled by a temperature microcontroller. A controlled and measured quantity of air was supplied. The airflow rate was regulated and measured by using a rotameter. The output syngas was channelled through a rubber tube which was passed through a pair of the filter. The filters trapped excess tar and moisture of syngas. A filtered syngas was pumped into the online gas analyzer to analyze its composition. A flare point was introduced between filter and gas pumping point to act as a buffer. Furthermore, flammability of the syngas was checked at flare point. Experiments were conducted for two varying parameters which were biomass blending ratio and airflow rate. The blends of W20:CS80, W50:CS50, and W80:CS20 were used and airflow rate was ranged from 2.0 L/min to 4.0 L/min. All experiments were carried out at an optimum temperature of 800°C. The details of experimental plan are presented in Table 2.

Table 2. Experimental Conditions for Co-gasification of Wood/Coconut Shells

Experiment	Temperature (°C)	Airflow Rate (L/min)	Blending Ratio
1	800	3.0	W20:CS80
2	800	3.0	W50:CS50
3	800	3.0	W80:CS20
4	800	2.0	W20:CS80
5	800	2.5	W20:CS80
6	800	3.5	W20:CS80
7	800	4.0	W20:CS80

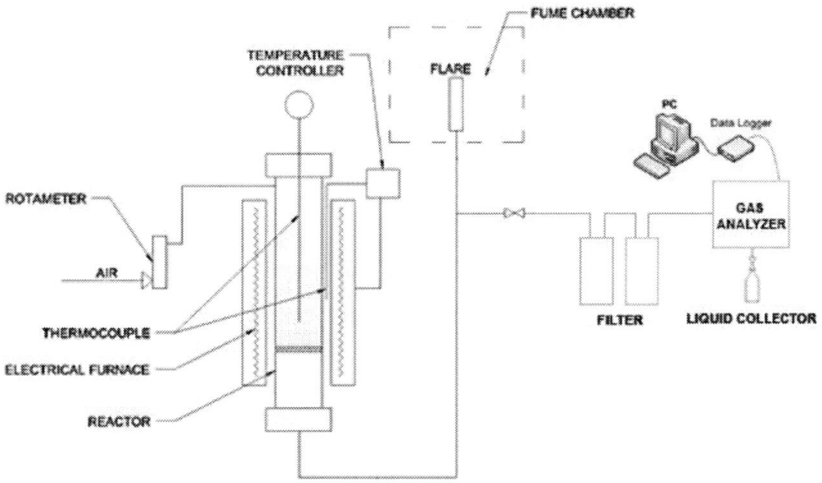

Fig. 1. Schematic diagram of experimental setup

2.3 Experimental procedure

The residual of ash and char of the previous experiments in gasifier was cleaned before to start the experiment. The gas piping was checked properly to ensure that there was no blockage in it. The gasifier temperature was set and monitored by the temperature microcontroller. The electric ceramic furnace was switched on 30 minutes before to start the experiment. As the desired temperature was obtained a pre-measured quantity of blended feedstock of W/CS was fed from the top of the gasifier. A control and measured air was supplied to the gasifier as gasification medium. The gas analyzer was connected to the computer via its interface Ethernet. A continues syngas was pumped to analyzer through diaphragm pump. The online analyzer software was used to record the syngas composition with an interval of 1 second. The syngas sampled was analyzed until all the feedstock was consumed. The interface of the gas analyzer was stopped and data was saved in a computer. The air supply was stopped and electric ceramic furnace was switched off so that gasifier was allowed to cool down. The residual of co-gasification such as ash and char from the gasifier was collected and was measured for further calculations.

The quality of syngas can be determined by the gas composition and its heating value. The heating value of a syngas is defined as the quantity of heat being released from combustion of a unit volume of syngas under the normal conditions. Higher heating value (HHV) also known as gross calorific value is the total energy released from syngas and can be determined by Eq. (1) [4, 6].

$$\mathrm{HHV}_{\mathrm{Syngas}} = 12.63 \times X_{co} + 12.74 \times X_{H_2} + 39.82 \times X_{CH_4} \quad (1)$$

where HHV of CO, H_2, and CH_4 are 12.63 MJ/Nm^3, 12.74 MJ/Nm^3, and 39.82 MJ/ Nm^3 respectively, and X_{CO}, X_{H2}, and X_{CH4} represent the volumetric percentage of CO, H_2 and CH_4 in dry syngas respectively [10].

3 Results and discussion

3.1 Effect of blending ratio on gas composition and syngas heating value

Effects of biomass blending ratios on of gas profile of different syngas components and its higher heating value (HHV) are illustrated in Figure 2. Experiments were conducted using W20:CS80, W50:CS50, and W80:CS20 blends at 3.0 LPM airflow rate and 800°C temperature. At the start of the co-gasification process, the yields of CO and CH_4 gases were high and concentration of H_2 was low. The CO with 44 Vol.% was quite higher among other gaseous components However, its production was continuously seen decreasing until 320 seconds of the process after that its get stable to the 10-15 Vol.%. The CH_4 was also higher in yield at the start of a process which was around 30 Vol.% but in a later stage, after 120 seconds it sharply decreased to reach to 5.0 Vol.% and gets stable after 320 seconds. The H_2 yield was around 10 Vol.% for W50:CS50 and W80:CS20 blends after the 380 seconds of the process. The H_2 starts decreasing after 520 seconds and it drop to 3.0 Vol. %. The H_2 yield from W20:CS80 was quite high as compared to the other blends it reached its maximum value 20 Vol.% at 400 seconds of the process after that it starts decreasing. At 600 seconds and onward of the process, all blends the gas components were dropped below 3.0 Vol.% for H_2 and CH_4 whereas CO was below 10 Vol.% in each experiment data was recorded for 1000 seconds. The decreased in gas composition after 600 seconds was subjected to consumption of feedstock during the process. Syngas HHV was calculated from the real time gas composition by Equation 1. HHV was high at the start of the process due to the higher values of CO and CH_4. The maximum value achieved around 18 MJ/Nm^3. The HHV started decreasing after 200 seconds of the process and until 600 seconds it dropped to 2.0 Vol.%. Among the other blends W20:CS80 blend has higher HHV as it shown in the Figure 2d a less steep curve of HHV for W20:CS80 blend. Overall W20:CS80 blend was better in gas composition and HHV as compared to other blends. Therefore, the later experiments of airflow rate were carried out using W20:CS80 blend.

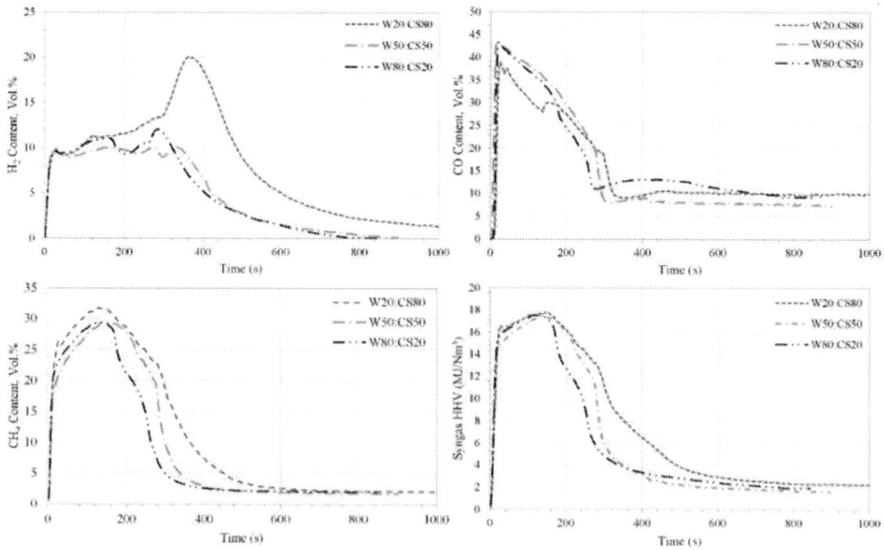

Fig. 2. Effect of biomass blending ratio on (a) H_2, (b) CO, (c) CH_4, and (d) syngas higher heating value during the co-gasification of W/CS.

3.2 Effect of airflow rate on gas composition and syngas heating value

The effect of airflow rate on the gas profile of different syngas components and syngas HHV is depicted in Figure 3. The blend of W20:CS80 was co-gasified at the 800°C temperature and airflow rate was varying from 2.0 to 4.0 LPM. The Figure 3a presented the H_2 profile at different airflow rate the H_2 yield was increased from 20 Vol.% to 30 Vol.% as the airflow rate decreased from 4.0 LPM to 2.0 LPM. Furthermore, the gas profile span also increased and shifted to the right side which shows that process was more stable at lower airflow rate range. The maximum H_2 was obtained at 2.5 LPM close to pyrolysis conditions because of low air supply the reaction gets more time to complete the reaction through water-gas shift reaction ($CO + H_2O \leftrightarrow CO_2 + H_2$) that is evidence of increasing H_2 and decreasing CO at 400 seconds of the process in Figure 3a and 3b. All gas components start decreasing after the 400 seconds because the feedstock was near to consume. The CH_4 was higher in yield at the start of the process (200 seconds) after that it starts decreasing. The highest CH_4 content was obtained at low airflow rate. The Figure 3d presented the HHV of syngas the HHV was also high at the start of the process due to higher production of all combustible gases at the start process. At low airflow rate range, the HHV was high due to the higher proportion of H_2 content. The maximum HHV value was 18.40 MJ/Nm^3 and it dropped to 3.0 MJ/Nm^3 at the end of the process.

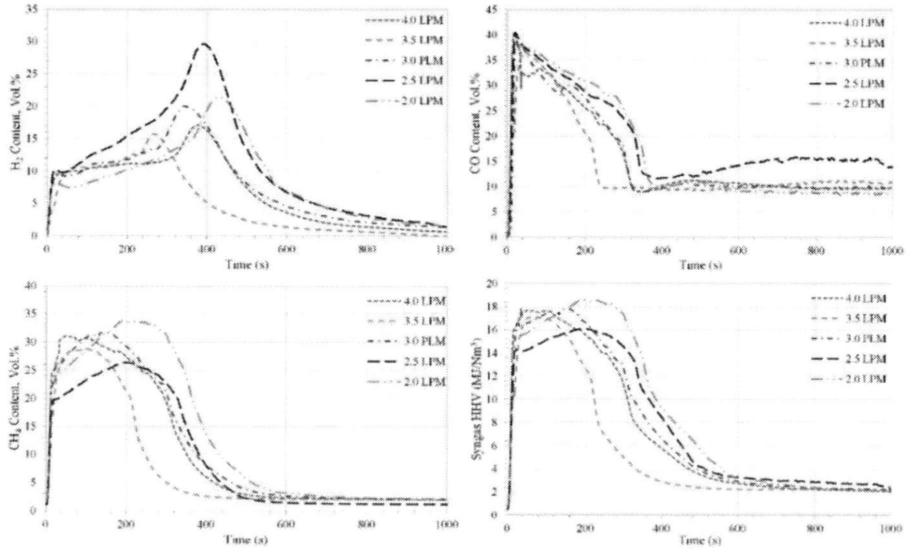

Fig. 3. Effect of airflow rate on (a) H_2, (b) CO, (c) CH_4, and (d) syngas higher heating value during the co-gasification of W/CS.

4 Conclusions

In this paper, the co-gasification of wood and coconut shells blends was carried out using an externally heated gasifier. The effect of varying biomass blending ratios and airflow rate on the gas profile of syngas components and its HHV was investigated. The W20:CS80 blend is observed to have higher H_2 as compared to other blends. The H_2 yield has a negative correlation with airflow rate. A 50 % increase in H_2 yield was observed when the airflow rate decreased from 4.0 LPM to 2.0 LPM. The CO has a constant decreasing trend in biomass blending ratio and airflow rate. The CH_4 yield increased at low airflow rate. The HHV was higher at the start of process and also at low airflow rate. It is concluded that low airflow rate and a higher proportion of coconut shells in blend can improve the syngas quality and process can be more stable with a lower airflow rate.

DOI: 10.1051/matecconf/201713103015

References

1 M. Inayat, S. A. Sulaiman, A. Abd Jamil, F. M. Guangul, and S. M. Atnaw, "The study of temperature profile and syngas flare in co-gasification of biomass feedstock in throated downdraft gasifier," in *ICGSCE 2014: Proceedings of the International Conference on Global Sustainability and Chemical Engineering*, M. A. Hashim, Ed., ed Singapore: Springer Singapore, 2015, pp. 203-10.

2 S. M. Shafie, T. M. I. Mahlia, H. H. Masjuki, and A. Ahmad-Yazid, "A review on electricity generation based on biomass residue in Malaysia," Renewable and Sustainable Energy Reviews 2012;**16**(8): 5879-89

3 H. W. Kong, "Current status of biomass utilizations in Malaysia," Forest Research Institute Malaysia 2000.

4 S. A. Sulaiman, R. Roslan, M. Inayat, and M. Yasin Naz, "Effect of blending ratio and catalyst loading on co-gasification of wood chips and coconut waste," Journal of the Energy Institute

5 B. Buragohain, P. Mahanta, and V. Moholkar, S., "Investigations in gasification of biomass mixtures using thermodynamic equilibrium and semi–equilibrium models," International Journal of Energy and Environment 2011;**2**(3): 551-78

6 M. Inayat, S. A. Sulaiman, A. Kumar, and F. M. Guangul, "Effect of fuel particle size and blending ratio on syngas production and performance of co-gasification," Journal of Mechanical Engineering and Sciences 2016;**10**(2): 2187-99

7 S. A. Sulaiman, M. Inayat, H. Basri, F. M. Guangul, and S. M. Atnaw, "Effect of blending ratio on temperature profile and syngas composition of woody biomass co-gasification," Journal of Mechanical Engineering and Sciences 2016;**10**(2): 2176-86

8 M. N. Z. Moni, S. A. Sulaiman, Y. S. Raja, K. Karunamurthy, M. Inayat, and M. A. Bou-Rabee, "Investigation of the relationship between moisture content and density of selected Malaysian biomass " Journal of Mechanical Engineering and Sciences 2016;**10**(2): 2111-25

9 S. A. Sulaiman, H. S. Bamufleh, S. N. A. Tamili, M. Inayat, and M. Yasin Naz, "Characterization of date palm fronds as a fuel for energy production," Bulletin of the Chemical Society of Ethiopia 2016;**30**(3): 465-72

10 P. Basu, *Biomass gasification and pyrolysis: practical design and theory*: Academic press, 2010.

Pre-treatment of oil palm fronds biomass for gasification

Shaharin Anwar Sulaiman[1,*], *Nor Hazwani* Mat Razali[2], *Mohamad Nazmi* Zaidi Bin Moni[1], *and Muddasser* Inayat[1]

[1]Department of Mechanical Engineering, Universiti Teknologi PETRONAS, 32610 Bandar Seri Iskandar, Perak Darul Ridzuan, Malaysia
[2]Electrical & Instrumentation, PETRONAS Level 2, East Coast Regional Office Complex, 24300 Kertih, Kemaman, Terengganu, Malaysia

Abstract. Oil Palm Fronds (OPF) has been proven as one of the potential types of biomass feedstock for power generation. The low ash content and high calorific value are making OPF an attractive source for gasification. The objective of this study is to investigate the effects of pre-treatments of OPF residual on gasification. The pre-treatments included the briquetting process and extensive drying of OPF which are studied separately. In briquetting process, the OPF were mixed with some portions of paper as an additives, leaflets, and water, to form a soupy slurry. The extensive drying of OPF needs to cut down OPF in 4-6 cm particle size and left to dry in the oven at 150°C for 24 hours. Gasification process was carried out at the end of each of the pre-treated processes. It was found that the average gas composition obtained from briquetting process was 8.07%, 2.06%, 0.54%, and 11.02% for CO, H_2, CH_4, and CO_2 respectively. A good composition of syngas was produced from extensive dried OPF, as 16.48%, 4.03%, 0.91%, and 11.15% for CO, H_2, CH_4, and CO_2 contents respectively. It can be concluded that pre-treatments improved the physical characteristics of biomass. The bulk density of biomass can be increased by briquetting but the stability of the structure is depending on the composition of briquette formulation. Furthermore, the stability of gasification process also depended on briquette density, mechanical strength, and formulation.

1 Introduction

Oil palm frond (OPF) is an agricultural by-product that is abundantly available in Malaysia. Its low ash content and high calorific value have made OPF a good source of biomass energy via gasification [1]. However, utilization of OPF as a feedstock for bioenergy is limited [2]. Oil palm industry generates numerous type of biomasses, which are different in particle size and density. Among these, small particle sizes and low-density biomasses are not suitable for gasification. Because low bulk density biomass often encounters with bridging in gasification process [3]. Furthermore, low densities of feedstock are also difficult to transport, store, and interface with bio-refinery feed systems [4]. Bridging in gasification process significant

* Corresponding author: shaharin@utp.edu.my

effect on pressure drop, performance, and its slow down the movement of the biomass down to the gasifier system. Insufficient flow of biomass under gravity which eventually affect the on gas quality, and low heating value of produce gas [5]. To improve the low-density of feedstock through the densification processes such as; pelletization and briquette are effective techniques which make low-density feedstock suitable for the gasification process [4]. The aim of this study is to investigate the effect of pre-treatments of OPF residual on the gasification process.

2 Materials and methods

2.1 Characterization and feedstock preparation for briquetting

The ultimate analysis of OPF residual was carried out using CHNS analyzer as per ASTM D3176-09 standard procedure [6] to determine the percentage of carbon (C), hydrogen (H), nitrogen (N), sulphur (S), and oxygen (O) contents in feedstock that are important for stoichiometric air calculation of gasification process. The results of ultimate analysis of oil palm fronds residual are presented in Table 1.

Table 1. Ultimate Analysis of Oil Palm Fronds Residual

Feedstock	**C (%)**	**H (%)**	**N (%)**	**S (%)**	**O*(%)**
OPF	41.33	5.30	0.22	0.22	52.93
OPF Leaflet	41.20	4.81	2.89	0.34	50.76

The raw OPF was collected from oil palm plantation and shred in the shredder. The shredded feedstock obtained from shredder was not uniform in particle size. Therefore, sieving was done to separate different particle sizes of the OPF feedstock. The shredded feedstock of OPF obtained after shredding and sieving operation is shown in Figure 1. The first sieving process involved with 10 mm sieve size. The shredded OPF was poured onto the 10 mm sieve. The second sieving process involved with 5 mm sieving size. Briquette was formed from the mixture of the OPF fibre, OPF sawdust, and diced type OPF. The briquette process was done in two batches. In the first batch, sawdust, fibre, leaflet of OPF, and paper were used. While in the second batch of briquette, diced OPF was used instead of fibre, and a paper was neglected for the second batch. Approximated 4-5 litters of water was added to the mixture to form the soupy slurry. The mixture was then left for about an hour to allow it blend well.

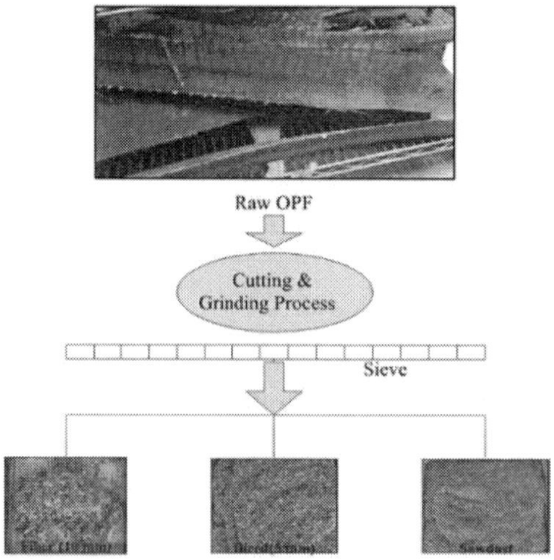

Fig. 1. Feedstock preparation for briquette formation

2.1.1 Forming briquette

The briquetting process was started after about an hour of soaking time. About 250 grams of the mixture was taken each time to fill the briquette mould. The mixture needed to be distributed evenly in the mould. It was to create the same pressure distribution when pressing was started. The pressing method was applied manually using the hand. The handle of the briquetting machine had to be pressed downward to remove the water content of the mould. The weight of the die used to press the mould was 8.0 N. The mould was left for 5.0 minutes under pressure to make sure all the water content was removed from the mixture. After 5.0 minutes, the briquette can be taken out. This process needs to be done carefully to make sure the structure of the briquette did not destroy because of the soft structure of the briquette. The handle of the level needs to be pulled upward to remove the briquette from the mould and left it for drying.

2.1.2 Extensive drying of OPF

OPF extensive drying was done to investigate the effect of moisture removal on syngas composition for gasification process. The drying was done at the temperature of 150°C for 24 hours using the electric oven. The particle size was 4-6 cm was used for extensive drying. Before the extensive drying, the raw OPF was naturally dried under the sun for 1 to 2 days after chopped down.

2.1.3 Gasification procedure

A downdraft and updraft gasifiers were used for gasification. OPF briquette and extensively dried OPF was used as a solid fuel for gasification. Around 6 kg of feedstock was fed into the gasifier for each run. A controlled air was supplied to the gasifier as a medium of gasification at a rate of 180-220 LPM by using a blower. The temperature of different zones of gasifier was monitor and recorded by Type K thermocouples and data logger profiles. Syngas compositions recorded at 1-minute interval along the 60-minutes of gasifier operation using X-STREAM online gas analyzer.

3 Results and discussion

3.1 Briquette properties

Briquettes produced from different OPF wastes, paper, and oily sludge at varying blending ratio and formulation of briquettes are presented in Table 2. More amount of OPF diced in blend producing more rigid, denser and stronger briquettes. The presence of oil palm leaflet in blend improved the surface contact of materials during press and shredded paper into the blending act as a binder material to help retain the moulded shape after briquetting operation. The physical shape was recorded as 10.5 cm in diameter and 2.5 cm in thickness, the inner diameter of the hole was 3.0 cm physical structure of the briquette is shown in Figure 2.

DOI: 10.1051/matecconf/201713103016

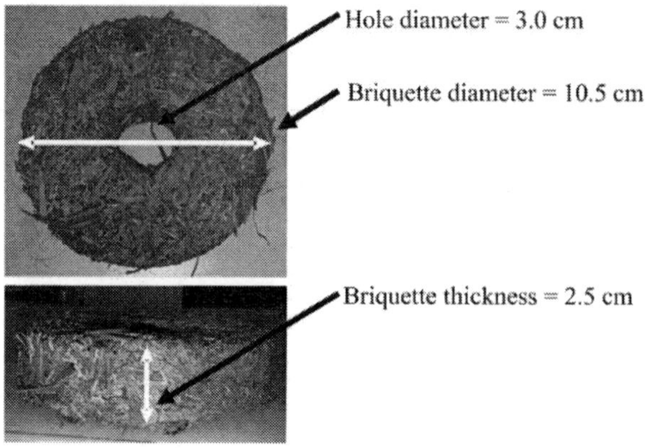

Fig. 2. Physical structure of a fully-dried briquette

The specific density of briquette on the dry basis was measured as 40 kg/m^3. There was a significant difference between the wet and dry densities of the raw materials. Approximate 80% moisture removed during the drying process. The bulk density of raw blended feedstock was measured by dividing the mass of biomass with the volume of the container used. The volume of the container was recorded as 9.5 cm x 9.5 cm x 11cm in dimensions. The total mass of raw blended feedstock that filled up the container was 40 grams. The approximate mass of one dried briquette was in the range of 35-55 grams each. The wet condition of the biomass might influence the mass range of dried briquettes. The mechanical strength of dried briquettes was determined the stability and toughness of its structure tested by performing a drop test. Dried briquettes were dropped from a height of 1-2 m as conducted by Sing and Aris [7] in their tests. Both briquettes sustained their shape with minor damages after impact. Two more drop tests followed with each from a height of 5.0 m and 10 m respectively. The 1st batch of briquette was sustained its structure whereas, the 2nd batch found few cracks on the briquette after it was dropped from a height of 5.0 m while still maintaining its shape. The structure of 2nd batch of briquette was damage significantly after a drop from 10 m height. Therefore, it's proven that feedstock nature and formulation of briquette have an influence on its mechanical strength and density.

Table 2. Briquette Formulation

Biomass Type	Oil Palm Based Biomass (%)				Non-Oil Palm Biomass (%)	
	Sawdust	Fiber	Leaflet	Diced	Paper	Oily Sludge
1st Batch of Briquette	37.00	14.80	7.50	N.A	3.70	37.00
2nd Batch of Briquette	41.66	N.A	16.67	41.67	N.A	N.A

3.2 Gasification of briquette

The gasification of fully dried briquette of the 1st batch was carried out in downdraft gasifier [8] and 2nd batch was gasified in updraft gasifier. In downdraft gasifier syngas was produced but due to leakage in experimental setup syngas was unable to analyze. Results of syngas profile and temperature produced from the 2nd batch of briquette are presented in Figure 3.

 DOI: 10.1051/matecconf/201713103016

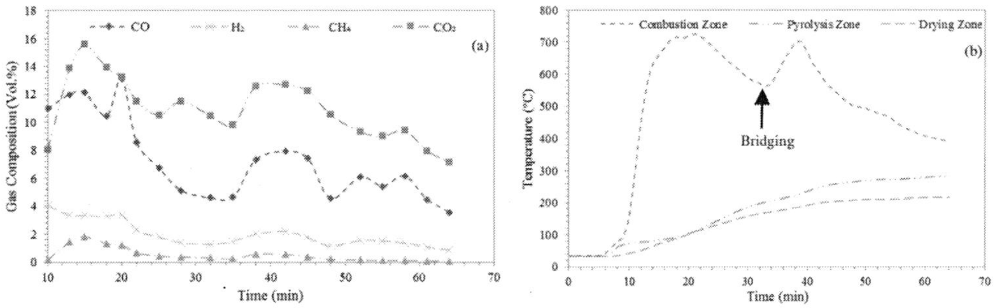

Fig. 3. (a) Syngas profile and (b) temperature profile produced during the gasification of 2nd batch of briquette

The average gas composition of process was recorded as 8.07%, 2.06%, 0.54%, and 11.02% for CO, H_2, CH_4, and CO_2 respectively. Comparison between gas compositions of briquette to typical OPF and extensive dried OPF as shown in Table 3. The Figure 3 exemplifies that the gases produced from briquettes were quite low due to the bridging that occurred during the experiment. The bridging occurred due to the low density of the particles from the briquette. As the structure of the briquette was not stable enough, as it was loaded into the gasifier, the briquette was broken into small particles. This contributed to the effect of low density to the operation. The low density of the biomass restricted the flow of the particles downward, and due to this, the gas cannot flow smoothly to the upper part of the gasifier [9]. By comparing the temperature profile of the process at the 32th minute, with the gas composition at the same duration of time, there is a drop in the production of gas and temperature. It occurred due to the bridging at that time. Onward from 35th minute of gasification, as the particles moving downward the gas production was increased.

Table 3. Gas composition produce from the pre-treatment methods and typical OPF gasification

Type of Feedstock	Average Gas composition (%)			
	CO	H_2	CH_4	CO_2
Typical Gasification of OPF	18.02	10.00	1.40	10.03
Gasification of Briquette	8.07	2.06	0.54	11.02
Gasification of Extensive Drying of OPF	16.48	4.03	0.91	11.15

The temperature profile of the gasifier during briquette gasification showed a maximum combustion zone temperature of 650-700°C. The low combustion zone temperature may be insufficient to break the heavy carbon rings in the syngas. Dynamic syngas carried tar formed during the gasification process. The evidence of tar formation was observed at some spots of gasifier pipeline. The drying and pyrolysis zone temperatures were in the range of 200-300°C and were satisfactory to remove excess moisture and volatiles from the briquettes simultaneously [10].

3.3 Gasification of extensive dried OPF

Gas profile and temperature profile of gasification of extensive dried OPF presented in Figure 4. Good compositions of syngas were produced during the gasification process. The average value of CO, H_2, CH_4, and CO_2 was 16.48%, 4.03%, 0.91%, and 11.15% respectively. The peak gas composition was achieved between 12-20 minutes of gasification operation. After the 20th minute, all syngas components started decreasing especially CH_4. After 28th minute concentration of CO_2 was start increasing and drop in others gaseous components which were sign of combustion start insisted of gasification at the end of process. This phenomenon caused due to the decrease in feedstock amount after being consumed in gasification. Whereas a constant amount of air was supplied to the system that shifted process of gasification toward the combustion due to change in air-fuel equivalence ratio that approaches to 1.0 [9]. The combustion zone temperature during the gasification of extensive dried OPF was found to be in the range of 450-650°C as shown in Figure 4b. The temperatures at drying and pyrolysis zones were found to be between 200-300°C respectively. No brigading was observed during the process.

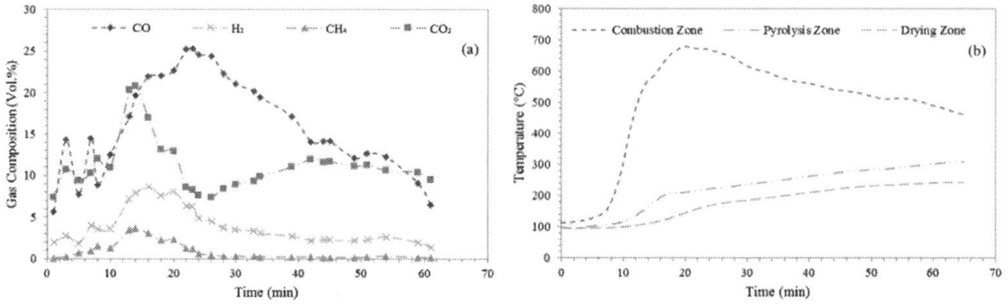

Fig. 4. (a) Syngas profile and (b) temperature profile produced during the gasification of extensive dried OPF

The comparison of the average gas composition of the two pre-treatment methods with the typical OPF gasification is presented in Table 3. The results of extensive drying method show a good potential in gasification operation as its average gas composition value are nearly approaching the average gas composition of typical OPF. For the briquette, the gas composition was lower due to the structure of the briquette that was not favourable and eventually creates bridging in the process due to the low density resulted from the broken.

4 Conclusions

Pre-treatment methods are very important in improving the physical conditions of the feedstock before being fed to the gasifier for the further gasification process. The experimental results show that bulk density of biomass can be increased by briquetting, however, the stability of the briquette structure depends on the formulation of the briquette. The extensive dried OPF improved the gasification process.

References

1. S.A. Sulaiman, S. Balamohan, M.N.Z. Moni, S.M. Atnaw, and A. O. Mohamed. Journal of Mechanical Engineering and Sciences **9**(1744-57 (2015)
2. F.M. Guangul, S.A. Sulaiman, M.N. Moni, S.M. Atnaw, and R.E. Konda. Asian Journal of Scientific Research **6**(2): 360-6 (2013)
3. M. Inayat, S. A. Sulaiman, A. Abd Jamil, F. M. Guangul, and S. M. Atnaw, "The study of temperature profile and syngas flare in co-gasification of biomass feedstock in throated downdraft gasifier," in *ICGSCE 2014: Proceedings of the International Conference on Global Sustainability and Chemical Engineering*, M. A. Hashim, Ed., ed Singapore: Springer Singapore, 2015, pp. 203-10.
4. J. S. Tumuluru, C. T. Wright, J. R. Hess, and K. L. Kenney. Biofuels, Bioproducts and Biorefining **5**(6): 683-707 (2011)
5. F.M. Guangul and S.A. Sulaiman. Asian Journal of Scientific Research **6**, 331-8 (2013)
6. ASTM, "ASTM Standard D3176-09 " in *Standard Practice for Ultimate Analysis of Coal and Coke*, ed. West Conshohocken: ASTM International, 2009.
7. C. Y. Sing and M. S. Aris. Journal of Applied Sciences **12**(24): 2621-5 (2012)

8. S. A. Sulaiman, R. Roslan, M. Inayat, and M. Yasin Naz. Journal of the Energy Institute 2017, in press
9. S. A. Sulaiman, M. Inayat, H. Basri, F. M. Guangul, and S. M. Atnaw. Journal of Mechanical Engineering and Sciences **10**(2): 2176-86 (2016)
10. Z. N. M. Moni and S. A. Sulaiman. Applied Mechanics & Materials **699**: 480-5 (2014)

Single and Multiple variables control using Tree Physiology Optimization

A.Hanif Halim*, and *I.* Ismail

Electrical and Electronic Engineering Department, Universiti Teknologi PETRONAS, 31750 Tronoh, Perak, Malaysia

Abstract. This paper presents the tuning of single-input single-output (SISO), and multiple-input multiple-output (MIMO) control system using Tree Physiology Optimization (TPO). TPO is a metaheuristic optimization algorithm that has a clustered diversification search strategy inspired from plant shoots growth. The clustered diversification is referred as tree branch and leaves. The exploration is amplified from roots growth counterparts. In the proposed method, each shoot from each branch search for possible solution in parallel and the fitness is evaluated based on all best values found by branch search. The proposed algorithm is also compared with deterministic gradient-free algorithm: Nelder-Mead simplex (NMS) and another metaheuristic algorithm: Particle Swarm Optimization (PSO). Results shown that TPO is able to find good PID parameters with lesser settling time for SISO and MIMO process. NMS is also able to find good PID parameters for SISO with lesser performance index, however not able to find better solution in MIMO control. PSO converged prematurely in SISO control and has high overshoot for MIMO control optimization.

1 Introduction

To date, the tuning of PID controller is still in extensive research. There are many PID controllers in practice that are not tuned properly [1]. Such tuning is normally carried out with trial and error method which is time consuming [2]. In addition, the multivariable process that involved controlling of many input and output variables are regularly practiced in industrial plant [2]. Therefore such high complexities of process need to be properly controlled. The multiple input-output control systems are also defined as MIMO control system. In contrast, the single input-output control system is denoted as SISO control. SISO PID control is available in most commercial controllers [3] and cited in many literatures as described in [4]. There are many tuning method proposed in the literatures. Among the most implemented method is metaheuristic algorithm. Metaheuristic algorithm is proven to be more efficient in computation time and performance [7].

*Corresponding author: abdul_g01772@ utp.edu.my

2 Optimization algorithm

There are essentially two types of optimization algorithms: deterministic and stochastic [8]. Deterministic algorithm has a repeatable path to find a solution by a given input. Deterministic algorithm is further divided into two approaches: gradient-based and gradient-free search method. Gradient-based algorithm uses gradient information to search for optimal solution, which may trapped into local optimum especially in multimodal problem. In contrast, gradient-free algorithm eliminate this disadvantage and able to search towards global optimum. Simplex algorithm is one example of gradient-free algorithm [10] and the algorithm is proven to be more effective than GA and Simulated Annealing (SA) [9]. Another section of optimization algorithm is the stochastic algorithm, which is also denoted as heuristic or metaheuristic algorithm. Metaheuristic algorithm has unique properties of search that incorporate two components: diversification and intensification [8]. Diversification is to generate a diverse solution in order to explore the search space. Intensification is to focus the search space in local region by exploiting the information of current good solution of respective region. Some examples of metaheuristic algorithms are GA, SA, PSO, FFA, and TPO. In this paper, three optimization algorithms for solving SISO and MIMO control system will be discussed. The algorithms are Nelder-Mead Simplex algorithm (NMS) [11], Particle Swarm Optimization (PSO) [12], and Tree Physiology Optimization (TPO) [16]. The characteristic of each algorithm is summarized in Table 1.

Table 1: Overview of proposed algorithm

Num	Algo. (year)	Main features
1	NMS (1965)	Triangular Search driven with 3 main steps: reflection, contraction and expansion [11,12]
2	PSO (1995)	Swarm-based algorithm. Solution is attracted to local and global best in each iteration [9, 13].
3	TPO (2013)	Plant growth inspired algorithm with shoots as search agent. Shoots consist of defined leaves and branches for parallel exploration. The search is amplified by root-shoot correlation [16 – 18].

The performance criterion for fitness evaluation of each algorithm is based on integral of absolute error (IAE) with response constraint as shown in (1).

$$J = \int |e|dt + \int_{0}^{t_{max}} (f_{up} - x_i) + \int_{t_1}^{t_{max}} (x_i - f_{low}) \tag{1}$$

Whereas f_{up} and f_{low} are continuous function that define the upper and lower border of shaded region as in Fig. 1.

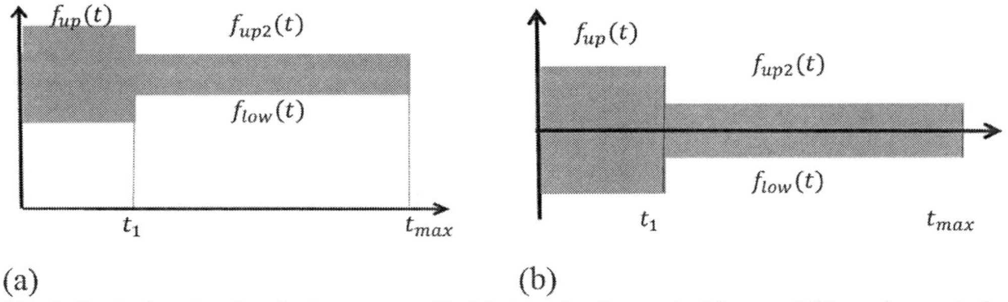

Fig. 1: Desired region for plant response with (a) step of active control loop and (b) passive control loop

The constraint f_{up} consists of two borders (f_{up} and f_{up2}). x_i is the plant response at each iteration. Based on the figure, the settings of $f_{up}(t), f_{low}(t), t_1$ and t_{max} are specified by user. The performance criteria of tuned control parameters are based on response constraint and IAE, since the response constraint may assist the estimation to desired response region. In this paper, the response setting is set with $f_{up} = 1.55$; $f_{up2} = 1.20$; $f_{low} = 0.8; t_1 = 1.6$; $t_{max} = 20$

3 Application on SISO optimization

The algorithm performance is compared with SISO of DC servo motor for position control as presented in [13] with transfer function defined in (2). The block diagram of closed loop PID control with online optimization is shown as in Fig. 2. The limitation of each controller parameter is set to [0,5].

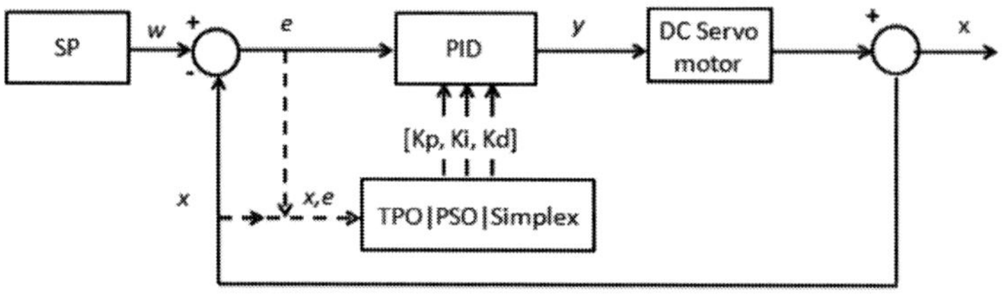

Fig. 2: Block diagram of closed loop controller tuning using proposed algorithm

$$G_{DC\ Servo\ Motor}(s) = \frac{0.1}{0.01s^3 + 0.09s^2 + 0.21s} \tag{2}$$

Based on Fig. 2, the PID parameters are updated by proposed the algorithm in each iteration. The control performance is measured based on plant output (x), that used to evaluate the constraint area and error value (e) that determine the IAE criteria.

3.1 Simulation result

Each algorithm is set according to their original setting. The number of iteration for PSO and NMS is set to 30. For TPO, the number of branches and shoots leaves is set to 10 and 20 respectively. The step response of each algorithm is depicted in Table 2 and Fig. 3. The best value is highlighted with bold. It is observed that NMS has the lowest J value (IAE with response constraint). All algorithms show merely the same rise time. TPO has the shortest settling time and converge in less than ±1% of set value as shown in Figure 3b. PSO has the lowest overshoot, however with longer time characteristics

Table 2: Optimum PID based on each method

Method	Kp	Ki	Kd	Rise time	Settling time	Overshoot %	*J*
NMS	4.701	0.0384	0.0301	**1.20**	5.00	23	**1.09**
PSO	3.577	0.158	0.005	1.22	7.60	**16**	1.27
TPO	3.156	0.001	0.001	1.21	**3.80**	20	1.21

(a)

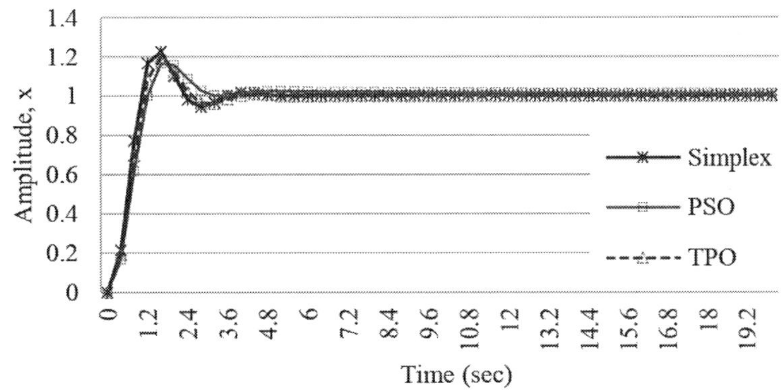

(b)

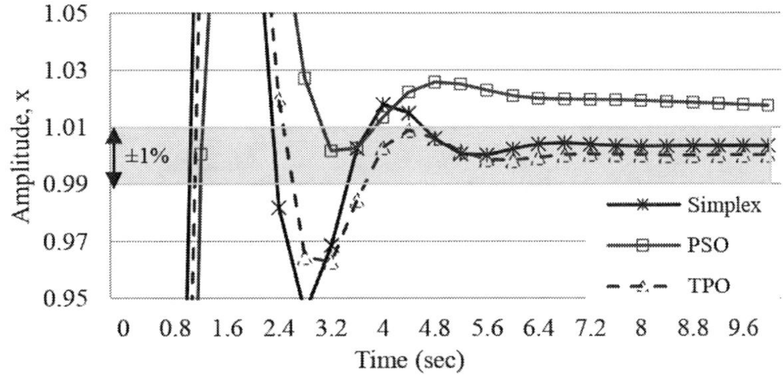

Fig. 3: (a) Optimized step response of each algorithm,
(b) Response of each method within ±1% tolerance

4 Application on MIMO optimization

For multiple variable processes, the performance of each algorithm is compared with Wood and Berry distillation column control [14]. The control strategy for multivariable system is presented by Wood and Berry, which is to reduce the interaction between the control of the overhead and bottom products [14]. The input-output variable for MIMO system is expressed with output: $Y(s) = \begin{bmatrix} Y_1(s) \\ Y_2(s) \end{bmatrix}$ and input: $U(s) = \begin{bmatrix} U_1(s) \\ U_2(s) \end{bmatrix}$. The inputs of this process are the reflux and steam flow rates with outputs of overhead and bottoms composition.

$$G(s) = \begin{bmatrix} G_{11}(s) & G_{12}(s) \\ G_{21}(s) & G_{22}(s) \end{bmatrix} = \begin{bmatrix} \dfrac{12.8e^{-s}}{1 + 16.7s} & \dfrac{-18.9e^{-3s}}{1 + 21s} \\ \dfrac{6.6e^{-7s}}{1 + 10.9s} & \dfrac{-19.4e^{-3s}}{1 + 14.4s} \end{bmatrix} \tag{3}$$

Two PID controllers with i^{th} vector representation are designed for this problem as shown in (4). The vector is updated in each number of iteration for each proposed optimization algorithm.

DOI: 10.1051/matecconf/201713103017

$$C_{1,2\,i} = \left[K_{P1_i}, K_{I1_i}, K_{D1_i}, K_{P2_i}, K_{I2_i}, K_{D2_i}\right] \tag{4}$$

Whereas $C_{1,2\,i}$ consists of C_1 that represents PID of K_{P1}, K_{I1}, K_{D1} and C_2 for K_{P2}, K_{I2}, K_{D2}. Figure 4b shows the MIMO control structure designed for this work.

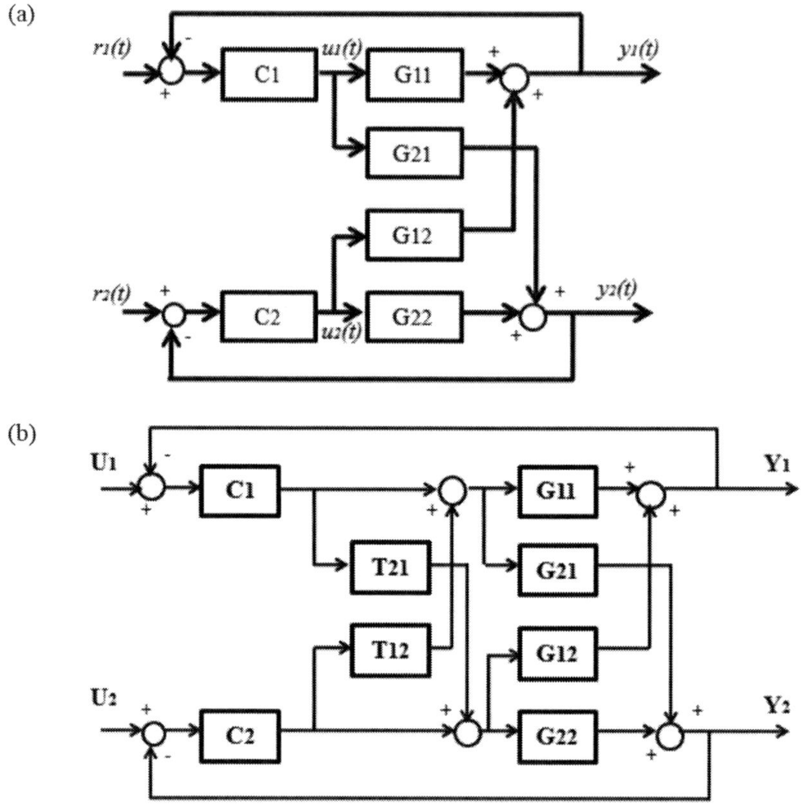

Fig. 4: (a) Block diagram of decentralized control of MIMO system,
(b) MIMO system with decoupler

In this paper, a simplified decoupling method is implemented. The best response is obtained by tuning of decoupler and PID controllers. A decoupling matrix, *T* is evaluated with transfer matrix. The controller parameters are tuned based on predefined elements of T. The elements of simplified decoupling are $T_{11} = T_{22} = 1$ and off-diagonal with, $T_{12} = -\frac{G_{12}}{G_{11}}$ and $T_{21} = -\frac{G_{21}}{G_{22}}$. The range of each controller parameter is set according to min-max value Kp: [-1.5,1.5], Ki: [-0.15,0.15] and Kd: [-0.2,0.2] and constraint setting with $f_{up} = 1.55$; $f_{up2} = 1.02$; $f_{low} = 0.8$; $t_1 = 20$; $t_{max} = 100$

4.1 Simulation result

The setting of each algorithm is set similar to SISO problem. The parameters of PID controller obtained by each algorithm and control loop are depicted in Table 2. The step response of each algorithm is shown in Table 3 and Fig.5.

DOI: 10.1051/matecconf/201713103017

Table 2: Optimum PID based on each method for MIMO control

Method	K_{P1}	K_{I1}	K_{D1}	K_{P2}	K_{I2}	K_{D3}
NMS	0.3010	0.0633	0.0143	-0.0383	-0.0278	-0.0502
PSO	0.3052	0.0230	0.0040	-0.1324	-0.0394	-0.0361
TPO	0.3266	0.0413	0.3662	-0.1277	-0.0140	-0.0733

Table 3: Response properties of each algorithm

	Step y_{11}				Step y_{22}			
Method	Rise time	Settl. time	% Os	J	Rise time	Settl. time	% Os	J
NMS	5.0	28	25.0	11.5	8.0	38	**2.7**	12.5
PSO	4.5	28	40.0	11.2	8.0	**22**	35.0	11.7
TPO	**4.0**	**20**	**20.0**	**7.0**	8.0	**22**	6.5	**10.3**

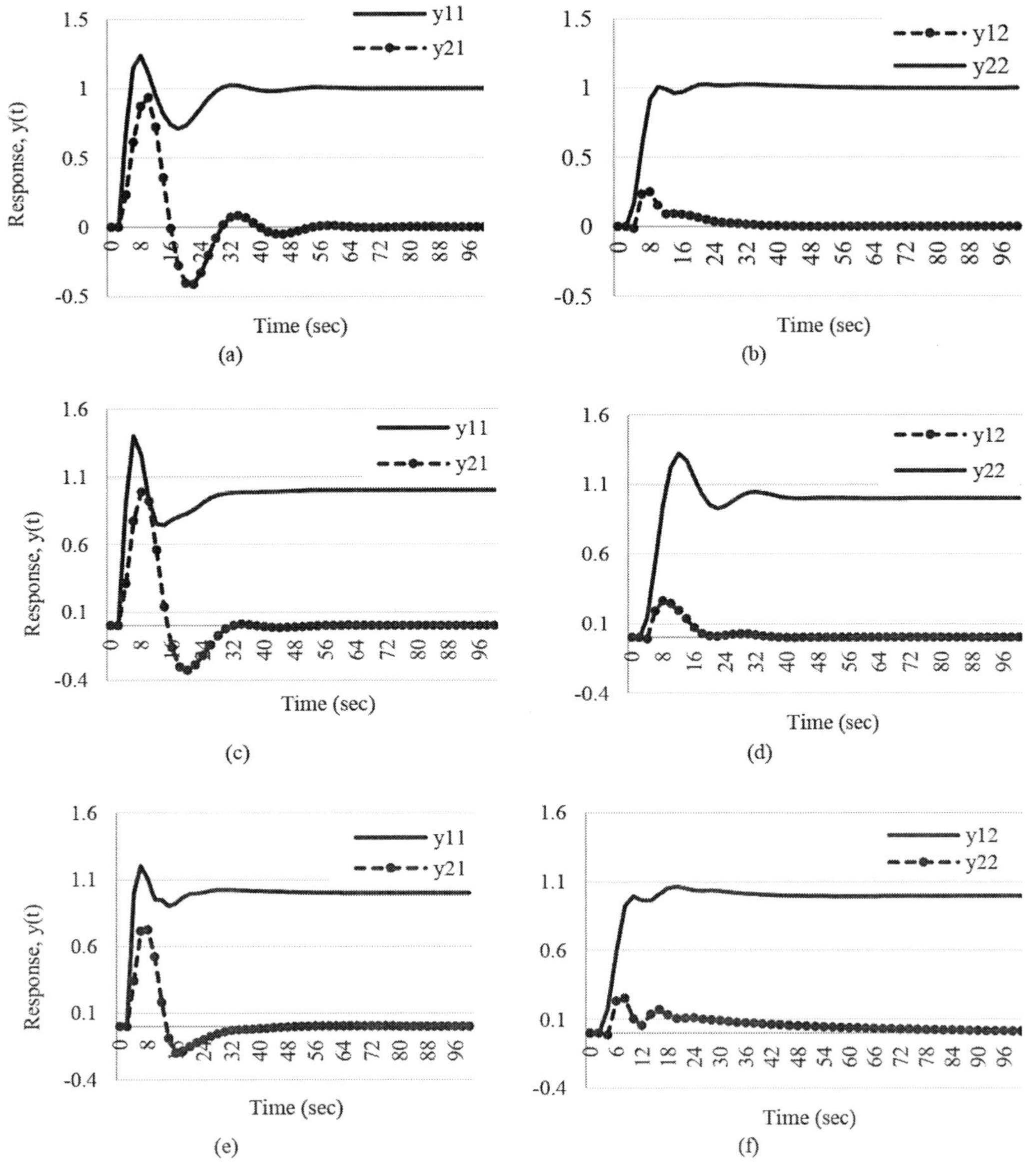

Fig. 5: Unit step response using NMS for (a) $y11$ and $y21$, (b) $y12$ and $y22$, PSO for (c) $y11$ and $y21$, (d) $y12$ and $y22$ TPO for (e) $y11$ and $y21$, (f) $y12$ and $y22$

Based on the response properties, PSO method has the highest overshoot. The performance index as in the table shows that TPO outperformed other algorithms in both steps y_{11} and y_{22} and followed by PSO

5 Conclusion

In this paper, the performance of step response for single and multiple variable PID controls for Nelder-Mead simplex, PSO and TPO are compared. For SISO model, TPO outperformed other algorithm in convergence with lesser than ±1% tolerance. The performance of the algorithm on MIMO control optimization is verified with Wood and Berry distillation column control. TPO shows the best performance index with lower IAE and integral response boundaries. It is also observed that NMS has lower overshoot in step response of y_{22}.. The optimum solution of TPO can be improved further by higher number of branches and shoots leaves, as current settings are only 10 branches search and 20 leaves search (which correspond to 10x20 searches). The search of each branch and leaves are running in parallel; this is to ensure a better diversification and hence also better intensification towards true solution, thus the computational cost is reduced significantly. NMS algorithm is able to search towards global optimum due to its triangularity search and driven with three simplex steps that compares and adapts with better performance during each step. The search of PSO algorithm is driven by the particle location and speed. However PSO search may end up with premature convergence as the algorithm is not able to search diversely in the search space as observed in SISO response. The findings from this paper suggest a necessity of comparison studies with more practical MIMO process and with increased complexity.

The research was supported by Universiti Teknologi PETRONAS. The author would like to thank the university organisation for the support. The author would also like to thank the reviewers for thoughtful comments that added to the efficacy of this paper.

References

1. Marcio F. Miranda, K. G. Vamvoudakis, "Online optimal auto-tuning of PID controllers for tracking in a special class of linear systems", *American Control Conference (ACC)*, Boston, pp. 5443-5448 (2016).
2. B. L. G. Costa, J.P.L.S., Almeida, B. A. Angelico, "Application of Optimzation Heuristics in Tuning Decentralized PID Controllers". *INDUSCON*, pp 1-8 (2012)
3. S. Boyd, M. Hast, K. Johan Åström, " MIMO PID tuning via iterated LMI restriction", Int. J. of Robust and Nonlinear Control, pp 1718-1731 (2015)
4. S. Sommer, A. Kienle., " Auto-Tuning of Multivariable PID Controllers Using Iterative Feedback Tuning", Automatisierungstechnik **60**, pp 20-26, Oldenbourg-Wissenschaftsverlag (2012)
5. K. J Åstrom, T. Hägglund, "*PID Controller: Theory, Design, and Tuning*", ISA Research Triangle Park, NC (1995)
6. W. L. Luyben., "Simple method for tuning SISO controllers in multivariables systems", Industrial & Engineering Chemistry Process Design and Development, **25**(3), pp 654-660 (1986)
7. O. Roeva, T. Slavov, " PID Controller Tuning based on Metaheuristic Algorithms for Bioprocess Control", Biotechnology & Biotechnological Equipment, **26**, 5, pp 3267-3277 (2014)

DOI: 10.1051/matecconf/201713103017

8. X. S. Yang, "*Nature-Inspired Metaheuristic Algorithms*" 2nd Edition, Luniver Press (2010)
9. J. Mange, S. Pace, A. Dunn, "Optimization of PID Controller Parameters for Automated Ground Vehicle Control on Dynamic Terrain", *Int'l Conf. Scientific Computing, CSC'15,* pp 3-8 (2015)
10. G. B. Dantzig, "Origins of Simplex Method", Technical Report SOL **87**-5, Systems Optimization Laboratory (1987)
11. J. A. Nelder., R. Mead, "A simplex method for function minimization", Computer Journal, **7**, pp 308-313 (1965)
12. J. Kennedy, R. Eberhardt, "Particle swarm optimization*", proc. of the IEEE Int. Conf. on Neural Networks*, Piscataway, NJ, pp 1942-1948 (1995)
13. Aishwarya Y., S. Wadhwani, A. K. Wadhwani, "Particle Swarm Optimization Based Intelligent Tuning of PID Controller for DC Servo Motor Control", Int. Journal of Scientific Engineering and Technology Research, IJSETR, ISSN 2319-8885, **03**, Issue 15, pp 3106-3110 (2014)
14. R. K. Wood, M. W. Berry, "Terminal composition of a binary distillation column", Chemical Engineering Science, **28**, pp. 1707-1717 (1973)
15. A. Hanif Halim, I. Ismail, "Nonlinear plant modeling using neuro-fuzzy system with Tree Physiology Optimization", *IEEE Student Conference on Research and Development (SCOReD)* (2013)
16. A. Hanif Halim, I. Ismail, " Online PID controller tuning using tree physiology optimization", *International Conference on Intelligent and Advanced Systems (ICIAS)*, pp 1- 5 (2016)
17. A. Hanif Halim, I. Ismail, "Comparative Study of Meta-heuristics Optimization Algorithm using Benchmark Function", International Journal of Electrical and Computer Engineering, IJECE, **7**, No 2, (2017)

Tree Physiology Optimization tuning rule for Proportional-Integral control

A.Hanif Halim*, and *I.* Ismail

Electrical and Electronic Engineering Department, Universiti Teknologi PETRONAS, 31750 Tronoh, Perak, Malaysia

Abstract. This paper presents a tuning correlation for Proportional-Integral (PI) controller parameters using Tree Physiology Optimization algorithm (TPO). TPO is a metaheuristic algorithm that has parallel search strategy inspired from plant growth system. The parallel search is referred to the plant branches and leaves that able to diversify the search space and therefore finding promising results in much faster iterations. The exploration is amplified from roots growth counterparts. In the proposed method, a tuning correlation of PI-controller using TPO is established and the performances of correlated parameters are compared with Ciancone correlation. The idea of correlation tuning is to reduce the optimization time consumption and engineering efforts in controller tuning. The results shown that the tuning correlation of TPO is able to find good PI parameters with lesser error.

1 Introduction

PI and PID controller is often used in industries and is acknowledged as an effective control system in numerous controlling applications. The unique properties of each controller part: proportional, integral and derivative made the controller as a standard controlling tool that should be able to reduce the errors with good set point tracking and disturbance rejection within a desirable tolerance. The tuning of control parameters is depending heavily on plant properties such as order of the system, plant gain, time constants and dead time [1]. In practice, other factor such as process interruption is also included in finding the optimum PID parameter. To date, there are numerous tuning rule and optimization approach of such problems. This includes the traditional methods: Ziegler-Nichols (ZN) method that based on integral of absolute error index (ISE) [2] Tyreus-Luyben (TL) [3], and Ciancone correlation [4]. Other proposed optimization approaches are using metaheuristic algorithm such as Particle Swarm Optimization (PSO), Genetic Algorithm (GA), and many more [5-9]. However, optimization search with certain defined of iteration number is time consuming and improper optimization parameter may result in poor controller performance [10]. In practice, it is sufficient to obtain the controller parameter that able to perform within defined response characteristic without performing an optimization task. In this paper, a tuning correlation of PI-controller parameter based on

*Corresponding author: abdul_g01772@ utp.edu.my

TPO algorithm is proposed. The correlation is defined by each controller parameters: proportional gain (Kp) and Integral time (Tn).

2 Proposed tuning for PI control

The proposed tuning rule for PI-control is obtained from TPO search algorithm. TPO is a metaheuristic algorithm developed by A. Hanif Halim and I. Ismail [12]. The algorithm is inspired from a plant growth that consists of two sections: shoots and roots. The shoots section expands their branches and leaves towards light and convert light into Carbon by photosynthesis process. The Carbon gain is then supplied downwards and utilized by the root system for root-growth in search for nutrients. The nutrient gain is then supplied upwards and used by the shoot system for shoots-growth in extension towards light. This relationship of both plant parts towards better growth had inspired the idea of TPO algorithm. The search of TPO has its uniqueness due to the branches system with many shoots that able to search towards optimum in parallel. This is synonym to the normal tree physiological concept. This unique attribute show a better probability of solution due to the ability of more search effort that driven by the best solution found within branch and between branches. The search become broader as the random motion included in each shoot propagation. Several studies on TPO performance and comparison with other algorithm have shown the ability of the algorithm to find the global optimum efficiently. In online PID controller optimization, TPO outperformed other algorithms with faster response and lesser overshoot [9]. TPO is also performed significantly better with lesser computation time than other meta-heuristic algorithm for nonlinear optimization [11, 12]. The idea of tree physiological system is devised into four equations: shoots propagation, carbon gain, root elongation and nutrient uptake. The shoot propagation towards light is shown as in (1):

$$S_i^{kj} = S_i^{kj} + \left(S_{gbest} - S_i^{kj}\right) + \beta N_i^{kj} \tag{1}$$

With S_i^{kj} as the current shoot at ith iteration, of kth shoot and jth branch, S_{gbest} is the global best shoot from all branches, N_i^{kj} is the current nutrient supply from root, and β is the dynamic factor of shoots extension. Each shoot will undergo the photosynthesis process that converts light into carbon. The carbon gain is equivalent to deviation of individual branch best with individual shoot as in (2):

$$C_i^{kj} = \theta^i \left(S_{popbest} - S_i^{kj}\right) \tag{2}$$

C_i^{kj} is current carbon gain, $S_{popbest}$ is best shoot from current branch. The parameter θ is a power law formula that reduces the randomness as iteration increases in a monotonically decreasing function. C_i amplifies the root elongation in search for more nutrients (3):

$$r_i^{kj} = r_i^{kj} + \alpha\ C_i^{kj}\ \epsilon \tag{3}$$

r_i^{kj} is current root, α is an absorption constant, ε is the random numbers drawn from uniform distribution. The nutrient uptake can be assumed as a proportional factor to the root elongation. It is expressed in the same formulation as carbon gain

$$N_i^{kj} = \theta\left(r_i^{kj} - r_{i0}^{kj}\right) \tag{4}$$

These formulations run iteratively until a significant optimum is found by the shoots counterparts. The computation of search process depends on the number of shoots, number of branches and number of iteration. The correlation of optimized PI controller parameter with TPO algorithm is carried out with first-order-with-dead-time transfer function as in (5), since many processes can be represented by this function [10].

$$G_s = \frac{K_C e^{-\theta s}}{1 + \tau s} \tag{5}$$

The correlation is established with parameter from fraction-dead-time that is defined with $\theta/(\theta + \tau)$ [15].

3 Optimization formulation

The optimum controller parameter determined from TPO is carried out by varying θ, τ: $[2,16]$ and K_C: $[0.2,3.2]$. The performance criterion to evaluate the performance of the algorithm is based on integral of absolute error (IAE) with conditional programming of response constraint as shown in (1) and Fig. 1.

$$IAE = \int |e| dt \tag{6}$$

Based on Fig. 1, f_{up} and f_{low} are the upper and lower border of shaded region. f_{up} consists of two borders (f_{up} and f_{up2}). y_i is the plant response at each iteration.

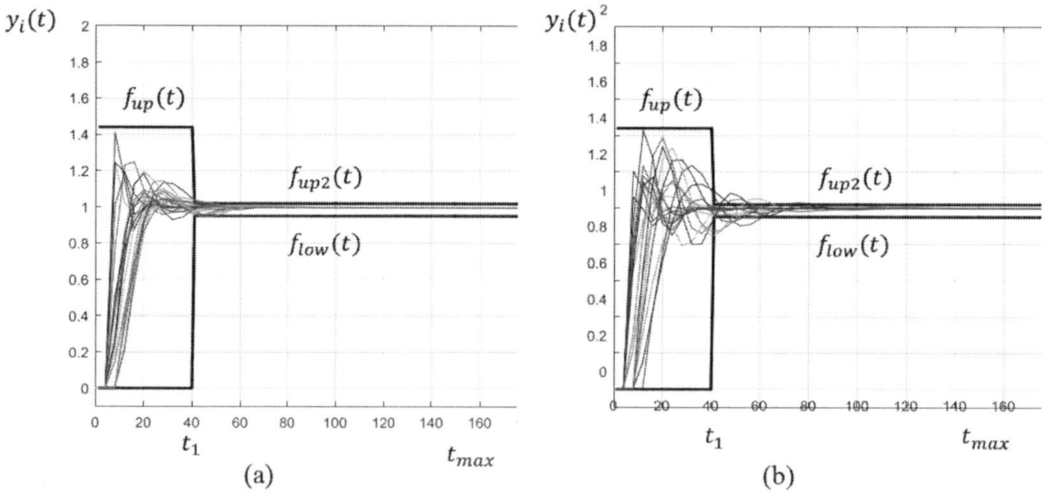

Fig. 1: Desired region for plant response with (a) lower Kc (b) higher Kc

As observed in the figure, the settings of $f_{up}(t)$, $f_{low}(t)$, and t_1 can be specified by user. In this paper, the response setting is set with $f_{up} = 1.44$; $f_{up2} = 1.08$; $f_{low} = 0.85$; $t_1 = 20\%$; $t_{max} = 100\%$. The algorithm run iteratively until it covers all the response value defined for θ, τ, Kc. The parameter setting for TPO in this study is set according to Table 1.

Table 1: Parameter setting for TPO

Number of leaves (population)	10
Number of branch (Sub-population)	5
Number of iteration	10
Dynamic factor, β	50
Randomization factor, θ	0.9
Absorption factor, α	0.3

The defined number of branch and leaves show that TPO algorithm is able to search in parallel with possibility solution of 5 X 10 in each iteration. A regression correlation between proportional gain and integral time with fractional dead-time is analyzed and depicted in Fig. 2 and Fig. 3 respectively.

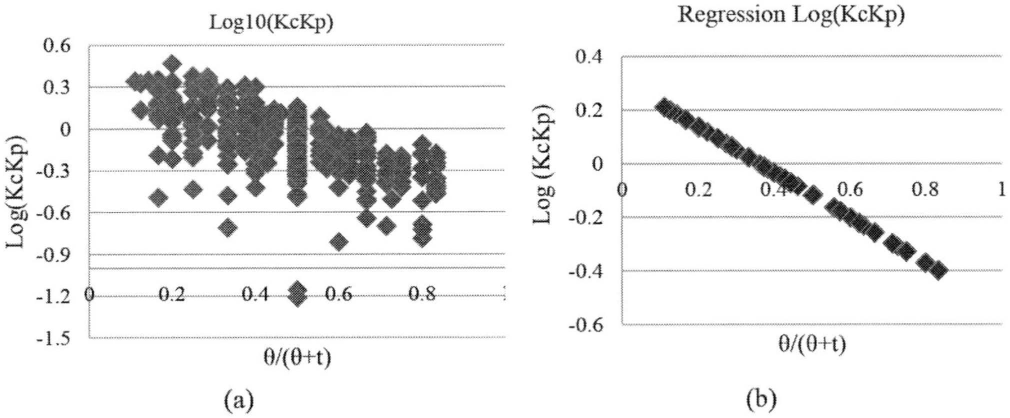

Fig. 2: Evaluated TPO results for Kp (a) and regression of Kp (b)

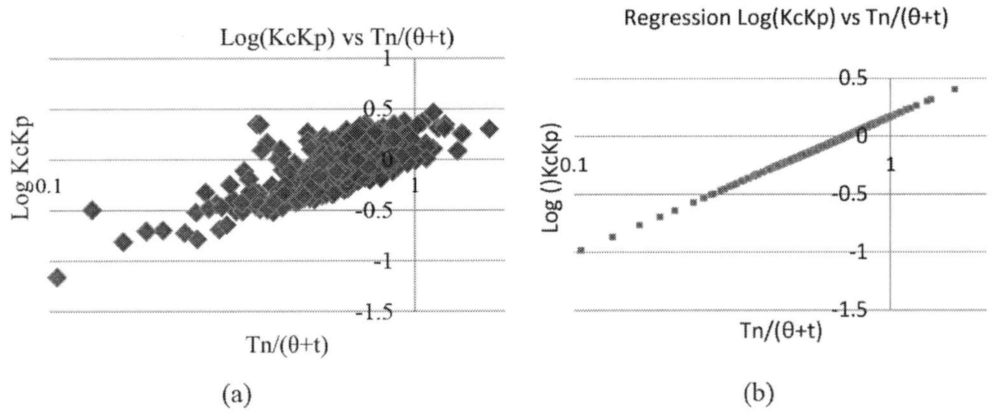

Fig. 3: Evaluated TPO results for Tn (a) and regression of Tn (b)

4 Performance comparison

The correlation performance is compared with Ciancone correlation as defined in [10]. Fig. 4 shows the comparison of step response from TPO and Ciancone correlation. The summary results of each method by different plant parameters are tabulated in Table 2.

DOI: 10.1051/matecconf/201713103018

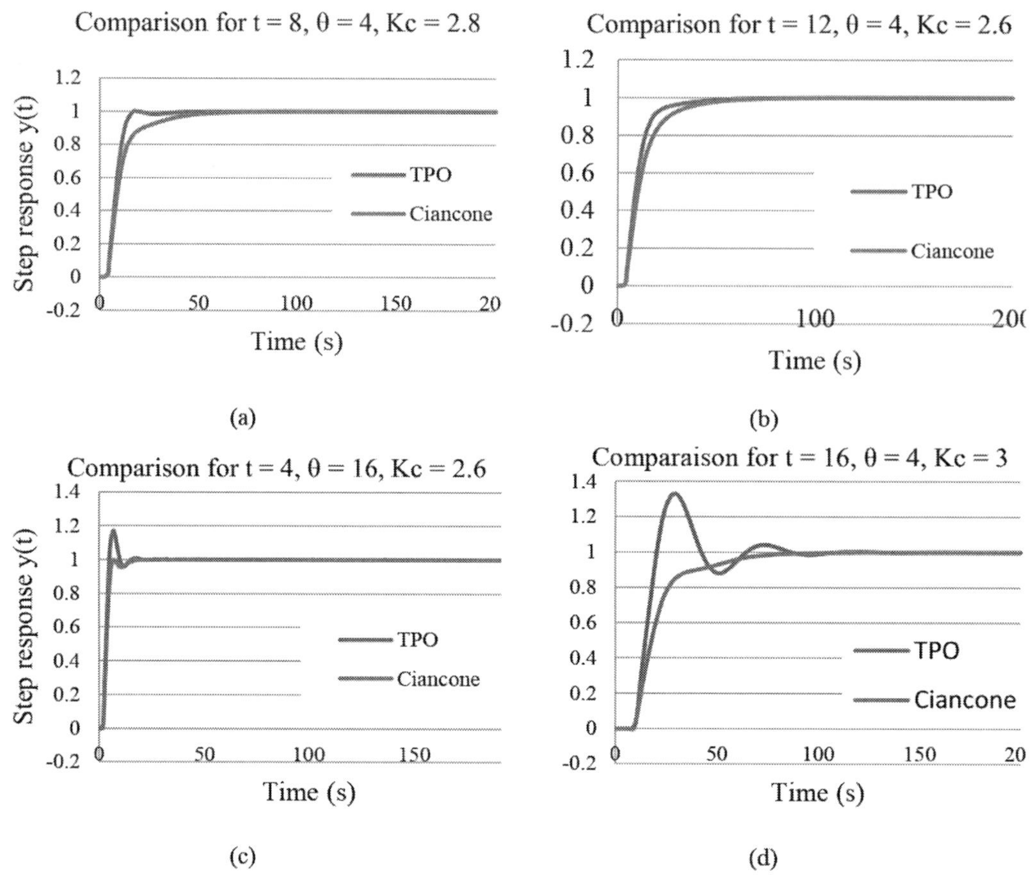

Fig. 4: Comparison of step response for TPO and Ciancone correlation by different time constant and plant gain

Table 2: Optimum PI of each method

Method	Kp	Ki	IAE
t = 8, θ = 4, Kc = 2.8			
Ciancone	0.3214	0.0298	11.9846
TPO	3.577	0.158	8.6475
t = 12, θ = 4, Kc = 2.6			
Ciancone	0.388	0.0276	13.935
TPO	0.477	0.034	11.312
t = 4, θ = 16, Kc = 2.6			
Ciancone	0.147	0.0144	26.783
TPO	0.163	0.0239	37.431
t = 16, θ = 4, Kc = 3.0			
Ciancone	0.286	0.0317	22.543
TPO	0.393	0.0604	22.065

Findings from Fig. 4 and Table 2 shows that TPO has slightly better results in several cases. Ciancone correlation resulted a step performance without overshoot compared to TPO. The performance of TPO can be significantly improved by increasing the number of search space, which is the number of branches and leaves. This can lead the search agents to discover more promising solution as the search space become broader. Another possibility of improvement is by tightening the constraints of step response as in Fig. 1.

5 Conclusion

In this paper, the correlation of optimum PI controller parameters and first-order-with-dead-time transfer function is established. The correlation is carried out by running TPO algorithm for wide range of dead time, time constant and plant gain. Based on the regression model of each controller parameter with fraction dead-time, a correlation for optimum step response with defined constraints can be evaluated. The performance of evaluated TPO is compared with Ciancone correlation by selecting several step properties. Based on the comparison, Ciancone correlation did shows good results without overshoot, but with slightly higher absolute error (IAE) for three cases. The performance of TPO can be further improved by increasing the number branches and leaves for broader exploration search, and also by tightening the constraints for step response properties. The findings from this paper suggest a necessity of more comparison study by increasing the search capability of TPO and to cover wider range of each plant parameters that include dead-time, time constant and plant gain. The findings also implies a necessity of finding a tuning correlation of set point tracking for PID controller and for both type (PI and PID controller) with disturbance rejection.

The author would like to thank Universiti Teknologi PETRONAS for the support of the paper.

References

1. K. Ogata, "*Modern Control Engineering*",5th Edition Prentice Hall (2010).
2. J. G. Ziegler, N. B. Nichols, "*PID Controller: Theory, Design, and Tuning*", ISA Research Triangle Park (1995).
3. L. Luyben., "Simple method for tuning SISO controllers in multivariables systems", Industrial & Engineering Chemistry Process Design and Development, **25**(3), pp 654-660 (1986)
4. Marlin T. E., "*Process Control*", 2nd Edition, McGraw-Hill (2000)
5. P. G. Medewar, R. K. Munje, "PSO based PID controller tuning for PMDC motor", *Int. Conf. on Energy Systems and Application* (2015)
6. P. Ouyang, Vangjel Pano, "Comparative Study of DE, PSO and GA for tuning Position Domain PID Controller", *ROBIO* (2014)
7. Kelvinder S., Pandian V., Irraivan E., Ramani K., "PID Tuning of Servo Motor using Bat Algorithm, 19th *Int. Conf. on Knowledge Based and Intelligent Information and Engineering Systems* (2015)
8. Omar B., "Continuous Firefly Algorithm for Optimal Tuning of PID Controller in AVR System", Journal of Electrical Engineering, **65**, No. 1, pp 44 – 49 (2014)
9. A. Hanif Halim, I. Ismail, " Online PID controller tuning using tree physiology optimization", *ICIAS*, pp 1- 5 (2016)
10. T. Marlin, "*Process Control, Designing Processes and Control Systems for Dynamic Performance*", 2nd Edition, McGraw-Hill (2015)
11. A. Hanif Halim, I. Ismail, "Comparative Study of Meta-heuristics Optimization Algorithm using Benchmark Function", IJECE, **7**, No 2 (2017)
12. A. Hanif Halim, I. Ismail, "Nonlinear plant modeling using neuro-fuzzy system with Tree Physiology Optimization", *(SCOReD)* (2013)

A Preliminary Study on the Thermal Performance of a Ventilated Honey Cassette for Stingless Bees

Firdaus Basrawi [1,*], *A Rahman* A Hamid [2], *Rizduan* Bahari [2], *Mohd Najib* Mohd Noordin [3], and *Mohd Hazwan* Yusof [1]

[1]Meliponini Engineering Laboratory (MepEL), Energy Sustainability Focus Group (ESFG), Faculty of Mechanical Engineering, Universiti Malaysia Pahang
[2]Qlulut Global, NO.48 Lrg 5 Taman Kampong, Padang Damai, Tanjung Pasir, Jalan Bk Goh,25200 Kuantan, Pahang
[3]RISDA, Kuala Kangsar District 33000 Kuala Kangsar, Perak

Abstract. Stingless bees are very sensitive to the changes of surrounding temperature. A report stated that fertility rate in broodcell is 0% when the broodcell temperature is higher than 34°C or lower than 26°C. In addition, propolis made honey pot in a honey cassette also could melt when temperature is high. Therefore, the objective of this research is to investigate the temperature profile of a ventilated honey cassette exposed to outdoor conditions, and to evaluate the temperature regulation in the hive using the ventilated honey cassette. To achieve these objectives, two hives with conventional and ventilated honey cassettes were exposed under sun light in cloudy and sunny day. Temperature inside each hive was measured at 3 points and was compared. It was found that there is no significant different between the hives when both hives were exposed under direct sunlight in a cloudy day. However, two significant improvements were found for ventilated hive in sunny day. It could help to reduce temperature at wall of honey cassette consistently below 33°C. This could avoid the melting of propolis around the ventilated wall area. Furthermore, it could facilitate in better temperature reduction as compared to the conventional honey cassette. However, further study when there is a colony inside the hives must also be conducted to validate the results.

1 Introduction

Stingless bees predate the stinging honey bee, and was found in the tropical earth for over 65 million years. They have diversify species in tropical and subtropical areas. Diversity of stingless bee, especially in Indo-Malayan region makes classifying them is difficult [1]. Although, stingless bees produce honey as honey bee, produced honey is less, and they do not use water to cool their nest [2]. Temperature of hive for more than 34°C causes 0% fertility rate because of the death of pupae in broodcell [3]. Another study suggested that the

* Corresponding author: mfirdausb@ump.edu.my

temperature inside a hive must be around 31-32°C for Trigona denoiti [4]. Thus, in natural hives, they live in shaded area inside hole in trees that could provide insulation, and this makes temperature inside the hive is relatively constant [5]. Hence, cooling methods by photovoltaic-thermoelectric and green roof were proposed to prevent overheating of hives when subjected to direct sunlight conditions.

For artificial hives, it is common to have separated compartment for bees to store honey and beebread. This part is usually called as topping or honey cassette. Temperature regulation is also important for this part because, when the temperature is too high, propolis made honey pot could melt and honey cassette will be flooded with honey. Air ventilation for this part is also important because there is activity of fanning the filled honey pot to reduce water content in honey. Therefore, if this space is ventilated, it could help in the honey drying process.

One of the possible cooling and ventilation technique is to provide holes to the honey cassette with wire mesh to avoid treats from other insects. From literature reviewed, there is no study on the ventilated honey cassette, and therefore the objectives of this study are:

- to investigate the temperature profile of the ventilated honey cassette exposed to outdoor conditions, and
- to evaluate the temperature regulation in the hive using ventilated honey cassette.

The study was done by exposing hives with different honey cassettes to outdoor condition. One of the hive was equipped with ventilated honey cassette, another one was equipped with a conventional honey cassette. The temperature profile inside the hives was recorded and analysed to evaluate the feasibility of ventilated honey cassette for stingless bee hives.

2 Materials and Method

An experiment was conducted at Universiti Malaysia Pahang, Pekan campus (coordinates: 3.5437° N, 103.4288° E, GMT + 8). Figure 1 shows the overall experimental setup. Two hives were exposed to outdoor conditions without any shades. They have the same broodcell and roof part with identical material, size, and colour. This is to ensure only honey cassettes part is the affecting parameter.

Fig. 1. Overall setup, with thermocouples measuring temperatures inside honey cassette and brood cell compartment of each hives, connected to a temperature data logger.

One hive was fitted with a ventilated topping as shown in Figure 2, and another one with the conventional one. Both honey cassettes have the same dimensions in terms of inner area. All cracks and gaps were seal with a tape to ensure air could only enters through entrance to simulate the real condition. It should be noted that there is no bees colony inside the hives.

Fig. 2. Hive with ventilated honey cassette.

Temperature were recorded at 3 points for each hive as shown in Figure 3, which are inner wall of both honey cassettes, middle hole of the honey cassette, inside broodcell. The measurement was recorded by a data logger (Lutron BTM-4208SD) shown in Figure 4, which was then transferred into computer for further analysis. K-type thermocouple was used as the sensors.

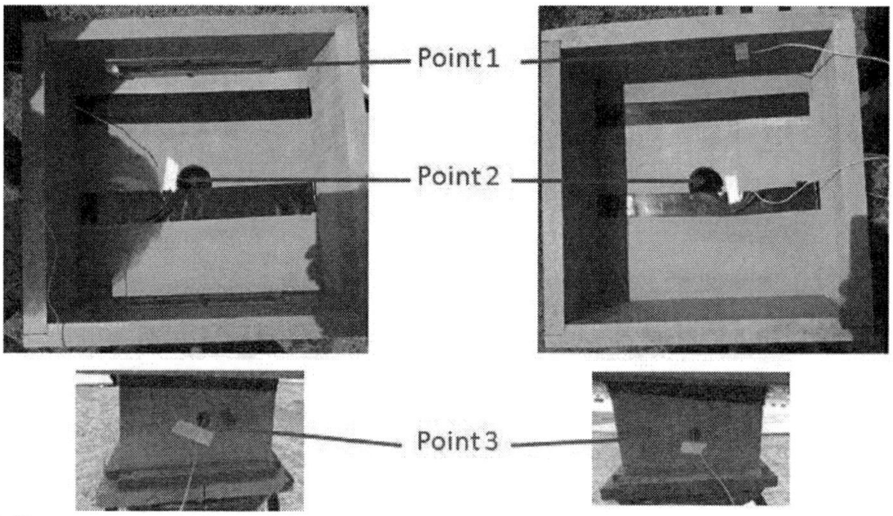

Fig. 3. Temperature recording points.

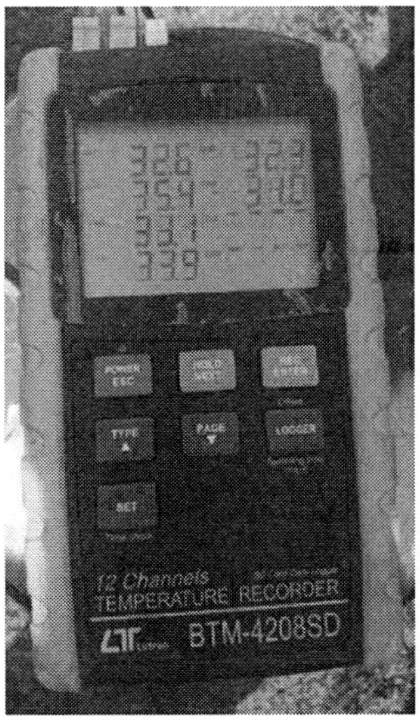

Fig. 4. Temperature recorder used.

Experiments were conducted 2 times, one time in a sunny day, and one time in a cloudy day. In addition, sampling time was 1 minute. Since this is a preliminary study, environment data including irradiance, humidity, ambient temperature, and wind speed were not recorded.

3 Results and Discussion

First experiment was conducted in a cloudy day, and the result is shown in Figure 5. Red lines show temperature for ventilated hive, whereas gray lines show temperature for conventional hive. Different line style show different point of measurement as shown in the figure. It was found that temperature at all points started to rapidly increase around 11:00 until 12:00, then temperature constantly increased until in the end of the experiment. Temperature varies in the range of 27-31°C, and this range is comfortable for bees. There is no significant difference in temperature between conventional and ventilated hive in all points. Thus, it can be concluded that the ventilated honey cassette does not improve thermal regulations under cloudy condition.

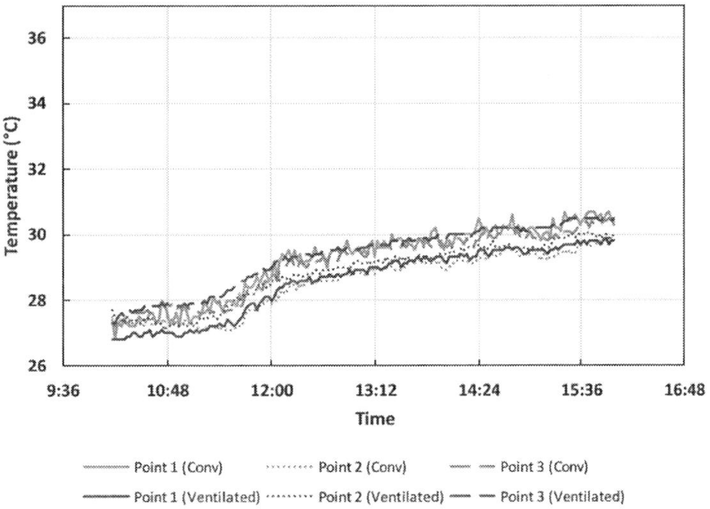

Fig. 5. Results of temperature for both hive in cloudy day.

 DOI: 10.1051/matecconf/201713104001

Experiment was repeated in sunny day, and the results is shown in Figure 6. It was found that the temperature for conventional hive is in the range of 32-36.5°C, whereas temperature range for ventilated hive is lower and more stable in the range of 32.5-34.5°C. If colony was alocated in the hive with the conventional honey cassette, pupae in the broodcell may die because the temperature is obviously above 34°C.

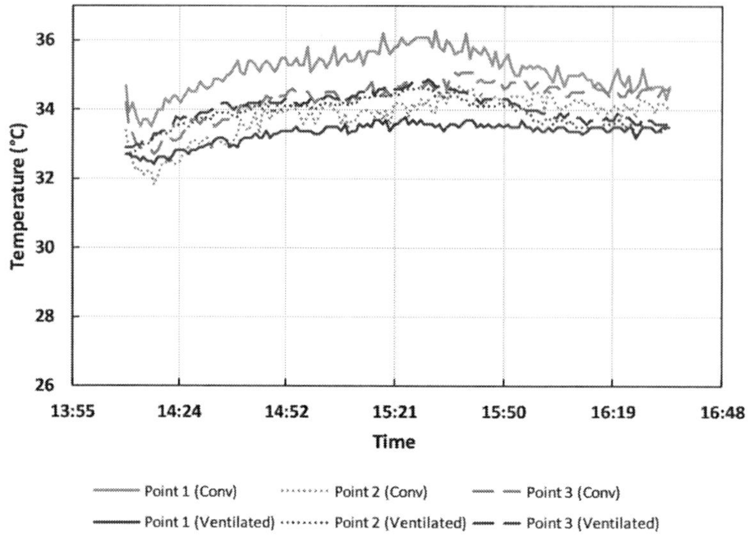

Fig. 6. Results of temperature for both hive in sunny day.

There are two significant improvements found in the hive with the ventilated honey cassette: (a) temperature at Point 1 for ventilated honey cassette was the lowest temperature, and it consistently below 33°C, and this could help to avoid for the melting of propolis around the ventilated wall, (b) Better temperature drop started around 15:30 for the ventilated honey cassette as compared to the conventional honey cassette. These suggest that providing ventilation to the hive could reduce temperature around the wall of honey cassette, and further enhance cooling process inside the honey cassette. On top of that, the ventilated space could also help in drying honey in pots, and therefore fanning process for drying honey by bees could be reduced. In addition, combining a greenroof with the ventilated honey cassette will be the best option because temperature could be further reduced for 3.3-6.5°C, on top of the existence of honey drying process by the ventilation. This could reduce the temperature inside hive to optimum temperature at 31-32°C as suggested by Fletcher et. al. However, further study when there is a colony inside the hives must also be conducted to validate the results.

4 Conclusion

Thermal performance of a ventilated honey cassette was tested and compared with a conventional honey cassette under direct sunlight. It was found that there is no significant difference between the hives when both hives were exposed under direct sunlight in a cloudy day. However, two significant improvement were found for ventilated hive in sunny day.

1) It could help to reduce temperature at wall of honey cassette consistently below 33°C. This could avoid the melting of propolis around the ventilated wall area.
2) It could facilitates in better temperature reduction as compared to the conventional honey cassette.

DOI: 10.1051/matecconf/201713104001

However, further study when there is a colony inside the hives must also be conducted to validate the results.

References

1. Schwarz, F., The Indo-Malayan Species of Trigona, *Am. Museum Nat. Hist.* **76** (1939) 83–141.
2. Roubik, D. W., Stingless bee nesting biology, *Apidologie* **37** (2006) 124–143.
3. Vollet-Neto, A., Menezes, C., and Imperatriz-Fonseca, V. L., Behavioural and developmental responses of a stingless bee (Scaptotrigona depilis) to nest overheating, *Apidologie* **46** (2015) 455–464.
4. Fletcher, D. J. C., and Crewe, R. M., Nest structure and thermoregulation in the stingless bee, Trigona denoiti, Hymenoptera: Apidae, *J. Entomol. Soc. South Africa* **44** (1981) 183–196.
5. Sung, I. H., Yamane, S., and Hozumi, S., Thermal characteristics of nests of the Taiwanese stingless bee Trigona ventralis hoozana (Hymenoptera: Apidae), *Zool. Stud.* **47** (2008) 417–428.

Proximate Analysis and Calorific Value Prediction using Linear Correlation Model for Torrefied Palm Oil Wastes

Siti Raishan Mohd Rashid[1,*], *Suriyati* Saleh[1] , and *Noor Asma Fazli Abdul* Samad[1]

[1]Faculty of Chemical & Natural Resources Engineering, University Malaysia Pahang, Lebuhraya Tun Razak, 26300 Gambang, Pahang, Malaysia

Abstract. Torrefaction is one of the pretreatment processes to upgrade the chemical and physical properties of biomass for power production. In this study, four types of palm oil wastes were selected. The wastes, which consisted of oil palm frond (OPF), palm kernel shell (PKS), palm mesocarp fibre (PMF), and empty fruit bunch (EFB), were subjected to torrefaction process at different temperatures of 240°C, 270°C, 300°C, and 330°C for 30 min residence time. Based on the analysis of torrefied palm oil wastes, it was observed that there is a linear relationship between the properties of torrefied palm oil wastes and torrefaction temperature. Based on this, linear correlation model as a function of mass loss was developed to predict the energy yield (EY), calorific value (HHV), and proximate analysis. A reliable correlation model ($R^2 > 0.90$) for predicting calorific value (HHV), fixed carbon (FC), and volatile matter (VM) was obtained, indicating the developed linear model is indeed reliable. Meanwhile, an acceptable coefficient of determination ($R^2 \approx 0.75$) was obtained when the linear model is used to estimate the energy yield (EY) and ash content (ASH). These developed linear correlation models are cost effective and can be used as a tool to predict the properties of palm oil wastes and to assess the suitability of biomass in torrefaction process.

1 Introduction

In Malaysia, palm oil industry is a major contributor in economic growth. Around 22.31 billion USD was generated from the palm oil industry in 2014, and it has been identified as the fourth largest source of national income [1]. Usually, the main products from palm oil industry are Crude Palm Oil (CPO), Palm Kernel Oil (PKO), and Palm Kernel Cake (PKC); these products are the common ingredients for a wide variety of food, feed, and non-food products [2]. Therefore, the demands for these products are continuous and expected to increase due to the population growth. In order to fulfil this demand, significant efforts have been made in terms of plantation expansion, thus increasing the palm oil cultivation. As a consequence, substantial amount of palm oil wastes such as empty fruit bunches (EFB), palm kernel shell (PKS), palm mesocarp fiber (PMF), and oil palm frond (OPF) are generated after

* Corresponding author: raishanrashid@gmail.com

oil palm harvesting and palm oil processing. Current practices are to utilise these palm oil wastes for pelletisation and as a feed for boiler to generate electricity, while some of the wastes are just left to rot in the plantation for mulching and nutrient-recycling purposes [1]. Alternatively, these wastes can be transformed into value-added products or biofuel for co-firing processes at palm oil mills [3]. However, the main problem is that the energy produced from co-firing processes that use these palm oil wastes is relatively low and infeasible. This is due to the unfavourable characteristics of palm oil wastes such as high moisture content, low calorific value, and hygroscopic nature. In order to overcome this limitation, the palm oil wastes need to undergo a pretreatment process in order to improve their properties.

One of the common methods to upgrade the properties of palm oil wastes is by using torrefaction. Torrefaction is a thermal treatment process in the temperature range of 200°C–300°C for a residence time between 15 and 60 min under inert condition. A vast study of torrefaction process has been implemented for various types of biomass and the resulting product from torrefaction is proven to exhibit brittle behaviour, high energy content, hydrophobic nature, and reduced mechanical strength [3–6]. In spite of that, the correlation analysis of the quality of torrefied product is rather limited. This correlation analysis is indeed important as a tool to assess the suitability of biomass in torrefaction process. This is due to the fact that although the torrefaction process is able to increase some of the properties, the increment for certain biomass is deemed low and insignificant. Therefore, there is a need for a simple correlation analysis to help in the biomass selection for torrefaction purposes. Through this correlation, the properties such as HHV and proximate analysis can be predicted and thus, the appropriate biomass can be selected for torrefaction purpose. It will reduce the needs to perform a number of experiments involving biomass and save a lot of time and energy. In this work, the objective is to develop a correlation analysis for predicting the properties of torrefied palm oil wastes. The torrefaction experiment of palm oil wastes is performed at temperatures of 240°C, 270°C, 300°C, and 330°C for 30 min residence time. The torrefied palm oil wastes are then measured in terms of mass yield, energy yield, high heating value (HHV), and proximate analysis. Subsequently, the correlation model is developed for predicting the energy yield, HHV, and proximate analysis using the mass loss. The validity of this model is then evaluated using the coefficient of determination (R^2).

2 Materials and methods

2.1 Palm oil wastes preparation

Four types of palm oil wastes used in this study were oil palm frond (OPF), palm kernel shell (PKS), empty fruit bunches (EFB), and palm mesocarp fiber (PMF). The palm oil wastes were obtained from Lepar Hilir Palm Oil Mills, Kuantan, Pahang. Before experimental implementation, all of the palm oil wastes were oven-dried at 105°C for 4 hours to remove unbound water and to avoid the degradation of palm oil wastes. Afterwards, the samples underwent grinding and sieving process where only particles in the range of 0.5–1.0 mm were collected. The prepared samples were placed in an airtight container until the experiments are carried out.

2.2 Torrefaction experiment

Torrefaction of palm oil wastes was performed using a vertical-stainless steel reactor with 39.7 cm long and 1.9 cm internal diameter. An amount of 2–3 g of biomass samples were approximately measured and placed in the reactor for torrefaction process. The reactor was flushed with nitrogen gas at a flow rate of 10 mL/min for 5 min as a carrier gas to create an inert environment. Afterwards, the biomass was heated to the desired torrefaction temperature (240°C, 270°C, 300°C, and 330°C) for 30 min residence time by an electric furnace. Various temperatures for torrefaction were applied to study the effect of physical and chemical changes behaviour in biomass at low and high torrefaction temperature. After the torrefaction process, the furnace was switched off and the reactor was allowed to cool to ambient temperature. The torrefied samples were then weighed to measure the mass loss of torrefied biomass. The torrefied samples were stored in an airtight container to avoid moisture from being trapped in the sample. For each temperature, the experiments were repeated for three times to enhance data reliability.

2.3 Property analysis of torrefied palm oil waste

The mass yield and energy yield were calculated by using Eqs. (1) and (2):

$$Mass\ Yield\ (\%) = \frac{Mass\ of\ torrefied\ biomass}{Mass\ of\ raw\ biomass} \times 100\% \quad (1)$$

$$Energy\ Yield\ (\%) = Mass\ yield\ \times \frac{Higher\ heating\ value\ of\ torrefied\ biomass}{Higher\ heating\ value\ of\ raw\ biomass} \quad (2)$$

Proximate analysis was conducted according to the American Society for Testing and Material (ASTM) standards. The volatile matter and ash content were calculated using Eq (3) – Eq (6) by referring to ASTM E871 [7], ASTM E872-82 [8], and E1755-01 [9] respectively while the fixed carbon content was calculated using Eq (7). The calorific analysis in the form of High Heating Value (HHV) was conducted using a bomb calorimeter.

$$Moisture\ content\ (\%) = \frac{m_{initial} - m_{final}}{m_{initial} - m_{crucible}} \times 100\% = B \quad (3)$$

$$Mass\ Loss\ (\%) = \frac{m_{initial} - m_{final}}{m_{initial} - m_{crucible}} \times 100\% = A \quad (4)$$

$$Ash\ content\ (\%) = \frac{m_{ash} - m_{final}}{m_{initial} - m_{crucible}} \times 100\% \quad (5)$$

$$Volatile\ Matter\ (\%) = A - B \quad (6)$$

$$Fixed\ Carbon = 100\% - Volatile\ matter - Ash\ content - Moisture\ content \quad (7)$$

3 Results and discussions

The summary of properties of raw and torrefied palm oil wastes in terms of mass yield (MY), energy yield (EY), high heating value (HHV), fixed carbon (FC), volatile matter (VM), and ash content (ASH) is shown in Table 1. Some of the data such as PKS, EFB, and OPF are collected from our previous work [3,6]. As shown in Table 1, the mass yield for all palm oil wastes decreases as the torrefaction temperature is increased. This is due to the decomposition of palm oil wastes, particularly moisture content and hemicellulose constituent, into volatile components. However, the HHV value for all palm oil wastes shows an increasing trend when the torrefaction temperature is increased. As shown in Eq. (2), the HHV value affects the value of the energy yield but since the mass yield is decreased dominantly, in the end, the energy yield also decreases. Meanwhile, the fixed carbon and ash content for palm oil wastes increase, but the volatile matter shows the opposite trend. Overall, it can be concluded that all of the properties for palm oil wastes either decrease or increase linearly as the temperature is increased, which is in accord with other published literature [5,10]. Therefore a linear correlation model ($y = mx + c$) has been developed in this work

DOI: 10.1051/matecconf/201713104002

due to its simplicity and based on the observed linear trends. Here, the mass loss is used as the variable (x) needed to predict the other properties (y) of torrefied palm oil wastes. The mass loss is determined by subtracting mass yield (MY) from 100%.

Table 1. Properties for raw and torrefied PMF, OPF, PKS, and EFB

Biomass	Condition	MY (%)	EY (%)	HHV (MJ/kg)	FC (%)	VM (%)	ASH (%)	Ref.
PMF	Raw	100.00	100.00	16.94	18.42	75.23	6.35	This work
	240	84.21	94.54	18.05	27.84	66.07	6.08	
	270	75.61	90.87	19.17	28.73	64.60	6.67	
	300	68.01	92.46	21.49	33.33	59.58	7.10	
	330	50.83	90.54	22.91	35.41	56.43	8.16	
PKS	Raw	100.00	100.00	16.15	13.69	79.46	6.85	[3]
	240	84.43	102.26	19.68	26.45	66.27	7.28	
	270	77.64	104.70	21.91	26.17	65.70	8.13	
	300	68.26	99.29	23.64	29.34	58.55	12.11	
	330	54.79	85.84	25.46	36.61	49.68	13.71	
EFB	Raw	100.00	100.00	15.49	18.24	77.18	4.58	[3]
	240	82.84	83.36	15.59	30.79	62.51	6.70	
	270	74.63	86.69	17.99	37.83	54.50	7.67	
	300	59.83	83.37	19.60	43.86	48.44	7.70	
	330	44.74	77.02	22.07	51.49	36.63	11.88	
OPF	Raw	100.00	100.00	17.74	14.29	82.30	3.41	[6]
	240	87.76	98.00	19.82	31.93	64.86	3.20	
	270	77.52	94.34	21.60	39.23	56.15	4.62	
	300	56.98	78.27	23.79	49.70	45.54	4.76	
	330	42.10	83.24	25.83	55.89	38.95	5.15	

Sum of squared residuals (SSR) is calculated in order to measure the discrepancy between the experimental data and estimation model as shown in Eq (8). The y_{exp} value is obtained from experimental data and y_{pred} is obtained from the linear correlation model. A small SSR value indicates a tight fit of the model to the data while large value of SSR indicates the unreliability of the estimation model. Number of samples used to predict the energy properties for torrefaction process are 20, (N=20). In addition the 95% confidence interval is also determined for coefficients of m and c in the linear model.

$$SSR = \sum_{i=1}^{20}\left(y_{exp} - y_{pred}\right)^2 \quad (8)$$

The linear correlation models as functions of mass loss for predicting the calorific value (HHV), energy yield (EY), fixed carbon (FC), volatile matter (VM), and ash content (ASH) for palm oil wastes are shown in Figs. 1(a)-(e). As shown in Figs. 1 (a)-(e), the linear correlation model for predicting HHV, FC, and VM shows $R^2 = 0.92$, $R^2 = 0.91$, and $R^2 = 0.93$ respectively, indicating that a reliable correlation has been achieved. The experimental values for HHV, FC, and VM at different torrefaction temperatures (240°C, 270°C, 300°C, and 330°C) are very close to the linear line of correlation model which confirms that the linear model as a function of mass loss is a good indicator to represent the torrefaction process. Meanwhile, only $R^2 = 0.74$ and $R^2 = 0.77$ coefficients of determination have been obtained from linear correlation model for energy yield and ash content. Although low coefficients of determination are obtained, the linear model can be used to predict the energy yield and ash content, however, further comparison with other sources is necessary to improve the reliability of the predicted data. Furthermore, the upper bound and lower bound obtained from 95% confidence interval are also included in Figs. 1(a)-(e) where most of the properties data are lies within the bounds except energy yields and ash content which explains

DOI: 10.1051/matecconf/201713104002

the low coefficients of determination is obtained for both properties. Overall, it can be concluded that all of the linear models can be used to predict the necessary properties data within the context of palm oil wastes.

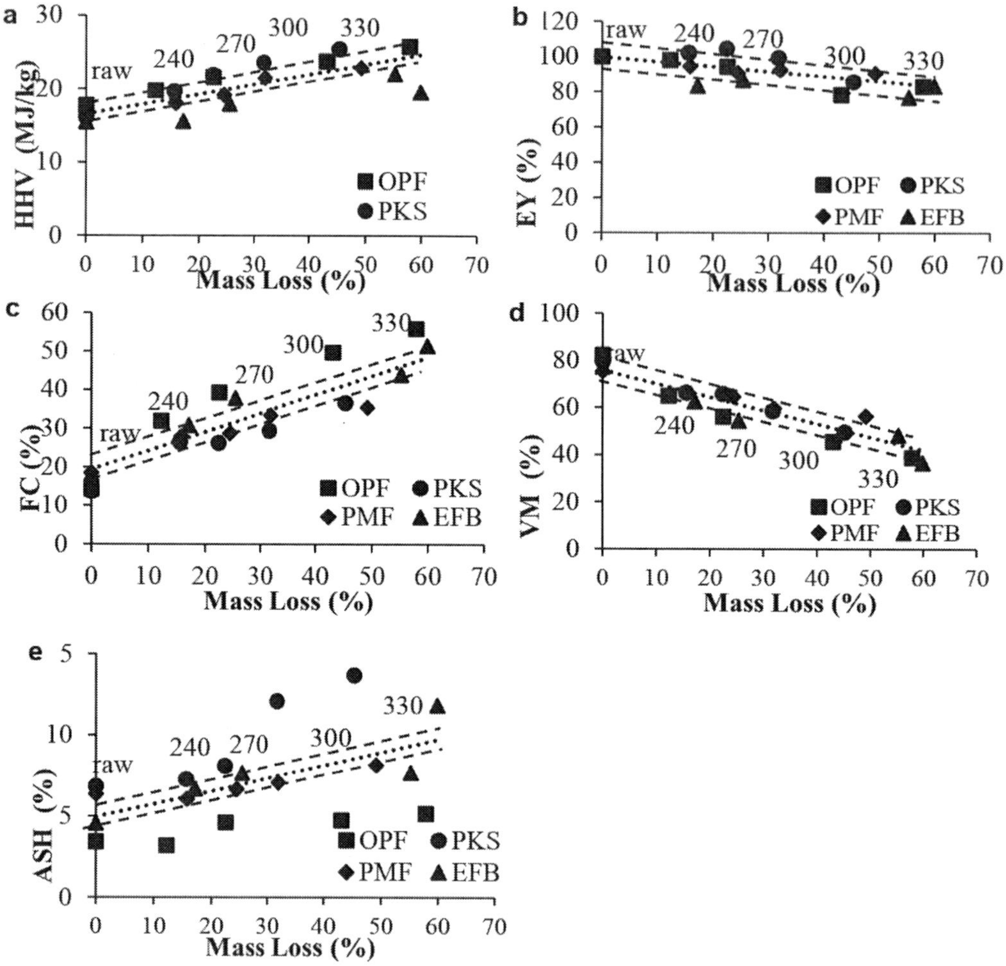

Fig. 1. Linear correlation model for (a) Calorific value (HHV); (b) Energy yield (EY); (c) Fixed carbon (FC); (d) Volatile matter (VM); (e) Ash content (ASH) as a function of mass loss.

The summary of simple linear correlation model with confidence interval for four types of biomass in terms of calorific value (HHV), energy yield (EY), fixed carbon (FC), volatile matter (VM), and ash content (ASH) as a function of mass loss is shown in Table 2. The SSR calculated for all properties are quite low (less than 5) indicating the predicted value using the model is very close to the experimental data for all data and thus the linear model is able to predict the desired properties. This model can be subsequently used for a preliminary assessment to determine the practicality of biomass to undergo torrefaction process. In this case, the values of calorific value (HHV), energy yield (EY), fixed carbon (FC), volatile matter (VM), and ash content (ASH) can be predicted by using the mass loss data only, without the needs to perform extensive experimental work and sample analysis. This is essential for example in gasification application where the simulation can be done using the predicted properties such as proximate analysis and HHV of torrefied biomass in order to evaluate the capability of torrefied biomass for producing synthesis gas and gasification efficiency [11]. This correlation model also helps in reducing the number of experiments to be implemented and thus, saves energy and reduces cost.

Table 2. Linear correlation model as a function of mass loss (ML) for palm oil wastes.

Properties	Linear Correlation Model	R^2	(SSR)
HHV (MJ/kg)	HHV = (0.135 ± 0.00675) x (ML) + (16.564 ± 0.8282)	$R^2 = 0.92$	0.89
EY (%)	EY = (-0.2767 ± 0.1384) x (ML) + (99.786 ± 4.989)	$R^2 = 0.74$	3.41
FC (%)	FC =(0.4826 ± 0.02413) x (ML) + (19.301 ± 0.9651)	$R^2 = 0.91$	2.72
VM (%)	VM = (-0.5618 ± 0.02809) x (ML) + (75.754 ± 3.7877)	$R^2 = 0.93$	0.74
ASH (%)	ASH = (0.0792 ± 0.00396) x (ML) + (4.9451 ± 0.24726)	$R^2 = 0.77$	4.09

4 Conclusions

In this study, a linear correlation model has been developed for predicting the calorific value (HHV), energy yield (EY), fixed carbon (FC), volatile matter (VM), and ash content (ASH) specifically for torrefied palm oil wastes. High coefficients of determination ($R^2 > 0.90$) for linear correlation models for predicting HHV, FC, and VM have been obtained, which are applicable for palm oil wastes. This proves that the reliable model has been obtained and the mass loss is a suitable variable to predict the HHV, FC, and VM. However, moderate coefficients of determination ($R^2 < 0.90$) are obtained for linear correlation model for predicting the EY and ash content, thus, cautionary use of the correlation model is necessary. Overall, the linear correlation model developed in this study is a useful model to be used as an assessment tool for suitability of palm oil wastes for torrefaction process.

Acknowledgements

This work was financially supported by Doctoral Research Scheme (DRS) under Universiti Malaysia Pahang.

References

1. M.F. Awalludin, O. Sulaiman, R. Hashim and W.N.A.W. Nadhari, Renew. Sustainable Energy Rev. **50**, 1469-1484 (2015)
2. D. Zwart, J Oil Palm Environ **4**, 41-46 (2013)
3. N.H.H.M. Harun, F.R.A.A. Wahid, S. Saleh and N.A.F.A. Samad, Chem. Eng. Trans. **56**, 1195-1200 (2017)
4. W-H Chen, J. Peng and X.T. Bi, Renew. Sustainable Energy Rev. **44**, 847-866 (2015)
5. K.M. Lu, W.J. Lee, W-H Chen, S.H. Liu and T.C. Lin, Bioresour. Technol. **123**, 98-105 (2012)
6. F.R.A.A. Wahid, N.H.H.M. Harun, S.R.M. Rashid, N.A.F.A. Samad and S. Saleh, Chem. Eng. Trans. **56**, 199-204 (2017)
7. ASTM E871-82, www.astm.org (2013)
8. ASTM E872-82, www.astm.org (2013)
9. ASTM E1755-01, www.astm.org (2015)
10. G. Almeida, J.O. Brito and P. Perre, Bioresour. Technol. **101(24)**, 9778-9784 (2010)
11. M.B. Muslim, S. Saleh, N.A.F.A. Samad, Chem. Eng. Trans. **56**, 1495-1500 (2017)

Silicon carbide- from synthesis to application: a review

Mohd Aizat Hafiz Mohd Sohor[1*], *Mazli* Mustapha[1], and *Jundika* Chandra Kurnia[1]

[1]Mechanical Engineering Department, Universiti Teknologi PETRONAS (UTP), 32610 Seri Iskandar, Perak, Malaysia

Abstract. This paper provides a review of the synthesis techniques that are used to produce nanostructured silicon carbide. The synthesis methods, consisting of carbothermal reduction, chemical vapor deposition, laser ablation, sol-gel and microwave heating are described. The silicon carbide properties and application are also explained. The paper then discusses the limitations of previous studies which involved complicated equipment and processes, that limit their further application and act as a barrier to further research and development in many fields.

1 Introduction

In recent years, nanotechnology has been increasingly matured following intensive efforts by materials scientists, physicists, chemists and engineers who have been inspired by its importance in both basic research and technological applications. The borders between various subjects such as physics, chemistry, materials science, biology, become indistinguishable on the nanometer scale. Thus, breakthrough in any of these areas in nanoscale could have an impact on the others. In the case of nanoelectronics, the related semiconductor research has been the most important driving force [1].

In this modern day, nanotechnology has stimulated attention of many researchers worldwide and is said to be revolutionary, replacing conventional materials science. This interesting scientific research area became a phenomenon in the past decades when Iijima had set a remarkable milestone with his first discovery of carbon nanotubes (CNTs) in 1991 [2], [3]. This was followed by exploration and production of identical geometrical compound namely SiC, BN, GaN and MoS_2 to name a few [4]. Significant number of research works were recorded with large number of articles published internationally [5]. This shows nanotechnology has become a widespread relevant materials science study which has been explored by researchers and scientists alike.

Nanostructure material can be classified into its functionality and dimensionality. However, only low dimensionality like zero dimensional (0D), one dimensional (1D), two dimensional (2D) and three dimensional (3D) have taken the interest of researchers to explore. Since the majority of research works related to nanostructure are focusing on 1D [5], the various types of 1D-nanostructure have been established with many synthesization techniques developed from time to time including the sol–gel [12], vapor–liquid–solid [8],

* Corresponding author: mohd.aizat_g03388@utp.edu.my

vapor– solid [2], laser ablation [6] and chemical vapor deposition (CVD) methods [9]. Among the various types of 1D nanostructure materials available, their specific names were classified based on their unique shapes and sizes such as nanowires, nanorods, nanocables, nanotubes and nanobelts [7], [8].

Different dimensions and shapes of nanostructure allow them to be applied in various applications. When a material is reduced to the range of nanometer, their properties changed significantly due to quantum confinement effect. 0D nanostructure is widely studied for its potential application in multifunctional nanoparticle probe. The charge carriers in 2D nanostructures are restricted to the direction perpendicular to the layer and can move freely in the layer plane. These 2D nanostructures, especially nanoporous materials have been studied for their applications in biosensors, nanophotonic and nanoelectronic applications. 1D nanostructures are known to exhibit quantum confinement effect like 0D nanostructures but for 1D nanostructures, the charge carriers can only travel in one direction along the wire, thus, making them useful as interconnects and critical devices in nanoelectronic and nano-optoelectronic applications[10]–[12].

2 Silicon carbide

Silicon carbide (SiC) is a non-oxide ceramic engineering material that has gathered a considerable amount of interest. It has a wide range of industrial applications [24] which plays a major role in high voltage, temperature and frequency semiconductor device applications.

SiC compound consists of a Silicon (Si) atom and four Carbon (C) atoms which are covalently bonded between two of them. The four C atoms surrounded the Si atoms through strong tetrahedral of sp3 combination bonds. Close-packed stacking structure between Si and C atoms can be considered in one period SiC crystallographic structure. Polytypes in SiC are formed when there are differences formed in crystallographic. More than 200 polytypes [13] or 250 polytypes in some reports claimed have been identified but only a few studied on them. Major polytypes of SiC are α-SiC and β-SiC, which are classified as hexagonal polytypes and cubical polytypes respectively. The most common polytype occured in α-SiC is 2H-SiC, 4H-SiC and 6H-SiC. Meanwhile, 3C-SiC owned zinc blend structure is the most synthesized polytype in β-SiC regime [14]. Another famous α-SiC polymorph type of SiC is 15R-SiC. The letter H, C and R in their name denoted as pure stacking of hexagonal, cubic, and rhombohedral respectively [3], [15].

2.1 Synthesize method

Up to the present, there are various fabrication methods that have been successfully conducted by the previous researchers. The synthesized products included micro and nano sizes with different dimensionality. However, with the recent advancement of nanotechnology, the race towards producing the nano level size SiC has become a phenomenon with the introduction of different synthesize techniques.

Composite powders which consist of a mixture in the Si-C-N system were produced by using a self-propagating high temperature synthesis (SHS). This process involved combustion reaction where it was started by an outer energy source in the powdery bed. Then, a reaction took place and followed by the propagation of high temperature thermal wave through the bed. Since it did not involve dilution, SHS was a simpler and more economical method. Four different compositions were prepared from the mixture of raw materials namely carbon black and silicon powder with C/Si vol.% ratio of 10, 20, 30, and 40.5 This mixture was then ball milled together with the grinding media which consisted of Si_3N_4 and isopropanol is added to wet homogenized them. The combustion process was performed under a nitrogen pressure of 3 MPa in high pressure reactor. The result indicated that carbon content had influenced the temperature profile of the reaction front. As the ratio of carbon increased, the temperature profile became higher and narrower. The specific surface area was also increased in the product with higher carbon content. XRD analysis on the combustion products has revealed two phases; β-SiC and β-Si_3N_4. In addition, a small discrepancy of the SiC content was detected when compared to the predicted ones. This research had concluded that composite (i.e SiC-Si_3N_4) in the Si-C-N system with controlled morphology and high purity could be attained using SHS without inert solution[16].

Through metal catalyst free thermal evaporation technique, the SiC nanowires were fabricated from Si substrate. The experimental process began with the placement of Si wafer and graphite powder side by side in the columnar graphite crucible. The system was set up in a vacuum furnace. Prior to heating, the surface of Si was washed with acetone and ethanol and then air-dried. Argon gas was supplied to the vacuum furnace after evacuation to 10 Pa via rotary pump. The heating temperature was increased to 1000 °C at 10 °C/min of heating rate then gradually increased to 1500 °C at 3 °C/min and kept to 3 hours. SEM and field emission scanning electron microscopy (FESEM) analysis clearly indicated the formation of straight and curved nanowires with the size of several hundred microns and 50-100 nm in diameter. The structure was then analyzed by energy-dispersive X-ray spectroscopy (EDX). With composition of a single-crystalline SiC core wrapped in thin amorphous SiO_2 shell, the nanowire has been categorized as SiC–SiO_2 core–shell structure. Vapor-solid process has been suggested and proposed to be the growth mechanism for the overall experiment [12].

The complex structure namely SiC nanowires decorated with SiC polycrystalline nanoparticles were grown via chemical vapor deposition (CVD) using methyltrichlorosilane (MTS) as starting material. The method involved the cutting of specimen (C/C) into two, cleaned with ethanol and dried. The specimen was then hung in the CVD furnace with inert Ar atmosphere and temperature raised to 1000 °C for 2 hours. The as-product was analyzed in transmission electron microscopy (TEM) revealing two compositions labeled as nanowire and nanoparticle. EDX analysis had been conducted on both structures which suggested the presence of Si and C with a small amount of O. This was due to the oxidation of SiC to form SiO_X during cooling process. The combination results of HRTEM and EDX safely concluded the appearance of many small single crystals with different orientations to form polycrystalline SiC and single crystalline SiC into SiC nanowires stem. However, the shift signal in Raman scattering indicated many stacking faults in the products. To conclude, the SiC nanowires decorated with SiC polycrystalline nanoparticles with diameters ranging from 50 nm to 200 nm and length up to several hundred micrometers was successfully synthesized through CVD with MTS precursor [17].

The inverted 3C-SiC nanoneedles with hexagonal structure were rapidly synthesized via a two-crucible heating technique. This study was conducted under high frequency and there was no catalyst involved. In this experiment, two different diameters hollow attached cylindrical graphite crucibles were used whereby the mixture of silicon monoxide and carbon coke was in smaller crucible and carbon fiber in the other. Then, the two crucibles were placed inside a quartz tube before heating to 1800 °C about 5 minutes dwelling time from room temperature under argon atmosphere. XRD pattern identified the structures as 3C-SiC in consistent with the standard. SEM images revealed hexagonal, spindle and thick portion of grown nanoneedle structure on the carbon fiber in the [1 1 1] orientation. The average length for each structure measured about 130 nm. The reaction process of SiO and carbon on the carbon fiber was believed to be the initial point forming SiC nuclei to SiC nanoneedle. Thus, with this synthesize effort, it has broaden the types of 1D nanostructure SiC [18].

Sol-gel method is one of the more versatile methods to synthesize SiC which offers several advantages including production of good chemical homogeneity, smaller particle size, and high-purity materials [19]. The sol-gel synthesis of nano-sized 3C-SiC has been conducted using sol-gel technique silicon-bearing gel and nano-sized carbon particles obtained from soot. The preparation of silica bearing sol started by using alkaline ethylene glycol and fumed silica which then yielded in a silicon bearing gel. At a temperature about 1400 °C to 1580 °C, the major amount of 3C-SiC was observed with minor trace of 2H-SiC. The different ranges of nanosize structural dimension was observed at 1400 °C and 1580 °C with 30–50 nm (particles) and 15-17 nm (crystallite) respectively [15].

Carbothermal reduction process involved long hours of heating raw materials at extremely high temperatures around 2200 - 3300 °C. Quartz sand and petroleum coke were used as precursor materials for this technique [20]. The summary of latest carbothermal reduction synthesis methods were summarized in **Table 1.** There are few publications that conducted the synthesization of nanostructured 2H-SiC and most of them carried out other polytypes product mainly 3C-SiC. However, the required synthesizing parameters for 2H-SiC produced were not fully analyzed by Zhang [9]. Complex carbothermal reduction method were employed by Zhu et al, Omidi et al, Mostaghi et al and Zhang et al including combination of sol-gel technique and TEOS infiltration. The carbothermal reduction experiment grown major phase of β -SiC and only small amount of α-SiC traced. The growth of α-SiC was claimed to have initiated beyond 1400 °C in temperature.

Table 1. Summary of recent carbothermal reduction synthesis method

Product	Size	Method	T (°C)	Reference
6H-SiC	Diameter ranging from 60 to 160 nm and length up to tens of microns	Carbothermal Reduction infiltrated TEOS	>1300	[14]
SiC	Ultrafine globular particles	Sol-gel technique and carbothermal reduction	1450	[19]
SiC nanoparticles	5-25 nm	Carbothermic reduction	1450	[21]
2H-SiC nanoribbon	Tens to hundreds of microns in length, a few microns in width and tens of nanometers in thickness	Carbothermic reduction	1400	[9]
β-SiC nanowires and nanocables	20 μm in length	Carbothermal reduction	1800	[22]

3 Silicon carbide properties and applications

Silicon carbide has been considered as a promising candidate material in industrial application due to its superior properties in mechanical, optical, chemical and electronic fields. In terms of size, SiC is light weight and normally has low dimensionality so it is commonly available in nano size devices [8]. This type of material achieved high evaluations in terms of mechanical strength [19], hardness [21], thermal stability [15] and conductivity [23], good thermal shock resistance [24] with low thermal expansion coefficient [9] and good wear resistance [21], [23]. Meanwhile, in view of chemical properties, SiC displayed chemical and oxidation resistance due to its inertness [25]. The strong bond of Si-C protected the structure from undergoing chemical reaction. The advancement of synthesization techniques has grown over time making SiC one of the more affordable materials in bulk volume [3].

On the other hand, in electronic properties and application, it is well known as a wide band gap material. 2H-SiC possessed the largest band gap among them compared to other polytypes like 3C-SiC, 4H-SiC, and 6H-SiC [3], [8], [12], [13], [17]. The simulation prediction done by Monte Carlo showed higher electron mobility in 2H-SiC than other polytypes [25]. Another most important parameter is thermal conductivity. As one material being exposed to heat, the changes of physical properties will happen. Thus, the carrier mobility will be lower as the temperature increases [26]. In harsh environment application, it is a necessity for SiC to possess high thermal conductivity compared to other metals [18]. In fact, most SiC has proven as a promising material which has higher thermal conductivity compared to copper (4.0 W/cm-K) that was set as benchmarking material [26].

Other SiC electronic properties are high breakdown electric field [2], high thermal and electrical conductivity [35], low quantum confinement [3], high ionic mobility [36] and high saturation electron drift velocity of the latter [24]. These electronic properties give major contribution to be the best selected material in electronic devices application such as power electronics, field electron emission, sensor, MOSFET, MESFET and so on. Besides, SiC appeared to be applied as blue and ultraviolet (UV) emitting diodes due to ability to emit high-intensity and stable UV and blue green lights [38].

On the other hand, since no cooling and transformer are needed, nanotechnology-based SiC become a more economical and effective material in engineering and industrial applications [26], [35]. SiC also owns diverse properties in electronic and optical fields. Thus, the application of SiC in optoelectronics, microelectronics, nanodevices, nanocomposites, hydrophobic devices and biomedical engineering application is vital and useful [1], [27].

4 Conclusion

This paper has reviewed several synthesis techniques that have been used to produce silicon carbide nanostructured material. Carbothermal reduction is the most commonly used technique due to its simple setup and economical process. Other techniques that can be used are chemical vapor deposition, laser ablation, sol-gel and microwave heating. All the techniques discussed have their own advantages to improve the formation of various types of silicon carbide.

We gratefully acknowledge the support provided by Universiti Teknologi PETRONAS and Ministry of Higher Education Malaysia (MOHE) of Malaysia for this work under Fundamental Research Grant Scheme (FRGS) no. 0153AB-K82.

References

1. P. Neudeck, R. Okojie, and L. Chen, *Proc. IEEE*, **90**, no. 6, pp. 1065–1076, (2002)
2. P. Kang, B. Zhang, G. Wu, J. Su, and H. Gou, *Mater. Lett.*, **65**, no. 23–24, pp. 3461–3464, (2011)
3. R. Wu, K. Zhou, C. Y. Yue, J. Wei, and Y. Pan, *Progress in Materials Science*, **72**, Elsevier Ltd, pp. 1–60, (2015)
4. H. Li, J. Wu, and Z. M. Wang, *Mater. Sci. 187*, (2013)
5. J. Fan and P. K. Chu, *Silicon Carbide Nanostructures*. (2014)
6. R. Wu, G. Yang, Y. Pan, J. Chen, R. Zhai, L. Wu, and J. Lin, *J. Alloys Compd.*, **453**, no. 1–2, pp. 241–246, (2008)
7. J. S. Lee, Y. K. Byeun, S. H. Lee, and S. C. Choi, *J. Alloys Compd.*, **456**, no. 1–2, pp. 257–263, (2008)
8. L. Z. Cao, H. Jiang, H. Song, X. Liu, W. G. Guo, S. Z. Yu, Z. M. Li, and G. Q. Miao, *Solid State Commun.*, **150**, no. 15–16, pp. 794–798, (2010)
9. H. Zhang, W. Ding, K. He, and M. Li, *Nanoscale Res. Lett.*, **5**, no. 8, pp. 1264–1271, (2010)
10. Y. Yao, S. T. Lee, and F. H. Li, *Chem. Phys. Lett.*, **381**, no. 5–6, pp. 628–633, (2003)
11. Y. L. Chiew and K. Y. Cheong, *Materials Science and Engineering B: Solid-State Materials for Advanced Technology*, **176**, no. 13. Elsevier B.V., pp. 951–964, (2011)

12. K. Chen, Z. Huang, J. Huang, M. Fang, Y. G. Liu, H. Ji, and L. Yin, *Ceram. Int.*, **39**, no. 2, pp. 1957–1962, (2013)
13. A. Ito, H. Kanno, and T. Goto, *J. Eur. Ceram. Soc.*, **35**, no. 16, pp. 4611–4615, (2015)
14. J. Zhu, J. Jia, F. L. Kwong, and D. H. L. Ng, *Diam. Relat. Mater.*, vol. 33, pp. 5–11, (2013)
15. M. Sharma, R. K. Upadhyay, S. S. Amritphale, and N. Chandra, *Mater. Lett.*, **65**, no. 14, pp. 2161–2164, (2011)
16. M. Soszyński, O. Łabędź, and A. Huczko, *J. Cryst. Growth*, **401**, pp. 445–448, (2014)
17. X. Qiang, H. Li, Y. Zhang, S. Tian, and J. Wei, *Mater. Lett.*, **107**, pp. 315–317, (2013)
18. [B. Qian, H. Li, Z. Yang, Y. Zhang, Y. Su, H. Wei, and Y. Zhang, *J. Cryst. Growth*, **338**, no. 1, pp. 6–11, (2012)
19. Z. Omidi, A. Ghasemi, and S. R. Bakhshi, *Ceram. Int.*, **41**, no. 4, pp. 5779–5784, (2015)
20. S. Cetinkaya and S. Eroglu, *J. Eur. Ceram. Soc.*, **31**, no. 5, pp. 869–876, (2011)
21. B. M. Moshtaghioun, A. Monshi, M. H. Abbasi, and F. Karimzadeh, *Int. J. Refract. Met. Hard Mater.*, **29**, no. 6, pp. 645–650, (2011)
22. L. D. Zhang, G. W. Meng, and F. Phillipp, *Mater. Sci. Eng. A*, **286**, pp. 34–38, (2000)
23. X. W. Zhigang Sun, Xiaojing Qiao, Qingguo Ren, Wangchang Li, *Int. J. Nanotechnol.*, **12**, pp. 896–908, (2015)
24. K. Zekentes and K. Rogdakis, *J. Phys. D. Appl. Phys.*, **44**, no. 13, p. 133001, (2011)
25. K. Bertilsson, E. Dubaric, H. E. Nilsson, M. Hjelm, and C. S. Petersson, *Diam. Relat. Mater.*, **10**, no. 3–7, pp. 1283–1286, (2001)
26. A. Stephen E, Saddow and Anant, *Advances in Silicon Carbide Processing and Applications*, no. 1. (2014)

MATEC Web of Conferences **131**, 04004 (2017) DOI: 10.1051/matecconf/201713104004
UTP-UMP SES 2017

Numerical investigation on effect of leaflet thickness on structural stresses developed in a bileaflet mechanical heart valve for its sustainable manufacturing

Syed Samar Abbas[1], *Mohammad* Shakir Nasif[1, *], *Mior Azman* Meor Said[1], *Rafat* Al-Waked[2], and *Saleem* Khalefa Kadhim[3]

[1]Mechanical Engineering Department, Universiti Teknologi PETRONAS, 32610, Seri Iskandar, Malaysia.
[2]Mechanical and Maintenance Engineering Department, School of Applied Technical Sciences, German Jordanian University, Amman 11180, Jordan.
[3]Control and System Engineering Department, University of Technology, Baghdad, Iraq

Abstract. Flow induced structural stresses can cause mechanical prosthetic aortic valve to fail due to yielding. In this study, we have performed the structural analysis, especially the effect of leaflet thickness on equivalent stresses developed in a Bileaflet mechanical heart valve (BMHV) due to blood flow through it has been investigated. The leaflet thickness varies from 0.5mm to 0.7mm, by 0.1mm. A fluid-structure interaction approach based on Arbitrary Lagrangian Eulerian (ALE) technique has been employed with the aid of an user defined function (UDF). Results of the analysis show that high von Mises stresses are developed in BMHV with leaflet thickness of 0.5mm and 0.6mm, being 75% and 13% higher than allowable equivalent stress respectively. Such thinner leaflets are therefore, not sustainable to be replaced with diseased aortic valve.

1 Introduction

The native aortic valve may be diseased due to inborn heart defects, rheumatic fever, bacterial infection etc. and thus can cause stenosis or valvular insufficiency [1]. A prosthetic replacement to such malfunctioning valve can be made with a 'Bileaflet Mechanical Heart Valve' (BMHV), which consists of two leaflets, housing and suture ring as shown in Fig. 1 [2].

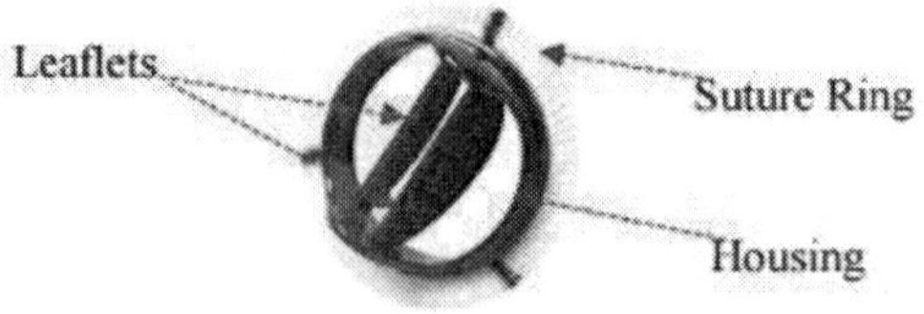

Fig. 1. A typical BMHV

* Corresponding author : mohammad.nasif@utp.edu.my

To have a deep insight into the fluid dynamics of prosthetic valves, numerical methods have evolved as a great tool in the past half century, originating with the pioneering work of Peskin et al. [3]. In his work, a two-dimensional numerical model of mitral valve was presented. The numerical simulation of prosthetic valves is a complex Fluid-structure Interaction (FSI) problem, which has to deal with kinematical description of the fluid domain. Based on this, the numerical models either employ a fixed or a moving grid approach. Due to their high accuracy, moving grid approaches are preferred over fixed grid methods [4,5]. Moving grid approaches are based on Arbitrary Lagrangian Eulerian (ALE) method, in which the fluid grid moves in an arbitrary way and it doesn't have to conform with fluid velocity. With ALE, greater distortions of the continuum can be handled better than would be allowed by a purely Lagrangian method, with more resolution than that afforded by a purely Eulerian approach [6]. Having applied ALE to 2-D geometry of BMHV, Dumont et al. [7] obtained velocity profiles for systolic phase of cardiac cycle and validated the results experimentally. Nobili et al. [8] and Yeh et al. [9] developed a 3-D model of the valve, and employed ALE to study hemolytic properties of blood flow during different phases of cardiac cycle. Their results were found in good agreement with experimental measures. It's important to note that due to large structural displacements involved while simulating BMHV, subsequent remeshing and smoothing is required to maintain a good quality mesh [10].

The structural strength of BMHVs is an important matter while designing the valve as structural stresses are developed in the valve due to blood flow that may cause valve failure. During its lifetime operation, the valve is exposed to blood pressure force [11], force due to regurgitant flow at the end of systole [12] and due to vortices formed in the wake region downstream the BMHV [13]. High equivalent stresses may cause yielding and thus failure of the structure. Kwon et al. [14] performed structural analysis of such valves with variable leaflet thicknesses by considering the effect of blood pressure force only. It was reported in his study that considerably thin valve (with thickness 0.5mm) undergo failure due to yielding. Some other studies have also considered the effect of leaflet thickness variation over the structural strength of valve experimentally [15,16] and have reported that the structural strength decreases with decreasing thickness. In another numerical study by Kwon et al. [17], a blood pressure force of 104 KPa was applied, even though the physiological blood pressure exerts a force of 16 KPa only [14]. The excess force was applied to account for lifelong and continuous exposure of leaflets to the blood pressure force.

Considering the effect of varying leaflet thickness and consequently, that of hinges, we have performed structural analysis of a BMHV, subjected to not only the blood pressure force, but also to the force due to recirculation of blood and regurgitant flow. The thickness of the leaflet has been varied between 0.5mm to 0.7mm [14], and hence the thickness of the hinges, by an interval of 0.1mm. A fluid-structure interaction model based on ALE has been employed. The von Mises stresses induced in the structure are computed and compared for various thickness. To the best of our knowledge, no previous study has performed such analysis while considering Fluid-structure interaction of blood flow through a BMHV.

2 Methodology

The geometric model of the aortic root containing BMHV has been developed using the experimental design presented by Reul et al [18], as shown in Fig 2. Aortic root is the initial tract of aorta, and is characterized by three bulbous sinuses of Valsalva [19]. BMHV is located at the initial verge of aortic root and consists of two leaflets made up of pyrolytic carbon, which fully closes at 25° while fully opens at 85°. The leaflets rotate about struts attached to their housing, which is made up of titanium. Leaflet with a thickness of 0.5mm

DOI: 10.1051/matecconf/201713104004

can be seen in Fig 3. The thickness is varied from 0.5mm to 0.7mm and complete cardiac cycle has been simulated for all the aforementioned thicknesses of the leaflet. Simulations have been carried out using finite volume code ANSYS Fluent coupled with finite element code Static Structure. The number of cells used in Static Structure and Fluent are 151,426 and 723,771, respectively and maximum mesh skewness of 0.76 was maintained.

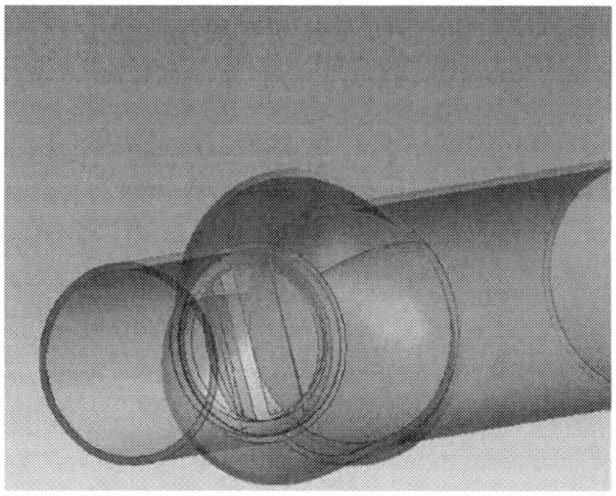

Fig. 2. Aortic root geometry

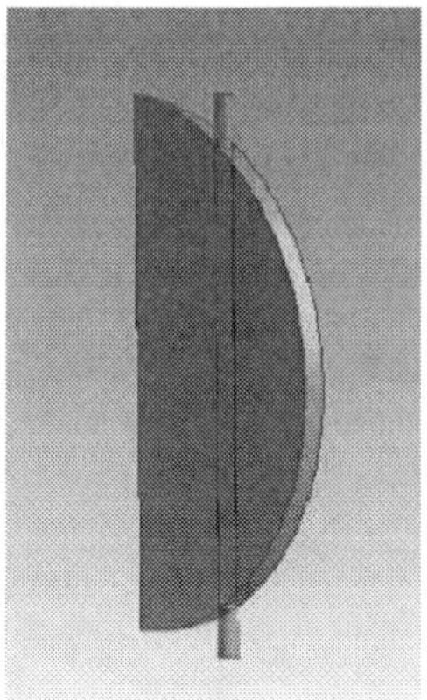

Fig. 3. Leaflet with thickness 0.5mm

ANSYS Fluent has been used as fluid solver with an in-house written user defined function to account for the movement of the two leaflets. The code was written using C++ programming language and was then incorporated into ANSYS Fluent. Dynamic mesh module of ANSYS Fluent has been used in conjunction with remeshing.

The set of Navier Stokes equations that the software solves for the fluid domain are as written in equations (1) and (2):

$$\nabla . v = 0 \tag{1}$$

$$p\left(\frac{\partial v}{\partial t} + v.\ \nabla v\right) = \ \nabla . \sigma + p.f \tag{2}$$

Where, p is the fluid density, v is the fluid velocity, σ is the Cauchy Stress tensor and f the body force.

A no-slip boundary condition has been applied to wall boundary. Blood has been modelledas laminar, Newtonian fluid, since the Newtonian behaviour of blood in large vessels like aorta has been widely agreed [18]. The viscosity of blood becomes asymptotically constant at high shear rates, which is the case for aorta. The blood's dynamic viscosity and blood density used in the model are 3.5 cP, and 1060 kg/m3 respectively.

A physiological pressure waveform, which was approximated using the Fourier series, has been incorporated as the inlet and outlet boundary condition, as mentioned in equations (3) and (4) respectively [8]. The governing equations for inlet and outlet pressures are listed below:

$$p(inlet) = 44.24 - 50.76\cos wt + 44.4\sin wt - 0.4108\cos 2wt - 19.45\sin 2wt - 0.3481\cos 3wt + 6.66\sin 3wt + 8.032\cos 4wt + 2.597\sin 4wt + 0.3551\cos 5wt + 2.903\sin 5wt \quad (3)$$

where w represents a factor in the pulsed wave equation and is equalled to 8.72.

$$p(outlet) = 96.71 - 18.66\cos wt + 2.59\sin wt + 0.182\cos 2wt - 8.67\sin 2wt + 4.41\cos 3wt + 0.46\sin 3wt + 0.059\cos 4wt + 1.39\sin 4wt - 2.101\cos 5wt + 1.19\sin 5wt \quad (4)$$

where w is equal to 8.438.

3 Results and Discussion

The von Mises stresses developed in the valve structures of leaflet's thickness 0.5mm, 0.6mm and 0.7mm are presented in Fig. 4 and Fig. 5 respectively at fully closed position of the leaflets (25^0, time 0.8sec), while the stresses for other significant positions of the leaflet as start to open (30^0, time 0.008sec), fully open (85^0, time 0.057sec) and start to close position (63^0, time 0.237sec) are listed in Table 1. It has been highlighted in previous studies that the valve structure is subjected to maximum stress at fully closed position and our results conform with this statement. Higher von Mises stresses than allowable range are developed in leaflets with thickness of 0.5mm and 0.6mm both at fully closed positions of the leaflets. This is contrary to results of Kwon et al [14,17] in regard that in their study, it was reported that von Mises stresses higher than the allowable range are developed in leaflets of thickness 0.5mm only. This is majorly due to the reason that they ignored recirculation and regurgitant flow in their study, subjecting the valve to blood pressure force only and therefore, the equivalent stresses in leaflets with thickness of 0.6mm were under-estimated.

It is the hinge region that develops the maximum von Mises stresses for all thicknesses. The maximum von Mises stress developed in leaflets with thickness 0.5mm is 56 MPa, being 1.5 times higher than those developed in leaflet thickness of 0.6mm (being 36 MPa) while 5.5 times higher than those developed in leaflet thickness of 0.7mm (being 11 MPa). These stresses occurred at the leaflet pin for a short time, however it may cause valve failure in the long run due to yielding as the allowable stress is 32 MPa [22].

The von Mises stresses developed in leaflet thickness of 0.5mm at start to open (30^0, time 0.008sec) and start to close position (63^0, time 0.237sec) are also higher than the allowable range, being 45 MPa and 35 MPa respectively (Fig. 4). Other than mentioned above, for all leaflet's thicknesses and positions, the stresses are within the allowable range. It is worth mentioning that for leaflet thickness of 0.7mm, the von Mises stresses developed during the whole cardiac cycle are well below the allowable range, with maximum equivalent stress being 11 MPa only.

(a) Fully closed (b) Start to close (c) Fully open

Fig. 4. von Mises stress profiles for different positions of leaflets with thickness 0.5mm.

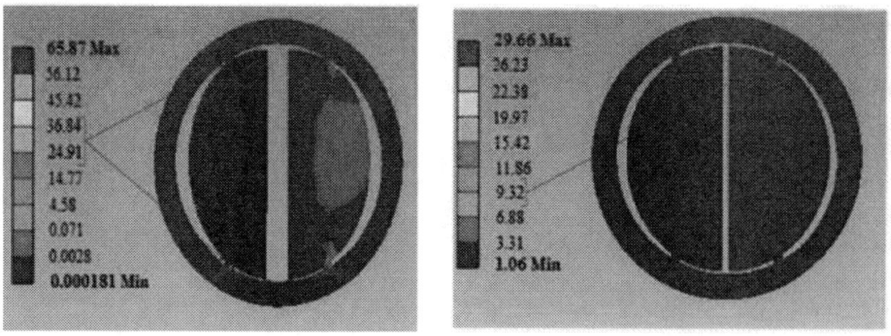

Fig. 5. von Mises stress profile for 0.6mm and 0.7mm thick leaflets

Table 1. von Mises stresses for variable leaflet thickness at different physiological positions of the valve leaflet

	0.5 mm	**0.6 mm**	**0.7 mm**
Leaflet's Position (θ°)	**von-Mises stress (MPa)**	**von-Mises stress (MPa)**	**von-Mises stress (MPa)**
25	56	36	11
30	45	30.5	9
63	35	24	5.5
85	14	11.5	3.5

3 Conclusion

von Mises stresses developed in a Bileaflet mechanical heart valve with variable leaflet's thicknesses have been presented in this paper, when it is subjected to physiological blood flow. The thickness of the leaflet varies from 0.5mm to 0.7mm, by an interval of 0.1mm. It has been observed that high von Mises stresses than the allowable range are developed in thinner leaflets of thickness 0.5mm and 0.6mm. The hinge region of the leaflets has been observed to be subjected to high equivalent stresses. Such leaflets can fail due to yielding and therefore, are not suitable for sustainable design of BMHV. A 0.7mm thick leaflet has been found to develop very low equivalent stresses than the allowable range, and is therefore recommended for sustainable manufacturing of BMHV.

References

1. T.H. Hong, C.R. Choi, C.N. Kim, Eng. Technol. **25** pp. 679-684 (World Academy of Science, Engineering and Technology, France, 2009)
2. F. Zhou, Y.Y. Cui, L.L. Wu and J. Yang, L. Liu, M.F. Maitz, I.G. Brown and N. Huang, Jr. of Med. and Bio. Engg., **36** pp. 110-120 (Biomedical Engineering Society of the R.O.C., Taiwan, 2016)
3. C.S. Peskin, Jr. of Comp. Phys.,**25** pp. 220-252 (Elsevier, United States, 1977)
4. E. Kuhl, H. Askes, P. Steinmann Comput. Methods Appl. Mech. Engrg. **193** pp. 4207-4222(Elsevier, United States, 2004)
5. C. Michler, S.J. Hulshoff, E.H. Brummelen, R. Borst, Computers & Fluids, **33(5)** pp. 839-848 (Elsevier, United Kingdom, 2004)
6. J. Donea, S. Giuliani, J.P. Halleux, *Comput. Methods Appl. Mech. Engrg.*, pp. 689-723 (Elsevier, United States, 1982)

DOI: 10.1051/matecconf/201713104004

7. K. Dumont, J.M. Stijnen , J. Vierendeels , F.N. van de Vosse , P.R. Verdonck , Comput Methods Biomech Biomed Engin. **3** pp. 139-146 (Taylor and Francis, United Kingdom, 2004)
8. M. Nobili, U. Morbiducci, R. Ponzinic, C.D. Gaudiod, A. Balduccid, M. Grigioni, F.M. Montevecchib and A. Redaelli, Journal of Biomechanics, **41** pp. 2539–2550 (Elsevier BV, Netherland, 2008)
9. H.H. Yeh, D. Grecov and S. Karri, Journal of Medical and Biological Engineering, **34(5)** pp. 482-486 (Biomedical Engineering Society of the R.O.C., Taiwan, 2014)
10. V.T. Nguyen, Y.T. Kuan, P.Y. Chen, L Ge, F Sotiropoulos, A.P. Yoganathan and H.H. Leo, Cardiovascular Engineering and Technology, **3** pp. 88-100 (Springer, United States, 2012)
11. Y. J. Kwon, C. N. Kim, J. W Lee, KSME International Journal, **17** pp. 1316–1323. (2003)
12. J. Mazumdar, *An introduction to Mathematical physiology and biology* (Cambridge University Press, Adelaide, 1989)
13. T. Nakamura, S. Kaneko, *Flow Induced Vibrations* (Elsevier Science, Tokyo, 2008)
14. Y.J. Kwon, Journal of Mechanical Science and Technology **22(11)** pp. 2038-2047 (Korean Society of Mechanical Engineers, South Korea, 2008)
15. D. N. Ghista, A. P. Rao, J. of Biomechanics, **5(3)** pp. 295–296. (Elsevier BV, Netherland, 1972)
16. A. Cataloglu, R. E. Clark, P. L. Gould, Jr. of Biomechanics, **10(3)** pp. 153–158 (Elsevier BV, Netherland, 1977).
17. Y. J. Kwon, JSME International Journal Series A, **46(4)** pp.550–558. (Japan Society of Mechanical Engineers, Japan, 2003)
18. H. Reul, A. Vahlbruch , M. Giersiepen ,V. Hirtz and S. Effert , Jr. Biomech. **23** pp. 181-191 (Elsevier BV, Netherland, 1990)
19. M.D. de Tullio, G. Pedrizzetti, and R. Verzicco, Acta Mech. **216** pp. 146-163 (Springer Verlag, Germany, 2010)
20. S. Annerel, T. Claessens, P. Van Ransbeeck, P. Segers, Vierendeels J, *Aortic Valve* pp. 29-39 (Intech, Croatia, 2011)
21. A.C. Guyton and J. Hall, *Textbook of medical physiology* (SAUNDERS ELSEVIER, Philadelphia,1994)
22. M.S. Nasif, S.K. Kadhim, H.H. Al-Kayiem, R Al Waked, ARPN Jr. of Engg. and App. Sci., **11** pp. 11971-1197 (Asian Research Publishing Network (ARPN), Pakistan, 2016)

Engineering economic analysis of meliponiculture in Malaysia considering current market price

Firdaus Basrawi[1*], *Asnul Hadi* Ahmad [1], *Daing Mohamad Nafiz* Daing Idris[1], *Mohd Rashidi* Maarof[1], *MRR* Chand[1] *and Ahmad Syazwan* Ramli[1]

[1] Meliponini Engineering Laboratory (MePEL), Energy Sustainability Focus Group (ESFG), Faculty of Mechanical Engineering, Universiti Malaysia Pahang.

Abstract. Stingless bees (kelulut) keeping is now a trend in Malaysia. However, since demand for the source of colony in log is increasing, the log price is rapidly increasing. But, there is no data reported on the economic viability of meliponiculture in the current market price. Thus, the objective of this study is to clarify the economic viability of investment in meliponiculture in the current market price by engineering economic perspective. Investment in meliponiculture was analysed using Equivalent Annual Uniform Cost (EAUC), Internal Rate of Return (IRR) and Breakeven Analysis. A small start-up with 30 units of logs or hives was considered in the analysis. All raw data was acquainted from current Malaysian market price, but only revenue from honey was considered. It was found that EAUC indicated that the annual worth of the log system is 23% better than the hive system. However, IRR calculation indicated that both the log and the hive systems offer margin exceeding 55% which is a very good return in general investment. In addition, it was also found that the log system had breakeven after 8th month, whereas the hive was 13th month. Better economic value could be obtained if revenue from by-products are considered. Thus, it can be concluded that meliponiculture is still very economically viable in Malaysia market trend, and the hive systems could be a better choice if splitting colony, maintenance, safety and aesthetics points of view are considered.

1 Introduction

Bee keeping is a traditional culture in getting honey in tropical climate country. For Malaysia, it is recorded that bee keeping has established since Malacca Sultanate era (Hassan, Z., 2003). However, it is believed that bee keeping has been practiced earlier than that in Malaysia. Bee keepers preferred Apis bees than stingless bees because the honey production for stingless bee is too little, and it is well known that the stingless bees honey could not be commercialized in a large-scale (Jaafar, 2012; Ismail, 2012). However, stingless bees honey is getting high demand and has high commercial value in Malaysia because it is an alternative to the widespread adulterated honey in the country, and also because of its better health benefit than Apis honey (Ismail, 2014; Idris, 2013; Jaafar, 2011b). In addition, high commercial value of by-products from meliponiculture including propolis and bee bread, and government support through funding and research in public institutions also contribute to the grow of industrialization of meliponiculture. Bee keepers

* Corresponding author: mfirdausb@ump.edu.my

usually obtain a colony from Orang Asli to start the meliponiculture activity. The cost to have a log with topping or honey cassette was about RM100-150/log in 2010-2012, but now the price has increased to RM350-700/log.

Many new investors think about higher capital cost, and other challenges due to stolen logs, threats by overheat, animals and insects (Sinar Harian, 2016; Sinar Harian, 2015). There are also perception that it is currently not worthy to invest in meliponiculture under current market price. However, there is no clear data on the economic analysis of meliponiculture under current market. Thus, the objective of this study is to investigate the economic viability of investment in meliponiculture in the current market price by engineering economic perspective. Investment in meliponiculture was analyzed using Equivalent Annual Uniform Cost (EAUC), Internal Rate of Return (IRR) and Breakeven Analysis. A small start-up with 30 units of log or hive were considered in the calculation. EAUC, IRR and Breakeven Analysis was applied to determine the comparison of annual operating cost, the exact operational margin, and the breakeven time consensus depending on fixed and variable cost.

2 Material and method

The first common evaluation in engineering economic is by Equivalent Annual Uniform Cost (EAUC). It is used to determine the equal opportunity of investment suitability. The objective is to convert money into an equivalent uniform cost in a year. The main advantage of EAUC is that it could compare different investments that have different lifetime. This is because it is an equivalent annual cost over the lifetime of the project. The parameter used in the calculation is shown in Table 1. Data used are value in the current market. It should be noted that hive system includes a mounting system, and data shown for labor, maintenance, rental, revenue are annual data.

Table 1. The parameter considered in EAUC method

Type	Log	Hive
Purchasing cost (RM) Logs Fence	450 x 30 = 13,500 5,000	90 X 30 = 27,000 5,000
Site Cleaning	1,000	1,000
Annual maintenance cost (RM	1,000	5,00
Annual labour cost (RM)	9,000	9,000
Rent (RM)	3,6000	3,6000
Sale value (RM)	32,400	32,400
Life (years)	10	10

EAUC can be calculated by the following equation.

$$AW_{Log} = (AP, i, \%LT) - A_{operation} + A_{revenue} \tag{1}$$

where AWlog is Annual Worth [RM], A/P is finding Annual operation cost and revenue in terms of present value [RM], i is expected dividend/ interest/ margin of the project in decimal, LT is lifetime of the investment [year], Aoperation is per annual expenditure due to operation [RM], and Arevenue is per annual revenue [RM] after sales.

Then, Internal Rate Return (IRR) was calculated to evaluate the relevance of the investment. IRR indicates the exact rate of return that is received on an investment. IRR is also defined as the investment rate that makes the Present Worth of all expenditures is equal to the Present Worth of all income, i.e., the Net Present Value (NPV) equals zero. To calculate the IRR, the NPV of all values are calculated as a function of the interest rate, and then interest rate that makes the NPV is equal to zero is calculated. IRR can be calculated by the following equation;

$$PW_{Loss} = PW_{Gain} \tag{2}$$

$$PW_{Loss} = A_{operation}(P/A, i, LT) - A_{revenue}(P/A, i, LT) \tag{3}$$

where *PW* is Present Worth [RM], $A_{operation}$ is per annual expenditure due to operation in negative value [RM], and $A_{revenue}$ is per annual revenue [RM] after sales, *P/A* is finding Present worth of money, *P* with given Annual operation and revenue cost [RM], *i* is expected dividend/ interest/ margin of the project, *LT* is lifetime of the investment [year].

Table 2. Schedule for breakeven analysis

Cost	Type	Log [RM]	Hive [RM]
Fixed	Logs	450 x 30 = 13,500	90 X 30 = 27,000
Fixed	Fence	5,000	5,000
Fixed	Site Cleaning	1,000	1,000
Fixed	Annual maintenance cost One-Off	1,000	5,00
Fixed	Annual labour cost (RM) RM 750.00 Monthly	9,000	9,000
Fixed	Rent (RM) RM 300.00 Monthly	3,6000	3,6000
Revenue	Sale value [RM / kg]	300	300
Variable Cost	Bottle [RM/ 100 ml]	2	2
Variable Cost	Harvesting and Bottling (RM 100/ 9 kg which produce 90 bottle of 100 ml capacity) 100 /9 = RM 11.11/ kg	11.11	11.11

Finally breakeven calculation was carried out to determine the payback period of the investment. Payback period is the period of time required for the revenue from an investment to equal the cost of the investment. The breakeven calculation based on given schedule shown in Table 2. It should be noted that more detail parameters were considered as compared to Table 1.
Payback Period in month PBP_{month} could be obtained by the following equation;

$$PBP_{month} = \frac{m_{BE}}{m_{month}} \tag{4}$$

Where m_{BE} of honey needed to achieve breakeven [kg], m_{month} is estimated mass of honey produced monthly [kg], and it is equal to [9 kg/month] for 30 logs or hives. m_{BE} could be calculated by Equation (5), and to achieve breakeven, the Total contribution $T.Cont$ must be equal to Fixed Cost (FC) as shown in equation (6).

$$m_{BE} = \frac{T.Cont}{Cont_{perkg}} \tag{5}$$

$$T.Cont = FC \tag{6}$$

where $T.Cont$ is in [RM], and $Cont_{perkg}$ is Contribution per-kg [RM/kg] is calculated by difference between revenue total variable cost per-kg shown in Equation (7).

$$Cont_{perkg} = REV_{perkg} - TVC_{perkg} \tag{7}$$

where Rev_{perkg} is revenue per-kg obtained from sales of honey [RM/kg], and TVC_{perkg} is total variable cost per-kg of bottling and operational cost [RM/kg]. It should be noted that it takes time for colony to produce honey after honey cassette is placed to the log, or after colony is transferred to hive system. Thus, expected 1st batch of the honey production is after four month with 0.3 [kg-honey/log/month] produced.

3 Result

3.1 Equivalent annual uniform cost

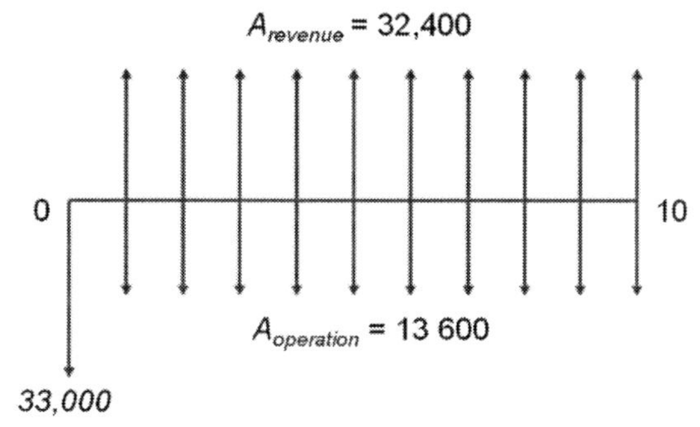

AW_{log} = -33,000 (A/P, 20%,10) -13,600 + 32,400
AW_{log} = -33,000 (0.2385) -13,600 + 32,400
AW_{log} = 10,929.50

Fig. 1. EAUC for log system

 DOI: 10.1051/matecconf/201713104005

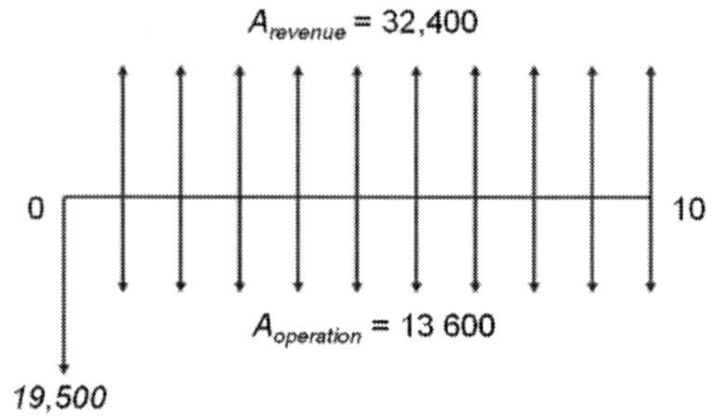

AW_{log} = -19,500 (A/P, 20%,10) -13,600 + 32,400
AW_{log} = -19,500 (0.2385) -13,600 + 32,400
AW_{log} = 14,149.25

Fig. 2. EAUC for Hive system

Results of EAUC for log system and hive system is shown in Fig. 1 and Fig. 2, respectively. It shows that uniform revenue and variable operational cost is same because same production and operation was estimated for both systems. Since initial cost of log type is 40% cheaper, log system has 22.7% higher EAUC value than hive system for 10 years of lifecycle considered. However, the calculation never discusses the ease of maintenance and operation, splitting colony benefit of hive.

3.2 Internal rate of return

Results of IRR for log and hive systems is shown in Table 3 and Table 4, respectively. It was found that IRR for the log system is more than 69%, whereas IRR for the hive system is 58.4%. This indicates that both investments shows very high return and worth to invest. Even the lower IRR of hive system is actually very good for investment, although higher initial investment is required. Thus, both investment opportunity are very promising even in the current market price. If production of propolis, bee bread, and by-products are considered, even higher return is expected from the investment.

$$PW_{Loss} = PW_{Gain}$$

$$19{,}500 = 13{,}600(P/A, i, 10) + 32{,}400(P/A, i, 10)$$

$$19{,}500 + 13{,}600(P/A, i, 10) + 32{,}400(P/A, i, 10) = 0$$

Table 3. IRR and NPV for log system

When i [%]	**P/A (10)**	**NPV**
i =20%	4.192	-59309.6
i= 60%	1.652	-11557.6

Then,

$$\frac{i - 20}{60 - 20} = \frac{0 - (-59309.6)}{-11557.56 - 59309.6}$$

$$i_{log} = 69.7\%$$

$$PW_{Loss} = PW_{Gain}$$

$$33{,}300 = 13{,}600(P/A, i, 10) + 32{,}400(P/A, i, 10)$$

$$33{,}300 + 13{,}600(P/A, i, 10) + 32{,}400(P/A, i, 10) = 0$$

Table 4. IRR and NPV for hive system

When i [%]	**P/A (10)**	**NPV**
i =20%	4.192	-45809.6
i= 60%	1.652	1942.6

Then,

$$\frac{i-20}{60-20} = \frac{0-(-45809.6)}{1942.6-(-45809.6}$$

$$i_{hive} = 58.4\%$$

3.3 Breakeven and payback period

Finally is the results of breakeven and payback period. Fig. 3 shows the breakeven of log and hive systems. As shown in Fig. 3, both systems starts to produce honey in 4th month of the life cycle. Since estimated production is same at 9kg/month, payback period for log system is at 8.35 months, whereas for hive system is 13.56 months. Although hive system has longer period, but it is still can get payback within 14 months which is considered as a short payback period for an investment. Thus, meliponiculture is in general it is too good as an investment with high IRR and short payback period. If bee bread, propolis and by-products are considered, the payback period could be even shorter. Both systems is therefore economically viable enough to be invested in Malaysia even in the current market price.

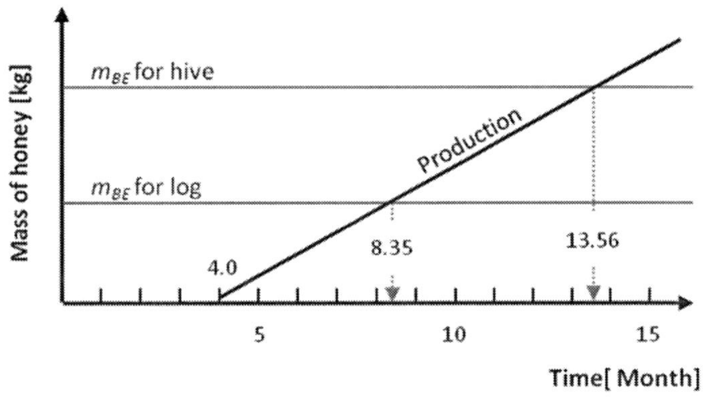

Fig. 2. Breakeven comparison of both system

4 Conclusion

Investment in meliponiculture by log system and hive system with only honey sales as revenue was analyzed by engineering economic analysis. It was found that hive system has 40% higher capital cost, and log system has 22.7% higher EAUC. IRR of log system is 70%, whereas hive system is 58%, which is both has high return. Breakeven analysis shows that log system has 8.35 months of payback period, whereas hive system has 13.3 months of payback period. Although log system shows better results, if easiness of maintenance, colony splitting, and risks are considered, hive system could be a better choice. Both investment is economically viable in the current market price, and if sales of bee bread, propolis and by-products are considered, even better economic results could be obtained.

The authors would like to thank the Ministry of Education Malaysia for providing Research Grant under FRGS, RDU140119.

References

1. Hassan, Z. (2003). Aktiviti Keusahawanan Masyarakat Melayu pada Zaman Kegemilangan Kesultanan Melayu Melaka. In Seminar Za'ba Mengenai Alam Melayu (Nilai-nilai Peradaban Melayu). 1-21.
2. Idris, S. H. (2013). Tingkatkan Penghasilan Madu Lebah Tempatan. Agrobiz, 6.
3. Ismail, M. M. (2012). Honey Marketing in Malaysia, Agricultural Marketing Issues of Selected. Commodities.
4. Ismail, M. M. (2014). Competitiveness of Beekeeping Industry in Malaysia. Inaugural Lecture Series, Universiti Putra Malaysia.
5. Jaafar, M. F. (2011b). Penternakan Kelulut (Teknologi MARDI). Agromedia, 54-57.
6. Jaafar, M. F. (2012). Manual Teknologi Penternakan Lebah Kelulut. MARDI.
7. Sinar Harian (2016). Lalat askar hitam ancam kelulut. 23 July 2016.
8. Sinar Harian (2015). Sarang lebah kelulut dicuri. 27 April 2016.

MATEC Web of Conferences **131**, 04006 (2017) DOI: 10.1051/matecconf/201713104006
UTP-UMP SES 2017

Integrating Life Cycle Costing (LCC) and Life Cycle Assessment (LCA) Model for Selection of Centralized Chilled Water Generation – Review Paper

Ali Akbar[1] and *Ainul Akmar Binti* Mokhtar[2]

[12] Mechanical Engineering Department, Universiti Teknologi PETRONAS, Bander Seri Iskandar, Perak, Malaysia

Abstract. This paper presents a review of literature relating to Life Cycle Costing (LCC) and Life Cycle Assessments (LCA) in Gas District Cooling plant. The gas district cooling plant uses either vapour compression process or absorption process for the chilled water generation. Both processes have impacts on the economic aspects as well as the contribution to CO_2 emissions; thus, integration of the cost and environmental analysis of both processes are necessary. An extensive body of literature exists on both subjects, however, there is very limited number of available literature on the integration of LCC and LCA. The purpose of this review is to find the most suitable optimization model which can integrate the LCC and LCA to provide the best decision in choosing the most cost-effective and environment-friendly system. This review assesses the literature from thirty-four journals, research papers, and case reports from the year 1995 to 2017. The result of this review shows that the goal programming methodology is the appropriate method for integrating LCC and LCA because it provides the optimal decision in choosing the most cost-effective, environment-friendly system for the chilled water generation.

1 Introduction

PETRONAS, the national oil corporation of Malaysia created the terminology Gas District Cooling (GDC). PETRONAS originally co-owned the plants. It is a cogeneration plant system that produces chilled water and electrical power simultaneously. Majid et al. [1] highlighted the various systems around the world which have a resemblance to GDC but they are known as district cooling, district heating, and district energy, it depends on the nature of their operations of which some do not produce electricity. Adzuieen et al.[2] discussed that the natural gas is used as the fuel for gas turbines for the GDC system, which generates electricity. The waste heat from the turbine outlet is used as the primary energy source in steam absorption chillers to produce cooling water for air-conditioning systems. Here, the heat is used to vaporize the cooling-liquid solution in the generator of steam absorption chillers. The free heat or energy from the turbine outlets is utilized by such an integrated energy system, which would otherwise be released to the environment and wasted [3]. Since GDC uses mainly the steam absorption chiller, which uses only water and lithium bromide so the environment is better protected with the use of GDC system [4]. The usage electric chillers, which use refrigerants such as R134a, are minimized in GDC plants, whereas in conventional systems, these chillers are the primary equipment used for generation of chilled water [5].

Gilani et al. [6, 7] indicated that most of the GDC plants used vapor compression and absorption processes for chilled water generation. These processes differ in terms of the source of energy, operation, and heat emission. Since these processes play an important role in enhancing the efficiency of the GDC plant but also contributes to CO_2 emissions to the environment. Many analyses have been done by different authors on gas district cooling (GDC) plant at University Teknologi PETRONAS (UTP) covering life cycle costing (LCC) of chilled water generation. However, there is no any integrative study was done which can integrate and simultaneously assess the cost and CO_2 emission of both the processes used in chilled water generation at GDC-UTP. So the purpose of this review is to find the most suitable optimization model which can integrate the LCC and LCA simultaneously to provide the best decision in choosing the most cost-effective and environment-friendly system for chilled water generation.

This review will assess the literature from maximum possible journals, research papers, and case reports available from the previous years. This review is initiated by categorizing the literature based on the year of publication, followed by the categorization based on the methodology used by literature review. Some of the categories were used to classify the LCC, some are in LCA, and the others are in the integration of LCC and LCA.

The next step is to decide how to collect the material and to choose the case studies suitable for the review. After that, the articles chosen will be classified according to the different categories on LCC, LCA and on their integration also.

Based on our literature review, the literature in the field of life cycle costing and life cycle assessment was quite fragmented. Therefore, we decided to use several databases instead of individual journals to identify LCC case studies. The databases were chosen according to our access to different databases. Due to the contents of these databases in Google Scholar, it was not possible to define article sections to narrow down the search, so we decided to limit the search only to article title.

1.1 Life Cycle Costing (LCC)

Life Cycle Costing (LCC) evolved in the 1960s, during the development of United States Department of Defense to enhance their cost-effectiveness [8]. Burhani et al. [9] emphasized that the LCC is the summation of costs estimates associated with the total product life cycle activities. It attempts to identify the best value for investment consumption (i.e. the lowest long-term cost that addresses the required performance) [4].

In these days, LCC is used widely in industries as decision-making tools or as management tools [10]. The analyst should assure that the level of details integrated into LCC is sound with the level of investment decision under consideration. There comes a point of diminishing returns as more and more cost factors are incorporated into LCC [11]. As Sullivan et al. [12] discussed the relative influence of individual factors on the LCC analysis results may change from primary to minor to inconsequential. For example, minor differences in future costs have a significant effect on discounted present value. Including such factors as this will complicate the analysis without producing actual improvement in analysis outcomes. Including all factors in every analysis is frequently not progressive [13]. In conducting an LCC, analysts should investigate all aspects of inclusion and clarify the motive for eliminating factors [14]. In simple terms, LCC is an analysis technique that guides to better investment outcomes [15].

1.2 Life Cycle Assessment (LCA)

Yusnee et al. [16] and Yang et al. [17] stress that the life cycle assessment (LCA) is the investigation and evaluation of the environmental impacts of a given product or service. The main objective of LCA is to identify and improve the environmental effects and aspects of products at various points in their life cycle. It is an integrative tool for the quantitative evaluation of the environmental impacts of a product or service system through all phases of its life. LCA takes a whole life cycle, from the initial stage to the disposal stage of any product/equipment. In industry, LCA is typically used for the design or improvement of products and processes [18]. This may involve identifying environmental hotspots in production processes as well as considering upstream or downstream processes such as energy sources, product additives, end-of-life management options or by-products consumed in the product use phase [19]. LCA results can also be used as part of the environmental management system to prove continuous improvement or to provide information for eco-labeling or environmental rating schemes [20]. LCA quantifies whether a product is environmentally friendly or not. In another way, LCA qualifies which product is greener [21]. LCA analyzes the life cycle of a product in environmental terms on the basis of emission released and consumption of raw materials. This actually can support future decision-making.

1.3 Combining Life Cycle Costing And Life Cycle Assessment

When combining economic and ecological dimensions to GDC, they can be alleged to balance each other in overall development. Lower energy utilization equals lower running costs [22]. The existing literature presents some examples of connecting the economic and ecological dimensions of life cycles. Attempted in bringing together the economic LCC with the environmental LCA was shown in the case of municipal waste management systems [23]. Heijungs et al. [24] introduced a common unit to measure both total costing and environmental impacts. However this technique enabled the analysis, but issues of rational methodology between economic assessments and environmental assessments remained same. Junnila et al. [20, 25] presented a method of calculating the economic and environmental life cycle costs in one assessment model and also proposed another method in which LCA and LCC were combined together to address the life cycle green costs of buildings. A life cycle management approach was proposed in this study for large-scale development resorts, where LCC and LCA were linked together in order to create designs that provide environmental sustainability with low operational costs [26]. This study combines LCC and LCA in a case of GDC plant, aiming to classify the actual technologies that could elaborate the maximum sustainable viability and assess the emissions and correspondent mitigation potentials associated with the different technologies [27].

Considering a life cycle perspective can support sustainable decision-making, with LCC and LCA being the leading tools for addressing economic and environmental burdens, respectively [28].

1.4 Multi-Objective Goal Programming

Due to the presence of two or more conflicting objectives at the same time, it is often impossible to model the problem as a single-objective programming problem [29]. To overcome this problem the need to use the Multi-objective goal programming approach cannot be neglected. Multi-objective goal programming (also known as a multi-objective optimization) is an area of multiple criteria decision making, that is concerned with mathematical optimization problems involving more than one objective function to be optimized simultaneously [30]. The concept of multi-objective optimization was originally defined by [31]. Since then, it has experienced enormous progress and has been successfully applied in different fields of science. Jones et al. [32] developed a multi-objective model by using goal programming methodology for dynamic cell formation problem. The first objective function of the developed model seeks to minimize total cost of machine procurement, machine relocation, inter-cell moves, overtime utilization, worker hiring/laying-off, and worker moves between cells; while the second objective function maximizes labor utilization of the cellular manufacturing system. As in 2016, Shao et al. [33] addressed the problem of representing the continuous but non-convex set of non-dominated points of a multi-objective linear programming by a finite subset of points. Thus it is recommended that a multi-objective linear programming is also required for GDC-UTP which will integrate the LCC and LCA and will be helpful to provide the best decision among choosing the most economical and environment-friendly system for chilled water generation

1.5 Conclusion and Overview

The purpose of this study is to examine both the environmental and economic life cycle results in order to add valuable information for decision-makers and future researchers. Despite the similarity of their names, LCC and LCA have major methodological differences. It is found that the process scope of the LCC includes only those processes imposing direct economic costs (or benefits) upon the decision maker. As in an LCA, we might neglect costs which are expected to be equivalent among alternatives. The LCC takes careful account of the timing of the cost flows, while LCA neglects flow timing. The LCC includes cost risks, and how they are altered or avoided as a function of product/process design investment options, but the LCA considers flows of pollutants and resources from each of the modeled processes.

However, this study combines LCC and LCA, aiming to clarify the decision situation which decides the suitability of the process or equipment and to classify the actual technologies that could elaborate the maximum sustainable viability. Whereas, findings in literature review indicated that the Goal programming approach can support sustainable decision-making, by integrating the LCC and LCA. Furthermore, the goal programming approach will also help to develop criteria for selecting the most cost-effective and environment-friendly system for chilled water generation.

References

1. A. Majid and M. Amin, (2011).
2. N. Adzuieen, M. Amin and A. Majid, (2013).
3. A. Nordin, N. Buyamin, M. A. A. Majid and S. A. S. Ariffin, Int. J. Comp. Elec. Eng. **5** (3), 284 (2013).
4. S. Amear, S. Ariffin, A. Nordin, N. Buyamin, M. Amin and A. Majid, Asian J. Sci. Res. 6 (2), 299 (2013).
5. M. A. A. Majid, S. A. Sulaiman, I. Ibrahim and Z. Baharddin, J. Appl. Sci. **12** (24), 2636 (2012).
6. S. Gilani, M. Amin and C. Rangkuti, Dist. Energy **94** (3), 14-18 (2008).
7. S. I.-u.-H. Gilani, M. A. A. Majid and C. Rangkuti, (2006).
8. M. J. Epstein, Report No. 0786302305, 1996.
9. N. R. A. Burhani, A. Majid and M. Amin, (2012).
10. P. Gluch and H. Baumann, Build. Environ. **39** (5), 571-580 (2004).

11. S. J. Kirk and A. J. Dell'Isola, LCC. Design. Prof. (1995).
12. W. G. Sullivan, E. M. Wicks and C. P. Koelling, (Singapore. Prentice Hall, Inc, 2012).
13. R. Flanagan, G. Norman, J. Meadows and G. Robinson, Practice. BSP Professional Books, Oxford (1989).
14. Y. Qian, Z. Huang and Y. Jiang, Chem. Ind. Eng. Progress **25** (2), 126 (2006).
15. D. Langdon, Literature Review, Davis Langdon Management Consulting (2007).
16. N. I. B. M. Yusnee, 2010.
17. D. Yang, L. Fan, F. Shi, Q. Liu and Y. Wang, Resources, Conserv. Recycl. (2016).
18. E. Martínez, F. Sanz, S. Pellegrini, E. Jiménez and J. Blanco, Renew. Energy **34** (3), 667-673 (2009).
19. H. W. Kua and S. Kamath, J. Clean Prd **81**, 190-200 (2014).
20. S. Junnila, A. Horvath and A. A. Guggemos, J. Infrastruct. Syst. **12** (1), 10-17 (2006).
21. S. Xing, Z. Xu and G. Jun, Energy Build. **40** (7), 1188-1193 (2008).
22. T. Nauclér and P.-A. Enkvist, McKinsey & Company 192 (2009).
23. M. C. Reich, J. Clean Prd **13** (3), 253-263 (2005).
24. R. Heijungs, E. Settanni and J. Guinée, Int. J. LCA **18** (9), 1722-1733 (2013).
25. S. Junnila, Int. J. LCA **13** (5), 432 (2008).
26. K. L. Brown, D. C. Greer and B. Schwegler, in Towards Life Cycle Sustainability Management (Springer, 2011), pp. 229-238.
27. M. Ristimäki, A. Säynäjoki, J. Heinonen and S. Junnila, Energy **63**, 168-179 (2013).
28. R. Kannan, C. Tso, R. Osman and H. Ho, Energy Convers. Manage. **45** (18), 3093-3107 (2004).
29. J. Razmi, E. Jafarian and S. H. Amin, Exp. Sys. Applic. **65**, 181-193 (2016).
30. C.-L. Hwang and A. S. M. Masud, Multi. Obj. dec. mak. Method. Applic. (Springer Science & Business Media, 2012).
31. S. Mirjalili, S. Saremi, S. M. Mirjalili and L. d. S. Coelho, Exp. Sys Applic. **47**, 106-119 (2016).
32. D. Jones, H. Florentino, D. Cantane and R. Oliveira, Eur. J. O.R. **255** (3), 845-855 (2016).
33. L. Shao and M. Ehrgott, Eur. J. O.R. (2016).

Performance measure of eco-process innovation: insights from a literature review

Suziyana Mat Dahan [1,*], *Sha'ri* Mohd Yusof [2], and *Mohd Yusof* Taib [3]

[1]Universiti Malaysia Pahang, Faculty of Industrial Management, Lebuhraya Tun Razak, 26300 Gambang, Pahang, Malaysia.
[2] Universiti Teknologi Malaysia, Razak School of Engineering and Advanced Technology, 54100 Kuala Lumpur, Malaysia.
[3] Universiti Malaysia Pahang, Faculty of Mechanical Engineering, 26600 Pekan, Pahang, Malaysia.

Abstract. Eco-process innovation has been recognised as one of the important strategies for mitigating the growing environmental challenges. Its concept has succeeded in drawing the interests of many scholars worldwide. The aim of this paper is to review the literature to clarify how one actually measure eco-process innovation performance. Critical analysis of literature has been performed in this study. Database searches were mainly relied to compile the literature. In relation to eco-process innovation assessment, results revealed that most prior works focused on the economic and environmental performance with the exclusion of social performance and very limited attempts done in analysing real operational data. It was identified that there is a need for empirical investigations on measuring the social performance of eco-process innovation along with the economic and environmental performance, and on development of operational measuring instrument as these study area have not been well explored.

1 Introduction

The environmental conservation agenda has emerged as essential concern amongst manufacturers around the globe due to growing unfavourable environmental impacts of their operations. As major consumers of energy and natural resources ([1]), and main potential producers of pollutions ([2], [1]), manufacturing firms are forced by various factors to seek for secret recipes of mitigating the environmental risks. Eco-innovation is one of the strategic green solutions to achieve sustainable development in manufacturing industries in response to the increasing economic, environmental and economic pressure ([3], [4], [5]). It is evident that manufacturers have focused on improving production processes (i.e. eco-process innovation) to ensure the optimal use of resources and minimal generation of waste and pollution in their production processes ([6]). This will change the perspective of seeing the manufacturing activities as causing the negative environmental

* Corresponding author: suziedahan@gmail.com

impacts to the one that perceives manufacturers as an entity which fulfilling society's needs and increasing their standard of living, and improving quality of environment.

The evaluation of eco-innovation implementation among manufacturers is essential to see if their performances are at par or corrective actions are required to reach the balance priority of economic, environmental and social targets ([7]). However, the existence of various versions of assessment indicators has limited the value of assessment results to the firm. Literature showed that limited knowledge has been established to create comprehensive performance measures of eco-innovation ([8]). Hence, the development of better instrument of assessing eco-process innovation performance is in need. Considering the highlighted necessity to establish comprehensive assessment indicators, it is important to ask: What are the main dimensions of eco-process innovation assessment in previous published works? How does one actually measure eco-process innovation performance? The purpose of this paper is to review the literature on measuring aspects and approaches which have been the focus of previous researchers.

This study offers some notable contributions to the literature by pointing out the most important gaps in assessing eco-process innovation performance at firm-level. It proposes a future study agenda for measuring eco-process innovation performance. This paper is structured as follows: Section 2 consists of methods applied to compile relevant literatures for analysis. Section 3 summarises the literatures on emergence of sustainable development, sustainable manufacturing and eco-innovation concepts, and definition and nature of eco-innovation. Section 4 discusses the main findings from the literature review and Section 5 concludes with future research agenda on measures of eco-process innovation performance.

2Research methods

A literature review was conducted to explore how eco-process innovation implementation has been measured, which involved the critical analysis of previous published studies. In compiling the literature, database searches were mainly relied whereby the published studies were identified from the Web of Science, Scopus and Google Scholar databases to ensure the analysis is inclusive of all relevant literature and not limited to the top journal publications only such as implemented by [9]. "eco-process innovation" OR "process eco-innovation" OR "green process innovation" OR "environmental process innovation" OR "sustainable process innovation" OR "green process improvement" AND "manufacturing"' were the search strings employed in finding by topic, title, abstract and keyword. The compiled literature was assessed and downsized based on their main focus in applying indicators for measuring eco-process innovation, in the context of manufacturing firms (i.e. micro-level), full text accessibility and included only empirical studies.

3The emerging concepts

3.1 Sustainable development, sustainable manufacturing and eco-innovation

With the global concerns over the worsening environmental degradation associated with the rapid expansion of industrial activities, there is growing consciousness to move gradually towards the concept of sustainable development. They started to realise the importance of establishing a healthy and sustainable business environment where economic wealth creation, environmental preservation and social wellbeing priority are placed on equal footing in deciding their operational actions ([6]). Manufacturers are playing more significant role to deal with growing environmental challenges by taking more comprehensive and integrated approaches ([1]). The resultant difficulties from economic development such as resource exhaustion (economic), environmental pollution (environment), and social injustice (social) have led to the introduction of sustainable development approach ([10]).

The first use of phrase "sustainable development" could be traced back to the 1980 "World Conservation Strategy", the United Nation Environment Programme (UNEP) and The World Wildlife Fund (WWF) ([10]). The definition then noted in publication of Brundtland Report in 1987 and the subsequent Earth Summits in Rio de Janeiro in 1992 and Johannesburg in 2002, where sustainability and environmental issues gained the participants' attention ([11], [12]). Sustainable development was referred to development that meets the needs of the present generation without compromising the ability of future generations to meet their own needs ([13]). The idea of sustainable development stressed the essential of balancing the Triple Bottom Line (TBL) – economy, environmental and society, in value creation activities in order to pursue the green and sustainable growth ([14]). It is an effort to preserve the earth as a place where human and nature can live together in harmony, while permitting the meeting of current and future generations' needs. In order to ensure the survival of the planet, the economic, ecological and social aspects should be put in a balanced priority. The use of sustainability concept within a more confined context such as production and manufacturing area has promoted the idea of sustainable manufacturing.

Sustainable manufacturing which is also known as environmentally responsible manufacturing or green manufacturing emerged from the United Nations Conference on Environment and Development (UNCED) held in Rio de Janeiro in 1992 ([1], [15]). It was defined as manufacturing practices that do not harm the environment during any part of the manufacturing process ([16]). The concept emphasises on designing and delivering products that minimise negative effects on the environment through their production, use, and disposal. The focus of the idea is on the use of processes that do not pollute the environment or harm consumers, employees, or other members of the community. This includes the initiatives of recycling, conservation, waste management, water supply, environmental protection, regulatory compliance, pollution control etcetera. Manufacturers are focusing on integrating these multiple green methods in their attempts to gain extensive benefits of sustainable manufacturing approach including minimisation of resources and energy consumption, reduction of waste and environmental pollution, and less safety and health threats as much as possible ([12]).

Recently, caused by the necessity to achieve sustainability, firms have been compelled to implement innovation to enjoy the privileges of putting a balanced priority to the three pillars of sustainable development ([17], [18], [10]). As such, the concept of eco-innovation was put forward. It received worldwide attention, was perceived as a bridge ([11], [19], [12]), a critical driver and provides promising way of leading manufacturing industry towards sustainable manufacturing ([20]). This require the manufacturers to adopt a holistic and integrative strategy, encompassing the identification of any potential for eco-innovation implementation at any stage of the manufacturing system, either technical or non-technical solutions, and extends beyond the conventional firm's boundaries which entails shifts in current socio-cultural norms and institutional structures. Thus, each change in ecological solutions introduced by the firm for instance from end-of-pipe technique (elimination of pollution at the point of discharge) to cleaner production technology (reducing energy and materials consumption in the production process) is considered as improvement facilitated by eco-innovation practice ([21], [1]). In other words, eco-innovation is the process of improving manufacturing sustainability ([17]). The adoption of eco-innovation concept plays essential roles in directing the manufacturing sector towards sustainable manufacturing.

The concepts of manufacturing sustainability and eco-innovation are not similar but closely related, and associated with many conceptual overlaps ([21], [1]). Figure 1 illustrates the overlaps that exist between the concepts of sustainable manufacturing and eco-innovation. The evolutionary stages of various sustainable manufacturing approaches are shown by the "waves" spreading towards the right-hand corner of the figure, connected and can only be realised through a combination of the eco-innovation targets on the left and mechanisms at the bottom. As the sustainable manufacturing practices evolve, the eco-innovation processes such as process modification, product design, business model alternatives and the creation of new methods, procedures and arrangements should also advance to induce greater economic, environmental and social gains out of the initiatives. Closed-loop production practice (removal of product disposal stage through wastes treatment and use them as new resources for production), for example, is associated with the application of alternative business model, but when changed to the adoption of industrial ecology approach, will require manufacturer to consider the creation of entirely new production system which also involves the introduction of new organisational and institutional structures.

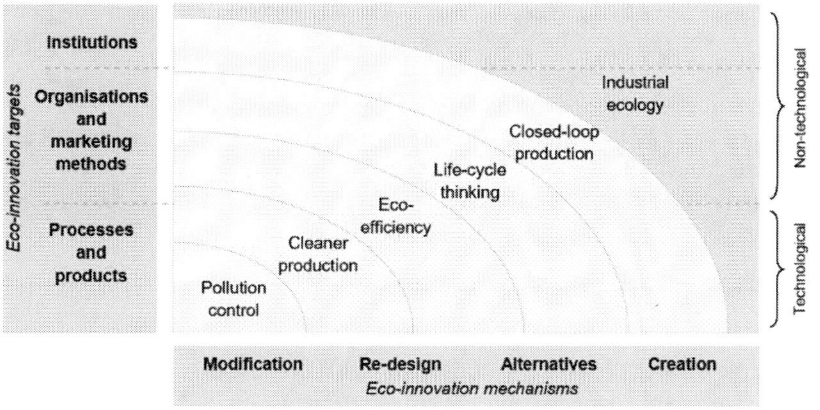

Fig. 1. Conceptual relations between sustainable manufacturing and eco-innovation.
Source: [21]

3.2 Definition and nature of eco-innovation

Understanding the meaning which carried by the term 'eco-innovation' is challenging because the definition keeps on changing and evolving since its first appearance in a book titled *Driving Eco-innovation: A Breakthrough Discipline for Innovation and Sustainability* by [22] ([23], [24], [25]). Since then, scholars have defined eco-innovation differently according to various perspectives and contexts the term is used. Economists, ecologists and sociologists, just to name a few disciplines, all have their preferred perspective. For the purpose of this study, eco-innovation refers to strategic implementation of any form of technical (product and process) and non-technical (organisational, marketing, institutional and system) changes that are either new to the world or new to the firm (through adaptation or adoption approach), with the intention of achieving a balanced priority of economic returns, environmental preservation and society well-being resulting in sustainable economic, environmental, social and institutional benefits. The adapted definition characterises key concerns of recent definitions proposed by scholars such as [26], [27], [28] and [29].

In a broader perspective of the nature of innovations, eco-innovation comprises of activities namely technical eco-innovation and non-technical eco-innovation ([30], [31], [32]). Technical eco-innovations take place in products and processes and involve technologies aimed at improving the environmental performance of products and processes. Non-technical eco-innovations include people-oriented changes which normally occur in marketing (related to implementation of more green product design and packaging, pricing, distribution and promotion activities) ([33]), organizational (shift to a new or improved ecological-oriented managerial arrangement such as Environmental Management System (EMS)) ([34]), institutional (includes improvement of values, norms, culture and behaviour) ([1]) and system (systemic level transformation among networked organizations which involve combination of all other types of eco-innovation – product, process, marketing, organizational and institutional) ([35]). The concern of this study is on the technological eco-innovation which taking place in the production processes. [34] and [12] used the term "eco-process innovation" to describe this type of eco-innovation, and the similar term is also used throughout the discussion of this paper.

Eco-process innovation refers to the changes of existing production methods or addition of new processes to minimise the environmental impacts ([34]). Its practice encompasses activities such as utilising closed looping for solvents, material recycling and consumption of less resource ([36]). Eco-process innovation is directly related to operations activities and concerned with process upgrades or introduction of new techniques and technology into production operations, which improve production efficiency, thereby lead to cost reduction ([34], [8]). In summary, for the purpose of this study, any changes or improvement made on production processes which aimed at or resulted in the reduction of unfavourable environmental impacts is considered as eco-process innovation.

4Results

There have been limited studies of measuring eco-process innovation performance carried out at the firm level. In total, 41 full papers, all empirical were extracted and reviewed to meet the aim of this paper which shown in Table 1. If looking at the measuring aspects, most prior works focuses on the economic and environmental performance without the assessment of social performance. Limited empirical studies have been found evaluating the social aspect of eco-process innovation performance. This is due to the complex nature of evaluating the social performance such as healthcare, safety, workers' satisfaction etcetera, and therefore in need of attention from scholars to develop an appropriate assessment method ([37]). It is noted that only [38], [12] and [39] carried out studies on evaluating the social performance of eco-process innovation practice. In their survey, [38] investigated the relationship between organisations' sustainability performance management practices and sustainability performance (economic, environmental, and social aspect). Whereas [12] study focuses on the influence of green manufacturing and eco-innovation on firm economic, environmental, and social performance. [39], however, performed survey investigation on the effect of eco-product and eco-process innovation on job satisfaction through job intensity (social aspect) with no assessment done on the economic aspect. Apart from these studies, the literature clearly visualise that the investigation on social performance of eco-process innovation implementation has not been well explored by previous researchers.

In terms of measuring approach, all scholars except for [40], [41] and [42], have conducted survey study (i.e. perceptual approach) to validate their proposed eco-process innovation measuring indicators. Results revealed that there were limited attempts to develop eco-process innovation measuring tool capable of quantifying technical and operational data of real production process (i.e. operational approach). [40] developed and tested a model of linking lean system and eco-innovation implementation in production process of a metal working firm. Their Life Cycle Analysis (LCA) showed a significant improvement of incomes, sustainability and social responsibility performance. [41] employed simulation of manufacturing systems method called Discrete Event Simulation (DES) for assessing lean and eco-process improvements. They developed model of manufacturing system which then verified and validated using the empirical data collected through interview of the expert consultants, shop floor measurement and historical data of an automotive manufacturing firm. Their cost-time profile analysis showed reduction in cost as impact of a tailored combination of lean and eco-process innovation. Using the similar method, [42] evaluated the energy efficiency in a vehicle assembly line. Results of the simulation model indicated improvements for energy saving and cost reduction objectives. Even though [40], [41] and [42] have undertook objective analysis of real operational data, but their assessment have been found to only cover the economic and environmental aspects of eco-process innovation. This demonstrates that the development of instrument which can assess the quantifiable measures of eco-process innovation performance has not been well addressed and therefore deserves a specific attention.

Table 1. Studies on eco-process innovation measure at firm level.

No.	Author	Measure Approach		Measure Aspect			Analysis Method	[Country] Sector
		Perceptual	Operational	Economic	Environmental	Social		
1.	[43]	√			√		Statistical	[Taiwan] Information & electronics
2.	[44]	√			√		Statistical	[Taiwan] Information & electronics
3.	[45]	√		√	√		Statistical	[Netherlands] Petrochemicals production
4.	[46]	√			√		Statistical	[Taiwan] Manufacturing & service
5.	[47]	√			√		Statistical	[Taiwan] Manufacturing
6.	[48]	√		√	√		Statistical	[Russia] Manufacturing
7.	[38]	√		√	√	√	Statistical	[Australia] Manufacturing
8.	[36]	√		√	√		Statistical	[Taiwan] IT, electronics, telecommunication

 DOI: 10.1051/matecconf/201713104007

								& others
9.	[49]	√			√		Statistical	[German] Manufacturing & service
10.	[50]	√			√		Statistical	[Italy] Manufacturing
11.	[51]	√			√		Statistical	[Africa] Manufacturing & service
12.	[52]	√			√		Statistical	[Taiwan] Information & electronics
13.	[53]	√			√		Fuzzy set theory, Analytical Network Process (ANP) & entropy weight	[Taiwan] Electrical & Electronics (E&E)
14.	[54]	√			√		Statistical	[China] Manufacturing
15.	[40]		√	√	√		LCA	[Spain] Metal working
16.	[41]		√	√	√		DES	[USA] Automotive
17.	[12]	√		√	√	√	Statistical	[Turkey] Automotive, chemistry & electronic
18.	[55]	√			√		Statistical	[China] E&E
19.	[56]	√			√		Statistical	[Ireland] Manufacturing & service
20.	[57]	√			√		Statistical	[Ireland] Manufacturing & service
21.	[58]	√			√		Statistical	[German] Manufacturing & service
22.	[59]	√			√		Statistical	[China] Manufacturing
23.	[34]	√		√	√		Statistical	[Taiwan] Information technology, electronics, telecommunication & others

 DOI: 10.1051/matecconf/201713104007

24.	[8]	√		√	√		Statistical	[China] Manufacturing
25.	[60]	√			√		Statistical	[French] Manufacturing
26.	[61]	√		√	√		Statistical	[Taiwan] Information technology, electronics, telecommunication & others
27.	[62]	√		√	√		Statistical	[Malaysia] Manufacturing
28.	[39]	√			√	√	Statistical	[Malaysia] (E&E)
29.	[63]	√		√			Statistical	[Malaysia] Food & beverage, rubber & plastics, chemical, E&E & metal working
30.	[64]	√		√	√		Statistical	[Spain] Manufacturing & service
31.	[65]	√		√	√		Statistical	[Brazil] Textile
32.	[66]	√			√		Statistical	[Europe] Manufacturing
33.	[67]	√			√		Statistical	[China] Manufacturing
34.	[68]	√			√		Statistical	[China] Manufacturing
35.	[69]	√			√		Statistical	[Brazil] (E&E)
36.	[42]		√	√	√		DES	[Malaysia] Automotive
37.	[70]	√		√	√		Statistical	[Slovenia] Manufacturing & service
38.	[71]	√		√	√		Statistical	[China] Manufacturing
39.	[72]	√			√		Statistical	[Slovenia] Manufacturing & service
40.	[73]	√			√		Statistical	[Malaysia] Chemical
41.	[74]	√			√		Statistical	[Malaysia] Green technology

5Conclusions

This study drew propositions for further empirical research on measures of eco-process innovation performance. It is argued that eco-innovation could contribute to the three pillars of sustainability performance. Eco-innovation is the process towards improved sustainability, thus measured with the focus on the resulting impact of the innovation process ([17]). The measurement should entail the three pillars of sustainability, namely economic, environmental and social aspects in accordance with eco-innovation definition ([29], [28], [27]). However, this study provides evidence that most empirical research on eco-process innovation assessment address the economic and environmental impacts with no consideration of social performance.

If looking at the measuring approach adopted in previous studies, most scholars have proposed and tested questionnaire-based survey, which noted as associated with respondents' generic and bias responses. Survey questionnaire only provides general information about eco-innovation practice in which technical and operational information of products and processes, and micro-level eco-efficiency aspects of innovation are not able to be assessed ([75], [76]). Moreover, provided data could be seriously biased since responses are self-declared by respondents and tend to give favourable answers to show socially desirable image of themselves or their firms ([76]). [34] added that studies which involve the analysis of objective data would lead to the actual eco-innovation performance and overcome biases. However, the analysis of real production system (i.e. objective data analysis) in measuring eco-process innovation performance has not been fully addressed in the literature and there is a need of extending its application in other quantifiable impacts.

Hence, it is the intention of our future study to develop an instrument of assessing eco-process innovation practice with the measurement indicators of economic, environmental and social performances. The developed set of performance indicators will be validated using operational data of a selected manufacturer.

The authors would like to acknowledge the Ministry of Education Malaysia (MOE), Universiti Malaysia Pahang and Universiti Teknologi Malaysia for their support in this research.

References

1. Organisation for Economic Co-operation and Development (OECD), *Eco-Innovation in Industry: Enabling Green Growth* (OECD Publishing, Paris, 2009)
2. O. C. Aja, H. H. Al-Kayiem, M. G. Zewge, and M. S. Joo, "Overview of Hazardous Waste Management Status in Malaysia", *Manag. Hazard. Wastes*, October (2016)
3. Y. Fernando, X. W. Wen, and M. S. Shaharudin, "Does a Firm's Innovation Category Matter in Practising Eco-innovation? Evidence from the Lens of Malaysia Companies Practicing Green Technology", *J. Manuf. Technol. Manag.*, **27**, 2 (2016)
4. Organisation for Economic Cooperation and Development (OECD), *Green Growth Indicators* (OECD Publishing, Paris, 2014)
5. V. Rizos, A. Behrens, and I. Taranic, "Measuring progress in eco-innovation", ISBN: 978-94-6138-466-9 (2015)
6. Organisation for Economic Co-operation and Development (OECD), "Sustainable Manufacturing and Eco-innovation: First Steps in Building a Common Analytical Framework", (Rochester, New York, 2008)
7. Eco-Innovation Observatory (EIO), "Europe in Transition : Paving the Way to a Green Economy through Eco-innovation", (Brussels, 2013)

8. Y. Dong, X. Wang, J. Jin, Y. Qiao, and L. Shi, "Effects of Eco-innovation Typology on its Performance: Empirical Evidence from Chinese Enterprises", *J. Eng. Technol. Manag.*, **34** (2014)
9. T. Schiederig, F. Tietze, and C. Herstatt, "Green Innovation in Technology and Innovation Management - An Exploratory Literature Review", *R&D Manag.*, **42**, 2 (2012)
10. J.H. Jo, T. Roh, S. Kim, Y.C. Youn, M. Park, K. Han, and E. Jang, "Eco-Innovation for Sustainability: Evidence from 49 Countries in Asia and Europe", *Sustainability*, **7**, 12 (2015)
11. D. Angelo, J. C. Jabbour, and S. V. Galina, "Environmental Innovation: in Search of a Meaning", *World J. Entrep. Manag. Sustain. Dev.*, **8**, 2 (2012)
12. B. Sezen and S. Y. Çankaya, "Effects of Green Manufacturing and Eco-innovation on Sustainability Performance", *Procedia - Soc. Behav. Sci.*, **99** (2013)
13. World Commission on Environment and Development (WCED), "Report of the World Commission on Environment and Development : Our Common Future", (1987)
14. International Union for Conservation of Nature (IUCN), "World Conservation Strategy: Living Resource Conservation for Sustainable Development" (IUCN, Geneva, 1980)
15. K. Rennings, "Redefining Innovation: Eco-innovation Research and The Contribution from Ecological Economics", *Ecol. Econ.*, **32**, 2 (2000)
16. S. Gupta, G. S. Dangayach, A. K. Singh, and P. N. Rao, "Analytic Hierarchy Process (AHP) Model for Evaluating Sustainable Manufacturing Practices in Indian Electrical Panel Industries", *Procedia - Soc. Behav. Sci.*, **189** (2015)
17. G. Huppes, R. Kleijn, R. Huele, R. Ekins, B. Shaw, M. Esders, and S. Schaltegger, "Measuring Eco-Innovation: Framework and Typology of Indicators Based on Caausal Chains" (2008)
18. I. S. Jawahir and A. D. Jayal, "Product and Process Innovation for Modeling of Sustainable Machining Processes", *8th Global Conference on Sustainable Manufacturing*, **2** (2011)
19. J. Carrillo-Hermosilla, D. R. Pablo, and K. Totti, "Diversity of Eco-innovations: Reflections from Selected Case Studies", *J. Clean. Prod.*, **18**, 10–11 (2010)
20. A. N. Sarkar, "Promotion of Eco-innovation to Leverage Sustainable Development of Eco-industry and Green Growth", *Eur. J. Sustain. Dev.*, **2**, 1 (2013)
21. T. Machiba, "Eco-innovation for Enabling Resource Efficiency and Green Growth: Development of an Analytical Framework and Preliminary Analysis of Industry and Policy Practices", *Int. Econ. Econ. Policy*, **7**, 2 (2010)
22. P. Fussler, C., & James, *Driving Eco-innovation: A Breakthrough Discipline for Innovation and Sustainability* (Pitman Publishing, Washington DC, 1996)
23. V. Oltra, R. Kemp, and F. P. Vries, "Environmental Innovation" (2009)
24. S. Ozusaglam, "Essays on Eco-Innovation", Université de Strasbourg (2014)
25 A. F. Xavier, "The Eco-innovation Concepts Through a Strategic Perspective", *International Association for Management of Technology* (2015)
26. B. Bossink, *Eco-Innovation and Sustainability Management* (Taylor & Francis Group, New York, 2012)
27. S. P. Ganapathy, J. Natarajan, A. Gunasekaran, and N. Subramanian, "Influence of Eco-innovation on Indian Manufacturing Sector Sustainable Performance", *Int. J.*

Sustain. Dev. World Ecol., **21**, 3 (2014)

28. S. Scarpellini, J. A. Usón, M. M. Fondevila, A. A. Usón, and E. L. Sastresa, "Eco-innovation Indicators for Sustainable Development: The Role of the Technology Institutes", *Int. J. Innov. Sustain. Dev.*, **10**, 1 (2016)
29. E. Calik and F. Bardudeen, "A Measurement Scale to Evaluate Sustainable Innovation Performance in Manufacturing Organizations", *Procedia CIRP*, **40** (2016)
30. M. Frondel, J. Horbach, and K. Rennings, "End-of-Pipe or Cleaner Production? An Empirical Comparison of Environmental Innovation Decisions Across OECD Countries" (2007)
31. Organisation for Economic Cooperation and Development (OECD), "Sustainable Manufacturing and Eco-Innovation: Framework, Practices and Measurement", (2009)
32. H. S. Özer, "A Review of the Literature on Process Innovation in Remanufacturin", *Int. Rev. Manag. Mark.*, **2**, 3 (2012)
33. N. A. Abu Seman, N. Zakuan, A. Jusoh, M. A. Shoki, A. Z. Bahari, N. Zaidin, and M. M. Z. Saman, "The Development of Green Innovation Measurement Based on Inter Rater Agreement Approach : A Preliminary Stud", *Adv. Mater. Res.*, **903** (2014)
34. C. C. J. Cheng, C. Yang, and C. Sheu, "The Link Between Eco-innovation and Business Performance: A Taiwanese Industry Context", *J. Clean. Prod.*, **64** (2014)
35. Eco-Innovation Observatory (EIO), "Introducing Eco-innovation : from Incremental Changes to Systemic Transformations" (EIO, 2011)
36. C. C. Cheng and E. C. Shiu, "Validation of a Proposed Instrument for Measuring Eco-innovation: An Implementation Perspective", *Technovation*, **32**, 6 (2012)
37. H. Zhang, J. Calvo-amodio, and K. R. Haapala, "A Conceptual Model for Assisting Sustainable Manufacturing Through System Dynamics", *J. Manuf. Syst.*, **32** (2013)
38. D. Gadenne, L. Mia, J. Sands, L. Winata, and G. Hooi, "The Influence of Sustainability Performance Management Pactices on Organisational Sustainability Performance", *J. Account. Organ. Chang.*, **8**, 2 (2012)
39. M. Iranmanesh, S. Zailani, S. Moeinzadeh, and D. Nikbin, "Effect of Green Innovation on Job Satisfaction of Electronic and Electrical Manufacturers' Employees Through Job Intensity: Personal Innovativeness as Moderator", *Rev. Manag. Sci.* (2015)
40. S. Aguado, R. Alvarez, and R. Domingo, "Model of Efficient and Sustainable Improvements in a Lean Production System Through Processes of Environmental Innovation", *J. Clean. Prod.*, **47** (2013)
41. N. Diaz-Elsayed, A. Jondral, S. Greinacher, D. Dornfeld, and G. Lanza, "Assessment of lean and green strategies by simulation of manufacturing systems in discrete production environments", *CIRP Ann. - Manuf. Technol.*, **62**, 1 (2013)
42. O. Abduaziz, S. M. Atnaw, J. K. Cheng, and L. Singh, "Improving Energy Efficiency for the Vehicle Assembly Industry: A Discrete Event Simulation Approach", in *IOP Conference Series: Materials Science and Engineering* (2016)
43. Y. S. Chen, S. B. Lai, and C. T. Wen, "The Influence of Green Innovation Performance on Corporate Advantage in Taiwan", *J. Bus. Ethics*, **67**, 4 (2006)
44. Y. S. Chen, "The Driver of Green Innovation and Green Image - Green Core Competence", *J. Bus. Ethics*, **81**, 3 (2008)
45. T. Ren, "Barriers and Drivers for Process Innovation in the Petrochemical Industry: A Case Study", *J. Eng. Technol. Manag.*, **26**, 4 (2009)

 DOI: 10.1051/matecconf/201713104007

46. T. Y. Chiou, H. K. Chan, F. Lettice, and S. H. Chung, "The Influence of Greening The Suppliers and Green Innovation on Environmental Performance and Competitive Advantage in Taiwan", *Transp. Res. Part E Logist. Transp. Rev.*, **47**, 6 (2011)
47. C. H. Chang, "The Influence of Corporate Environmental Ethics on Competitive Advantage: The Mediation Role of Green Innovation", *J. Bus. Ethics*, **104**, 3 (2011)
48. A. Grobecker, J. Wolf, and R. N. Germain, "The Antecedents and Outcomes of Environmental Innovation: Lessons from Russia", *EBS Bus. Sch. Res. Pap. Ser.*, **12**, 6 (2012)
49. J. Horbach, C. Rammer, and K. Rennings, "Determinants of Eco-innovations by Type of Environmental Impact — The Role of Regulatory Push/Pull, Technology Push and Market Pull", *Ecol. Econ.*, 78 (2012)
50. D. Antonioli, S. Mancinelli, and M. Mazzanti, "Is Environmental Innovation Embedded within High-performance Organisational Changes? The Role of Human Resource Management and Complementarity in Green Business Strategies", *Res. Policy*, **42**, 4 (2013)
51. U. Van den Berg, J. P. Labuschagne, and H. Van den Berg, "The Effects of Greening The Supplier and Innovation on Environmental Performance and Competitive Advantage", *J. Transp. Supply Chain Manag.*, **7**, 1 (2013)
52. Y. S. Chen and K. C. Chang, "The Nonlinear Effect of Green Innovation on The Corporate Competitive Advantage", *Qual. Quant.*, **47**, 1 (2013)
53. M. L. Tseng, R. Wang, A. S. F. Chiu, Y. Geng, and Y. H. Lin, "Improving Performance of Green Innovation Practices Under Uncertainty", *J. Clean. Prod.*, **40** (2013)
54. K. Bi, Q. Bao, and D. I. Feng, "Effect Analysis of Informatization Level to Green Process Innovation in Manufacturing Enterprises", *Theor. Appl. Informatio Technol.*, **47**, 2 (2013)
55. S. K. S. Wong, "Environmental requirements, knowledge sharing and green innovation: Empirical evidence from the electronics industry in China", *Bus. Strateg. Environ.*, **22**, 5 (2013)
56. A. Muscio, G. Nardone, and A. Stasi, "Drivers of Eco-Innovation in the Italian Wine Industry", *Proceedings in Food System Dynamics* (2013)
57. J. Doran and G. Ryan, "Eco-Innovation – Does Additional Engagement Lead to Additional Rewards?", *Int. J. Soc. Econ.*, **41**, 11 (2014)
58. S. Rexhäuser and C. Rammer, "Environmental Innovations and Firm Profitability: Unmasking the Porter Hypothesis", *Environ. Resour. Econ.*, **57**, 1 (2014)
59. H. Lin, S. X. Zeng, H. Y. Ma, G. Y. Qi, and V. W. Y. Tam, "Can Political Capital Drive Corporate Green Innovation? Lessons From China", *J. Clean. Prod.*, **64** (2014)
60. V. Chassagnon and N. Haned, "The Relevance of Innovation Leadership for Environmental Benefits: A Firm-level Empirical Analysis on French Firms", *Technol. Forecast. Soc. Change*, **91** (2015)
61. J.-W. Huang and B. Yong-Hui Li, "Green Innovation and Performance: The View of Organizational Capability and Social Reciprocity", *J. Bus. Ethics* (2015)
62. S. M. Masoumik, S. H. Abdul-Rashid, and E. U. Olugu, "Importance-performance Analysis of Green Strategy Adoption within the Malaysian Manufacturing Industry", *Procedia CIRP*, **26** (2015)
63. M. Abdullah, S. Zailani, M. Iranmanesh, and K. Jayaraman, "Barriers to Green Innovation Initiatives Among Manufacturers: The Malaysian Case", *Rev. Manag. Sci.*,

10, 4 (2015)

64. M. Segarra-Oña, Á. Peiró-Signes, and J. Mondéjar-Jiménez, "Twisting the twist: how manufacturing & knowledge-intensive firms excel over manufacturing & operational and all service sectors in their eco-innovative orientation", *J. Clean. Prod.* (2016)
65. M. V. De Oliveira Brasil, M. C. Sá de Abreu, J. C. L. Da Silva Filho, and A. L. Leocádio, "Relationship Between Eco-innovations and The Impact on Business Performance: An Empirical Survey Research on The Brazilian Textile Industry", *Rev. Adm. (Environmental Manag.)*, **51**, 3 (2016)
66. N. Kawai, R. Strange, and A. Zucchella, "Stakeholder Pressures, EMS Implementation and Green Innovation in MNC Overseas Subsidiaries", *Dep. Econ. Manag. DEM Work. Pap. Ser.*, **121** (2016)
67. X. Peng and Y. Liu, "Behind Eco-innovation: Managerial Environmental Awareness and External Resource Acquisition", *J. Clean. Prod.*, **139** (2016)
68. T. Kong, T. Feng, and C. Ye, "Advanced Manufacturing Technologies and Green Innovation: The Role of Internal Environmental Collaboration", *Sustainability*, **8**, 10 (2016)
69. D. L. Arenhardt, L. F. Battistella, and M. Z. Grohmann, "The Influence of the Green Innovation in The Search of Competitive Advantage of Enterprises of The Electrical and Electronic Brazilian Sectors", *Int. J. Innov. Manag.*, **20**, 1 (2016)
70. J. Hojnik and M. Ruzzier, "The Driving Forces of Process Eco-innovation and Its Impact on Performance : Insights from Slovenia", *J. Clean. Prod.*, **133** (2016)
71. X. Xie, J. Huo, G. Qi, and K. X. Zhu, "Green Process Innovation and Financial Performance in Emerging Economies: Moderating Effects of Absorptive Capacity and Green Subsidies", *IEEE Trans. Eng. Manag.*, **63**, 1 (2016)
72. J. Hojnik, M. Ruzzier, and B. Antončič, "Drivers of Eco-innovation : Empirical Evidence from Slovenia", *Int. J. Entrep. Innov. Manag.*, **21**, 4 (2017)
73. J. S. Keshminder and V. Chandran, "Eco-Innovation in the Chemical Manufacturing Firms : Insights for Policy Response", *Institutions Econ.*, 9, 1 (2017)
74. Y. Fernando and X. W. Wen, "The Impact of Eco-innovation Drivers on Environmental Performance : Empirical Results from the Green Technology Sector in Malaysia", *Sustain. Prod. Consum.*, **12**, (2017)
75. M. Miedzinski and A. Reid, "Systemic Eco-innovation Report for Europe INNOVA Sectoral Innovation Watch Project", (2008)
76. P. Berrone, A. Fosfuri, L. Gelabert, and L. R. Gomez-Mejia, "Necessity as the Mother of Green Inventions: Institutional Pressures and Environmental Innovations", *Strateg. Manag. J.*, **34** (2013)

Pure intelligent monitoring system for steam economizer trips

Firas Basim Ismail[1*], *Khalid* Hamzah Abed[1], *Deshvin* Singh[1], and *Mohammad* Shakir Nasif[2]

[1]Power Generation Unit, Institute of Power Engineering (IPE), Universiti Tenaga Nasional, 43000 Kajang, Selangor, Malaysia.
[2]Mechanical Engineering Department, Universiti Teknologi Petronas, 32610 Tronoh, Malaysia.

Abstract. Steam economizer represents one of the main equipment in the power plant. Some steam economizer's behavior lead to failure and shutdown in the entire power plant. This will lead to increase in operating and maintenance cost. By detecting the cause in the early stages maintain normal and safe operational conditions of power plant. However, these methodologies are hard to be achieved due to certain boundaries such as system learning ability and the weakness of the system beyond its domain of expertise. The best solution for these problems, an intelligent modeling system specialized in steam economizer trips have been proposed and coded within MATLAB environment to be as a potential solution to insure a fault detection and diagnosis system (FDD). An integrated plant data preparation framework for 10 trips was studied as framework variables. The most influential operational variables have been trained and validated by adopting Artificial Neural Network (ANN). The Extreme Learning Machine (ELM) neural network methodology has been proposed as a major computational intelligent tool in the system. It is shown that ANN can be implemented for monitoring any process faults in thermal power plants. Better speed of learning algorithms by using the Extreme Learning Machine has been approved as well.

1 Introduction

Steam economizer is one of the main component in a power plant. Steams Economizers uses water as a heat transfer average tool and frequently also as the final point to reject heat into the atmosphere by evaporating inside cooling towers. Waterside problems develop in cooling water systems from scaling, corrosion, dirt and dust accumulation, and biological growth [1]. All these problems develop depending on the quality of water supply. The problems stated are commonly related with copper alloy type of steam economizer tubes. Rough solids in the cooling water could cause excess force inside the tubes. These forces create highly turbulent conditions at tube inlets, which results in destruction of protective passive films on the tube, causing a turbulent flow situation. Changes in flow direction and unfavourable design configurations of the waterbox are other factors that can contribute to high turbulence [2].

* Corresponding author: Firas@uniten.edu.my

When there is a tube failure, abnormal changes can be observed in the related economizer variables such as economizer inlet pressure, economizer outlet temperature and feed water flow rate. As the tube failure is not detected and the unit continues operating, the readings of the variables deviates from its normal range. This deviation causes trips to occur when it reaches a certain maximum point. This causes frequent economizer trips which eventually leads to a power plant shutdown. Frequent shutdowns in a plant causes a significant increase in its operating and maintenance cost [3].

Conventional detection methods in power plants gives information on a particular problem only once serious damage has been done. Consequently prediction of economizer trips is important to maintain a normal and safe operational conditions. This is because when a trip is detected early, preventive measures can be taken by the plant operator to avoid further damage and shutdown. Eventually operating loss can be reduced by minimising repair cost and secondary damages [4]. Early detection is possible using intelligent monitoring system. Measurements from several sensors provide information on the operating system and its failures. This information can be used by an intelligent monitoring system for fault diagnosis [5]. One of the most commonly used intelligent systems for early fault detection is the Artificial Neural Network. The ANN approach is able to developed input-output simulation and forecasting models in situations that do not require modelling of the internal structure of the watershed [6]. The most commonly used family of feed-forward networks is a layered network in which neurons are organized into layers with connections strictly in one direction from one layer to another [7]. The most common type of feed-forward networks is the multilayer preceptor (MLP). An MLP, which has three types of layers: an input layer, an output layer and a hidden layer. Each neuron is in the hidden layer including its input signals after weighting them with the strengths of the relating connections from the input layer. The application of ANN in fault detection and diagnosis is based on model approximation and pattern recognition [8-10]. The pattern recognition method is a convenient approach to resolve the fault identification problem, for instance in determining the size of the fault [11]. Beale et al. provided a simple ANN model of steam cycle has been performed in 2002. The aim of that ANN model was just to show the capability of ANNs to predict the power output. That ANN model had just a few input parameters, and thereby the accuracy of the ANN was not so high [12]. In all proposed solutions, there has not been any intelligent monitoring system established particularly for detecting economizer trip in thermal power plants. Therefore it is crucial to develop a reliable intelligent monitoring system for prediction of thermal economizer trip. Mathematically simulated data were used by several researches instead of real plant data for training and validation, which is inaccurate for decision making. In most applications, the raw data are used directly for the training process. However this data is not always complete and the quality is limited. Therefore a data preparation framework is important for a thorough data analysis [13]. In this study, we concentrate on the steam economizer with 10 variables; only two of them were optimized and validated. The trips can be minimized using the fault detection and diagnostic (FDD) method and more specifically an artificial neural network (ANN) was the core diagnostic part in the system. This will have the effects on the power plant great performance and help the desperate industry to achieve a better outcomes and its production.

2 Description of power plant

This research was done based on real data collected from Manjung power plant in Malaysia. The unit consists of single reheat, controlled circulation type sub-critical pressure boiler. With a power generation of 700 MW(e), the combustion system consists of a single furnace with balanced draught and tangential firing.

In this research the power plant being considered consists of three sub-critical pressure boilers. Each boiler is a single reheat and controlled circulation type. The boiler is fired with pulverized coal to produce steam for the continuous generation of 700 MW(e). The combustion system consists of a single furnace with balanced draught. Using direct tangential firing located in the furnace corners, the maximum heat input that can be attained is 40% of the Boiler Maximum Continuous Rate (BMCR). The capacity of steam generating for the unit is 172 bar with steam outlet temperature of 540°C. The main auxiliary equipment includes two forced draft fans, two induced draft fans and two primary air fans which are centrifugal fans with control vanes at the inlet. Other components of the auxiliary equipment include two steam air preheaters, one piece of soot-blowing equipment, two electrostatic precipitators. The plant consists of low nitrogen oxide combustion burner system known as Over Fire Air (OFA) ports. The flue gas desulphurisation plant is controls the sulphur dioxide emission whereas the dust in the flue gas at the boiler outlet is removed by an Electro-Static Precipitator (ESP).Remote controlled fuel oil burners with high energy ignitions are used to start-up the boiler. Pulverized coal is able to undergo combustion at low firing rates using these burners.

3 Data preparation framework

Real-world data is often incomplete, inconsistent or lacking in certain behaviors or trends, and is likely to contain many errors. With proper data preparation framework, such issues can be resolved. The series of steps involved in this stage are:

3.1 Data Capturing

Data was captured from MNJ Power plant, Malaysia. More than 1000 variables were included in the data file. However, we have considered the most influent operational variables which can led to steam economizer trip. The time interval range when the fault has occurred is specified. Therefore, it is easy to recognize the variables through a sensitivity measure. The scope for this project is the boiler section whereby the unit is shut down due to the leakage of the boiler tubes. Based on the real data collected from the selected thermal power plant, the unit has been shut down from 25th April 2008 until 31 April 2008 and it is approximately about 5 days according to the outage.

3.2 Data cleaning

Large spreadsheet of data collected from sensors in power plants contains invalid values also known as noise. To avoid any error in the input data, the data has to undergo a few processes. The low pass filter method with moving average window was applied for this stage. The low pass filter with window size of 3 is applied on the data. The result of applying the low pass filter of type moving average window is acceptable in reducing the noise component that is affecting the raw data. However, small noise is still active. Probably, increasing the window size from 3 to larger sizes like 5 or 7 can have better performance in removing this noise component. But it is better to leave this option to be selected according the first impression obtained by the data visualization stage. Our choice is 3 because we aim at demonstrating the ANN decision validity even in case of further noise interfering with the data.

3.3 Data interpolation

Piecewise polynomial interpolation was chosen. It has the capability of optimizing the smoothness of the fitted curve while passing through all the data points. Also, using it has the advantage of reducing the order of degree polynomial function because in each subinterval different polynomial is selected.

3.4 Data normalization and segmentation

Data were subject to data normalization in order to limit the values of the data between [-1 +1]. This step is needed because the neural network is trained on the relative change in the input not on the absolute values. Data was partitioned into two groups: 50% training and 50% checking. The only difference between the partitioned data and the raw data of the variable is the size of the data. This gives the opportunity to the ANN to capture all the information from the data and to reflect this information on its weights.

After undergoing the data processing procedure, the data was shortlisted into 10 important variables based on plant operator experience as listed in table 1. The variables are shortlisted based on the critical sensors that contributed to the trip of that particular unit of boiler. Among all those 10 variables, there are several variables that had been identified contributed to the trip before the real shutdown. It is possible to reduce the number of variables to the most influential variables based on the data visualization of these 10 variables. Therefore variables number 3 and 5 were selected. These variables are very important because they caused immediate trip to the boiler.

Table 1. Selected variables for fault interpretation

No	KKS	Description
1	JMJG_U2_HAJ01CP501XV1	RH LT R/HTR inlet pressure
2	JMJG_U2_HAC01CT501X3V	Economizer inlet temperature
3	JMJG_U2_HLA14CP501XV1	RH Hot sec air pressure to burners [H12]
4	JMJG_U2_HAC01CF501X5V	Steam flow
5	JMJG_U2_HLA15CF501FL	Combustion air flow
6	JMJG_U2_HAC61CF501X5V	Feed water flow
7	JMJG_U2_HAD01CP501AXV1	Boiler drum pressure
8	JMJG_U2_HAH92CP501X3V	SH steam pressure
9	JMJG_U2_HAH92CT501X3V	SH steam temperature
10	JMJG_U2_HAC01CF501X5V	Total combined steam flow

4 Design implementation and results of ims

An efficient learning algorithm for single-hidden layer feed-forward neural network (SLFNs) called ELM is tested for training the data in steam economizer. Three different activation functions were applied to the weighted sum of its input. Training algorithms, learning rate, momentum coefficient, activation functions and number of neurons are the main neural network (NN) topology taken into consideration. A single hidden layer is used to be able to approximate arbitrarily with any system with continuous mapping from one finite space to another. The performance of each training results was indicated by their RMSE values. Based on this performance, the system is trained and validated. The variables selected are used as inputs for the IMS. The output is set to '0' or '1', representing normal economizer operation and faulty economizer operation respectively. In the first stage of designing the IMS, preliminary training is done to determine the optimal NN topology combination. Faulty data set is used as input to the IMS. Once optimal topology is obtained, the next stage involves basic training using 50% of the data. The RMSE results of the basic training process is compared with the calculated RMSE values of the optimal NN topology combination. This is to ensure a smaller or equal calculated RMSE values compared to the preliminary training process. The next stage involves the validation process using the remaining 50% of the data set. The optimal NN topology combination was applied to validate the real data set in order to determine the rapidness of the proposed IMS. Validation is also done for performance of the system during normal economizer operations. The capability of the IMS for specific trips was explored using the performance results obtained. It can observed from the table 2 the RMSE output for different activation function and different number of neurons in ANN hidden layer. Comparing the RMSE values in the table, the ANN topology with 20 neuron in its hidden layer performed better than the others. The activation functions performed differently for different number of neurons.

Table 2. RMSE values for different activation functions and number of neurons combination

Activation Function	5	10	20
sigmoid	0.2344	0.2031	0.1194
sin	0.3210	0.1393	0.1295
purelin	0.2869	0.2301	0.1900

The trained output of the proposed monitoring system is set to '0' and '1' for normal economizer operation and faulty economizer operation respectively. In between 0 to 1, the lower threshold represents the point where economizer operation is normal whereas the upper threshold indicate the point of faulty operation.

The output of the system is displayed in figure 1. From the figure, we know that ELM performed better than conventional ANN algorithms. It has one-iteration learning, which gives this algorithm the capability of being executed for real time diagnostic. Furthermore ELM has a capability of universal approximation and is much simpler. The RMSE results obtained were very satisfying even with low number of neurons. The particular thermal economizer trip was successfully predicted by the proposed system.

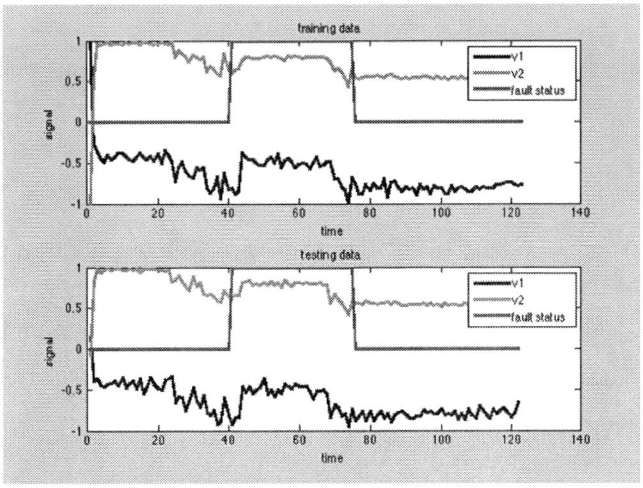

Fig. 1. ANN training data, and testing results for ELM.

5 Conclusion

In this study, the early detection of economizer trips using intelligent monitoring system has been confirmed. Real data from power plant were used for training and validation purposes instead of mathematically simulated data. An integrated framework form economizer trip detection has been proposed in this study. The best ANN topology combination was achieved with RMSE of 0.1194. The single layer feed-forward neural network was applied with ELM training algorithm in which it performed better than the conventional algorithms. The best activation function is the sigmoid function with 20 neurons in the hidden layer. ANN plays an important role in monitoring and analyzing the power plant's trips and faults. Using the monitoring system developed, trips can be detected earlier in order to maintain a continuous and safe plant operation. Further development of the proposed system can be considered in which online real time data is used for prediction of thermal power plant economizer tube leak trip.

The authors would like to express our gratitude to the power generation unit, institute of power engineering Universiti Tenaga Nasional (UNITEN) TNB Janamanjung and Ministry of Higher Education (MOHE), Malaysia for the opportunity to sponsorship this corresponding research.

References

1. Md. Mujibur and A. Kamal, *Failure analysis of high temperature superheater tube of a pulverized coal-fired power station,* Int. Scientific Conf. (2011)
2. Tallman P., Smith J., ASME Power (2004)
3. D. Fred, A. Dave, and E. Loren, *Detection of tube leaks and their location using input/loss method* Int. Joint Power Generation Conf. (2004)
4. D. Flynn, Thermal power plant simulation and control (The Institution of Electrical Engineers, 2003)
5. L. Fortuna, S. Graziani, A. Rizzo and M.G. Xibilia, Advances in Industrial Control (2007)
6. S.K. Chang and P.L. Hsu, *Fault detection observer design for linear systems with unknown inputs,* European Control Conference, **4**, pp.1975-1980 (1993)
7. A.K. Jain, J. Mao and K.M. Mohiuddin, Computer, **29**, pp. 31-44 (1996)
8. A. Lipnickas, Advanced information and knowledge processing, pp. 209-230 (2006)
9. Simani S., Fantuzzi C. and Patton R.J., Advances in Industrial Control, **1** (2003)
10. L. Zhou, R. E. Dickinson, Y. Tian, X. Zeng, Y. Dai, Z.-L. Yang, C. B. Schaaf, F. Gao, Y. Jin, A. Strahler, R. B. Myneni, H. Yu, W. Wu, M. Shaikh, Journal Of Geophysical Research, **108** (2003)
11. Simani S., Fantuzzi C. and Patton R.J., Advances in Industrial Control (2003)
12. Beale M.H., Hagan M.T., and Demuth, H.B., MathWorks (2014)
13. Messai, A. Mellit and I. Abdellani, Progress in Nuclear Energy, pp.8-21 (2015)

Numerical investigation on the effect of channelled and unchannelled screens on smoke contamination in atriums upper balconies with open upstand

Mohammed Mutafi[1], *Mohammad* Nasif[1*], *William* Pao[1] and *Feras* Ismail[2]

[1]Universiti Teknologi PETRONAS, 32610 Seri Iskandar, Perak Darul Ridzuan, Malaysia.

[2]Power Generation Research Centre, Collage of Engineering University Tenaga Nasional, Selangor, Malaysia.

Abstract. Channel screens are used to restrict the smoke from spreading laterally under the first balcony. Effect of channel screens on smoke contamination in atrium upper balconies with the presence of balcony open upstand and 1m deep down stand at the fire compartment is investigated in this paper. Fire dynamic simulator (FDS6) which is a computational fluid dynamic (CFD) software was utilized to conduct the simulation. A comparison between the results of smoke contamination with the presence and omission of channeling screens showed that the height of smoke contamination in atrium upper balconies increased by 90% in case of no channel screens at the fire compartment are installed. A new correlation obtained from this work is able to predict the height of smoke contamination in atrium upper balconies with the presence of 1m deep down stand, balcony open upstand and no channel screens.

1 Introduction

Smoke toxic gases are found to be the main cause of death in a fire incident. Smoke in general has three different hazards which are visibility, thermal hazards and toxicity. In United States, it was reported that most fire victims died because of the inhalation of toxic gases of the smoke [1].

The word atrium is used to describe a building with a large space at the center and contains multiple floor. This modern design of building has been widely used in airports, hotels, hospitals and shopping malls. The open space of the atrium allow the smoke to spread easily during a fire incident inside an atrium which may endanger the life of occupants. When fire takes place in an atrium compartment, It was expected that the bulk amount of smoke leaving the fire compartment rises vertically towards the atrium ceiling. However, it was found by previous research that sometimes smoke directed inward the atrium upper balconies rather than completely rising upward. Consequently, smoke contamination occurs at that particular level. This phenomenal causes risk to the occupant's life.

Cox [2] studied the reason that make the smoke curl into the balcony of the atrium. He found that the raising plume entrains more air as it moves upward toward the ceiling and because of the less supply of air from the side of the plume that is facing the atrium, a low pressure region is created inside the balcony causing the smoke to curl into the balcony which cause a smoke contamination at that specific balcony. Previous researches [3-4] found that

*Corresponding author: mohammad.nasif@utp.edu.my

DOI: 10.1051/matecconf/201713104009

it's essential to study the effect of building structure on smoke contamination occurrence. Hence, the effects of balcony breadth, width of fire compartment opening and the depth of down stand were studied. Moreover, the effect of the channelling screen which is installed beneath the first balcony and used to restrict the lateral spread of the smoke was investigated only with the presence of balcony solid upstand. Therefore, the effect of channel screen in case of using balcony open upstand is investigated in this paper.

In this study, the results of the smoke contamination occurrence in case of using channel screen is compared to the results when no channelling screens are used.

2 Model development

The geometry used in this study is a full scale three levels atrium modeled using fire dynamic simulator (FDS) which is a computational fluid dynamic (CFD) software that models fluid flow driven by fire. It solves numerically a form of the Naiver-Stokes equations of a low speed fluid flow, thermally – driven flow with an emphasis on smoke and heat transport from fires [7].

A numerical investigation of the effects of different building structure parameters is performed in this work using (FDS). The balcony breadths used in this research are 5m, 3m, 2m and 1.5m whereas the values of fire compartment opening width are 10m, 8m, 6m, 4m and 2m. As for the heat release rate, the values of 1.6 MW, 3.2 MW and 4.7 MW were selected for this study to be consistent with the recommended design fire range in an atrium [3] and corresponded to those used by [9]. A 1m down stand and an open balcony upstand have been used for all cases. The dimensions of the atrium used in this research are like the atrium dimensions used in Tan experiment but for full scale [3] Upstand structure modelled as shown in figure 1(c).

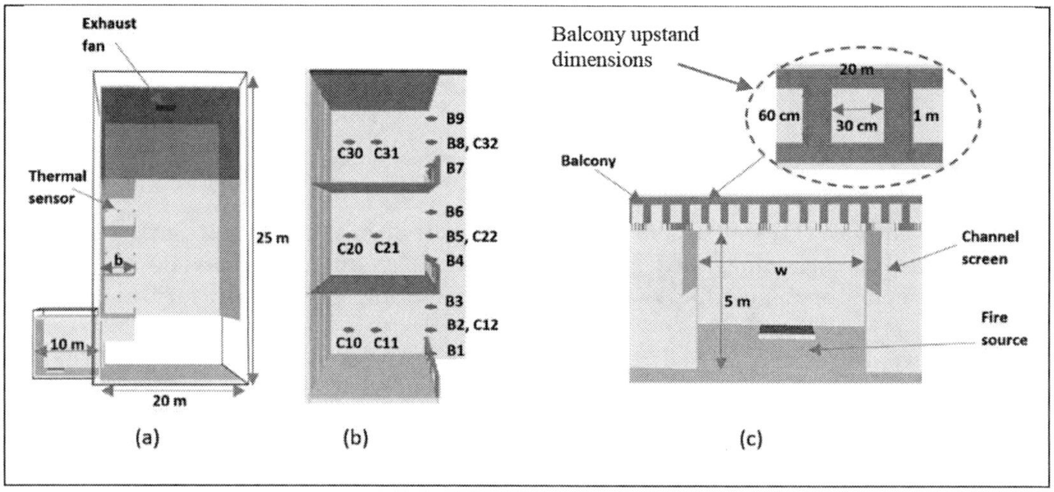

Fig. 1. Schematic drawing of atrium and fire compartment (a) side view of atrium, (b) Location of temperature sensor, (c) Front view of fire compartment with the balcony upstand

$\frac{D^*_{spill}}{\Delta x}$ ratio is used in (FDS) in order to determine the grid size in modelling [5]. The criterion that was utilized in this study for choosing an appropriate mesh size was set by Harrison [7]. The criterion for the size of grid cells is shown in equation (1) and (2).

$$n^*_{spill} = \frac{D^*_{spill}}{\Delta x} \geq 0.9 \quad (1)$$

$$D^*_{spill} = \left(\frac{Q_c/_{w}}{\rho C_p T\sqrt{g}}\right)^{\frac{2}{3}} \quad (2)$$

Where D^*_{spill} is characteristic length of plume for determining the grid size (m), n^*_{spill} is coefficient for determining the grid size, ΔX is grid dimension in x-axis, Qc is convective heat flow layer below spill edge, ρ is air density at ambient condition, Cp is a specific heat, T is absolute temperature and g is gravity acceleration. Hence, a grid size of 200 mm met the criteria mentioned above and it was selected for all simulations. The total number of mesh elements in the atrium space and fire compartment is 2332800 and 81000 respectively. Values adopted to the simulation like the balcony breadth (b), fire compartment opening width (w) and heat release rate (HRR) were varied for each case. However, down stand depth was kept constant as 1m for all cases. The maximum heat release rate used in this study is 4.8MW which is considered acceptable according to past studies performed for shopping mall fires with sprinklers fire protection system. The parameters used in the simulation are listed in Table 1.

Table 1. Values adopted to simulation

Case	Balcony Breadth b (m)	Fire compartment opening width w (m)	Heat Release Rate (kW)
1	5	6	1581
2	3	4	4740
3	2	4	3160
4	2	2	1581
5	1.5	4	1581
6	1.5	2	4740
7	3	6	3160

3 Results and discussion

In this comparison study, 7 cases were selected and each case was simulated twice with a channel screen and without channel screen. The comparison is based on the smoke temperature, smoke contamination occurrence and correlation.

3.1 Comparison of smoke temperature

Tan [3] recognized that smoke contamination occurrence can be indicated by the smoke temperature at that balcony. Tan found out that when the smoke temperature inside the balcony is ≥ 10 °C above the ambient temperature which was assumed to be **20** °C, the smoke contamination occurs in that balcony. [8] and [9] used this criterion to determine the smoke contamination inside the atrium's balconies. To measure the temperature in the balcony, 15 thermocouples were installed in (FDS6) at the edge and along the breadth of the balcony. The sensors position is similar to Tan's experiment. Figure 2 (a) illustrates the smoke temperature at the edge of the balcony for both cases with channel screen and without channel screen. (b) and (c) shows the temperature along the balcony breadth in case of using channel screen and not using it, respectively.

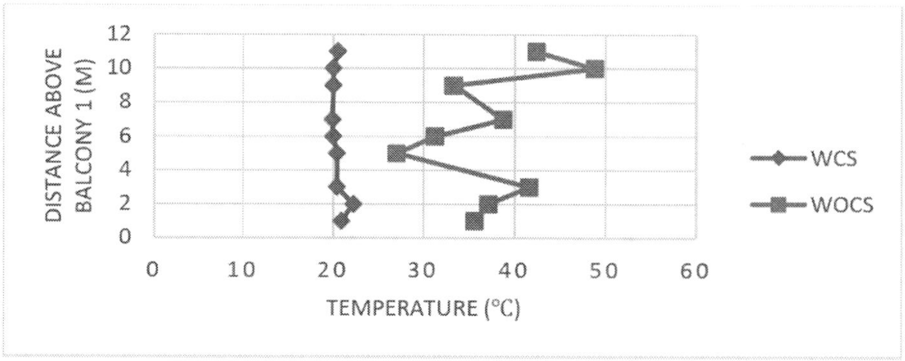

(a)

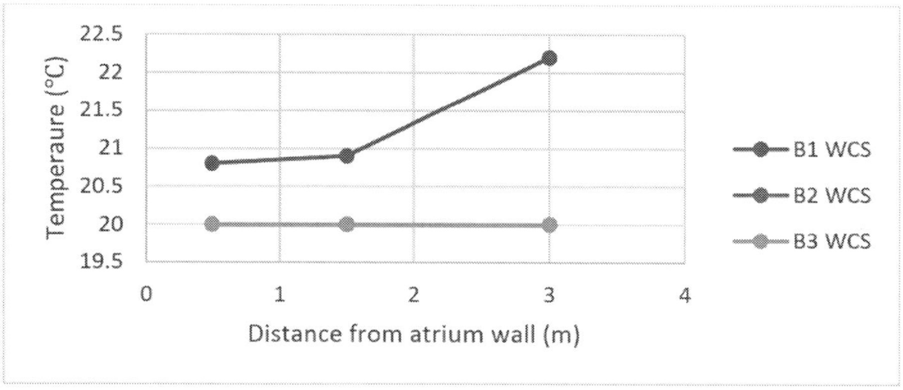

(b)

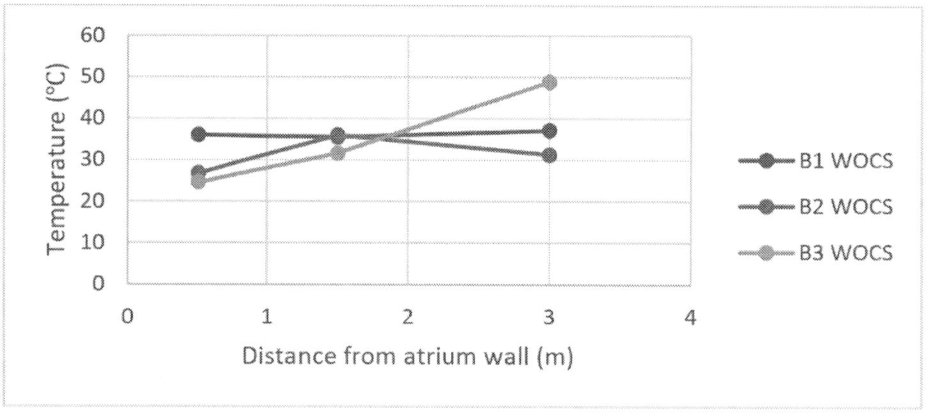

(c)

Fig. 2. Smoke temperature (a) along the balcony edges, (b) along the balcony breadth in case of using channel screen (WCS) and (c) along the balcony breadth in case of not using channel screen (WOCS).

DOI: 10.1051/matecconf/201713104009

Case number 2 was selected to show the temperature difference along the balcony edges and inside each balcony. It can be seen in figure 2 (a) that the temperature along the balcony edges is higher for (WOCS) compared with (WCS). The temperature difference is illustrated in Table 2.

Table 2. Comparison of smoke temperature along balcony edges

Sensor symbol	Sensor position above balcony 1 (m)	Temperature		Temperature difference (%)
		WCS	WOCS	
B1	1	20.9	35.6	41.3
B2	2	22.2	37.1	40.2
B3	3	20.4	41.6	50.9
B4	5	20.4	27	24.4
B5	6	20	31.3	36.1
B6	7	19.9	38.7	48.6
B7	9	20	33.3	39.9
B8	10	20	48.8	59.0
B9	11	20.5	42.4	51.7

3.2 Comparison of correlation

The developed correlation to predict the height of smoke contamination is shown in figure 3. The correlation has four parameters that are smoke plume depth below the spill edge (d) which is taken from FDS, smoke contamination height (H) which is found from a visual observation of the smoke temperature contours, balcony breadth (b) and fire compartment opening width (w), the last two are structural dimensions of the atrium. Comparison of correlations illustrated that (WOCS) configuration has more smoke contamination indicated by a lower smoke height in the correlation.

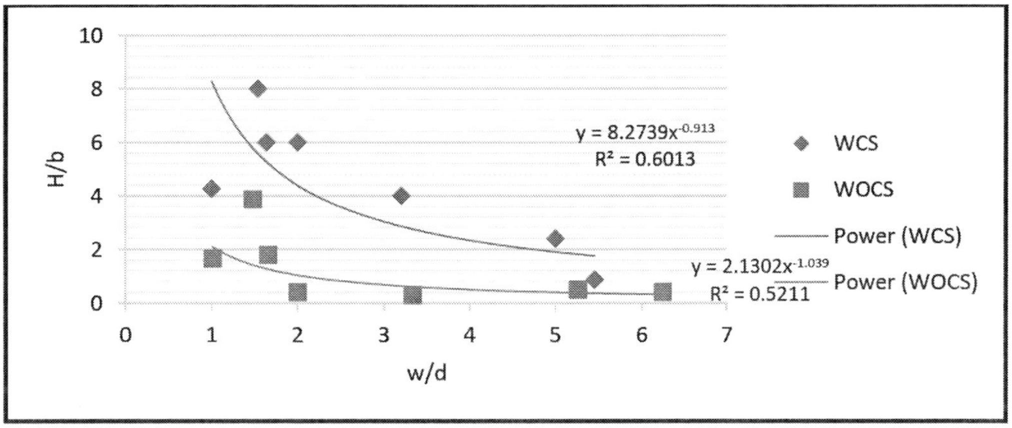

Fig. 3. Correlation to predict the smoke contamination height

3.3 Comparison of smoke contamination height

Table 3 shows the difference in occurrence of smoke contamination in case of using channel screen and not using it. It can be seen that the height of smoke contamination increased significantly for WOCS atrium compared with WCS atrium. Hence, the smoke contamination is more when no channel screens are used. On the other hand, the depth of the smoke plume is higher for WCS atrium because the smoke has been restricted from moving in a lateral direction and when the smoke hits the obstruction it losses buoyancy and becomes stagnant.

Table 3. Comparison of height and depth of smoke contamination

Case	b(m)	w(m)	HRR(kW)	WCS		WOCS		Comparison	
				H	d	H	d	ΔH (%)	Δd (%)
1	5	6	1581	12	1.2	2.5	1.14	79.2	5
2	3	4	4740	12	1.25	0.89	1.2	92.6	4
3	2	4	3160	12	2	0.8	2.12	93.3	0.0
4	2	2	1581	12	1.22	3.6	1.206	70	1.1
5	1.5	4	1581	12	2.6	5.8	2.3	51.7	11.5
6	1.5	2	4740	6.4	2	2.5	1.98	60.9	1
7	3	6	3160	2.6	1.1	1.25	0.96	51.9	12.7

4 Conclusion

In this work, CFD simulation using FDS 6 software was conducted to investigate the effect of installing channel screens in the shopping mall's shops. 7 cases were selected in this study. The structural dimension of the atrium such as the balcony breadth and the fire compartment opening width were varying in each case. As well as the heat release rate. However, the depth of the down stand was kept constant and the balcony open up stand was used for all cases. It was found that, the omission of channel screens results in a greater lateral extent of smoke and hence, more smoke contamination in atrium upper balconies.

References

1. R. j.Hall, *Burns, toxic gases, and other hazards associated with fires: Deaths and injuries in fire and non-fire situations.* NFPA, Fire Analysis and Research Division, Quincy, MA, (2001).
2. Cox G. Compartment fire modelling. Combustion fundamentals of fire, 329-404 (1995)
3. R. Harrison "Entrainment of air into thermal spill plumes," Ph.D. thesis, University of Canterbury, (2009).
4. S. A. Hasnain, M. S. Nasif, W. Pao and R. Al-Waked *Numerical investigation of smoke contamination in atrium upper balconies at different down stand depths,* Building Simulation (2017)
5. S. A. Hasnain, M. S. Nasif, W. Pao and R. Al-Waked, *Effect of Down Stand Depth Variation on Atriums Upper Balconies Smoke Contamination* MATEC Web of Conferences, (2014)
6. Ho C. L., Li K. Y. and Spearpoint M. J. *Numerical simulation of scale-model smoke contamination of upper atrium levels by a channelled balcony spill plume*, Fire and Materials 37(8) 581-96 (2013)
7. K. McGrattan, S. Hostikka, J. Floyd, H. Baum, R. Rehm, W. Mell and R . McDermott *Fire dynamics simulator (version 5), user's guide* 1019(5) 1-186. (2010)
8. H. P.Morgan, B. K. Ghosh, G. Garrad, R. Pamlitschka, J. C. De Smedt and L. R. Schoonbaert *Design Methodologies for Smoke and Heat Exhaust Ventilation*. Construction Research Communication Ltd, Londres, R, U. (1999)
9. F.Tan , R. Harrison and M. J. Spearpoint *Physical scale modeling of smoke contamination in upper balconies by a channeled balcony spill plume in an atrium* Fire Sciences 1;28(4) 313-36.(2010)

Energy efficient of the residential buildings based climatic condition using experimental design: a case study in malaysia

Seyed Mojib Zahraee[1], *Morteza* Khalaji Assadi[1,*], *Sajjad* Bayat[2], and *Shokoufeh* Bakhoda[1]

[1]Department of Mechanical Engineering, Universiti Teknologi PETRONAS, Bandar Seri Iskandar 32610, Perak, Malaysia

[2]Faculty of Mechanical Engineering, Department of Materials, Manufacturing & Industrial Engineering, Universiti Teknologi Malaysia, 81310 Skudai, Johor, Malaysia

Abstract. In recent years, energy consumption has become a critical issue in the developed and developing countries. Residential buildings are one of the most users of energy in the construction sector that use the highest share of energy. This paper aims at evaluating the effect of four factors that are temperature, humidity, airflow and pressure on the cooling load in the residential buildings. To achieve this goal, statistical experimental design is used to determine the optimum setting of factors that result in optimum energy usage. Simulation software and energy analysis is used to simulate a two-storey building in Malaysia as the case of study. Final results showed that the temperature, humidity and interaction between them have the most significant effect on the energy cooling load. Moreover, to obtain the minimum value of cooling load the temperature and humidity should be equal to A=20 Celsius degree and B=60% respectively. In addition, the other two insignificant factors, airflow and pressure should be placed at the high level which are equal to C=3 cubic meters per hour, and D=6 Pascal (P) respectively.

1 Introduction

Energy optimization plays an important role in the new world. There is a global concern due to a probable lack of energy in the near future as well as some environmental effects like global warming [1]. Energy is becoming more and more costly and greenhouse gas emission is the disastrous effect of global warming that threat the whole human life [1]. One of the most cost-effective measures to minimize of carbon dioxide emission is to improve the energy efficiency of buildings [2]. Hence, energy efficiency is a key factor that should has been considered as an effective solution [1]. Among the widest range of energy consumers, residential buildings consume the largest amount of energy most of which is consumed by air conditioning systems in tropical countries [3]. In tropical regions with naturally hot and humid weather, a high amount of energy has to be consumed to provide a tolerable environment. In Malaysia it is claimed that air conditioning and refrigerators consumed nearly 70% of the electricity in buildings [3]. In this paper one residential building in Malaysia was simulated using Revit Architecture as a case study. The statistical method,

*Corresponding author : morteza.assadi@utp.edu.my

Design of Experiments (DOE) was applied to assess the effect of four climatic factors that are temperature, humidity, airflow and pressure on the cooling load and energy saving.

2 Literature review

There are some investigations about the energy efficiency assessment in the residential buildings that are discussed here briefly. One case study in in the humid and hot climate in Saudi Arabia used Visual DOE4 software to assess the energy consumption of a five storey office building [4]. The results indicated that increasing the insulation thickness does not have any considerable impact on energy efficiency. Another study in Singapore applied thermal analysis software (TAS) to investigate the effect of some microclimatic criteria on minimization of the heat in terms of passive climate control in residential buildings that are ventilated naturally [5]. In another study in Saudi Arabia, the impact of three factor on energy saving was investigated. The results claimed that these factors can enhance the thermal comfort and energy usage [6]. Recently, some investigations were done by using combined computer simulation and statistical analysis for enhancing the performance of construction process such as concrete pouring process and energy consumption in buildings [7, 8]. In one study, Experimental design (DOE) approach was implemented to find the best combination of some selected factors in a two-storey building which located in Malaysia. Final result indicated that changing ceilings and ceiling materials play a significant role to decrease energy consumption [3]. Additionally, in another study, statistical Taguchi method was proposed to find the optimum value of the main parameters of buildings that are window, ceiling, and wall along with the effect of uncontrollable factors such as humidity, temperature, and air flow in residential buildings[9]. In another research, building simulation and DOE were combined to evaluate the effect of main climate factors on energy saving and cooling load. Final result showed that the temperature and humidity have the most significant effect on the energy saving [10]. Therefore, this paper aims at evaluating the effect of four factors that are temperature, humidity, airflow and pressure on the cooling load in the residential buildings. To achieve this goal, building simulation and Design of Experiments (DOE) is applied to find an optimum setting of factors by doing the sensitivity analysis.

3 Materials and method

3.1 Simulation model and experimental design

In this study, a two-story building separated into 2 uniform apartments (each level 1 unit) was selected Malaysia. Total building area is 676 m^2. The house is separated into eleven zones with separate thermal properties for each level. Revit Architecture software is one of the most useful dynamic simulation tools that is used to simulate the building. In order to simulate, the CAD drawings are imported to Revit Architecture and simulated there by particular parametric design principles. The specifications were re-assigned to Ecotect software and a final sketch-up was imported to Energy Plus software.
DOE is a statistical technique first suggested by R.A. Fisher in the 1920's. DOE assess the effect of main factors and interaction between them on the responses variable [11]. In order to implement DOE, the following steps are followed. The steps are: 1. Selection of the factors and their levels, 2. choosing a response variable, 3. Determination of experimental design 4. Running the experiments 5. Analysis of results 6. Conclusion and recommendations.

3.2 Choosing factors and response variable

As mentioned earlier, four climatic factors were chosen to examine their effects on the cooling load in the selected case. Table 1 shows the variation range or level of factors is indicated that each factor has a high (+) and low (-) level. Cooling load was considered as a response variable. Due to small number of factors, full factorial design (2^n) is used. In factorial design, all possible combinations of factors are considered in an experiment, which is replicated two times in order to decrease the error. In addition, three center points are considered to examine the curvature of experiment.

Table 1. Factors and levels

Factor Name	Low Level (-)	Center (0)	High Level (+)
A= Temperature (Celsius)	20	23	26
B= Humidity (Percent)	60	70	80
C= Air Flow (M^3 / hr)	1	2	3
D= Pressure (Pascal)	2	4	6

4 Result and discussion

4.1 Performing simulation experiment

Table 2 shows the result of 35 experiments that is run using the simulation software.

Table 2. Result of simulation experiment

Run Order	Temperature	Humidity	Air flow	Pressure	Response (Cooling load) Replication 1	Replication 2
1	20	60	3	2	199400	199350
2	26	60	1	6	225120	224140
3	26	80	3	2	260160	289400
4	20	80	3	6	219450	225300
5	26	80	1	2	255508	278570
6	26	60	3	2	215600	210470
7	20	80	1	6	213050	235780
8	20	60	1	6	199150	199400
9	20	80	3	2	214179	236100
10	26	60	1	2	214370	199160
11	20	60	1	2	198370	198240
12	26	80	3	6	287150	278358
13	20	80	1	2	220346	250558
14	20	60	3	6	199200	198240
15	26	80	1	6	276900	287220
16	26	60	3	6	223470	235480
17	23	70	2	4	297800 220670	297670

4.2 Determination of significant factors

In this study, Minitab software was used. P-value is a significant parameter that is usually used to identify the statistically significant factors, because it has influenced the final responses. It should be noted that, whenever P-values are less than 0.05, they should be considered as a significant. In contrast, when P-values are more than 0.05, they should be assumed as an insignificant factor [11]. Table 3 shows main effects of A (Temperature), B (Humidity) and two-way interaction (AB) are significant factors.

The normal probability plot of effects shows that the significant factors are far from the line [11]. Figure 1 shows main effects of A (Temperature), B (Humidity) and two-way interaction (AB) are significant factors. Pareto chart is also used to compare the sequence of the statistical significance of the both main and interaction effects from highest to lowest effects. The factors that exceed the reverence line are considered to be significant with the confident level of 95% (Figure 2).

Table 3. ANOVA table for cooling load

Source	Degree of Freedom	Seq SS	Adj MS	F	P-Value
A	1	10734531395	10734531395	24.05	0.00
B	1	15644197391	15644197391	35.04	0.00
C	1	1238345	1238345	0.00	0.959
D	1	11451702	11451702	0.03	0.875
AB	1	2758629212	2758629212	6.18	0.023
AC	1	662889563	662889563	1.48	0.239
AD	1	952693988	952693988	2.13	0.161
BC	1	365614601	365614601	0.82	0.377
BD	1	28658128	28658128	0.06	0.803
CD	1	107600783	107600783	0.24	0.629
ABC	1	17560220	17560220	0.04	0.845
ABD	1	107982534	107982534	0.24	0.629
ACD	1	303386186	303386186	0.68	0.421
BCD	1	15988099	15988099	0.04	0.852
ABCD	1	169321403	169321403	0.38	0.546
Curvature	1	5578800419	5578800419	12.5	0.2
Residual error	18	8035451463	446413970		
Total	34	45495905430			

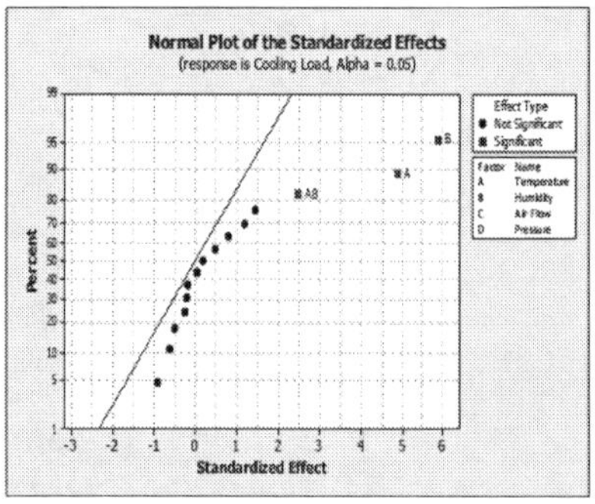

Fig. 1. Normal probability plot

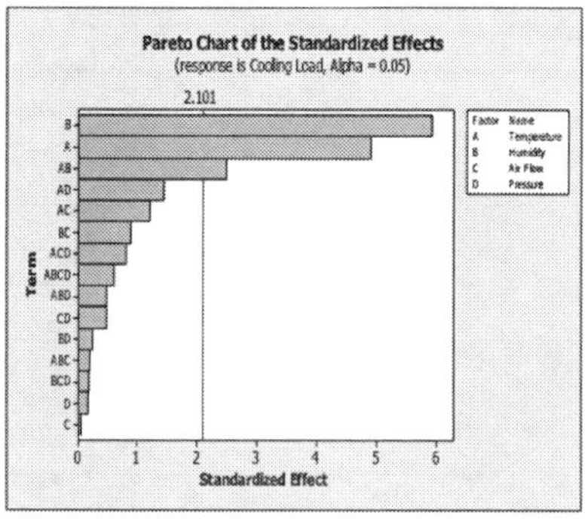

Fig. 2. Pareto chart

4.3 First order regression model

Regression model is a formula used to determine the predicted results of the experiment based on the significant factors of the experiment (Equation 1). Based on this experiment, the significant factors are main effect of Temperature (A), Humidity (B) and AB Interaction. In order to generate the regression model the constant coefficient value and all significant coefficient values should be considered. Table 4 shows the coefficients to develop the regression model.

$$Y = B_0 + B_1 x_1 + B_2 x_2 + B_{12} x_1 x_2 + \varepsilon \tag{1}$$

Y=226974 + 18315 (A) + 22111 (B) + 9285 (AB)

Table 4. Estimated effects and coefficients for cooling load

Term	Effect	Coefficient	SE Coefficient	T	P value
Constant		226947	3735	60.76	0.00
Temperature (A)	36631	18315	3735	4.90	0.00
Humidity (B)	44221	22111	3735	5.92	0.00
Temperature * Humidity (AB)	18570	9285	3735	2.49	0.023

In order to find the optimum setting of cooling load, sensitivity analysis is used by running and comparing different scenarios. Fig. 3 (a) shows that minimum cooling load and maximum desirability will be achieved when temperature and humidity placed on low level and two other factors placed on high level. In addition, Fig 3(b) claimed that if all the factors place on center points, the maximum desirability will be 25 %. The minimum desirability will be achieved when all the factors place on high level (Fig 3 (c)).

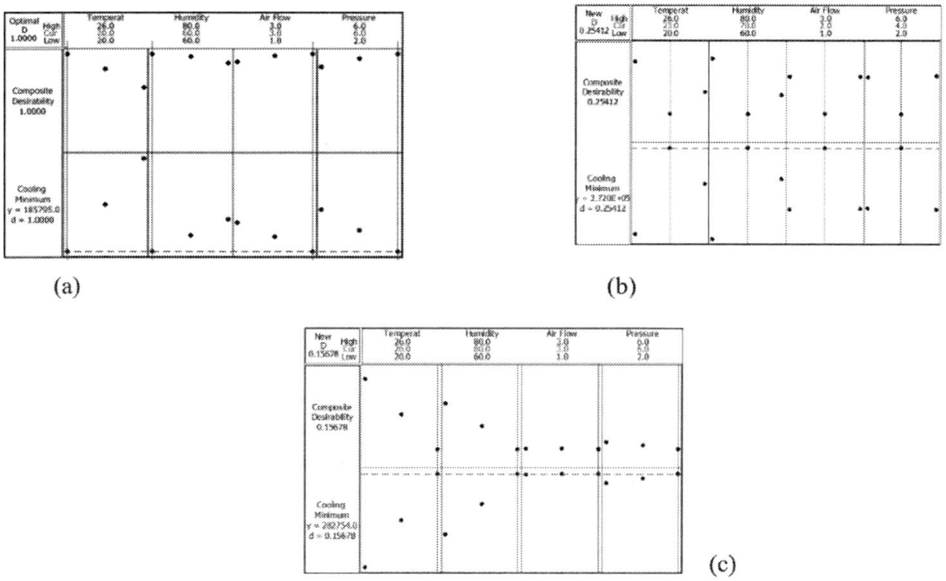

Fig. 3. Sensitivity Analysis for optimum setting

5 Conclusion

Energy efficient of building depends of some parameters such as climatic condition, location and physical material of buildings. In this paper, a model for assessing the effect of four climatic parameters which were temperature, humidity, airflow and pressure on cooling loads in residential tropical buildings was introduced. The design of experiments (DOE) was applied for evaluating the effect of factors on the energy saving and cooling load by running the simulation experiment. Finally, a sensitivity analysis is done to find the optimum setting of selected climate parameters in order minimize the cooling load. Final result claimed that maximum desirability of energy saving will be obtained when temperature and humidity is equal to A=20 Celsius degree and B=60% respectively. In addition, the other two insignificant factors, airflow and pressure should be placed at the high level which are equal to C=3 cubic Meters per hour and D=6 Pascal respectively.

We gratefully acknowledge the support provided by Universiti Teknologi PETRONAS and Ministry of Higher Education Malaysia (MOHE) of Malaysia for this work under Fundamental Research Grant Scheme (FRGS), cost center: 0153AB-k88.

References

[1] S. Zahraee, M. K. Assadi, and R. Saidur, "Application of artificial intelligence methods for hybrid energy system optimization," Renewable and Sustainable Energy Reviews, vol. **66**, pp. 617-630, (2016).

[2] H. Abdul-Rahman, C. Wang, and M. Y. Kho, "Potentials for sustainable improvement in building energy efficiency: Case studies in tropical zone," International Journal of Physical Sciences, vol. **6**, pp. 325-339, (2011).

[3] A. N. Sadeghifam, S. M. Zahraee, M. M. Meynagh, and I. Kiani, "Combined use of design of experiment and dynamic building simulation in assessment of energy efficiency in tropical residential buildings," Energy and Buildings, vol. **86**, pp. 525-533, (2015).

[4] I. Iqbal and M. S. Al-Homoud, "Parametric analysis of alternative energy conservation measures in an office building in hot and humid climate," Building and environment, vol. **42**, pp. 2166-2177, (2007).

[5] N. Wong and S. Li, "A study of the effectiveness of passive climate control in naturally ventilated residential buildings in Singapore," Building and Environment, vol. **42**, pp. 1395-1405, (2007).

[6] M. S. Al-Homoud, A. A. Abdou, and I. M. Budaiwi, "Assessment of monitored energy use and thermal comfort conditions in mosques in hot-humid climates," Energy and Buildings, vol. **41**, pp. 607-614, (2009).

[7] S. M. Zahraee, H. Milad, M. Y. Noordin, M. R. Jafri, and Z. Farhad, "Combined use of design of experiment and computer simulation for resources level determination in concrete pouring process," Jurnal Teknologi (Sciences and Engineering), vol. **64**, pp. 43-49, (2013).

[8] S. M. Zahraee, G. Rezaei, A. Shahpanah, A. Chegeni, and J. M. Rohani, "Performance Improvement of Concrete Pouring Process Based Resource Utilization Using Taguchi Method and Computer Simulation," Jurnal Teknologi, vol. **69**, pp. 17-24, (2014).

[9] S. M. Zahraee, M. Hatami, A. A. Bavafa, K. Ghafourian, and J. M. Rohani, Application of statistical taguchi method to optimize main elements in the residential buildings in Malaysia based energy consumption vol. **606**, (2014).

[10] S. M. Zahraee, A. Khademi, S. Khademi, A. Abdullah, and H. Ganjbakhsh, "Application of design experiments to evaluate the effectiveness of climate factors on energy saving in green residential buildings," Jurnal Teknologi (Sciences and Engineering), vol. **69**, pp. 107-111, (2014).

[11] D. Montgomery, "Basic experiment design for process improvement statistical quality control," ed: John Wiley and Sons, Inc., USA, (2009).

Effect of Stall Strip Position, Size and Geometry on the Lift Coefficient of NACA 001 Aerofoil

Wee Tee[1], *Ahmed Nurye* Oumer[1], *Ahmmad* Shukrie[1], *Azizuddin* AbdulAziz[1], and *Aklilu* Baheta[2]

[1]Faculty of Mechanical Engineering, Universiti Malaysia Pahang, 26600 Pekan, Malaysia
[2]Faculty of Mechanical Engineering, UTP, 32610 Seri Iskandar, Malaysia

Abstract. This study aims to determine the optimum position and geometry of stall strips (SS) to control sudden fall of lift in wind turbine blades. The type of airfoil used in this study is NACA 0015 with 150 mm of chord length. Total of five positions, two geometries and three sizes of SS configurations are simulated by using Ansys Fluent software. For position configuration, SS of size 2 mm is placed on the apex (POS-1), and on the upper and lower surfaces at distance of 0.65 mm (POS-4 and POS-2 respectively), and 2.45 mm (POS-5 and POS-3), respectively, from the leading edge. The shapes tested are dome and equilateral triangle. The results show that the addition of SS as a method of controlling sudden loss of lift decreases the maximum lift coefficient. Attachment of SS at the lower surface of the airfoil did not bring any significant effect to the lift and stall characteristics; while for the upper surface it reduces the sudden fall of lift but at the cost of big reduction in maximum lift coefficient. The optimum position and geometry of SS are POS-1 and triangle shape. Increasing in size of SS shows positive effect in control stalling effect.

1 Introduction

Aerodynamic ability is the main criteria in designing shape of wind turbine blades. The lift coefficient of an airfoil increases with the increase in angle of attack (AOA). Beyond critical AOA, lift coefficient drops significantly due to stalling effect. The design of shape of airfoil is essential to provide maximum lift and minimum drag of turbine blade without inducing major stalling effect, directly affecting performance of a wind turbine. Lift modifying devices have been used on wind turbine blades to improve the lift/drag ratio or otherwise modify the stalling characteristics of the blade. The addition of these devices on airfoil bring effects on lift and drag, and trade-offs between lift and drag are usually expected [1]. Stall strip (SS) is a lift modifying device usually triangle in shape, installed to the leading edge of airfoil in wind turbine and aircraft. Installation of stall strips serve the purpose of modifying the streamline flow around the airfoil so that the stalling effect is altered [2].

Various research works have been done on the application of stall strips for improved performance of turbine blades and airfoils [3-5]. Experiment on wind tunnel test of NACA 63-415 attached with stall strips was done by Christian et al. [6]. Stall strips are positioned at different location to analyze the effects on lift coefficient. His results show that aerodynamic coefficients are most significantly affected when stall strips are attached at stagnation position of -2°. Maximum lift decreased from 1.33 to 1.00 while minimum drag increased to 0.010 from 0.008. The effect of SS to the aerodynamic characteristic of airfoil can be ignored when mounted beyond stagnation point 10° AOA [6]. Zahle et al. [7] investigated the same type of airfoil by using 2D computational analysis. The simulation when compared to experimental dome by Christian, shows similar results. However, the author did not make any conclusion due to insufficient agreement with the experimental data. In separate study, Lewis et al. [8] conducted experiment to determine the optimum position and size of SS on NASA LS (1)-0417 MOD airfoil. From the experiment, SS were found to be most effective at position close to the base line. At this location, SS of height 3 mm provided drag control without inducing stalling effect. The author concluded that using SS to control stalling effect has a limitation which is not possible to increase the lift coefficient.

To the best knowledge of the authors this paper, the best location and geometry of the SS is not fully understood. Thus, the main objective of this project is to determine the optimum position and geometry of SS on the leading edge of airfoil to control sudden fall of lift in wind turbine blades. The additional objective is to study the effect of the size of SS towards the stalling characteristics of airfoil in wind turbine. The project is approached by 2D computational analysis method using ANSYS Fluent.

2 Aerofoil and Stall Strip Geometries

The type of airfoil used in the simulation was NACA 0015 with chord length of 150 mm. The geometries of stall strips tested were equilateral triangle and dome shapes. Each geometry had three different sizes namely 2 mm, 3 mm and 4 mm. The simulation was done by placing the stall strips at five different positions near to the leading edge of the airfoil. The SS of size 2 mm was tested at all position while SS of size 3 mm and 4 mm were tested at position POS-1 only. The chord-wise positions of the stall strips attached near the leading edge of the airfoil and their distance from the leading edge are shown in Table 1.

Table 1: Positon of stall strips on airfoil

Denotation	Chord-wise position (%C)	Distance from leading edge (mm)
POS-1	0.000	0
POS-2	0.435	0.65
POS-3	1.635	2.45
POS-5	1.635	2.45
POS-4	0.435	0.65

3 Numerical Details

3.1 Governing Equations

In this study, the flow field is assumed to be incompressible, steady and isothermal. Therefore, the governing equations for the continuity and momentum were solved together with turbulent equations. Thus, the governing equations are expressed as [9-13]:

$$\vec{\nabla} \cdot \vec{V} = 0 \quad (1)$$

$$\rho[(\vec{\nabla} \cdot \vec{V})\vec{V}] = -\vec{\nabla}P + \rho\vec{g} + \mu\vec{\nabla}^2\vec{V} \quad (2)$$

where $\vec{\nabla}$ is divergence operator, V is velocity vector of the fluid (m/s), ρ is density of the fluid (kg/m3), P is the hydrostatic pressure (Pa), g is gravitational acceleration (m/s^2) and μ in the fluid viscosity (kg/m.s).

3.2 Meshing and Boundary Conditions

The mesh was generated by using ANSYS ICEM CFD and C-type grid topology was used. Sizing method with biased behavior was used to create finer grids at locations close to the airfoil walls where superior accuracy is needed. The objective is to obtain accurate result near the airfoil and reduce computational time. The flow in the simulation was considered to be laminar with Reynolds number around 100,000. Air was forced to enter to the computational domain from the left side at constant velocity, while the top and bottom walls are assigned as symmetry planes. The right side of the domain and the edges of the airfoil were assigned as outlet and wall, respectively. The properties of air such as density and viscosity were obtained at room temperature.

Most of the time in numerical simulation, increase number of mesh elements lead to better accuracy of result. However, increasing of element quantity requires extensive of computational time. Thus, proper mesh size should be selected for any computational domain. For this, mesh independence study was carried out to determine the optimum number of elements that are sufficient to produce accurate results. In order to provide accurate results, the grids around the airfoil were refined. It was observed that the values of C_L are same for mesh sizes of 40,000 and above. Thus, mesh size of 40,000 is sufficient to deliver reliable results for the rest simulation works. Further increment of elements cost a lot of computational times and the improvement in accuracy is small-scale.

3.3 Model Validation

Before further analysis of the effect of stall strip on the performance of an airfoil, the accuracy and reliability of the developed model should be checked. For this, model validation was carried out to determine whether the developed model is suitable for analysis of drag and lift of airfoil without SS. The simulation results of clean NACA 0015 airfoil were compared with the results from published experimental data obtained using wind tunnel test [15]. Fig. 2 shows variation of lift coefficient, C_L at various angles of attack obtained from the experimental data of Robert et al. and current simulation model. As we can see from the figure, the lift coefficient values of both experiment and simulation model have similar pattern. The pick value of C_L for the experiment occurred slightly at lower AOA than the simulation one. The average percentage error between the current simulation results and experiment results is 6.1 %. Since the average error of lift coefficient is within the acceptable range, the results obtained from the simulation are considered to be correct.

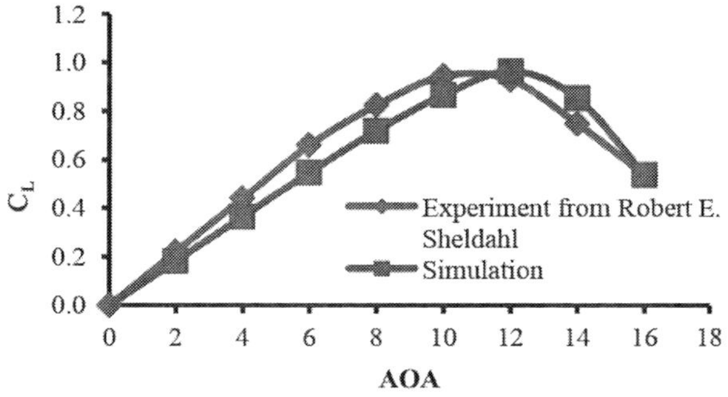

Fig. 1. Comparison of CL between simulation and experiment [15] at various AOA

4 Results And Discussions

4.1 Effect of SS Position on Lift

Shown in Figures 2a and 2b are variations of lift coefficient curves when dome and triangular shaped stall strips are added at five different positions. As can be clearly seen in Figure 2a, the stall strips placed at position POS-2 and POS-3 did not affect the lift and drag coefficients of NACA 0015 significantly. The highest lift coefficient was obtained for POS-3 around AOA 12° and this value is equal to 0.982; while the clean airfoil had 0.964. Placing of stall strips at POS-1 reduced the maximum lift coefficient to 0.84 and the AOA where this value obtained was at 11°. For position POS-4 and POS-5, the AOA for maximum C_L were around 8° to 10°, respectively, with maximum lift coefficient of 0.704 and 0.613 respectively. The lift coefficient trend rose back again after AOA 12°. At AOA 16°, the lift coefficient for all configurations converged to value between 0.482 and 0.562. The effect of the SS is largely dependent on the formation of separation bubble. Introducing free-stream turbulence most likely alters the formation of such a bubble by increasing the mixing of the fluid around the SS.

Similar trends of C_L were obtained for the case of triangular strips addition as shown in Figure 2b. Stall strips placed at POS-3 increased the maximum lift coefficient of NACA 0015 from 0.964 to 0.988. However the curve gradient from AOA 12° to AOA 14° was increased. It indicates enhanced effect in sudden loss of lift. In POS-2, the maximum lift coefficient was dropped by 4.97% and did not show any improvement in controlling stalling problem. Placement of stall strips at POS-4 and POS-5 positions decreased the critical AOA from 12° to 8° and 6° respectively. POS-4 and POS-5 showed positive effect in reducing stalling effect but maximum lift coefficient before stalling dropped drastically by 46.8% and 62.5%, respectively. For POS-1, the critical AOA was reduced to around 12° and magnitude in drop of lift coefficient beyond critical AOA was reduced.

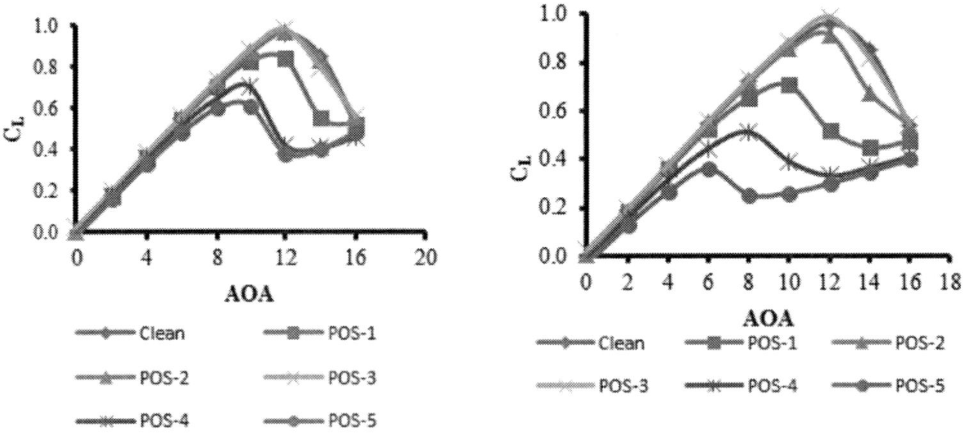

(a) **(b)**

Fig. 2. Variation of C_L curves when (a) dome shaped and (b) triangular shaped stall strips are added at five different positions.

4.2 Effect of SS Shape on Lift

The effect of stall strip shape towards the stalling characteristic of airfoil was brought to comparison. For this, lift coefficient results for clear airfoil and airfoils with doom shaped and triangular shaped stall strips attached at position POS-1 were compared as shown in

Figure 3. For all geometries, there was no noticeable difference in lift coefficient curves at low angle of attacks (i.e., AOA< 6°). However, the difference is visible at high AOA with maximum lift coefficient occurring at angle of attack of 13°, 12°, and 10° for the clean, doom, and triangular stall strips respectively. Overall, the triangular shaped stall strip showed less sudden loss of lift effect, lower maximum lift coefficient and lower critical AOA than dome shaped SS. The curve gradients at the stalling region of triangular shaped SS were lesser while that of dome shaped had no significant difference with clean airfoil. On the other hand, the triangular shaped stall strip showed better results in controlling sudden loss of lift of the airfoil. The difference in lift coefficient started to converge after AOA 12°. At AOA 16°, all geometries had very close values of lift coefficient.

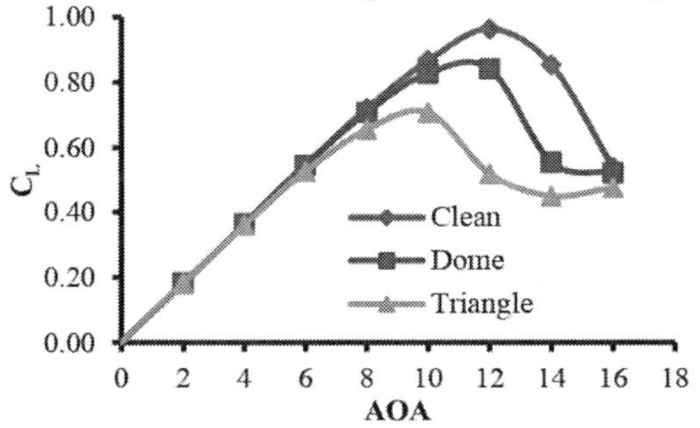

Fig. 3. Comparison of lift coefficient between clean airfoil and airfoil with dome and triangular shaped stall strips attached at positon POS-1

4.3 Effect of SS Size on Lift

Shown in Figures 4a and 4b are variation of lift coefficient with stall strip size when dome and triangular shaped stall strips, respectively, are added at position POS-1. As can be clearly seen in Figure 4a, the lift coefficient curves for 2 mm, 3 mm, and 4 mm stall strips remain almost the same with a slight increment from AOA 12° onwards. The average increment of lift coefficient for 3 mm and 4 mm stall strips compared to the 2 mm one were only 0.67% and 3.25%, respectively. It can be concluded that increasing in size of dome shaped stall strips did not improve the stalling behavior of the airfoil. Similarly, the effect of stall strip on the lift coefficient for triangular SS addition is not significant as shown in Figure 4b. However, the lift coefficient curves indicated that the increase in size of SS brings positive effect in controlling sudden stall effect. Increasing of size decreased the stalling effect as well as the maximum lift coefficient. Comparison between 4 mm SS with 2 mm SS showed that the maximum lift decreased by 8.43% and the fall of lift from AOA 10° to AOA 14° decreased by 29.8 %.

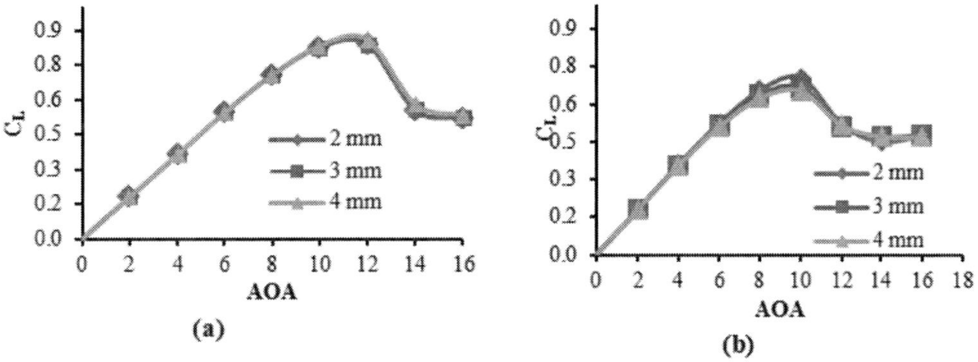

Fig. 4. Variation of lift coefficient with stall strip size when (a) dome and (b) triangular shaped stall strips are added at position POS-1

5 Conclusion

In this research, total of five positions, two geometries and three sizes of stall strip configurations added on NACA 0015 airfoil were simulated. The results led to the a conclusion that addition of stall strips as a method to control sudden fall of lift in wind turbine blades sacrificed the maximum lift coefficient. The optimum (suitable) geometry of stall strip appeared to be triangular shaped SS. With triangular shaped stall strips, position POS-1 had the optimum sudden fall control to reduction of lift coefficient ratio. On the other hand, different geometries of the stall strips showed different effects to the lift coefficient. For triangular stall strips, increasing in size of stall strips did slightly reduce the sudden fall of lift. While for dome shaped stall strips, the increment of size of stall strips did not change the stalling effect. The findings from this project may help in future design and development of stall strip in airfoils.

Acknowledgment

The authors would like to thank Universiti Malaysia Pahang for providing necessary assistance for providing financial support under grant number RDU150375.

References

1. H. Stiesdal, P. B. Enevoldsen, Google Patents (2006).
2. D. Crane, *Dictionary of Aeronautical Terms*. 3rd ed. (1997).
3. C. Amanullah, M. Arjomandi, and R. Kelso, Renew. Ener., **86**: 26-37, (2016).
4. J.W. Lee, J.S. Lee, and J. H. Han,. Jour. of Wind Eng.and Ind. Aerodyn., **110**: 62-69, (2012).
5. W. Geissler, G. Dietz, and H. Mai, Aerosp. Sci.e and Tech., **9**(5): p. 390-399.
6. B. Christian, F. Peter, J. Jeppe, A. Ioannis, Riso Nat. Lab., Roskilde, Denmark, (2000).
7. F. Zahle, N.N. Sørensen, and J. Johansen, DTU Orbit - The Res. Info. Sys. (2002).
8. R. I. Lewis, I. Potts, and A. A. Arain,. **15**: 40 (1995).
9. Y. A. Cengel, and J.M. Cimbala, New York: McGraw-Hill, (2013).
10. A. N. Oumer, A., N. Hamidi, and I.M. Sahat. *IOP Conf. Series: Mat.Sci. and Eng.(*2015)
11. A. N. Oumer, and O. Mamat, Int. Jour. of Comp. Sci. and Eng., **9**(3): 247-256 (2014).
12. A. N. Oumer, and O. Mamat, Compos. Part B: Eng.**43**(3): 1087-1094 (2012).
13. N. T. Rao., A. N. Oumer, and U.K. Jamaludin, The Jour. of Supercri. Flu., **116**: 132-147 (2016).
14. Syavash. *CFD online* Fluent airfoil help (2012);
Avail. from:http://www.cfd-online.com/Forums/fluent/107547-fluent-airfoil-help.html#post384545.
15. E. Robert. P. C. K. Sheldahl, 118, (1981).

Experimental investigation of zero energy office under natural and forced ventilation

Hussain H. Al-Kayiem[1], *Syed I. U.* Gilani[1,*] and *Mahmoud S.* Abdalfatah[2]

[1]Mech. Eng. Dept., Universiti Teknologi PETRONAS, 32610 Bandar Seri Iskandar, Malaysia
[2]Smart Tech Solution, 1/15A Third Settlement, Industrial Zone, New Cairo, Cairo, Egypt

Abstract. With the rapid rise of energy consumption in the world, lowering the energy needs of houses and buildings can have a significant effect on future of energy demand and consumption. This paper presents experimental investigation results on the effectiveness of the natural and forced ventilation in prototype of zero-energy office. The main aim of the different ventilation systems is to improve the thermal comfort within the office. The installed forced ventilation system achieved a 6°C drop in maximum temperature within the office and improved the thermal comfort by 46%.

1 Introduction

The residential sector accounts for approximately 18% of the global energy usage with an expected annual increase rate of 1.5% [1]. Hence, lowering the energy needs of the houses and buildings is vital to decrease the energy consumption and increase the dependence on renewable energy sources. It is found that up to 66% energy savings could be achieved in the household annual energy consumption by using net zero-energy building (nZEB) methods to optimize air flow through the usage of passive design techniques [2].

1.1 Concept of net zero-energy building

Many definitions have been put forward by researchers to define what exactly is nZEB and how to measure their performance. A proper technical consistent classification is required for countries and unions to align their ambitions and plans and set KPIs. This is what pushed the Federation of European Heating, Ventilation and Air Conditioning Association (REHVA) to decide on one technical definition to be used all across Europe. The research was led by [3] in 2013 to define nZEB as "A building that has very high energy performance. The nearly zero or very low amount of energy required should be covered to a very significant extent by energy from renewable sources, including energy from renewable sources produced on-site or nearby.

The definition adopted by the REHVA follows "avoided burden" logic. In this perspective the building is defined as nZEB because it avoids the need for primary energy that would otherwise have been needed to keep the building running.

- Corresponding author: syedihtsham@utp.edu.my

There are many methods by which the energy balance could be calculated within the avoided burden approach. [4] presents four methods for such calculations namely, a) Site balance b) Source energy balance c) Source energy balance with building embodied energy c) Source energy balance with RES embodied energy. A source energy balance is done at the primary energy level and termed ZEB source. This balance method is highly recognized by the international community.

nZEB must be optimized to consume very little energy so that a big part of the energy needed can be generated by the renewable sources installed within the building. Minimizing the energy usage of the building can be achieved using passive design techniques or energy efficient technologies or a hybrid of both. Passive design techniques mainly focus on using the building design to reduce the energy consumption (see figure 1). Passive design techniques include using the building's envelop, orientation and geometric and ratios to reduce the energy needed to power the building [5]. A major objective of passive design techniques is to use isolation to keep the building temperature stable and to use natural ventilation to provide the thermal comfort needed. On the other hand, using energy efficient technologies lowers the total energy usage by using higher efficiency HVAC systems, appliances and lighting. A hybrid solution aims at using both techniques to lower the energy needs of the building.

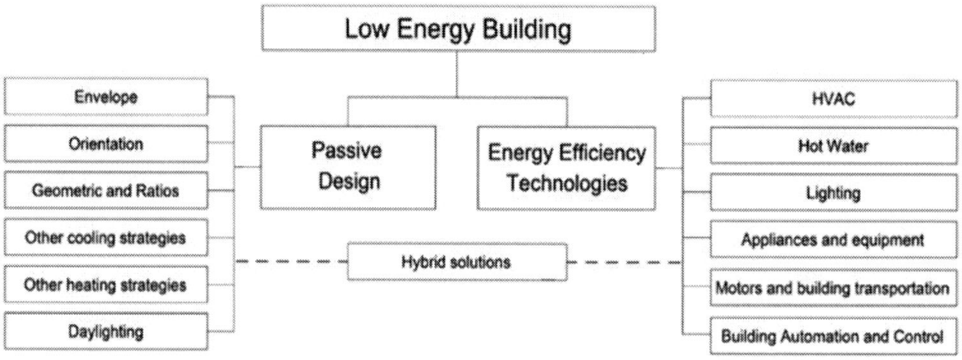

Fig. 1. Low energy building techniques. [5]

A buoyancy-driven natural ventilation system uses the difference in air density resulting from different air temperature to generate airflow. When the building's indoor temperature is higher than the outdoor temperature, the cold, high density, air from outside enters the building at lower floors. At the same time, the hot, low density, air at lower floors, inside the building rises up to higher floors where it exits. Thus the lower floor windows allow air to come in, which then exits from the upper floor creating airflow within the building; such phenomenon is called the "stack effect". The different in pressure between the two sides of the window, resulting from the difference in temperature and thus difference in density, controls whether air enters the building or flow out of it.

In a multi floor building, the top opening is made to be the larger than all the other windows in the building. This design assures that the neutral pressure plane is higher than the top floor, thus there is no recirculation of air within the building. Increasing the window to wall ratio to 0.24 can increase thermal condition indoor [6]. Furthermore, a recommended total area of inlet and outlet apertures is about 40% [7]. Figure 2, shows how the stack effect is implemented in a building with an atrium [8]. Airflow of cold air turns into hot air before exiting the building at the top chimney through the atrium. The design of the building and the implementation of the atrium can be adapted differently to suit the different conditions for different buildings.

 DOI: 10.1051/matecconf/201713104012

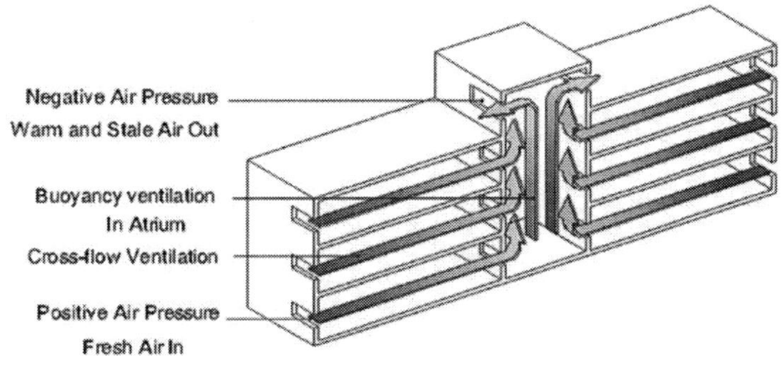

Fig. 2. Principles of stack effect in envelop of building [8]

1.2 Thermal comfort within the office

The ultimate objective of using the solar chimney is to achieve thermal comfort within the building while reducing the energy consumption. Although thermal comfort is a very subjective state, the acceptable levels of thermal comfort are specified in ASHRAE building standards. The standard way of measuring the thermal comfort within a building is by conducting a survey whether the condition within the building is comfortable or not. After extensive research and hundreds of conducted surveys, ASHRAE has released a method to calculate the Predicted Mean Vote (PMV). PMV is a standard measurement of thermal comfort within any building and ASHRAE [9] releases standard values of PMV measurement that offices and public buildings must adhere to. Furthermore, the PMV can be calculated based on the temperature of air within the building, the mean radiant temperature the occupants feel, air speed, relative humidity level, metabolic rate and clothing level. While the PMV model predicts thermal sensation well in buildings with HVAC systems, field studies in warm climates in buildings without air-conditioning have shown that it predicts a warmer thermal sensation than the occupants actually feel [10]. It's for this reason that Fanger et al. have introduced a way to adjust the PMV values from the standard ASHRAE calculation to be closer to the measured PMV in warm climates countries where the buildings are without air-conditioning [11].

The currently implemented Zero Energy office in UTP is using a natural ventilation system to achieve temperature comfort needed within the building. The house studied in this project uses a buoyancy-driven natural ventilation system. The problem at hand is to confirm whether the installed ventilation system is sufficient to achieve thermal comfort within the building.

The objective of the current work is to carry out monitoring and measurement on the office to evaluate and compare the effectiveness of different ventilation procedures at achieving thermal comfort within the office.

- The ventilation procedure either natural or forced.
- Monitoring the effect of the ventilation system on the level of thermal comfort within the office.
- Monitoring ΔT and ΔRH between indoors and outdoors.

In this project the adjusted PMV model developed by Fanger will be adapted. A recommended expectancy factor of 0.7 will be used. An expectancy factor of 0.7 has been adopted in Singapore and getting a very accurate expectation of PMV. Fanger's adjusted PMV model predicts the thermal comfort response of the occupants in hot weather accurately

DOI: 10.1051/matecconf/201713104012

2 METHODOLOGY

This project is concerned with the zero-energy office prototype. The building prototype consists of the office room, a hallway where cross wind flows and a third room that is of no interest to the project at hand. Figure 3 shows how natural ventilation happens in the office prototype.

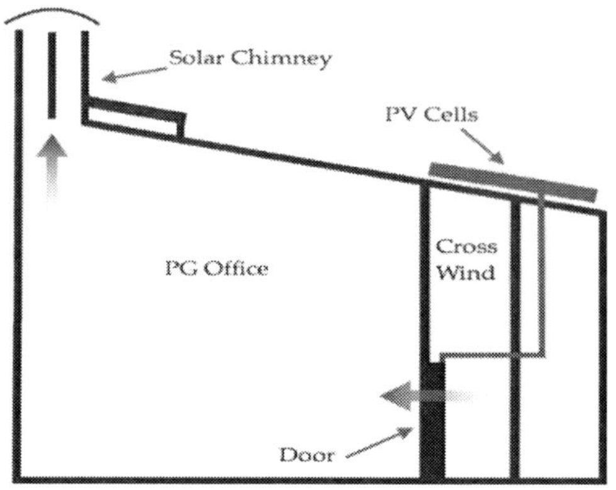

Fig. 3. Schematic of the Zero-energy office

The cold air enters the office from the open door. The hallway, where the cross winds acts, is the main source of fresh air in this case. The solar chimney at the top helps to heat up the air at the top levels thus acting as an area of low pressure to suck up the stagnant air to outside the building.

The methodology in this project is to experimentally measure the effectiveness of the natural ventilation system. The effectiveness is indicated by an increased thermal comfort. The experiments were designed to track the change in the temperature within the building during the day and during the night. Another factor to be experimentally evaluated is the change in humidity throughout the day. Lastly the airflow rate will be measured. With measurement results at hand, the effectiveness of the system at increasing the thermal comfort could be evaluated. A higher airflow rate between the inside and outside of the building is preferred. The temperature and humidity level in the building should always lie within the comfort levels throughout the day. If the system is not effective enough, a re-evaluation of the design of the natural ventilation system is needed. Such effort could be carried out separately in a different project. Figure 4 shows the measurement tools installation placement at the site.

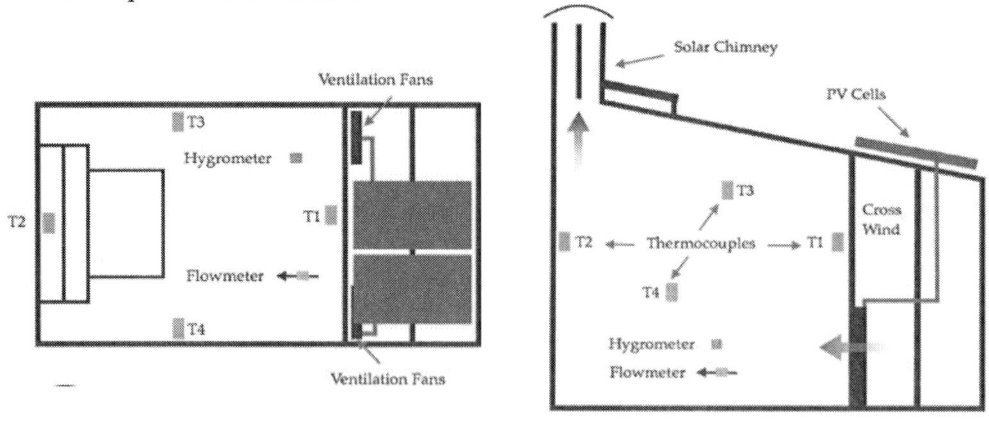

Fig. 4. Testing the effectiveness of natural and forced ventilation systems.

The two most critical variables, to calculate the thermal comfort, in accordance with Fanger's equation, are air temperature inside the office, radiant temperature sensed by the occupant and air flow within the office. Thermocouples T1 and T2 are used to measure the air temperature inside the office. The average of T1 and T2 was used as the average air temperature within the office. The average radiant temperature inside the office, the average of T3 and T4 was considered.

The forced ventilation system is comprised of four solar fans that are powered by PV cells. The fans are used to increase the air rate of change in the office. The forced flow of air caused by the fans will aid the ventilation system and stack effect. The usage of the forced ventilation system will only happen after measuring the effectiveness of the ventilation under the natural ventilation only. The experiment is to be carried out once without any active ventilation systems. The resulting data will be taken as a benchmark to measure the progress against. Secondly, data will be collected with active natural ventilation system exclusively. Lastly, the experiment is to be carried out with both natural and forced ventilation system working in tandem.

Three different cases have been experimented. Each experiment has been repeated 4 days and the mean is considered in the presentation and discussion of results.

Table 1. Identifications of the measurement cases

case	Ventilation status	Conditions and description of the case.
Case-1	Natural ventilation	No in flow. The building locked and sealed. Only inlet to the solar chimney is operated.
Case-2	Natural ventilation	Flow through the opened door. And exit to the chimney inlet.
Case-3	Forced ventilation	4 solar fans are operating to bring ambient air to inside the office. Exit to the solar chimney is opened.

3 Results and discussion

Case-1: No active ventilation system

In case-1, the door was closed to deactivate any possible ventilation. Thus follows the first set of data collected with a closed door, i.e., no active ventilation system. Table 2 shows the maximum and minimum thermal comfort points. The predicted adjusted thermal comfort PMV at highest temperature is 5.65, which indicates too hot conditions. At the lowest temperature, during the measurement period, the Adjusted Thermal Comfort PMV is 0.61 which indicating slightly warm. Results of case-1 show how the spaces inside the office can heats up to very high temperature without any ventilation system at work. This is the normal state at which the office would be at without any of the ZEB techniques applied to aid in the ventilation of the space. With a maximum radiant temperature of 49°C, the space had a very poor rating of thermal comfort of 5.65 which is considered too hot according to ASHRAE-55 standards. Since the ASHRAE-55 scale extends from -3 to +3, the resulting thermal comfort in this case is off the chart. The space is simply very uncomfortable to stay in at such state.

Table 2. Maximum and minimum thermal comfort as measured for case-1

Time	Average Air Temperature (°C)	Average Radiant Temperature (°C)	Adjusted Thermal Comfort PMV
5:15 PM	45.93	49.34	5.65 (Too Hot)
7:30 AM	28.47	27.27	0.61 (Slightly Warm)

 DOI: 10.1051/matecconf/201713104012

Case-2: Active natural ventilation system

In this case, the door was open to allow natural ventilation system to refresh the air within the office. The mean data of 4 days is shown in table 3, where only particulars of thermal comfort at maximum and minimum temperatures are presented. At the maximum point, the radiant temperature reached 48.16°C while the air temperature is top at 43.67°C. The following table shows the predicted thermal comfort based on maximum and minimum measurement for this set of data. In case-2, the doors were open thus the air was allowed to flow in the house and exit through the solar chimney at the top of the house. This natural ventilation system aids in getting the air to circulate through the office. This ventilation system is purely passive and costs nothing but the initial cost of setup for the solar chimney. Natural ventilation systems are valuable since they can improve thermal comfort, although marginally, without using any mechanical systems. The natural ventilation system dropped the temperature by 2°C and improved the thermal comfort level by 9%. Although the thermal comfort level is still far away from the recommended range by ASHRAE, the 9% increase at no additional operating cost and with only a solar chimney installed on top of the office is effective.

Table 3. Maximum and minimum thermal comfort as measured for case-2

Time	Average Air Temperature (°C)	Average Radiant Temperature (°C)	Adjusted Thermal Comfort PMV
6:00 PM	41.84	44.06	5.80 (Too Hot)
8:15 PM	33.99	32.89	1.9 (Slightly Warm)

Case-3: Active natural and forced ventilation systems

In this last experiment the ventilation fans were turned on. This was the natural ventilation system and the forced ventilation system was both working concurrently. The ventilation fans allow for more airflow within the office and hence lower temperatures are expected. Figure 11 shows the resulting data from the last experiment where the maximum ventilation occurred. The drop in the maximum temperature is noticeable clearly. The thermal comfort throughout the day was also improved. Following are the data for the thermal comfort.

Table 4. Maximum and minimum thermal comfort as measured for case-3

Time	Average Air Temperature (°C)	Average Radiant Temperature (°C)	Adjusted Thermal Comfort PMV
2:15 PM	38.89	40.39	3.47 (Too Hot)
8:15 PM	31.04	30.13	1.33 (Slightly Warm)

The results show progression improvement in thermal comfort from one experiment to another. In case-3 the ventilation fans were turned on for 24 hours a day. The ventilation fans run on 12VDC coming from the battery which is connected to the solar panels on top of the office. The four ventilation fans aided the air flow greatly throughout the office and helped improve thermal comfort significantly. Although the ventilation fans don't consume any grid energy, the energy embedded in fabricating the PV panels is considerable. The installation of the PV panels and the ventilation fans might be costly, yet the resulting data are encouraging. With the forced ventilation working side by side with natural ventilation the maximum temperature dropped by 6°C and the thermal comfort improved by 46%.

Such improvement in thermal comfort is highly significant. At best the thermal comfort was very close to accepted levels within the ASHRAE-55 standards while at worst, the PMV was slightly overshooting the scale at 3.47.

4 Conclusion

ZEB has been investigated experimentally at three different cases to evaluate the level of comfort. Solar fans have been used to impose forced ventilation and solar chimney has been used to enhance the natural ventilation. The ultimate aim of any ventilation system is to provide thermal comfort within the building. Hence, the effect of the installed systems on the thermal comfort levels within the building is the focus point of this project. With three experiment conducted under three different settings, the results show a promising improvement in thermal comfort levels.

While using the natural ventilation system exclusively decreased the maximum temperature by 2°C, using the forced ventilation system decreased the maximum temperature by 6°C and consequently improved the thermal comfort by 46%. Nevertheless, the resulting thermal comfort levels are not within the accepted range recommended by ASHRAE-55 standards. This calls for more passive and/or active ventilation techniques to improved thermal conditions in the office envelope. Live green wall is seems to be active technique to reduce the solar radiation absorption by the walls of the office. Since most of the heat is transferred to the office through the wide glass windows, it is recommended to shade the windows or cover them by insulation material.

The authors acknowledge Universiti Teknologi PETRONAS for supporting the work, logistically and financially. Also, the authors acknowledge the Ministry of Higher Education of Malaysia (MOHE) for allowing the use of the equipment under project FRGS/1/2017/STG02/UTP/01/1.

References

1 L. Pérez–Lombard, J. Ortiz, C. Pout, ENB **40**, 394–398. (2007)
2 Yuanming Kwan, Lisa Guan, Procedia Engineering, **121**, 604-611(2015)
3 J. Kurnitski, F. Allard, D. Braham, D. van Dijk, C. Feldman, J. Fox, REHVA nZEB. Technical Report, REHVA, (2013).
4 J. S. Bourrelle, I. Andresen, Arild Gustavsen, ENB **65**, Pages 84-92 (2013)
5 E. Rodriguez-Ubinas, C. Montero, M. Porteros, S. Vega, I. Navarro, M. Castillo-Cagigal, E. Matallanasc, A. Gutiérrez, ENB **83**, 10–22. (2014)
6 W. Liping, W. N. Hien, Building and Environment **42**. 4006–4015. (2007)
7 C. Tantasavasdi, J. Srebric, Q. Chen, ENB **33**. 815–824. (2001)
8 L. Moosavi, N. Mahyuddin, N. Ab Ghafar, M. A. Ismail, RSER **34**, Pages 654-670 (2014)
9 ASHRAE Handbook 2017—Fundamentals, Chapter 9, Thermal Comfort. (2017)
10 G. S. Brager and R. de Dear, ENB **27**. 83-96. (1998)
11 P. Ole Fanger, J. Toftum, ENB **34**, (6), 533-536 (2002)

Harnessing 'Blue energy': A Review on Techniques and Preliminary Analysis

Khairul Nisak Md Hasan[1,*], *Tian Xian* Khai[1], *Ramani* Kannan[1], and *Zainul Azhar* Zakaria[2]

[1]Electrical & Electronic Eng. Dept., Uni. Teknologi PETRONAS, 31750 Tronoh, Perak, Malaysia
[2]Universiti Kuala Lumpur, Malaysian Inst. of Marine Eng. Technology, Lumut, Perak, Malaysia

Abstract. Energy harvesting from the ocean called 'blue energy' began in the 1970's and should have reached its peak by today, but due to varying interests in the field and the growing potentials of other sources, its development was delayed. Recently, it receives more interest and a number of institutes has deployed their pilot plant. Blue energy is a type of renewable energy based on salinity gradient where power is generated from different salt concentration in saltwater (the ocean) and freshwater (river). When the mixing of seawater and freshwater occurs, an increase in the entropy of this system is observed and free energy is dissipated. This research aims to identify and compare existing techniques or methods of blue energy harvesting developed over time. Five different techniques were reviewed, looking at their principle of operation, configuration and performance. Based on the review, capacitive mixing method was selected for further analysis. Experiment was conducted to evaluate different factor including the concentration of sea water, volumes and type of electrodes. The highest output power obtained is 89.7 mW, while the average is about 30 mW.

1 Introduction

The world's increasing demand for energy encourages the use of renewable energy as an alternative to fossil fuels that are getting scarce. Renewable energy is also important to reduce pollution and carbon emission caused by the usage of fossil fuels as energy sources. There is another source of renewable energy called "Blue Energy" [1] or better known as salinity gradient power that has yet to be fully utilized. Salinity gradient power relies on the different salt concentration in saltwater and freshwater to generate energy. When the mixing of seawater and freshwater occurs, an increase in the entropy of this system is observed and free energy is dissipated. This renewable energy is clean and green as it does not produce any wastes. Globally, with an estimated 2.6 TW [2] of energy output from blue energy, this renewable energy will be able to provide a significant amount of the global energy needs.

In order to harness blue energy, many techniques with different power generation mechanism were developed. Five different techniques in extracting blue energy will be discussed, which are: Pressure Retarded Osmosis (PRO), Electrodialysis (RED), Capacitive Double Layer Expansion (CDLE), Capacitive Donnan Potential (CDP) and Mixing Entropy Battery (MEB). Three of the techniques, which are CDLE, CDP and MEB are under

* Corresponding author: khairulnisak.hasan@utp.edu.my

capacitive mixing (CAPMIX) category. The main problem faced by blue energy technology is the non-economical energy production based on current methods of harvesting blue energy. These methods such as PRO and RED, requires expensive materials especially the membrane. The relatively new methods, CAPMIX, was introduced aimed to eliminate the use of membrane. Besides the cost effectiveness, blue energy is still a developing technology that yet to acquire more attention along with other alternatives.

1.1 Review of Blue Energy Technology

1.1.1 Pressure-Retarded Osmosis (PRO)

PRO is a method that extracts osmosis pressure difference between seawater and freshwater. By utilizing a semi-permeable membrane, this pressure difference can be used in power generation. The amount of energy produced depends on both concentration of the solutions as well as the membranes used. In [6], power density between 0.11 – 1.22 W/m^2 was obtained. By using different type of membranes, which is cellulose acetate, [8] reported that it yields favourable results, with power density up to 2-5 W/m2 being obtained. This relationship can be explained by the osmotic potential difference equation developed by Van't Hoff, that is expressed by the activity of the salt in the two varying solutions:

$$\Delta\pi = RT(a_c - a_d) \tag{1}$$

Where $\Delta\pi$ is Osmotic Potential Difference, R is Molar Gas Constant (8.314 m^2 kg s-2 K-1 mol-1), T is temperature (°c), a_c and a_d is activity of the salt in concentrated and diluted solution, respectively. A pilot plant for power generation through PRO was commissioned at Tofte, Norway on 24th November 2009 with capacity of 4 kW. The membrane used by the pilot plant can only generate roughly 1 W/m² which is an improved performance from the previous membrane type [4], but it is still far lower than the targeted power density of 5 W/m². This means that that membrane was unable to reach an economical break-even for the power generated per square meter and cost of the membrane. As a result, an economical barrier prevents the PRO from being a cost-effective method of generating power. It is also reported that the PRO plant is facing fouling problem, which is very detrimental causing the performance to deteriorate over time [5]. The fouling problem is classified into pore narrowing, pore plugging and gel/cake formation. Review papers on PRO were reported in [5-8] which provides thorough review in PRO process, membrane technology development, current development up to 2015, the fouling and cleaning process and the viability of the technology itself. Improvement area for PRO focuses on to reduce cost of osmotic power to increase membrane power output to at least 5 W/m^2, improved membrane design to get higher membrane area, improve efficiency of the PRO process, scale-up the system to commercial production and handling the fouling and cleaning issue [7].

1.1.2 Reverse Electrodialysis (RED)

Reverse electrodialysis (RED) utilizes two different types of membranes: anion exchange membrane (AEM) and cation exchange membrane (CEM) are stacked alternatingly between each other and placed in between an anode and a cathode. By flushing salt water and fresh water thorough the space in between these membranes, electrochemical potential is produced due to concentration difference between the solutions and ion-selectivity of the membranes. Positive ions, Na+ will diffuse thorough the CEM while the Cl- will diffuse thorough the AEM producing ionic current, thus generating power. Available research in the area covers many factors including the RED simulation model, membranes, configuration, other factors affecting performance and even implementation of RED in micro and nano sizes. In a research by Z. Jia et al. [4], experimental results showed a power density of 0.93 W/m² based on artificial seawater and river water. It is not up to the targeted power density needed in order for RED to be economical in commercial uses. Currently, a pilot plant located at Afsluitdijk, the Netherlands is expected to produce 50 kW per hour, aiming to raise the output to 200 kW per plant. The progress of RED technology also evolved around the membrane technology. In [10], electro dialyzer equipped with a 0.19-mm spacer was tested with different concentration of dilute and power peak of 460 mW/m^2 was found at 0.54 cm/s linear flow velocity that gives an estimation of \$6.79/kWh for the investment cost. J.Vermaan in 2009 has evaluated 6 commercial membrane to compare their performance and found out the best membrane could produce up to 1.2 W/m^2 with thermodynamic efficiency is around 14% - 35% [13]. In [14], besides evaluating performance of commercially available membrane, they had tested a tailor made membrane with significant reduction in resistance and achieved 1.3 W/m^2. More recent type of membrane was reported such as pore-filling ion exchange membranes with open area spacer that achieved 2.4 w/m^2 [15], asymmetric heterogeneous membrane [16] which can be applied in nanofluidic application and membrane with chevron corrugation that reduce the pressure drop is able to produce up to 0.89w/m^2. Works on modelling and simulation of RED process including the cell pair model based on thermodynamic and transport equation was proposed by Veerman [17]. An enhanced model is proposed by Tedesco to improve it to a wider range of concentration [18] but lacking few non-ideal effects. The work by [19] proposed to improve this model by including non-ideal effect such as water transport through membranes, concentration polarisation and parasitic currents.

1.1.3 Capacitive Double Layer Expansion (CDLE)

CDLE is a method where two electrodes are immersed into an ionic solution to form a super capacitor. When the electrodes are connected to power supply and charged, electrical charge is stored in the electric double layers (EDL) between the carbon and solution. When the salinity of the solution decreases, the potential difference across EDLs increases with constant charge. This occurs due to counter-ions in the EDL moving away from the electrode using free energy of the solutions, hence the name of this technique.

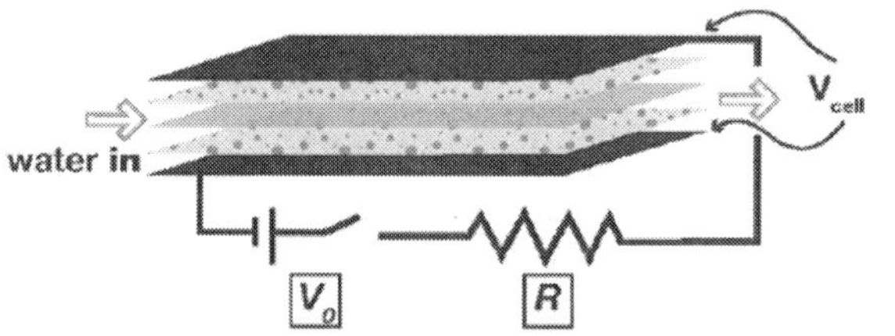

Fig. 1. Cell design of CDLE [4]

Fig. 1 shows the cell design of CDLE where the black layer indicates the electrodes, the grey layer is carbon and the middle layer is the solution. An experiment was done by D. Brogioli, et al. [20], in which the highest power density of 50 mW/m² was produced using A-PC-2 and NS30 as activated carbon materials. He stated that in order to improve the energy extraction thorough CDLE, different materials should be used for positive electrodes and negative electrodes. When two electrodes with different potential rises were used, the cell will experience a voltage rise, or called "zero charging" in which no external supply was needed for charging and discharging phase. This finding is supported by work by C.Hatzell which analyse the effect of strong acid functional group on electrode potential [21]. It was also reported that porous electrode produced more power as the total surface area are huge [22].

Further work by [23] reported that the porous electrode possesses an optimal charging potential and the pore size is an important parameter to optimize the power generated. This is also reflected in the study by [24] where the porous electrode is used and configuration of stacking them was evaluated. This technique has an advantage because it does not utilize membrane which has been a determinant factor for the technology to mature. It is expected that this technique or technology will be evolved faster.

1.1.4 Capacitive Donnan Potential (CDP)

In CDP, membranes are used to extract the salinity gradient energy. It is proposed by B.B Sales after the CDLE where an external supply is replaced by 'Donnan Potential' produced by adding ion-exchange membrane [25-26]. There are layers of AEM and CEM in between the layer of carbon and water flow. As a result, when water seawater flows through, ions from the concentrated seawater diffuse thorough the AEM and CEM into the electrodes. This movement of ions will produce current to an external load. After reaching equilibrium, freshwater will flow into the system and the ions inside the electrodes and membranes will diffuse out to form current in the opposite direction. The advantage of this technique is it does not require external power supply as the power is 'internally' generated [27]. Based on a research done by B.B. Sales et al. [28], maximum power density of 131.7 mW/m^2 was reached and an average power of 8.36 mW/m^2 was produced for one cycle of CDP technique. Optimization of distance between the electrodes in respect to the hydrodynamics of the flow of solution is recommended by them in order to improve this technique. Some works on CDP explaining the physics of the reaction, impact of electrode parameters, membrane and configuration are presented in [29].

1.1.5 Mixing entropy battery (MEB)

In MEB or sometimes known as faradaic pseudo capacitor [4], the energy is extracted from salinity difference between two solutions and stored inside the electrode. Through the use of battery-like electrodes, the ions are captured in the solution through redox reactions. The voltage of the electrodes used changes with respect to the solution and this will reflect chemical potential ions in the solution. The working principle of MEB can be divided into 4 steps. First, the electrodes are immersed into a low concentration solution such as river water, in which the battery is charged and the ions are removed from the electrodes. Step 2, the river water is exchanged with sea water which has higher concentration of ions and thus higher potential difference between the electrodes. The battery is discharged in step 3 and the ions are restored into the electrodes. Finally, the seawater is flushed out and replaced with river water which results in decrement of potential difference between the electrodes. In an experiment done by F. L. Mantia et al., they managed to produce power density of 138 mW/cm^2 [30]. The process was repeated for over 100 cycles (times) and the power density produced was stable. As mentioned, MEB might be a potential technique to harness blue energy due to the renewable nature of this technique; no loss of reagent even with repeated usage of the materials in MEB. Research conducted in [31-32] in evaluating different materials of electrodes and the chemical reaction formed that results in power generated. The electrolyte or the solution is also determinant factor in this technique. It has simple design but more studies are required in determining its reliability and the cost per power generated. Table 1 shows comparison among 5 techniques on their working principle (type), membrane characteristic, maximum power density reported to date and important remark.

 DOI: 10.1051/matecconf/201713104013

Table 1. Review of five salinity gradient techniques

Technique	Type	Membrane Characteristics	Max. Power density produced	Remarks
PRO	Osmotic pressure difference	Water selective	1 W/m²	Pilot plant
RED	Electrochemical reactions	Ionic selective	0.93 W/m²	Pilot plant
CDLE	Ultra Capacitor	None	50 mW/m²	Experimental scale
CDP	Ultra Capacitor	Ionic selective	131.7mW/m²	Experimental scale
MEB	Electrochemical reactions	None	138mW/m²	Experimental scale

2 Experimental Procedure

Analysis was carried out on the mixing entropy battery technique. It can produce more power for the same dimension of setup and did not require any membrane which reduced the cost and complexity. However, the technique requires nano-electrodes. The solution used in this work is the salty water. Experiment was carried out to analyse several factors including the concentration of sea water, volumes and type of electrodes on the power density produced.

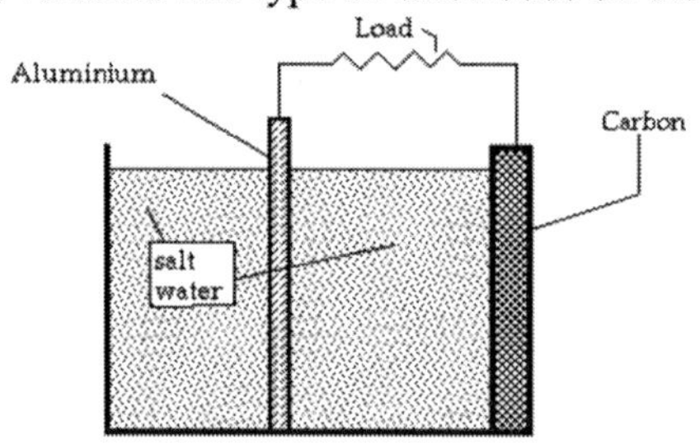

Fig. 2. Mixing entropy battery experimental setup

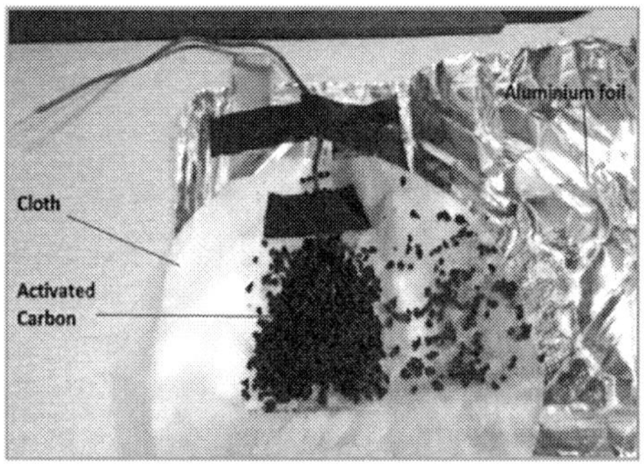

Fig. 3. Activated carbon in aluminium sheet

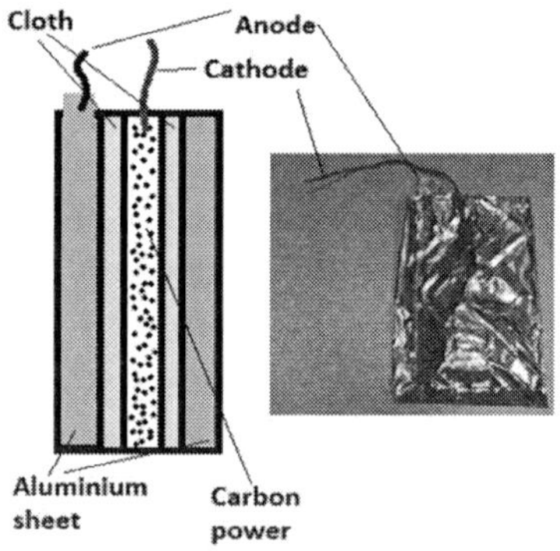

Fig. 4. Side view of the interiors

Figure 2 is a representative of how a mixing entropy experimental cell look like. Two electrodes, an aluminium electrode and a carbon electrode is prepared. A solution of salt water is also prepared by dissolving salt into water. When the aluminium electrode and carbon electrode are immersed into the salt water, a redox reaction occurs. The open circuit voltage Voc or the potential difference between the two electrodes is measured using multimeter. The current is measured by applying load or resistance in between the two terminals (both aluminium and carbon electrodes). In order to increase the voltage or current, multiple cells can be connected in series to increase the voltage or connected in parallel to increase the current. Activated carbon powder were filled into layers of cloth (as a hydrophobic membrane between carbon and aluminium) connected with a multi-cores wire and an aluminium foil was wrapped around the cloth as shown in Figure 3 and Figure 4. The aluminium foil acted as anode while the activated carbon acted as air cathode. No continuity was allowed between the anode and cathode to prevent the anode and cathode from short-circuiting. Readings are measured and recorded accordingly.

3 Result and Discussion

Multimeter was used to measure the voltage and current of the cell. Different parameters such as concentration of solute, volume of water and type of electrolytes were used as variables in the experiment. The readings for both voltage and current were tabulated in detail, but is not included due to space. For all experiment, reading was taken over 30 min at 5 min interval. In the first experiment, three different concentrations are tested for the same type of electrodes and volume. The electrode used is carbon-aluminium, volume of water is 1000 ml and type of electrolyte is NaCl. For the second experiment, volume is varying from 500 ml, 750 ml and 1000 ml. In the third experiment, different type of electrolytes tested are water, salty solution and water with vinegar. Overall, the power density obtained reduced over time with the highest reading was obtained for carbon-aluminium electrodes. Graphs of output power was plotted for the three different variables.

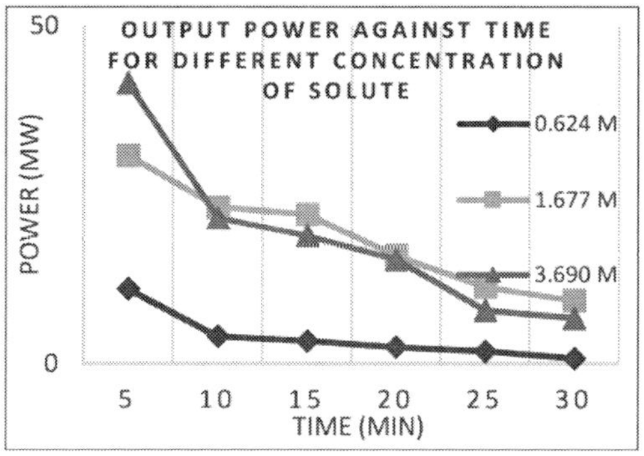

Fig. 5. Graph of output power against time for different solute concentration

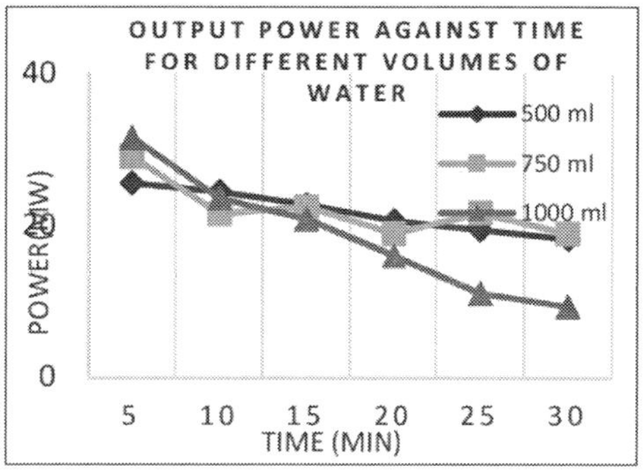

Fig. 6. Graph of output power against time for different volume of water

For experiment 1, it is worth to mention that more concentrations of solutes were tested before the three concentration is chosen to show the obvious finding. Voltage and current produced is measured accordingly. It is observed that the output voltage depreciated at higher rate in the solution with lower concentration. In Figure 5, it can be observed that higher concentration of solute provides higher output power. However, this is true only for the initial 5 minutes of operation. After 5 minutes, the output power for both 1.677 M and 3.69 M was almost equal. On the other hand, solute with lower concentration (0.624 M), consistently producing lower output. It can be concluded that concentration of solute has an optimal value which will produced maximum output power and increasing the concentration beyond this will not increase the output power. For this, the optimal concentration of solute for the cell would be 1.677 M. In experiment 2, three different volumes which having same solute concentration of 1.677 M were tested. The output power produced were consistent with the first experiment as plotted in Figure 6. It is observed that the volume of the water did not affect much on the output power. The purpose of third experiment is to analyse the output power produced by different type of electrolyte. Acetic acid+water is chosen to compare whether it can achieve more output power. Volume of the solution is 500 ml and concentration of NaCl solute is 1.677 M.

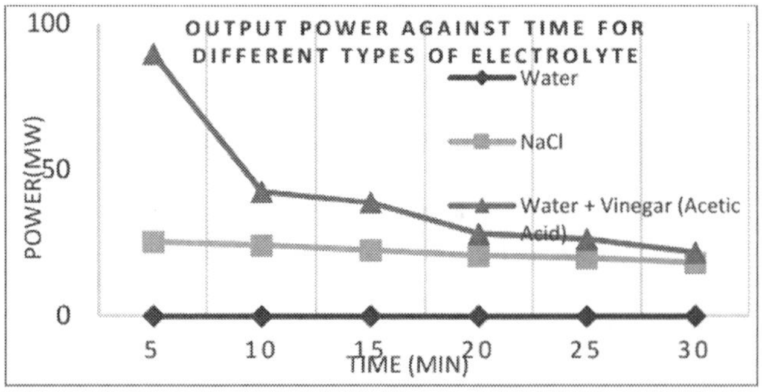

Fig. 7. Graph of output power against time for different types of electrolyte

Figure 7 represents the graph of output power against time for three different solutes, which were water, salt solution and water mixed with vinegar. The highest output power achievable was the mixture of water and vinegar. The salt solution provided the most stable output power among all three electrolytes. Water did not provide any output power due to the inability of water as an electrolyte.

3 Conclusion

Based on the comparison of techniques, mixing entropy battery was chosen for the experimental part of this research paper. Using dissimilar materials with different electrical potential for the anode and cathode electrodes such as aluminium and carbon, an experiment was conducted. Different variables such as volume of water, concentration of solute and type of electrolyte were tested in order to study their effects on the efficiency of the cell. It was found that volume of water does not affect the cell but the concentration of the solute is crucial for the cell. Besides that, the mixture of water and vinegar (acetic acid) was found to be the best electrolyte among all three different electrolytes used. The highest achievable output power was 89.7mW which is high enough to power low-power sensors. The output voltage and output current can be further increased or optimized. By using different configurations of placement for the electrodes, the surface area of the carbon can be increased which will increase the current obtained. Besides that, several cells can also be produced and connected in series to obtain higher voltage from the setup for higher voltage uses.

References

1. J. W. Post, J. Veerman, H. V. M. Hamelers, G. J. W. Euverink, S. J. Metz, K. Nymeijer, et al., Journal of Membrane Science, vol. 288, pp. 218-230, 2/1/ 2007.
2. G. L. Wick,W. R. Schmitt, Marine Technology Society Journal, vol. 11, pp. 16-21, 1977.
3. D. A. Vermaas, E. Guler, M. Saakes, and K. Nijmeijer, Energy Procedia, vol. 20, pp. 170-184, 2012.
4. Zhijun Jia, B. Wang, S. Song, and Y. Fan, Renewable and Sustainable Energy Reviews, vol. 31, pp. 91-100, 2014.
5. Qianhong She, Yew Kong Willis Wong, Shuaifei Zhao, Chuyang Y. Tang, Journal of Membrane Science 428 (2013) 181–189.
6. A. Achilli, T. Y. Cath, A. E. Childress, Journal of membrane science, vol. 343, pp. 42-52, 2009.
7. A. P. Straub, N. Y. Yip, and M. Elimelech, Environmental Science & Technology Letters, vol. 1, 2013.
8. Andrea Achilli, Amy E. Childress, Desalination 261 (2010) 205–211.

9. Ali Altaee, Adel Sharif, Desalination 356 (2015) 31–46.
10. Jihye Kim, Kwanho Jeong, Myoung Jun Park, Ho Kyong Shon and Joon Ha Kim, Energies 2015, 8, 11821-11845, 2015.
11. Anthony P. Straub, Akshay Deshmukh and Menachem Elimelech, Energy Environ. Sci., 2016, 9, 31--48.
12. M. Turek, B. Bandura, EuroMed 2006 conference on Desalination Strategies in South Mediterranean Countries, 21–25 May (2006).
13. J. Veerman,b, R.M. de Jong, M. Saakes, S.J. Metz , G.J. Harmsen, Journal of Membrane Science 343 (2009) 7–15, 2009.
14. Enver Güler, Rianne Elizen, David A. Vermaas, Michel Saakes, Kitty Nijmeijer, Journal of Membrane Science 446, 2013, 266–276
15. Han-Ki Kim,Mi-Soon Lee, Seo-Yoon Lee,a Young-Woo Choi, Nam-Jo Jeong and Chan-Soo Kim, J. Mater. Chem. A, 2015, 3, 16302–16306
16. Zhen Zhang, Xiang-Yu Kong, Kai Xiao, Qian Liu, Jie Ma, Ye Tian, Liping Wen, and Lei Jiang, Journal of the American Chemical Society 137 (46), 14765-14772 (2015)
17. J. Veerman, M. Saakes, S.J. Metz, G.J. Harmsen, Chemical Engineering Journal, Volume 166, Issue 1, 1 January 2011, Pages 256–268.
18. Tedesco, M., Cipollina, A., Tamburini, A., van Baak, W., Micale, G., Desalination Water Treat.49, 404–424, 2012.
19. Veerman, J., Saakes, M., Metz, S.J., Harmsen, G.J., J. Membr. Sci. 327, 136–144.
20. D. Brogioli, R. Ziano, R. A. Rica, D. Salerno, and F. Mantegazza, Journal of Colloid and Interface Science, vol. 407, pp. 457-466, 10/1/ (2013).
21. Marta C. Hatzell, M. Raju,Valerie J. Watson,Andrew G. Stack, Adri C. T. van Duin and Bruce E. Logan, Environ. Sci. Technol., 48, 14041–14048 (2014)
22. Raúl A. Ricaa, Roberto Zianoa, Domenico Salernoa, Francesco Mantegazzaa, Martin Z. Bazant, Doriano Brogioli, Electrochimica Acta 92 304–314 (2013).
23. Xian Kong,ab Alejandro Gallegos, Diannan Lu, Zheng Liu and Jianzhong Wu, Phys. Chem. Chem. Phys., (2015).
24. Guillermo R. Iglesias, Silvia Ahualli, María M. Fernandez, María L. Jimenez, Angel V. Delgado, Journal of Power Sources 318, (2016).
25. Raul A. Rica, Roberto Ziano, Domenico Salerno, Francesco Mantegazza, Rene van Roij and Doriano Brogioli, Entropy 2013.
26. B. B. Sales, F. Liu, O. Schaetzle, C. J. N. Buisman, and H. V. M. Hamelers, Electrochimica Acta, vol. 86, (2012).
27. B. B. Sales, M. Saakes, J. W. Post, C. J. N. Buisman, P. M. Biesheuvel, and H. V. M. Hamelers, Environmental Science & Technology, vol. 44, pp. 5661-5665, (2010).
28. M. F. M. Bijmans, O. S. Burheim, M. Bryjak, A. Delgado, P. Hack, F. Mantegazza, et al., Energy Procedia, vol. 20, pp. 108-115, (2012).
29. M. Marino, L. Misuri, M. L. Jiménez, S. Ahualli, O. Kozynchenko, S. Tennison, et al., Journal of Colloid and Interface Science, vol. 436, pp. 146-153, 12/15/ (2014).
30. F. La Mantia, M. Pasta, H. D. Deshazer, B. E. Logan, and Y. Cui, Nano Letters, vol. 11, pp. 1810-1813, 2011/04/13 (2011)
31. Marino, M.; Misuri, L.; Carati, A.; Brogioli, D., Energies 2014, 7, 3664-3683 (2014).
32. Kim T, Rahimi M, Logan BE, Gorski CA, ChemSusChem. May 10;9(9):981-8, (2016)

Remaining Useful Life Prediction of Gas Turbine Engine using Autoregressive Model

Shazaib Ahsan [1], and *Tamiru* Alemu Lemma [1,*]

[1]Universiti Teknologi PETRONAS, 32610 Seri Iskandar, Perak Darul Ridzuan, Malaysia.

Abstract. Gas turbine (GT) engines are known for their high availability and reliability and are extensively used for power generation, marine and aero-applications. Maintenance of such complex machines should be done proactively to reduce cost and sustain high availability of the GT. The aim of this paper is to explore the use of autoregressive (AR) models to predict remaining useful life (RUL) of a GT engine. The Turbofan Engine data from NASA benchmark data repository is used as case study. The parametric investigation is performed to check on any effect of changing model parameter on modelling accuracy. Results shows that a single sensory data cannot accurately predict RUL of GT and further research need to be carried out by incorporating multi-sensory data. Furthermore, the predictions made using AR model seems to give highly pessimistic values for RUL of GT.

1 Introduction

The GTs are widely used as prime movers to many applications such as aviation, marine, power generation, gas compression and pumping facilities. Due to higher power to weight ratio and quick starting characteristic, GT are preferred over steam turbines. Subject to the recent advancements in renewable energy, GT are now expected to operate more efficiently under part-load operating conditions. Either under part-load or peak load conditions, GTs are also required to operate flexibly with relatively higher number of firing start-ups and shut downs, which reduces the time between failures [1, 2]. Therefore, the maintenance strategy for such system should be robust and flexible in order to have improved reliability and better availability that would result in less downtime and operating cost [3]. Improper maintenance may lead to increased rate of deterioration. Due to that maintenance cost may also reach up to 35% of operating cost [4].

Maintenance strategies can be classified into three main categories namely breakdown maintenance, planned maintenance and condition based maintenance [2, 5]. In recent years a new concept of maintenance called "prognostics based maintenance" that would potentially minimize risk while maximizing useful life has been reported [1]. This approach is based on the condition monitoring data and involves prediction of likely fault cases instead of waiting until sudden failure occurs. As such, RUL is predicted after a fault is diagnosed. Accordingly, all the required planning for maintenance is completed well ahead of any sudden failure. This approach is reported to have saved the cost associated with unplanned maintenance and is considered superior to diagnostics alone [6].

* Corresponding author: tamiru.lemma@utp.edu.my

So far various techniques have been used for the prognostics of GTs. These include support vector machines (SVM) [1, 7-9], Bayesian forecasting [10-12], Kalman Filters [10], state-space models [12], artificial neural networks [13-16], independent component analysis [17], regression techniques [18], Dempster-Shafer regression [1] and one parameter double exponential smoothing [10]. These techniques can be classified into two types: (i) physics based modelling approaches, and (ii) data driven models [5, 19-21]. Physics based models are effective in dealing with novel faults but are considered difficult to implement to complex systems [1] since they require detail knowledge about the system. In the context of dynamic condition, it may require real-time parameter optimization, which is a limiting factor since GT is featured by many interacting parameters [21]. Under such circumstances, a model that relies on historical data for input and output sensors is preferred. The general steps for developing such kind of model entails the selection of a model structure followed by parameter estimation, simulation, and validation [22].

The purpose of this paper is to investigate the use of Auto-regressive model, which is one type of historical data based model, to GT remaining useful life prediction. A parametric study is carried out by varying model order and training algorithms.

2 Methodology

2.1 Time Series Models

In time series modeling, historical data representing the process is used for prediction. The general procedure is that, first part of the historical data is used to train a model and then the remaining data is used to validate the model. The proportion for the data, however, depends on the judgment of the model developer even though 70/30 or 60/40 proportions are widely reported in the open literature. Regardless, the parametric and time invariant version of time series models, which is the main target of the present paper, is given by

$$A(q^{-1})y(t) = \frac{B(q^{-1})}{F(q^{-1})}u(t) + \frac{C(q^{-1})}{D(q^{-1})}e(t) \quad (1)$$

Where, $y(t)$ is the predicted value; q^{-1} is the shift operator; $e(t)$ is the modeling and measurement error, $G(q^{-1}) = B(q^{-1})/F(q^{-1})$ and $H(q^{-1}) = C(q^{-1})/D(q^{-1})$ are transfer functions for the input and error data, respectively. Depending upon the forms of $A(q^{-1})$, $G(q^{-1})$ and $H(q^{-1})$, the resulting model can be Autoregressive Moving Average (AR), Autoregressive Integrated Moving Average (ARIMA), Autoregressive Moving Average with Exogenous Inputs (ARMAX) or Box-Jenkins (BJ) model. In the present paper, AR model, for which $B(q^{-1}) = 0$ and $C(q^{-1}) = D(q^{-1}) = F(q^{-1}) = 1$ is investigated.

Table 1: Summary of Performance Measures.

Performance Measure	Criteria	Formula
Fit Percent	Near 100, best	$100\left(1 - \frac{\lVert y_{measured} - y_{model}\rVert}{\lVert y_{measured} - \bar{y}_{measured}\rVert}\right)$
MSE	Near Zero, best	$\frac{(\lVert y_{model} - y_{measured}\rVert^2)}{N}$
FPE	Smaller Value, best	$det\left(\frac{1}{N}\, e^T.e\right).\left(\frac{1+\frac{n_p}{N}}{1-\frac{n_p}{N}}\right)$
AIC	Smaller Value, best	$N.log\left(det\left(\frac{1}{N}\, e^T.e\right)\right) + 2.n_p.N.(n_y.log(2\pi) + 1)$
BIC	Smaller Value, best	$N.log\left(det\left(\frac{1}{N}\, e^T.e\right)\right) + N.(n_y.log(2\pi) + 1) + n_p.log(N)$

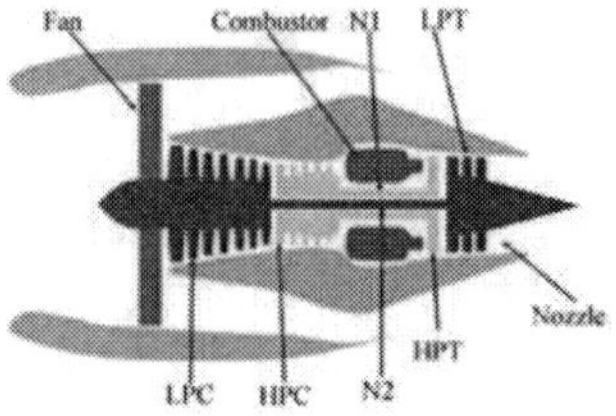

LPC: Low Pressure Compressor
HPC: High Pressure Compressor
N1: Shaft 1
N2: Shaft 2
LPT: Low Pressure Turbine
HPT: High Pressure Turbine

Fig. 1. Simplified Diagram of the Turbofan Engine GT [23].

2.2 Model performance Measures

The AR model parameters will be determined using optimization algorithms from MATLAB model identification toolbox. At each trail, quality of the model will be assessed using Fit Percent, MSE, FPE, AIC, and BIC. Fit Percent and MSE are measures of the actual quantity that is minimized during the parameter estimation whereas FPE, AIC and BIC are computed as properties of the output disturbance according to the relationship. For that purpose, we have summarized some of the model performance measures in Table 1.

3 Results and Discussions

3.1 NASA Benchmark Data

The data used for the case study is degradation data for Turbofan Engine which is made publicly available by NASA Prognostics Center of Excellence. Details on how the data was generated and the nature of the benchmark problem itself can be found from the publication by Saxena and Goebel [24]. An illustration of the Turbofan engine is provided as Fig. 1. The reader can refer to [25] for the list of sensors installed in the system. As highlighted in Table 2, the available data covers four operating regions.

Table 2: Benchmark data set details [24].

Data Set Name	Training Data set	Testing Data set	Operating Regions	Fault Modes
FD001	100	100	1	HPC Degradation
FD002	260	259	6	HPC Degradation
FD003	100	100	1	Fan Degradation & HPC Degradation
FD004	248	249	6	Fan Degradation & HPC Degradation

3.2 Effect of Optimization Algorithm

For modelling, we have tested five parameter optimization algorithms: burg's approach (burg), geometric mean approach (gl), least square (ls), Yule-Walker (YW) and forward approach (fb). For the best and optimum algorithm selection, a comparison was made using the performance measures provided in Table 1.

The data set for sensor number 2 was used for AR model performance investigation. For this particular sensor, there are 192 cycles before the turbofan is declared to be failed. From the 192 cycles, the first 134 cycle (70% of the data) are taken for model construction while the cycles from 135 to 192 (30% of the data) are used for model validation. Table 3 depicts the comparison of different optimization algorithms based on performance measures. From collective values of MSE, FPE and AIC, it can be concluded that 'burg' and 'gl' methods tend to provide best results.

Table 3: Comparison of different Performance Algorithm (PA) based on performance measures.

PA	Performance Measures				
	Fit Percent	**MSE**	**FPE**	**AIC**	**BIC**
burg	54.9	0.0	0.1	10.6	204.8
gl	54.9	0.0	0.1	10.6	204.8
ls	100.0	0.0	0.0	-6,632.1	-6,438.0
YW	-1,233.0	20.4	61.1	918.2	1,112.3
fb	60.9	0.0	0.1	-27.4	166.8

3.3 Effect of Changing Model Order

After the selection of optimum and best parameter optimization algorithms, we wish to find best and true model order number, again using the same set of performance measures. This has to be done to mitigate non-consistent estimation and increased variance [26]. The AR model was run for model order numbers in the range of 45 to 65 with the increment of 10. The two optimizers ('burg' and 'gl') demonstrated almost similar performance (ref. Table 4). The noticeable differences are only in terms of AICc and BIC. It can be concluded that as the model order increases, the performance of the model improves. The implication to that is that in the field, the designer need to choose between expensive model and good performance.

Table 4: Effect of changing model order number – 'burg' and 'gl'.

Model Order No.	Performance Measures							
	Fit Percent	**Loss Function**	**MSE**	**FPE**	**AIC**	**AICc**	**nAIC**	**BIC**
45	42.93	0.04	0.04	0.075	29.74	76.78	-2.62	160.14
55	48.28	0.03	0.03	0.073	23.35	102.32	-2.66	182.73
65	53.89	0.02	0.02	0.070	12.61	138.79	-2.74	200.97

3.3 Predicted vs Actual RUL comparison

For RUL investigation, AR models with 80 model parameters are produced for Sensor 2 and Sensor 7, respectively. The decision to choose the two sensors was made based on the condition that the data from the two sensors must follow different trends. Since the full information is known, the RUL from AR predictions are compared with the actual RUL. The predicted and actual RUL along with the corresponding errors are given in Table 5.

Table 5: RUL values and Error determined from AR(80 parameters) for sensors.

Sensor	RUL Values				Error in RUL Values		
	Pessimistic	**Actual**	**Optimistic**	**Observed**	**Pessimistic**	**Actual**	**Optimistic**
# 2	154	192	229	237	83	45	8
# 7	164	192	217	179	15	-13	-38

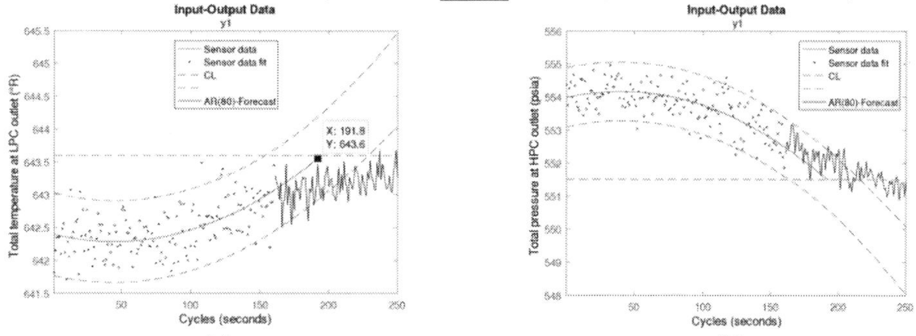

(a) (b)

Fig. 2. RUL Prediction using AR(80) of (a) increasing trend sensor# 2, (b) decreasing trend sensor# 7.

Plots of the data for sensor #2 and sensor # 7 are shown in Fig. 2. The corresponding RUL prediction by quadratic regression (ref. Table 5), indicates that the simple regression model provides optimistic and pessimistic RUL, with the actual RUL being always in between. AR model, on the other hand, tends to provide longer pessimistic RUL. Practically, the optimistic RUL urges maintenance preparation to be made as quickly as possible in anticipation of early failure. The RUL from AR model, on the other hand, shows longer RUL which might lead the maintenance team to believe that there is still ample time to do the preparation. Overall, the results imply that AR has a good potential for RUL prediction. One way to improve on the performance might be to use multiple input single output model structure.

4 Conclusion

The purpose of this paper was to investigate the use of AR models to RUL prediction. Accordingly, the Turbofan benchmark problem was studied with the hope of uncovering important observations. Five different optimization algorithms have been tested to assess AR model performance. At the end, it was possible to draw the following conclusions:

- Performance of AR model is weakly dependent on the type of parameter optimization algorithm.
- Increasing the model order increase the model accuracy. However, high model order is less preferred due to the need to have more data and high computational cost.
- AR model can be used for RUL prediction. However, the accuracy depends on the type of data. Too much noise in the data means, AR model highly likely to give longer optimistic RUL.

Future work will concentrate on expanding the AR model by considering data pre-processing and different model structure, perhaps Wiener model structure.

The Authors are thankful to Universiti Teknologi PETRONAS for the facilities provided. The study is partly funded by Ministry of Higher Education (MOHE), Malaysia under grant number FRGS/2/2014/TK01/UTP/03/2. Also, Authors are grateful to Saxena and Goebel for providing Turbofan engine simulation data at NASA Ames Research Center.

References

1. J. Xu, Y. Wang and L. Xu, *PHM-oriented integrated fusion prognostics for aircraft engines based on sensor data.* IEEE Sensors Journal, 2014. **14**(4): p. 1124-1132.
2. A. Heng, S. Zhang, A. C. Tan and J. Mathew, *Rotating machinery prognostics: State of the art, challenges and opportunities.* Mechanical systems and signal processing, 2009. **23**(3): p. 724-739.
3. M. Boyce, Gas turbine engineering handbook. 3rd, Third ed. (Gulf Professional Publishing, 2006).
4. T. A. Lemma, F. M. Hashim and M. A. B. A. Majid, *An integrated approach to industrial gas turbine diagnostics and reliability montoring.* 2006.
5. H. Hanachi, J. Liu, A. Banerjee, Y. Chen and A. Koul, *A physics-based modeling approach for performance monitoring in gas turbine engines.* IEEE Transactions on Reliability, 2015. **64**(1): p. 197-205.
6. A. K. Jardine, D. Lin and D. Banjevic, *A review on machinery diagnostics and prognostics implementing condition-based maintenance.* Mechanical systems and signal processing, 2006. **20**(7): p. 1483-1510.
7. W. Hu, J. Liu, J. Cui, Y. Gao, B. Cui and L. Jiang, *Fault diagnosis of gas turbine based on support vector machine.* in *Control and Decision Conference (2014 CCDC), The 26th Chinese.* 2014. IEEE.

8. P. G. Nieto, E. García-Gonzalo, F. S. Lasheras and F. J. de Cos Juez, *Hybrid PSO–SVM-based method for forecasting of the remaining useful life for aircraft engines and evaluation of its reliability.* Reliability Engineering & System Safety, 2015. **138**: p. 219-231.
9. D. Zhou, H. Zhang, and S. Weng, *A New Gas Path Fault Diagnostic Method of Gas Turbine Based on Support Vector Machine.* Journal of Engineering for Gas Turbines and Power, 2015. **137**(10): p. 102605-102605-6.
10. A. Cavarzere and M. Venturini, *Application of Forecasting Methodologies to Predict Gas Turbine Behavior Over Time.* Journal of Engineering for Gas Turbines and Power, 2012. **134**(1): p. 012401.
11. M. A. Zaidan, R. F. Harrison, A. R. Mills and P. J. Fleming, *Bayesian hierarchical models for aerospace gas turbine engine prognostics.* Expert Systems with Applications, 2015. **42**(1): p. 539-553.
12. J. Sun, H. Zuo, W. Wang and M. G. Pecht, *Application of a state space modeling technique to system prognostics based on a health index for condition-based maintenance.* Mechanical Systems and Signal Processing, 2012. **28**: p. 585-596.
13. M. Tahan, M. Muhammad, and Z.A. Karim, *A multi-nets ANN model for real-time performance-based automatic fault diagnosis of industrial gas turbine engines.* Journal of the Brazilian Society of Mechanical Sciences and Engineering, 2017: p. 1-12.
14. M.B. Rahmoune, A. Hafaifa, and M. Guemana, *Fault diagnosis in gas turbine based on neural networks: Vibrations Speed Application*, in *Advances in Acoustics and Vibration*. 2017, Springer. p. 1-11.
15. M.A. Ghafir, , Y. Li, and L. Wang, *Creep life prediction for aero gas turbine hot section component using artificial neural networks.* Journal of Engineering for Gas Turbines and Power, 2014. **136**(3): p. 031504.
16. M.F. Abdul Ghafir, Y.G. Li, and L. Wang, *Creep Life Prediction for Aero Gas Turbine Hot Section Component Using Artificial Neural Networks.* Journal of Engineering for Gas Turbines and Power, 2013. **136**(3): p. 031504-031504-9.
17. J. Ma, Q. Wang, G. Li and D. Zhou, *Reconstruction-based fault prognosis for flue gas turbines with independent component analysis.* Asia-Pacific Journal of Chemical Engineering, 2014. **9**(2): p. 205-213.
18. Y. Li, Y., *Gas turbine performance and health status estimation using adaptive gas path analysis.* Journal of Engineering for Gas Turbines and Power, 2010. **132**(4): p. 041701.
19. George, V., et al., *Intelligent fault diagnosis and prognosis for engineering systems*. 2006, John Wiley & Sons, Inc.
20. A.D. Fentaye, S.I.U.-H. Gilani, and A.T. Baheta. *Gas turbine gas path diagnostics: A review*. in *MATEC Web of Conferences*. 2016. EDP Sciences.
21. J. Yan, *Machinery prognostics and prognosis oriented maintenance management*. 2014: John Wiley & Sons.
22. S. Araghinejad, *Data-driven modeling: using MATLAB® in water resources and environmental engineering*. Vol. 67. 2013: Springer Science & Business Media.
23. D.K. Frederick, J.A. DeCastro, and J.S. Litt, *User's guide for the commercial modular aero-propulsion system simulation (C-MAPSS)*. 2007.
24. A. Saxena, K. Goebel, *Turbofan Engine Degradation Simulation Data Set.* NASA Ames Prognostics Data Repository, 2008.
25. A. Saxena, K. Goebel, *Damage propagation modeling for aircraft engine run-to-failure simulation*. in *Prognostics and Health Management, 2008. PHM 2008. International Conference on*. 2008. IEEE.
26. R. Shibata., *Selection of the Order of an Autoregressive Model by Akaike's Information Criterion.* Biometrika, 1976. **63**(1): p. 117-126.

MATEC Web of Conferences **131**, 04015 (2017) DOI: 10.1051/matecconf/201713104015
UTP-UMP SES 2017

Development of eco-friendly washing machine

Hamzah Sakidin[1*], *Pretesh* Sudhakaran[2], *Muhammad Syaamil* Saad[3], *Sarah Syamimi* Rozali[4], *Abdul Hadi* Nasarudin[5] and *Ahmed Abdullah* AlMashhor[6]

[1] Fundamental & Applied Sciences Department, Universiti Teknologi PETRONAS, 32610 Bandar Seri Iskandar, Perak

[2] Chemical Engineering Department, Universiti Teknologi PETRONAS, 32610 Bandar Seri Iskandar, Perak

[3] Chemical Engineering Department, Universiti Teknologi PETRONAS, 32610 Bandar Seri Iskandar, Perak

[4] Civil Engineering Department, Universiti Teknologi PETRONAS, 32610 Bandar Seri Iskandar, Perak

[5] Mechanical Engineering Department, Universiti Teknologi PETRONAS, 32610 Bandar Seri Iskandar, Perak

[6] Petroleum Engineering Department, Universiti Teknologi PETRONAS, 32610 Bandar Seri Iskandar, Perak

Abstract. A portable, non-electric washing machine powered by human leg identical to cycling is produced in order to fulfil the need of varsity students to wash their clothes in a cheap, eco-friendly and time saving and also to do physical body exercises. A black plastic bin has been used as the main body. At the bottom lies the rotating disk to provide agitation to the water and also cleaning the clothes. Shaft and gears are also been used in the rotating and moving mechanism that is connected to the pedal. This eco-friendly washing machine with an affordable price will benefit the lower income family and also for the college students.

1. Introduction

1.1 Background

Conventional washing machine generally is not eco-friendly. On average, 1kg of clothes would take up to 10 liters of water in a conventional washing machine. An official measurement of energy efficiency for washing machines called the "energy factor" and measures how many cubic feet of washing can be accomplished per kilowatt-hour of electricity consumed [1]. The usage a lot of water for a few clothes can be of huge wastage. Therefore, this project is aimed to track these necessities and fulfill the requirement with the most suitable product which is a portable non-electric powered washing machine.

*Corresponding author: hamzah.sakidin@utp.edu.my

1.2 Objectives

The objective of this project is given below:

1. To fulfil the need of varsity students to wash their clothes in a cheap, eco-friendly and time saving manner.
2. To do more physical body exercise while doing the laundry.
3. To give choice to lower income group for doing the laundry

2 Literature Review

Doing a laundry is an essential factor in human life. Since the first electric washing machine was invented by Alva J. Fisher in 1908, it had been benefit for people's lives [2]. However, the way they do it differs from one person to another according to their necessities.

Furthermore, according to a survey done all around Malaysia stated in Cuti-cuti Malaysia on 2014 which showed concerns raised against the use of public laundry services among young adults as many of them wished if they have a portable washing machines because they usually just wash their customer's clothes with less of detergent and use detergent of a low quality. They also put clothes over limit of the washing machine capacity, thus it causes the washing machines not to wash properly and cleanly. Another reason is related to workers mistakes in separating customer clothes in which result in missing some of these clothes [3].

In addition, designing a lightweight washing machine would settle the problem of homeowners with no extra room or small apartments such as studio apartments, and for people who travel frequently [4]. It is also suitable for operation in case of lack of water connection as well as for weekend riders and anyone who wants to wash quickly [6].

3 Methodology

Appropriate tools and selection of materials are important so that the prototype can be built smoothly and will not cause any trouble. An existing product is required to be modified to the extent of being able to insert a few elements for the mechanism to operate properly.

The planning and discussion of this project need to be done in a proper way to ensure the final prototype can be produced successfully.

The flow chart in Figure 1 shows the stages of the project from the initial stage to the final stage.

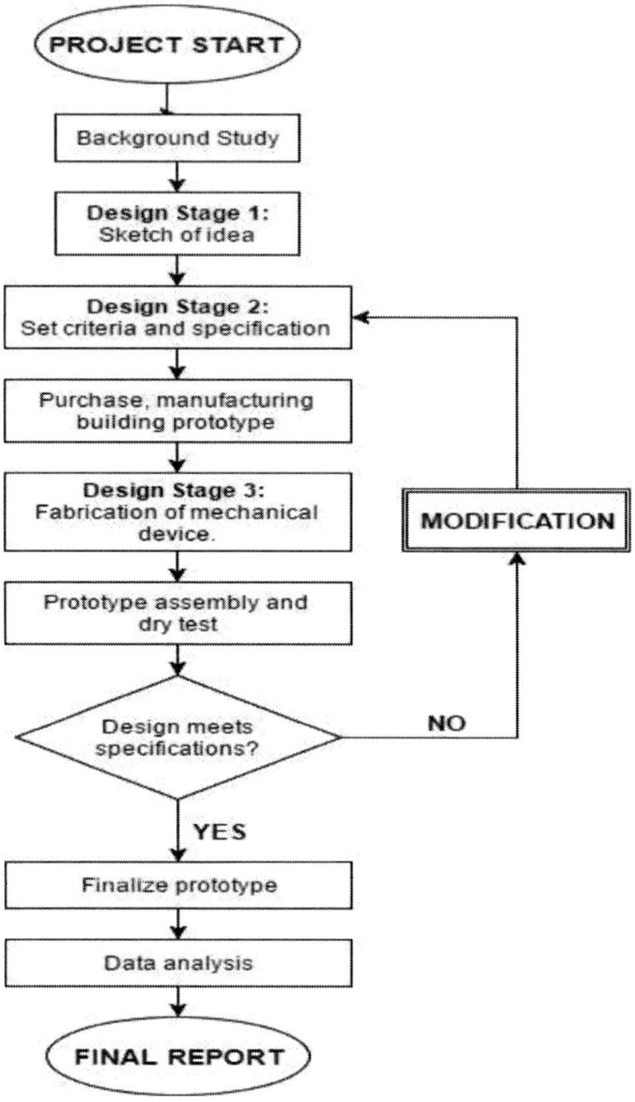

Fig. 1: Flow Chart of the project

3.1 Concept of design

This project is feasible and efficient design for our portable, non-electric washing machine. The main body of the washing machine consists of a cylindrical barrel or a bin. This barrel would be made of durable and sturdy plastic and polymer, like that of household pails.

This material is chosen for its overall economic value. A rotary base is attached to the bottom of the barrel; coaxially. This base is equipped with ridges or fins that functions as an agitator to promote the whirlpool motion of water in the device. The height of these ridges will be designed to be tall enough to create sufficient agitation and to avoid tangling up the clothes in the washing machine.

Chain and gears will provide the rotation motion for the barrel. This mechanism will be powered by a paddle located at the side of the barrel. The paddle will spin a horizontal gear that is perpendicularly aligned with a horizontal gear. This horizontal gear acts as the connection between the main peddle and the rotary base. As this gear spins with the kinetic energy supplied by peddling, the rotary base will spin along with it, thus agitating the contents in the barrel. To ensure the spaces in between the gears, rotary base, and the bottom of the barrel are waterproof, they will be sealed off with either silicon seal or a rubber seal.

The entire peddling mechanism will be isolated at the bottom of the barrel in a hollow plastic enclosure, protecting it from external forces or disturbances. This enclosure will also act as a foundation for our washing machine. The designed is capped off with a durable plastic cover, equipped with fitting sockets to ensure the barrel is sealed tight when the washing commences. The water in the barrel can be disposed off simply by tilting the device over once the washing is done.

3.2 Description of prototype material

Below is the description of materials to be used for prototype in the design:

Table 1: Materials description

Material	Description
Plastic bin with cover	Main prototype material. The clothes are loaded in the plastic bin, where the clothes are washed. The external tub covers the inner tub and supports it during various cycles of clothes washing.
Rotary disk with ridges	The rotary disk is located inside the tub of the washing machine. It is the important part of the washing machine that actually performs the cleaning operation of the clothes. During the washing cycle the rotary disk rotates continuously and produces strong rotating currents within the water due to which the clothes also rotate inside the tub. The rotation of the clothes within water containing the detergent enables the removal of the dirt particles from the fabric of the clothes. Thus the rotary disk produces most important function of rubbing the clothes with each other as well as with water.
Pedal	User pushes the pedal using foot powered. The pedal are connected to gear that will rotates the rotary disk and form kinetic energy.
Gears	Transfer kinetic energy from pedal foot powered to mechanical energy of the mechanism
Hollow Plastic Foundation	Separates the mechanical part of the washing including pedal, gears, shaft, bolts and nuts from washing space. It can prevent water from seeping into the mechanical part of the washing machine that can cause rust.
Shaft	Transfer power from gears for rotation of the rotary disk

The listed item in Table 1 are the things that used to build the prototype. A black plastic bin has been used as the main body. At the bottom lies the rotating disk to provide agitation.

DOI: 10.1051/matecconf/201713104015

to the water and also cleaning the clothes. Shaft and gears are also been used in the rotating and moving mechanism that is connected to the pedal.

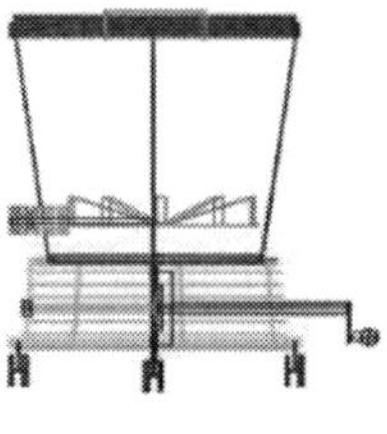

Fig 2a : Front view

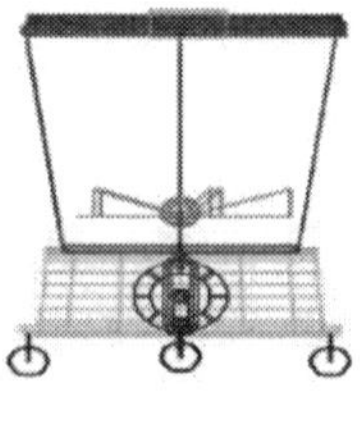

Fig 2b: Side view

Fig 3: Prototype

Figure 2a and Figure 2b show the schematic diagram of the eco-friendly washing machine. The prototype of the eco-friendly washing machine is given in Figure 3.

4 Analysis

4.1 Survey data

An online survey was conducted to 36 respondents from varsity students to identify the response for eco-friendly washing machine. Based on the survey, most of varsity students think that washing machine and laundryman service are expensive and inefficient. Washing machine and dryer services that are offered at their places need approximately RM 6 each time to operate and most of the students are doing laundry once a week. Mathematically, if we calculated it, one student would spend RM24 a month and RM 288 a year. Whereas, some of the self-service washing machine offered RM 10 cost each time to operate and that would have cost RM 480 a year.

Some students would prefer to wash their laundry manually by hand which would consumes energy and time. Therefore, there is a need to create a portable and a cheap alternative of doing laundry. Besides that, it would also save water and energy. It is foot pedal powered without using any electricity and would not be using too much water as the conservative washing machine nowadays. As for a small capacity of washing machine, this design prototype washing machine can be carried around with ease and very compact.

From the trial run, the consumption of water can be reduced from the usual 40 gallons to 9 gallons with this washing machine. These 9 gallons constitutes one wash for about 8 to 9 pieces of clothes. It shows that this prototype takes one gallon to wash one piece of clothing whereas a normal washing machine that can usually for around 20 pieces of clothing would use 2 gallons per clothing item. This is an improvement from the usual conventional electric washing machine.

Moreover, this prototype uses absolutely no form of electrical power as the hand powered washing machine which provides good exercise for the users. The mechanical gear system along with the hand paddle make up the operational system of our prototype, using the principles of kinetic energy. It registers 3 rotations of the rotary device per rotation of the handle. This gives us 225 rotations per minute of the device without water and 105 rpm with water. Although it is much lesser than that of a normal washing machine, it is standard for a hand powered one and is enough to complete the washing of clothes.

4.2 Costing

The economical of the project is important to ensure the items and parts are under the budget allocated. In this project, the economic analysis covers the overall cost of the prototype and also the real product. With variety of products in the market, the price range for every component in the project is quite high, therefore it is crucial to choose the affordable one, yet can maintain the quality of the prototype build by our team. The cost of the prototype can be shown in Table 2.

Table 2 : Costing for the prototype

No	Item	Quantity	Unit Price (RM)	Total Price (RM)
1	Barrel	1	10	10
2	Gear and had drill set	1	17	17
3	Metal bar including welding	-	50	50
4	Rotating base	1	50	50
5	Basin	1	6	6
6	Bicycle tyre rim	1	14	14
7	wheels	4	4.50 x 4	18
Total				**165**

Based on the costing above, may be the cost is quite expansive, however for the future purposes, the cost can be reduced after some modification on the prototype.

5 Conclusion

This project is a manual pedal-powered machine, thus it reduces electricity consumption in which decreases carbon footprint and that contributes to protecting our lovely globe from global warming.

More interestingly, this handy portable machine is designed to save money because all what it takes is just a good foot exercise in order to cut down on detergent usage and your electricity and water bills. Besides the materials selected to build the machine are cheap mainly plastic and steel, and the size is small as well as no electric parts or programming involved in the design process, increasing the chance of innovating a low-cost and competitive product.

Additionally, a normal electric washing takes 30-60 minutes per wash compared to the manual washing machine which only take 7 minutes to perform the duty.

The advantages of this washing machine are:

1. Portable
2. Can reduce carbon footage
3. Can reduce water usage
4. Good for hand muscle workout
5. Cheap

Overall, this washing machine do not require electrical components, so that it can last longer than traditional washing machines and never require maintenance or parts replacement.

The authors would like to acknowledge Universiti Teknologi PETRONAS (UTP) for financial assistance.

References

1 Foreman.G. (2011). "Your Washing Machine and the Environment". The Dollar Stretcher.com. [Online] Available: http://www.stretcher.com/stories/00/000911e.cfm

2. Barracuda (2015). "Who invented the Washing Machine?". Service Force. [Online] Avaiable: http://www.serviceforce.co.uk/blog/kitchen-appliances/who-invented-the-washing-machine/

3. Mamat, N. (2015). New Portable Washing Machine. [Online] Retrieved on May 27, 2015 [Online] Available : https://www.essays24.com/History-Other/New-Portable-Washing-Machine/62265.html

4 Z, Kara. (2014). Living Direct: The Benefits of Portable Washers. [Online] Retrieved on May 9, 2014. [Online] Avaiable : https://learn.livingdirect.com/portable-washer-benefits/

5. N, Radojicic. (2010). The Advantages of Portable Washing Machines. [Online] Available: http://ezinearticles.com/?The-Advantages-of-Portable-Washing-Machines&id=4551964

6. S. John. (2017). Portable Washing Machines: What are the Advantages and Disadvantages? [Online] Avaiable : http://www.blogher.com/portable-washing-machines-what-are-advantages-and disadvantages

Rational Cubic Spline Interpolation for Monotonic Interpolating Curve with C^2 Continuity

Samsul Ariffin Bin Abdul Karim[1,*]

[1]Fundamental and Applied Sciences Department, Universiti Teknologi PETRONAS, Bandar Seri Iskandar, 32610 Seri Iskandar, Perak DR, Malaysia

Abstract. Monotonicity preserving interpolation is very important in many sciences and engineering based problems. This paper discuss the monotonicity preserving interpolation for monotone data set by using C^2 rational cubic spline interpolant (cubic/quadratic) with three parameters. The data dependent sufficient conditions for the monotonicity are derived with two degree freedom. Numerical results suggests that the proposed C^2 rational cubic spline preserves the monotonicity of the data and outperform the performance of the other rational cubic spline schemes in term of visually pleasing.

1 Introduction

Visualizing the scientific data is important in computer graphics and geometric modeling areas. One of the requirements for scientific visualization is that the ability of the interpolating scheme to preserves the geometric properties of the data sets. For instance if the given data sets are monotone, then the resulting interpolating curves also must be monotonic on the entire given interval. Some examples of monotonic data includes the population at one country as well as the approximation of couples and quasi couples in statistics (Karim and Kong [13]). The usual C^2 cubic spline interpolation is very smooth but for shape preserving interpolation purpose, the cubic spline are unable to preserves the shape of the data on entire given interval resulting the interpolating curve may be oscillates on some interval. Fritsch and Carlson [8] constructed the monotonic cubic Hermite spline polynomial to preserve monotone data set. Their scheme have some drawback such as it require the modification of the first derivatives if shape violation is found. Fritsch and Carlson [8] scheme is not recommended for monotonicity preserving interpolation. Thus one of the alternative strategy was the used of rational cubic spline for monotonicity preservation.

There exist many research papers on the construction of rational cubic spline for monotonicity preserving interpolation. Some literature reviews are as follows: Delbourgo and Gregory [6,7] have discussed the monotonicity and convexity preservation by using rational cubic spline (quadratic denominator) with one parameter. Sarfraz [18], Sarfraz et al. [19] and Abbas et al. [1] also studied the use of rational cubic interpolant for preserving the monotone data. Meanwhile Hussain and Hussain [11] and Hussain et al. [12] studied

* Corresponding author: samsul_ariffin@utp.edu.my

 DOI: 10.1051/matecconf/201713104016

monotonicity preserving for curves and surfaces by utilizing rational cubic spline with quadratic denominator- without any degree freedom i.e. no free parameters for shape modification. In Sarfraz et al. [19] the rational cubic spline with quadratic denominator has been used for monotonicity with C^2 continuity – without any degree freedom. Sarfraz [18] also discussed the used of rational cubic spline (cubic/cubic) for monotonicity preserving with C^2 continuity but his schemes also not received much attention since there are no degree freedom. Similarly the rational cubic spline of Gregory [9] also did not have any degree freedom i.e. free parameters for shape modification. Abbas et al. [1] discussed the monotonicity by using C^2 rational cubic spline with two free parameters. This paper is a continuation from our previous works (Karim and Kong [14]). The rational cubic spline with C^2 continuity will used for monotonicity-preserving interpolation for functional data. The main contribution this paper is summarized as follows:

(i) Our scheme give comparable results compare with existing schemes. Indeed for some data set, our scheme is better.
(ii) Our scheme not involving any knots insertion as appear in the works of Lahtinen [15], Lam [16], Schumaker [21], Fristch and Carlson [8], Schmidt and Hess [20] and Butt and Brodlie [5].
(iii) Our scheme has two degree freedom meanwhile there are no degree freedom in the works of Sarfraz [18], Delbourgo and Gregory [6,7] and Gregory [9].

2 C^2 rational cubic spline interpolant

This section gives the definition for C^2 rational cubic spline interpolant with three parameters. The mathematical derivations are taken from Karim and Kong [14]. For simplicity, we assume that the data points $\{(x_i, f_i), i = 0,1,...,n\}$ are given such that $x_0 < x_1 < ... < x_n$. Let $h_i = x_{i+1} - x_i$, $\Delta_i = (f_{i+1} - f_i)/h_i$ and $\theta = (x - x_i)/h_i$ such that $0 \le \theta \le 1$. For $x \in [x_i, x_{i+1}], i = 0,1,2,...,n-1$ the rational cubic spline interpolant with three parameters is defined as follows:

$$s(x) \equiv s_i(x) = \frac{A_{i0}(1-\theta)^3 + A_{i1}\theta(1-\theta)^2 + A_{i2}\theta^2(1-\theta) + A_{i3}\theta^3}{(1-\theta)^2\alpha_i + \theta(1-\theta)(2\alpha_i\beta_i + \gamma_i) + \theta^2\beta_i} \tag{1}$$

with $A_{ij}, j = 0,1,2,3$ as:

$$\begin{cases} A_{i0} = \alpha_i f_i, \\ A_{i1} = (2\alpha_i\beta_i + \alpha_i + \gamma_i) f_i + \alpha_i h_i d_i, \\ A_{i2} = (2\alpha_i\beta_i + \beta_i + \gamma_i) f_{i+1} - \beta_i h_i d_{i+1}, \\ A_{i3} = \beta_i f_{i+1}. \end{cases} \tag{2}$$

C^2 continuity i.e. $s^{(2)}(x_{i+}) = s^{(2)}(x_{i-})$, $i = 1,2,...,n-1$, lead us to the following system of linear equations:

$$a_i d_{i-1} + b_i d_i + c_i d_{i+1} = e_i, \quad i = 1,2,...,n-1. \tag{3}$$

with

 DOI: 10.1051/matecconf/201713104016

$$a_i = h_i\alpha_{i-1}\alpha_i$$
$$b_i = h_i\alpha_i\left(\gamma_{i-1} + 2\alpha_{i-1}\beta_{i-1}\right) + h_{i-1}\beta_{i-1}\left(\gamma_i + 2\alpha_i\beta_i\right)$$
$$c_i = h_{i-1}\beta_{i-1}\beta_i$$
$$e_i = h_i\alpha_i\left(\gamma_{i-1} + \alpha_{i-1} + 2\alpha_{i-1}\beta_{i-1}\right)\Delta_{i-1} + h_{i-1}\beta_{i-1}\left(\gamma_i + \beta_i + 2\alpha_i\beta_i\right)\Delta_i$$

The system of equation in (3) gives $n-1$ linear equations for $n+1$ unknown derivative values. By imposing (4) and (5), then the system of linear equations in (4) is strictly tri-diagonal and has a unique solution for the unknown derivative parameters d_i, $i = 1,2,\ldots,n-1$ for all $\alpha_i, \beta_i > 0, \gamma_i \geq 0$. This solution can be found by using Thomas algorithm.

$$s^{(1)}\left(x_0\right) = d_0 \qquad (4)$$

$$s^{(1)}\left(x_n\right) = d_n \qquad (5)$$

3 Sufficient condition for monotonicity

The monotonicity preserving by using C^2 rational cubic spline will be further elaborate in this section. Given that the monotone data sets $\left(x_i, f_i\right), i = 0,1,\ldots,n$ such that $x_0 < x_1 < \ldots < x_n$. For a monotonic increasing (decreasing), the necessary condition for monotonicity is:

$$f_0 \leq f_1 \leq \ldots \leq f_n \text{ (or } f_0 \geq f_1 \geq \ldots \geq f_n \text{ for monotonic decreasing)} \qquad (6)$$

Condition (6) is equivalent to

$$\Delta_i \geq 0 \quad (\text{or } \Delta_i \leq 0 \text{ for monotonic decreasing data}) \qquad (7)$$

The necessary and sufficient condition for the C^2 monotonicity of rational cubic spline cubic interpolant will be derived by finding the conditions on parameter γ_i and it is data dependent. For monotonic increasing rational cubic interpolant $s(x)$ in (1), the first derivative must satisfy the following inequalities:

$$d_i \geq 0, i = 0,1,2,\ldots,n \qquad (8)$$

The rational cubic spline, $s(x)$ is monotonic increasing if and only if

$$s^{(1)}(x) \geq 0, \; x_0 \leq x \leq x_n \qquad (10)$$

The first derivative of the rational cubic spline interpolant $s(x)$ is given in Karim and Kong [13].

Theorem 1. Given a strictly monotonic increasing set of data satisfying (7), there exists a class of monotonic rational (of the form cubic/quadratic) interpolating spline $s(x) \in C^2\left[x_0, x_n\right]$ involving free parameters α_i and β_i provided that if it satisfy the following sufficient conditions:

$$\alpha_i > 0, \beta_i > 0,$$

$$\gamma_i = \delta_i + \text{Max}\left\{0, \beta_i\left(\frac{d_{i+1}}{\Delta_i} - 2\alpha_i\right), \alpha_i\left(\frac{d_i}{\Delta_i} - 2\beta_i\right)\right\}, \delta_i > 0, \Delta_i > 0. \qquad (11)$$

DOI: 10.1051/matecconf/201713104016

Remark 1: If the data are constant on certain interval $\Delta_i = 0$, then setting $d_i = d_{i+1} = 0$, will produce $s(x) = f_i = f_{i+1}$ is a constant on the interval $\left[x_i, x_{i+1}\right], i = 0,1,2,...,n-1$.

4 Numerical demonstrations

In order to illustrate the monotonicity preserving interpolation by using the proposed C^2 rational cubic spline interpolation (cubic/quadratic), choose Akima [4] data set i.e. wave distortion data in experiment conducted in laboratory.

Table 1. A monotone data from Akima [4]

x_i	0	2	3	5	6	8	9	11	12	14	15
f_i	10	10	10	10	10	10	10.5	15	50	60	85
d_i	0	0	0	0	0	0	1.083	24.083	25	18.333	31.667

Table 2. Numerical results

i	0	1	2	3	4	5	6	7	8	9	10
$d_i(C^1)$	0	0	0	0	0	0	1.083	24.083	25	18.333	31.667
Δ_i	0	0	0	0	0	0.5	2.25	35	5	25	
α_i	-	-	-	-	-	2	2	2	2	2	
β_i	-	-	-	-	-	2	2	2	2	2	
γ_i	-	-	-	-	-	0.25	13.66	0.25	0.25	0.25	
$d_i(C^2)$	0	0	0	0	0	0	0.747	17.498	25.388	14.815	31.667

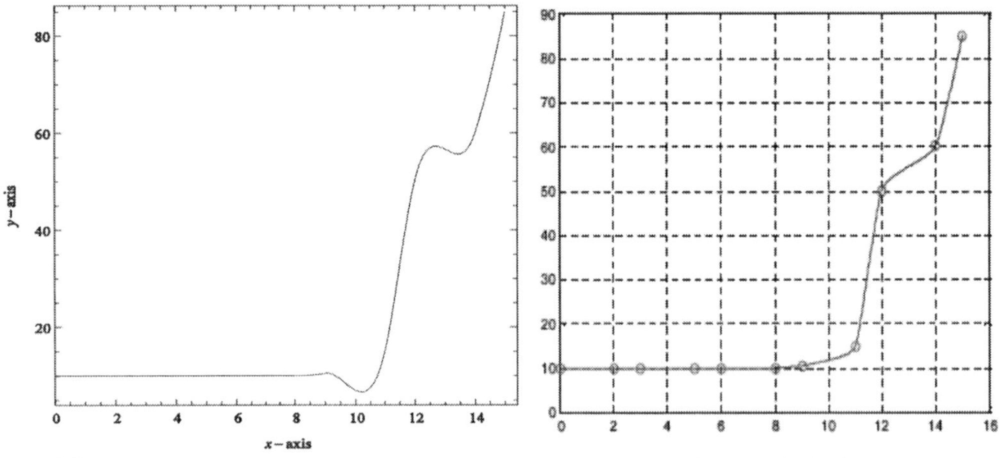

(a) Cubic Hermite spline

(b) From Sarfraz et al. [19]

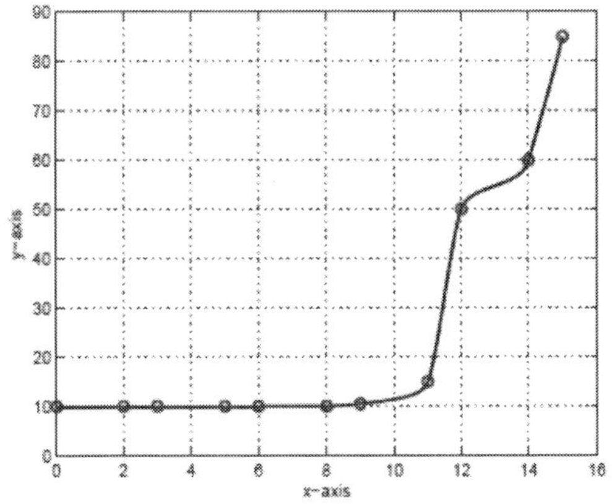

(c) From Abbas et al. [3]

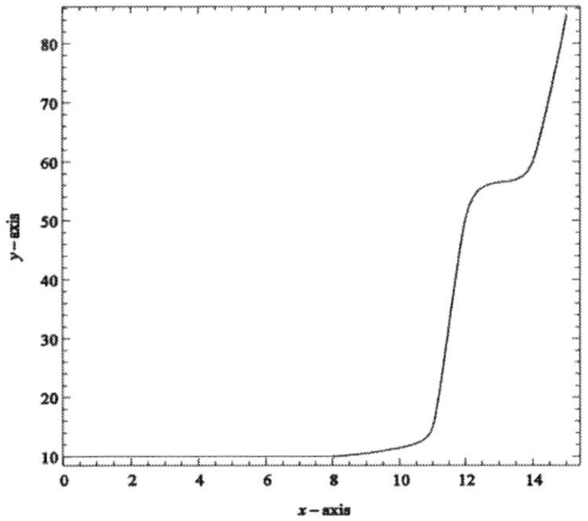

(d) The proposed scheme with $\alpha_i = \beta_i = 2$

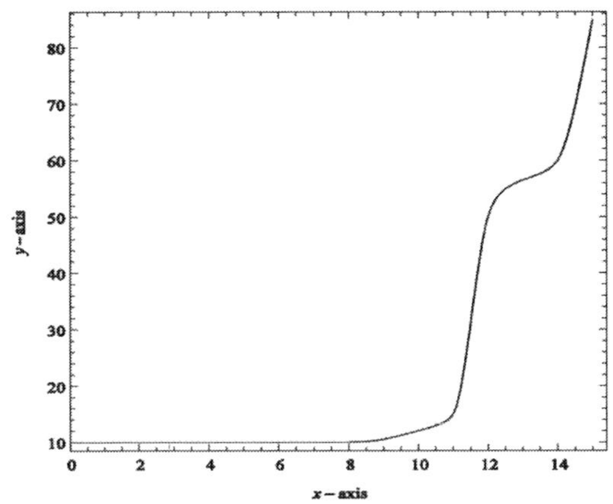

(e) The proposed scheme with $\alpha_i = \beta_i = 0.5$

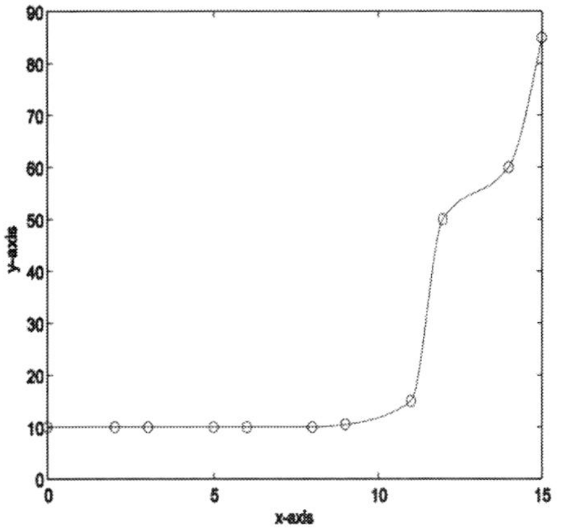

(f) PCHIP

Fig. 1. Comparison of interpolating curves

Fig. 1 shows the monotonic interpolating curves for monotone data sets given in Table 1. Figs. 1(a)) shows the default cubic Hermite spline polynomial. Figs. 1(b) and 1(c) show the monotonic interpolating by using the rational spline schemes by Sarfraz et al. [19] and Abbas et al. [1] respectively. Fig. 1(d) is obtained by using the data in Table 2. From Figs. 1(d) and 1(e) the proposed C^2 rational cubic spline give smooth results compare to the works of Sarfraz et al. [19] and Abbas [3]. Meanwhile from Fig. 1(f), shows that PCHIP also can produce the monotonic interpolating curves but not visually pleasing and some part of interpolating curves very tight and the smoothness is only C^1.

Discussions and Conclusion

From all numerical results, it can be seen clearly that the proposed C^2 rational cubic spline works very well and it is comparable with existing schemes such as Abbas [4], Abbas et al. [1] and Sarfraz et al. [19]. Future work will be focusing on other shape preserving criteria.

Acknowledgement to Universiti Teknologi PETRONAS (UTP) for the financial support received in the form of research grants: **Short Term Internal Research Funding (STIRF) No. 0153AA-D91 and YUTP: 0153AA-H24** including the Mathematica and MATLAB Softwares.

References

1. M., Abbas, A.A. Majid and J.M. Ali. Applied Mathematics and Computation, 219: pp. 2885-2895 (2012).
2. U., Bashir and J.M. Journal of Applied Mathematics, Vol. 2013, Article ID 531497, 10 pages, (2013).
3. M., Abbas. *Shape preserving data visualization for curves and surfaces using rational cubic functions [Ph.D. thesis]*, (School of Mathematical Sciences, Universiti Sains Malaysia, Penang,Malaysia, 2012).
4. H., Akima. J. Assoc. Comput. Mech. **17**:589-602 (1970).
5. S., Butt and K.W. Brodlie. Computer and Graphics 17 (1):55-64 (1993).
6. R., Delbourgo and J.A. Gregory. SIAM J. Sci. and Statist. Comput. **6**:967-976 (1985).
7. R., Delbourgo and J.A. Gregory. IMA Journal of Numerical Analysis, 5:397-406 (1985).
8. F.N., Fritsch and R.E. Carlson. SIAM J. Numer. Anal. **17**: 238-246 (1980).
9. J.A., Gregory. Computer Aided Design 18(1):53-57 (1986).
10. M.Z., Hussain and M. Sarfraz. International Journal of Computer Mathematics, Vol. 86, No. 3, March 2009, pp. 423-430 (2009).
11. M.Z., Hussain and M. Hussain. Applied Mathematics and Computation 190, 1353-1364 (2007).
12. M.Z., Hussain, M. Sarfraz and T.S. Shaikh. World Applied Sciences Journal 16(11): 1496-1508 (2012).
13. F.M., Ibraheem, Hussain and M.Z. Hussain, The Scientific World Journal. Volume 2014, Article ID 602453, 14 pages, (2014).
13. S.A.A., Karim and V.P. Kong. Research Journal of Applied Sciences (RJAS), Volume: 9, Issue: 4, Page No.: 214-223 (2014).
14. S.A.A., Karim and V.P. Kong. Journal of Applied Mathematics, Volume 2016 (2016), Article ID 4875358, 14 pages, (2016).
15. A., Lahtinen. Annals of Numerical Mathematics **3**:151-161 (1996).
16. M.H., Lam. Virginia Journal of Science. Volume **41(1)**: 1-13 (1990).
17. S., Liu, Che, Z. and Zhu, Y. Mathematical Problems in Engineering, Volume 2015, Article ID 983120, 20 pages (2015).
18. M., Sarfraz. Computers & Graphics **24(4)**:509-516 (2000).
19. M., Sarfraz, M.Z., Hussain, T.S., Shaikh and R., Iqbal. Data visualization using shape preserving C^2 rational spline. In Proceeding of 5th International Conference on Information Visualisation, 2011, pp. 528-533 (2011).
20. J.W., Schmidt and W., Hess. BIT 28, 340-352 (1988).
21. L.L., Schumaker. SIAM J. Numerical Analysis 20:854-864 (1983).

Optimal Long-Term Hydro Generation Scheduling of Small Hydropower Plant (SHP) using Metaheuristic Algorithm in Himreen Lake Dam

Ali Thaeer Hammid[1, 2, *], *Mohd Herwan* Sulaiman[1]
[1]*Faculty of Electrical & Electronics Engineering, University Malaysia Pahang, 26600 Pekan, Malaysia*
[2]*Department of Computer Engineering Techniques, Al Yarmouk University College, 32001 Ba'aqubah, Diyala, Iraq*

Abstract. This study focuses on the improvement of optimization model by applying particle swarm optimization and firefly algorithm methods to get a stable power production utility at its maximum level. Furthermore, it investigates on the minimization of utility loss in power production from the hydropower system, which is done by optimizing the variables of operation control in the hydropower plant at Lake Himreen - Diyala Dam. The variables mentioned are net turbine head, the rate of water flow and power production which had been gathered in the data during a research throughout a 10-year period. The results obtained from these two methods, namely Firefly Algorithm (FA) and Particle Swarm Optimization, (PSO) are compared. The inferences for general comparisons are created through several behavior indicators. The behavior indicators illustrate that FA's performance is better than PSO's performance, in some fields. At the end, the results show the strength of FA, as well as its efficiency and superiority.

1 Introduction

Hydropower golden age was set in the first half of the 20th century before oil control gained its dominance in the provision of energy. Several growth republics gradually began to get rid of traditional energy sources that existed in oil, coal, and natural gas [1-3]. In the present, as power demand keeps growing all day, more generation resources and various grid constructions are needed. While the network load is increasing, the generators that are attached to a hydro-turbine begin to be sluggish. This condition leads to the change of the output through the reduction of electrical power frequency. Consequently, the entire system of the hydro-turbine would collapse; this situation is called blackout or cascaded failure [4]. Usually, the observations made on the reliability of generation systems regard the resource of energy for the power production as continuously available. This indicates that, in general, even one cause of the power production deficit leads to the destruction of electricity generation in power plants. During hydropower production, when the reservoir is sufficient to generate the energy, the abundance of power production adheres to the rule of the rate of water flow, hence, this model of hydro-turbine is right [5].

The objective of this project is to employ both of Firefly Algorithm (FA) and Particle Swarm Optimization (PSO) into the hydropower plant located in Lake Himreen dam. This action is performed in order to create an effective solution to optimization power problem, as well as an actual solution to operational problems, which is suitable for both cases of irrigation provider and hydropower generation. Moreover, the objective of the mathematical function is to specify the maximum or the minimum values, which may be subjected to restrictions on its variables due to the diversity of real-life applications [6]. Additionally, the objective of power system production is to find out its restrictions and to perform an optimization by controlling specific variables of the power system. The test outcomes of both algorithm methods are obtained. Then, a comparison is made between the validation results of FA method and the validation results of PSO methods in order to display the possibilities and the superior points of the proposed algorithm. Firefly Algorithm (FA) would provide the solution to power system problems, which offers some superiority points and improved behavior to general results [7].

2 Small Hydropower Plants (SHP)

The hydro-turbine obtains mechanic hydropower and mechanically changes it into rotation power. Then, the turbine is connected to an electric power generator. Actually, the turbine efficiency depends on the turbine's power, its type, fluid percentage, etc. Kaplan turbine is observed to ensure that its efficiency reaches the maximum rate of water flow. Its efficiency will prove the appeal of this kind of turbine for a river with a variant in the management of the water flow rate [8, 9]. The unit of the power generator in a hydropower plant relies on four factors, which are gravity acceleration, the flow rate of water, the height of the water drop, and the general efficiency, as illustrated in Eq. (1)

$$P = g \bullet Q \bullet H \bullet \eta \qquad (1)$$

*Corresponding author: ali.thaeer.hammid@gmail.com

Where, P = generated power (KW); g= gravity acceleration (m2/s); Q = water flow rate (m3/s); H = Net head of Turbine (m); η = efficiency of hydropower plant.

3.1 Particle Swarm Optimization (PSO)

PSO mimics the action of bird grouping. Assume the general strategy; bird grouping is randomly looking for unknown regions as places for eating. However, there is only one area for eating left in the chosen regions. Even so, all of the birds could not pinpoint the location of the eating area. They only recognize the distance between the eating areas through the iteration on the distance amount.

3.2 PSO Operation

PSO is created by using a collection of particles which are as solutions. Afterwards, this operation looks for the ideal situation by performing updates on the power generators. During the update process, each particle targets for the best or superior values during every iteration. The first value is the best solution and it is known as fitness. Then, the fitness value is kept, and labeled as "p-best". The other best value is erased after the particle swarm optimizer is obtained by any of the particles within the population. The best value is a global best, and labeled as "g-best". Every particle within the population participates in this operation because it lives next to or beside the location of the operation. The best value is the local or personal best, labeled as "p-best". When two of the best values are discovered, the particle performs updates on its velocity (vi) and positions (xi), as illustrated mathematically in Eq. (2) and Eq. (3) [10].

$$v_i^{t+1} = wv_i^t + c_1 \times rand \times (pbest_i - x_i^t) + c_2 \times rand \times (gbest - x_i^t) \quad (2)$$

$$x_i^{t+1} = X_i^t + V_i^{t+1} \quad (3)$$

v_i^t is velocity particle, while i is at iteration t and it should be placed in field Eq. (4).

$$V_{min} \le V_i \le V_{max} \quad (4)$$

x_i^t is the present location of particle i at iteration t. Meanwhile, $pbest_i$ represents p-best in proxy i at iteration t, and *gbest* represents the best solution under this limit. Generally, the inertia's weight (w) is fixed depending on the formula in Eq. (5).

$$W = W_{max} - \left[\frac{W_{max} - W_{min}}{ITER_{max}}\right] \quad (5)$$

According to the formula above, w is the factor of inertia's weight. W_{max} is the highest amount of weighting factor, while W_{min} is the lowest amount of weighting factor. $ITER_{max}$ is the highest iterations number, and ITER is the current iteration number [11].

4.1 Fireflies' behavior and operation

Firefly Algorithm (FA) is an empirical algorithm, inspired from the flashing of light from fireflies. This algorithm was established by Xin-She Yang, at Cambridge University in 2007. A lot of fireflies create frequent light flashes and have a various light flashing dynamic. Firefly Algorithm is an effective local survey, in which the brightness of light flashes is related to its objective function. Parameters of fireflies' algorithm, such as the constancy of attractiveness and randomization, perform an essential task in specifying the optimum solution in the survey [12]. These parameters are significantly necessary for the computation of the convergence rapidity and FA algorithm's performance. FA has three main concepts that have been pursued from the fireflies' activities amongst one another.

- In general, all fireflies are unisexual. They will move individually towards fireflies which are brighter and more attractive, without relying on their sexual role.
- The grade of a firefly's attractiveness has a direct proportion to the brightness of its light. The brightness may be reduced due to an increase of the distance between fireflies. This is because their lights are absorbed by the air formed within the distance. It moves around randomly when it doesn't come across one that is brighter or more attractive.
- Therefore, the objective function is specified by the firefly's brightness or the intensity of the firefly's light [13].

4.2 Mathematical Model of Fireflies Algorithm (FA) Optimization

In a condition which displays the barriers of the highest FA optimization, the brightness (I) of a firefly is determined by two particular positions (x) which are marked as 'I (x) α f (x)'. This is because the distance between fireflies and their level of attractiveness (β) are proportional to each other. This depends on the distance of fireflies (r_{ij}) from the earth and the fireflies' variations (j^{th}). Light intensity reduces with the increase of space between the fireflies because the light is absorbed into the environment. Thus, 'β' changes with absorption rate, which inversely depends on the square law, as illustrated in Eq. (6):

$$I(r) = \frac{I_s}{r^2} \quad (6)$$

I(r) represents the light intensity within a space (r), and it has the same light intensity in the source space [14, 15]. Since a firefly's attractiveness is directly proportional to the light intensity, it is recognized and distinguishable amongst neighboring fireflies. Therefore, the relation between the level of firefly's attractiveness (β) and the size of space (r) is illustrated in Eq. (7):

$$\beta = \beta_0 e^{-\gamma r^2} \tag{7}$$

β0 represents a firefly's attractiveness level at r = 0. γ represents the constancy of environmental absorption, which the regulates the rate of light intensity reduction. The motion of the firefly which is attracted to the brighter firefly 'j' is labeled as 'i', as illustrated in Eq. (8):

$$X_i^{t+1} = X_i^t + \beta_0 e^{-\gamma r_{ij}^2}(X_j^t - X_i^t) + \alpha_t \varepsilon_i^t \tag{8}$$

The creation of the second part of the formula above is due to the Firefly's attraction towards the brighter firefly. Meanwhile, randomization (α_t) is included in the last third part of this formula. This symbol represents the parameter of randomization, and (ε^t_i) is the symbol of the dispersed vector numbers, achieved from the regular supply and delivery made at specific time 't' [16]. For both fireflies 'i' and 'j', the space (r) between them is calculated, as illustrated in Eq. (9):

$$r_{ij} = \sqrt{\sum_{k=1}^{d}\left(x_{i,k} - x_{j,k}\right)^2} = \|x_i - x_j\| \tag{9}$$

'r_{ij}' refers to the space between fireflies 'i' and 'j', and they are placed at 'x_i' and 'x_j' respectively. This space is known as the Cartesian space. Meanwhile, '$X_{i,K}$' is the kth element of the three-dimensional axis 'x_i' created by firefly 'i^{th}' under 2-D condition [17]. When most of the generation is almost finished, the fireflies are classified depending on their brightness, and the superior firefly group is discovered from each category. Finally, the fireflies start moving according to their groups. Additionally, the intensity of brightness of all fireflies is recently improved with the information provided in fitness function for each group. Moreover, when the procedures are performed on the groups, the firefly with the maximum level of brightness indicates the highest value of fitness. This is established as the optimal solution to optimization problems [18].

5 Results and discussion

In this study, only the constrained generated power was checked and taken into account, according to the main restrictions. There were various limitations on the operation of the power generator, on both maximum and minimum level. An observation was also conducted on two input parameters. This refers to the rate of water flow, ranging from 8 m^3/s to 175 m^3/s, and the net head of turbine, ranging from 14.17 to 29.84 meters. Moreover, based on the observation, the amount of power output production barely reached 44000 KW. Sometimes, the amount of power production dropped to zero and there was a wide variation of power production

Since the creation of FA and PSO depends on their historical data, the input and output data collected during this study were analyzed statistically. These two methods were built to be specialized in 4 inputs and 1 output, which originated from fitness function. Clearly, the evaluation of the plant behavior was based on some of the main variables. From the past up until now, there has been no rule for selecting the superior techniques from Swarm Intelligence (SI). The correct way of selecting the superior techniques is through trial and error. Therefore, the execution of each model will be checked so that the structure of its right behavior will be clearly seen. In this study, there were some attempts of optimizing power production using the R2015a version (The Mathworks Inc.) of the MATLAB software package. The FA and PSO techniques were modeled to accomplish the superior fitness level of fitness function, which was identified by Eq. (1)

The well-known optimization methods such as Particle Swarm Optimization (PSO) as well as another one is a Firefly Algorithm (FA) have also been considered here in the similar difficult so as to make a comparison between results with the two both diverse optimization methods to solve optimum power production problem. 500 is maximum iterations have been considered for FA. 3000 is maximum iterations have been considered for PSO. From power generated function that applied on PSO and Firefly algorithms, PSO is starting the power generation which has ranged from 41000 kW up to 49000 kW as illustrated in Figure 1. And the FA is starting the power generation which also has a range from 37600 kW up to 37700 kW as illustrated in Figure 2. In PSO method, the approximate maximum best fitness and approximate minimum best fitness were 49010 and 42737 respectively which equals to power generation function. In FA algorithm, the exact maximum best fitness and exact minimum best fitness were 37746 and 36676 respectively which equals to power generation function.

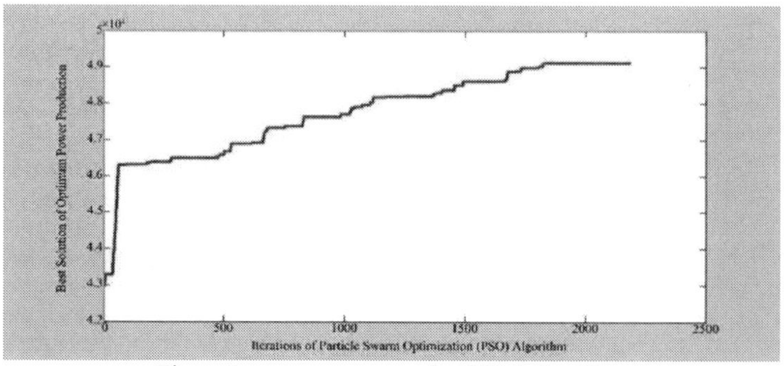

Figure 1. convergence graph of PSO algorithm

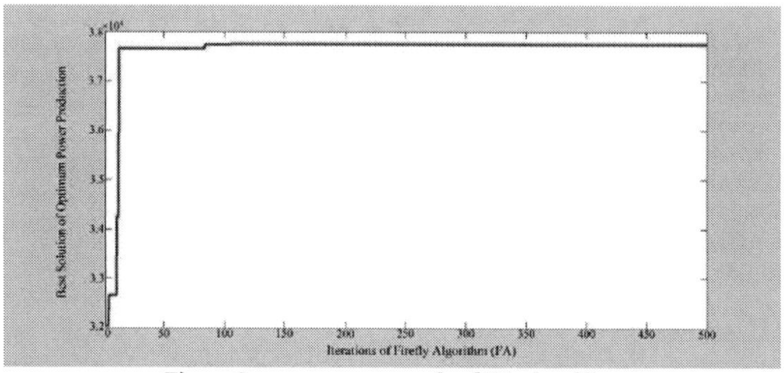

Figure 2. convergence graph of FA algorithm

In the main case, the best result achieved when 500 iterations are done by using the FA optimization method has been made a comparison with the PSO other optimization methods. Results in Table 2, illustrates that all power production rearranging in a founded solution of FA is less than the results obtained by basing solution of PSO. Moreover, the mathematical test function or fitness function value is the best solution that has been done from the power generation equation as illustrated in Eq. (1). Since it includes power sensitivity analysis in algorithms selection function, it covers all over probabilities and it contains all function parameters

It is clear that has the FA generated values of the fitness function included the lower standard deviation than the dependency runs compared of the PSO. Furthermore, the best values or maximum values of fitness functions for entire check test problems completely which is obtained by the FA are nearby the global optimal values, Thus FA might exactly get to the global optimal value of the power generation function in the check test table. Both of two algorithms have been shown for a similar number of population which is 3570. In the case of fitness function (power function), FA almost gets to the optimum point and it does not exceed the optimum point as illustrated in Figure 2. While, PSO is exceeding the observed maximum point, but in a much too long time as illustrated in Figure 1.

The performance of the FA convergence method which is illustrated in Figure 2 can be actually realized that throughout the algorithm the fitness value converges after 100 iterations turn out to be steady. On the contrary, the performance of the PSO convergence method which is illustrated in Figure 1 cannot realize any period for steady case due to increasing values throughout all periods of iteration the algorithm. As a result, a comparison must be getting into the relationship between the suitable convergence and computational time. Thus, there is a far away convergence of PSO method which may be a reason to obstruct the fitness function and lead it in local values. It can be prevented by swarm size increment but, the computation time will be greater than before also.

So as to estimate the algorithm accomplishment and illustration the advantage and efficiency of both suggested algorithm, the total execution and ran both of them are 10 times. Now, 'step numbers' means a whole number of time steps of optimization algorithm and 'iteration number' period is employed for getting sequentially earlier estimates the solution of our own specified problem.

6 Conclusions

In this research, a successful execution of Firefly Algorithm (FA) and Particle Swarm Optimization (PSO) has been accomplished on nonlinear power function. With this, the stability of power production utility at its maximum level was obtained, and the loss of power production in the hydropower system was minimized by optimizing the control of operation variables in the hydropower plant at Himreen Lake Dam - Diyala. The analysis of the daily data set was influenced by the net turbine head and rate of water flow. These elements of the hydropower plant functioned as the input and the power was produced as the output. The daily data set was modeled based on the actual observed daily data throughout 2005–2015. Thus, the experimental results were indicated by the compression FA with PSO, which is discovered, the better for this proposed work specifically. It's owing to the benefits of the meta-heuristic process of FA which is comprehensively cleaning the searching area to discover the optimal solution for combinatorial optimization correctly. The FA is not only containing the improvement operation related to the present area, but it's containing the enhancement amongst its specific another area from the earlier steps. Moreover, FA is a stochastic algorithm which depends on attracting and Light intensity of common performance of fireflies and its function has a unique behavior to solve the certain restricted of optimization problems.

References

1. A.T. Hammid. Applications of Tuning Control Actions for the Efficient Load/frequency Control in Steam Turbine. International Journal of Current Engineering and Technology, (2013). **3**(5): p. 1895-1898.
2. A.T. Hammid, et al., Load Frequency Control for Hydropower Plants using PID Controller. Journal of Telecommunication, Electronic and Computer Engineering (JTEC), (2016). **8**(10): p. 47-51.
3. A.T. Hammid, A.K. Bhardwaj, and S. Prakash., Design Remote Power Control I/O Data Acquisition System and Control on Home Automation. IJECCE, (2013). **4**(2): p. 528-535.
4. A.T. Hammid, M.H.B Sulaiman, and A.N. Abdalla., Prediction of small hydropower plant power production in Himreen Lake dam (HLD) using artificial neural network. Alexandria Engineering Journal, (2017).
5. S.H. Kiran, D.S. Sekhar, and C. Subramani., Performance of two modified optimization techniques for power system voltage stability problems. Alexandria Engineering Journal, (2016). **55**(3): p. 2525-2530.
6. M. Shalini, et al., Multi-objective parametric optimization of nano powder mixed electrical discharge machining of AlSiCp using response surface methodology and particle swarm optimization. Alexandria Engineering Journal, (2017).
7. M.S.R. Naidu, P.R. Kumar, and C. K, Shannon and Fuzzy entropy based evolutionary image thresholding for image segmentation. Alexandria Engineering Journal, (2017).
8. S.S. Nanivadekar and U.D. Kolekar., An approach involving dynamic group search optimization for allocating resources in OFDM-based cognitive radio system. Alexandria Engineering Journal, (2016).
9. L. Baohong, et al., Study on the optimal hydropower generation of Zhelin reservoir. Journal of Hydro-environment Research, (2013). **7**(4): p. 270-278.
10. L. Peng, et al., Short-term hydro generation scheduling of Xiluodu and Xiangjiaba cascade hydropower stations using improved binary-real coded bee colony optimization algorithm. Energy Conversion and Management, (2015). **91**(Supplement C): p. 19-31.
11. M.R. Ahmadi, et al., Comment on 'Differential evolution technique-based short-term economic generation scheduling of hydrothermal systems' by K.K. Mandal, N. Chakraborty Electric Power Systems Research 78 (2008) 1972–1979. Electric Power Systems Research, (2017). **147**(Supplement C): p. 310-312.
12. Tedjani, M., et al., Dynamical modeling of Li-ion batteries for electric vehicle applications based on hybrid Particle Swarm–Nelder–Mead (PSO–NM) optimization algorithm. Electric Power Systems Research, (2016). **131**(Supplement C): p. 195-204.
13. M. Li, et al., Short-term hydro generation scheduling of Three Gorges–Gezhouba cascaded hydropower plants using hybrid MACS-ADE approach. Energy Conversion and Management, (2013). **76**(Supplement C): p. 260-273.
14. H. Wang, et al., A hybrid multi-objective firefly algorithm for big data optimization. Applied Soft Computing, (2017).
15. Y. Zhang, S. Xian-fang, and G. Dun-wei., A return-cost-based binary firefly algorithm for feature selection. Information Sciences, (2017). **418**(Supplement C): p. 561-574.
16. H. Wang, et al., Firefly algorithm with neighborhood attraction. Information Sciences, (2017). **382**(Supplement C): p. 374-387.
17. D. Sánchez, P. Melin, and O. Castillo, Optimization of modular granular neural networks using a firefly algorithm for human recognition. Engineering Applications of Artificial Intelligence, (2017). **64**(Supplement C): p. 172-186.
18. H. Wang, et al., A hybrid multi-objective firefly algorithm for big data optimization. Applied Soft Computing, (2017).

AUTHOR INDEX

AUTHOR INDEX